XI Latin American Workshop on Nonlinear Phenomena 2009

(LAWNP 09)

Journal of Physics: Conference Series Volume 246

Buzios-RJ, Brazil
5-9 October 2009

ISBN: 978-1-61738-962-7
ISSN: 1742-6588

Printed from e-media with permission by:

Curran Associates, Inc.
57 Morehouse Lane
Red Hook, NY 12571

Some format issues inherent in the e-media version may also appear in this print version.

Printed by Curran Associates, Inc. (2010)

For permission requests, please contact the Institute of Physics
at the address below.

Institute of Physics
Dirac House, Temple Back
Bristol BS1 6BE UK

Phone: 44 1 17 929 7481
Fax: 44 1 17 920 0979

techtracking@iop.org

Additional copies of this publication are available from:

Curran Associates, Inc.
57 Morehouse Lane
Red Hook, NY 12571 USA
Phone: 845-758-0400
Fax: 845-758-2634
Email: curran@proceedings.com
Web: www.proceedings.com

TABLE OF CONTENTS

The XI Latin American Workshop on Nonlinear Phenomena (LAWNP) has been held in Búzios-RJ, Brazil, from 5–9 October 2009. This international conference is one in a series that have gathered biennially, over the past 21 years, physicists and other scientists who direct their work towards several aspects of nonlinear phenomena and complex systems. The main purpose of LAWNP meetings is to create a friendly and motivating environment, such that researchers from Latin America and from other parts of the globe can discuss not only their own latest results but also the trends and perspectives in this very interdisciplinary field of investigation. Hence, it constitutes a forum for promoting scientific collaboration and fomenting the emergence of new ideas, helping to advance the field.

The XI edition (LAWNP'09) has gathered more than 230 scientists and students (most from Latin America), covering all of the world (27 different countries from North and South America, Asia, Europe, and Oceania). In total there were 18 plenary lectures, 80 parallel talks, and 140 poster contributions. A stimulating round-table discussion also took place devoted to the present and future of the Latin American Institutions in Complex Phenomena (a summary can be found at http://lawnp09.fis.puc-rio.br, in the Round-Table report link). The 2009 workshop was devoted to a wide scope of themes and points of view, pursuing to include the latest trends and developments in the science of nonlinearity.

In this way, we have a great pleasure in publishing this Proceedings volume based on the high quality scientific works presented at LAWNP'09, covering already established methods as well as new approaches, discussing both theoretical and practical aspects, and addressing paradigmatic systems and also completely new problems, in nonlinearity and complexity. In fact, the present volume may be a very valuable reference for those interested in an overview on how nonlinear interactions can affect different phenomena in nature, addressing: classical and quantum chaos; instability and bifurcation; cooperative behavior; self-organization; pattern formation and synchronization; far-from-equilibrium and fluctuation dynamics; nonlinearity in fluid, plasmas, granular media, optics, and wave propagation; turbulence onset; and complexity in natural and social systems.

The success of the conference was possible thanks to the financial support from many agencies, especially the Brazilian agencies Capes and CNPq, and the international agencies, Binational Itaupú, ICTP-Trieste, and CAIS-Albuquerque. Equally very important was the support by the organizer's institutions PUC-Rio de Janeiro and UFPR-Curitiba. We also must thank *Journal of Physics: Conference Series*, for believing in the success and scientific quality of the conference, and to the journal staff, specially Anete Ashton, for the kind and prompt help during the whole production process of this publication.

Finally, and most important, we acknowledge all the participants of the LAWNP'09, whose interest and enthusiasm in advancing the science of nonlinearity constitutes the true moto making the present Proceedings a very valuable scientific contribution.

Celia Anteneodo (PUC-Rio, Brazil)
and
Marcos G E da Luz (UFPR-Curitiba, Brazil)
Conference Chairs

Some of the conference participants.

CAPES This issue was supported by CAPES (Agency for Evaluation and Support of Graduate Studies Programs), Brazilian govern entity devoted to the formation of human resources. CA would like to thank CAPES for financial support.

All papers published in this volume of *Journal of Physics: Conference Series* have been peer reviewed through processes administered by the proceedings Editors. Reviews were conducted by expert referees to the professional and scientific standards expected of a proceedings journal published by IOP Publishing.

Extreme fractal structures in chaotic mechanical systems: riddled basins of attraction

S. Camargo [1], S. R. Lopes [2] and R. L. Viana [2]

1. Instituto de Física, Universidade de São Paulo, 05315-970, São Paulo, SP, Brazil.
2. Departamento de Física, Universidade Federal do Paraná, Caixa Postal 19044, 81531-990, Curitiba, PR, Brazil.

E-mail: viana@fisica.ufpr.br

Abstract. Chaotic dynamical systems with certain phase space symmetries may exhibit riddled basins of attraction, which can be viewed as extreme fractal structures in the sense that, regardless of how small is the uncertainty in the determination of an initial condition, we cannot decrease the fraction of such points that are certain to converge to a given attractor. We investigate a mechanical system exhibiting riddled basins of attraction: a particle under a two-dimensional potential with friction and time-periodic forcing. The verification of riddling is made by checking its mathematical requirements through computation of finite-time Lyapunov exponents as well as by scaling laws describing the fine structure of basin filaments densely intertwined in phase space.

1. Introduction

The connection between chaos and fractal geometry is a cornerstone of nonlinear dynamics. Fractal sets appear quite often in nonlinear dynamics due to the stretch-and-fold construction leading to chaotic attractors [1]. Fractal basin boundaries have been observed in many dissipative dynamical systems with multiple attractors, leading to final-state sensitivity [2]. By the latter we mean that, even if a considerable improvement is made in the accuracy with which an initial condition is known, little is gained in diminishing the fraction of uncertain initial conditions, i.e. the relative number of initial conditions which we do not know *a priori* the final state they asymptote to [3].

Non-attracting fractal sets are also important for nonlinear dynamics, for they are the underlying structure of chaotic orbits. In dissipative dynamical systems, such sets are exemplified by the strange saddles, which are formed by the intersection of the stable and unstable manifolds of unstable hyperbolic fixed points, containing a dense orbit. In conservative dynamical systems, such non-attracting fractal sets are responsible for many phenomena like fractal escape boundaries [4], Wada basins [5], and chaotic scattering [6].

One epitome of the connection between chaos and fractals is the horseshoe set discovered by Smale in the 60's [7]. A horseshoe is a non-attracting invariant Cantor-like fractal set with (i) a countable set of periodic orbits; (ii) an uncountable set of bounded non-periodic orbits; (iii) a dense orbit [1]. The strange saddle, for example, being formed by an infinite number of transversal intersection of invariant manifolds, is a fractal horseshoe-like set. If an orbit starts off but near the strange saddle, it will wander erratically so as to closely approach a large number of unstable orbits embedded in the strange saddle, forming the basis of transient chaos [8].

An extreme form of fractality is represented by the so-called riddled sets. Such sets are so intertwined that they virtually defy any attempt to improve the final-state predictability. In other words, while for fractal basins a great reduction of the initial state uncertainty is necessary to yield a better predictability, for riddled basins no improvement at all would be achieved by means of this procedure. This is so because of the rather stringent mathematical requirements for riddled sets, particularly the existence of phase-space symmetries leading to invariant manifolds in which chaotic dynamics takes place, and with conspicuous transversal stability properties [9].

In spite of these mathematical requirements, it turns out that riddled sets have been observed and many dynamical systems and even in experiments. Basin riddling was named and described by Alexander *et al.* [10], who also presented a variety of illustrative examples chosen from discrete maps. Independently, however, riddling has been also noticed by Pikowsky and Grassberger [11]. Riddled basins have been described in mechanical systems, like coupled elastic arches [12], and were generally found to occur in spatially extended dynamical systems, like coupled oscillators (where the invariant set contains a chaotic synchronized state) [13, 14] and coupled circuits [15, 16]. Other applications include ecological population models [17, 18], learning dynamical systems [19], chemical reactions of the Belouzov-Zhabotinsky type [20], and in models of interdependent open economies [21].

In this paper we are to revisit an example first studied by Ott and co-workers [22, 23, 24], namely of a particle under a two-dimensional potential function, subjected to both friction and time-periodic forcing. In References [22, 23, 24] there was considered the case where the invariant set contains a chaotic attractor of the type exhibited by the forced Duffing equation, obtaining riddled basins of attraction for wide parameter intervals. On the other hand, the Duffing equation can be viewed as an approximation of the behavior of a nonlinear pendulum, so we have studied the full driven pendulum in order to verify that it also able to exhibit riddling.

The existence of riddling can be verified in two basic ways: the first is to check that all mathematical conditions for riddling are fulfilled by the system. This direct approach can be pursued by investigating the behavior of Lyapunov exponents along the directions transversal to the invariant set in which the chaotic attractor lies. The second approach handles scaling laws that relate the basin structures with the properties of finite-time Lyapunov exponents, and is based on a stochastic model developed by Ott *et al.* [24]. We have verified the presence of riddled basins using both approaches.

This paper is organized as follows: in Section II we outline the basic concepts of riddled basins of attraction, focusing on the computation of finite-time Lyapunov exponents to be used further. Section III presents the mechanical system to be investigated whereas Section IV contains the our results on riddled basins, using a characterization *via* two scaling laws. Our Conclusions are left to the final Section.

2. Riddled basins of attraction

The most common example of a riddled set is a riddled basin of attraction, which may appear in chaotic systems having certain symmetries and quite general topological and metric properties. A dynamical system may have a chaotic attractor A whose basin of attraction is riddled with "holes" (in a measure-theoretical sense) belonging to the basin of another (non necessarily chaotic) attractor B. In other words, riddled basins of attraction never exhibit a disk in phase space, because every point in the basin of attraction of attractor A has pieces of the basin of attraction of attractor B arbitrarily nearby [9, 10].

Riddling affects dramatically our ability of predicting what attractor the trajectory originating from a given initial condition asymptotes to. Let p be an arbitrary point belonging to the basin of a chaotic attractor A. If the basin of A is riddled by the basin of the other attractor B, then a small ball of radius ϵ centered at p has a nonzero fraction of its volume belonging to the basin of B, irrespective of how small the radius ϵ might be [Fig. 1]. Hence, if we regard this ϵ-ball as an uncertainty neighborhood related to the (numerical or experimental [1]) determination of the initial condition, the resulting trajectory always has a positive probability of falling into the basin of the other attractor. In other words, the probability of escaping from the basin of attractor A is nonzero for every uncertainty ϵ. Consequently, riddling implies an extreme form of final-state sensitivity.

In fractal basins of attraction, the uncertain fraction of initial conditions scales with the uncertainty radius as $f(\epsilon) \sim \epsilon^{\alpha}$, where $0 < \alpha < 1$ is the uncertainty exponent. If D is the dimension of the phase space and D_0 is the box-counting dimension of the fractal basin boundary, then it can be proved that $\alpha = D - D_H$ [3]. From this point of view, a riddled basin is associated with an exponent $\alpha = 0$ (within numerical accuracy), meaning that the uncertain fraction is always 100%, regardless of the value of ϵ.

In order to express the mathematical requirements for riddling, we first give the definition of attractor to be used throughout this work. Let H be the phase space in which the dynamical system **F** is defined (it is a discrete-time map, but continuous-time flows can be also described by **F**, if Poincaré sections are taken). A closed subset A $\in$ H is an attractor of **F** if it satisfies the following conditions:

- A has a basin of attraction, denoted $\beta(\mathsf{A})$, of positive Lebesgue measure (volume) in the phase space H.
- A is a compact set with a dense orbit. In the Milnor definition of attractor, the basin of attraction does not need to include the whole neighborhood of the attractor [25].

If the basin of attraction of A has positive Lebesgue measure, A is a weak Milnor attractor [26].

This measure-theoretical definition implies the following set of conditions under which riddled basins occur in a dynamical system [23]

1. There is an invariant subspace M $\in$ H whose dimension d_M is less than that of the phase space d_H.

2. The dynamics on the invariant subspace M has a chaotic attractor A;

3. There is another attractor B, chaotic or not, and not belonging to the invariant subspace M.

4. The attractor A is transversely stable in the phase space H, i.e., for typical orbits on the attractor the Lyapunov exponents for infinitesimal perturbations along the directions transversal to the invariant subspace M are all negative.

5. A set of unstable periodic orbits embedded in the chaotic attractor A becomes transversely unstable. As a consequence, at least one of the Lyapunov exponents along directions transverse to M, although negative for almost any orbit of A, experiences positive finite-time fluctuations.

Condition **1** is a consequence of the system having some symmetry which enables it to display an invariant subspace M, in the sense that, once an initial condition is exactly placed on M, the resulting trajectory cannot escape from M for further times. To have riddling, it is necessary to have a dense set of points with zero Lebesgue measure in the attractor lying in the invariant subspace which are transversely unstable, thus it is necessary that this attractor be chaotic (condition **2**). The existence of another attractor (condition **3**) is necessary for the basin of an attractor to be riddled with holes belonging to the basin of this second attractor.

[1] The uncertainty involved in numerical simulations is related to the truncation of numbers, whereas in experiments is due to the accuracy of the measurements.

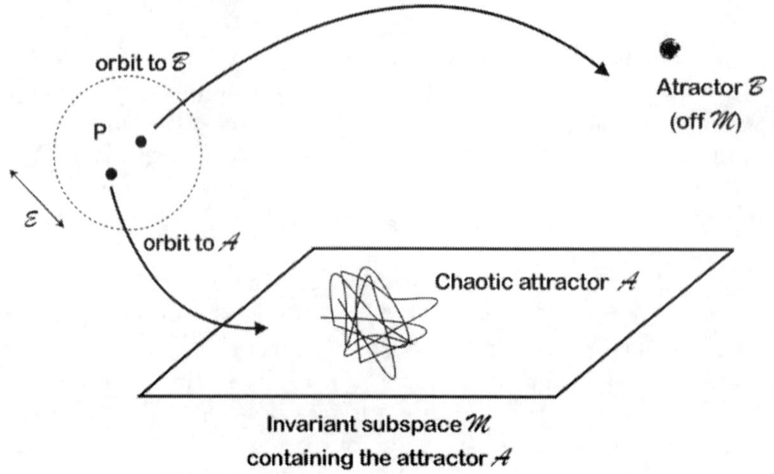

Figure 1. Schematic figure showing some of the requirements needed for riddling.

If the transverse Lyapunov exponents of typical orbits lying in the invariant subspace M are all negative (condition **4**), then A is an attractor in the weak Milnor sense, and its basin has positive Lebesgue measure. Condition **5** states that, while the invariant subspace M is still transversely stable, there will be trajectories on the attractor A that are transversely unstable. Condition **4** can be quantitatively checked by computing the maximal Lyapunov exponent along a transversal direction to M. Verifying condition **5**, on the other hand, would require the computation of finite-time Lyapunov exponents.

The finite-time Lyapunov exponents for a d_H-dimensional map $\mathbf{F}$ are defined as follows. Let n be a positive integer and $\mathbf{DF}^n(\mathbf{x}_0)$ be the Jacobian matrix of the n-times iterated map, with entries evaluated at an initial condition $\mathbf{x}_0 \in$ A. Supposing that the singular values of $\mathbf{DF}^n(\mathbf{x}_0)$ are ordered: $\xi_1(\mathbf{x}_0, n) \geq \xi_2(\mathbf{x}_0, n) \geq \ldots \geq \xi_{d_H}(\mathbf{x}_0, n)$, the k-th time-n Lyapunov exponent for the point $\mathbf{x}_0$ is defined as

$$\tilde{\lambda}_k(\mathbf{x}_0; n) = \frac{1}{n} \ln ||\mathbf{DF}^n(\mathbf{x}_0) \cdot \mathbf{v}_k||, \tag{1}$$

where $k = 1, 2, \cdots d_H$ and $\mathbf{v}_k$ is the singular vector related to $\xi_k(\mathbf{x}_0, n)$.

The infinite time limit of the above expression is the usual Lyapunov exponent $\lambda_k = \lim_{n\to\infty} \tilde{\lambda}_k(\mathbf{x}_0, n)$. Although the time-$n$ exponent $\tilde{\lambda}_k(\mathbf{x}_0, n)$ generally takes on a different value, depending on the chosen point, its infinite time limit takes on the same value for almost all $\mathbf{x}_0$ with respect to the natural ergodic measure of the invariant set A.

If the attractor in A is chaotic (or hyper-chaotic, in general), there are d_M infinite-time Lyapunov exponents related to directions in the invariant subspace M, of which many can be positive; and $d_H - d_M$ Lyapunov exponents along directions transverse to M. For the sake of determining the transversal stability of M it suffices to consider the largest one, $\lambda_\perp$. Hence, condition **4** implies that $\lambda_\perp$ takes on a negative value:

$$\lambda_\perp = \lim_{n\to\infty} \tilde{\lambda}_\perp(\mathbf{x}_0, n) < 0. \tag{2}$$

The possible existence of an infinite number of transversely unstable orbits embedded in a transversely stable attractor implies fluctuations of the finite-time largest transversal exponent, $\tilde{\lambda}_\perp(\mathbf{x}_0, n)$. Consider the probability distribution $P(\tilde{\lambda}_\perp(\mathbf{x}_0, n))$, from which we can obtain the average value of this exponent (assuming proper normalization):

$$\left\langle \tilde{\lambda}_\perp(n) \right\rangle = \int_{-\infty}^{+\infty} \tilde{\lambda}_\perp(n) P(\tilde{\lambda}_\perp(n)) d\tilde{\lambda}_\perp(n), \tag{3}$$

When n is large enough this probability distribution function is assumed to be Gaussian [27]:

$$P(\tilde{\lambda}_\perp(\mathbf{x}_0, n)) \approx \sqrt{\frac{nG''(\lambda_\perp)}{2\pi}} \exp\left[-\frac{nG''(\lambda_\perp)}{2} (\tilde{\lambda}_\perp(n) - \lambda_\perp)^2 \right], \tag{4}$$

where G is a function with the following convexity properties: $G(\lambda_\perp) = G'(\lambda_\perp) = 0$, $G''(\lambda_\perp) > 0$. Substituting (4) into (3) there results

$$< \tilde{\lambda}_\perp(\mathbf{x}_0, n) >= \lambda_\perp. \tag{5}$$

The standard deviation in the Gaussian approximation approaches zero with $n^{-1/2}$, when $n \gg 1$, we can write the corresponding variance as

$$\sigma^2 =< (\tilde{\lambda}_\perp(n) - < \lambda_\perp >)^2 >= \frac{2D}{n}, \tag{6}$$

such that the diffusion coefficient $D = 1/2G''(\lambda_\perp)$ is independent of n.

In terms of the distribution of finite-time transversal exponents, condition **5** for riddling implies that there is a positive fraction of positive values of $\tilde{\lambda}_\perp(n)$ for initial conditions $\mathbf{x}_0$ randomly chosen in the attractor A, i.e.,

$$\phi(n) = \int_0^\infty P(\tilde{\lambda}_\perp(n)) d\tilde{\lambda}_\perp(n) > 0. \tag{7}$$

3. A mechanical system with riddled basins

Let us consider a point particle of unit mass under the influence of a two-dimensional potential $V(x, y)$, viscous friction, and a time-periodic external forcing, whose equation of motion is

$$\ddot{\mathbf{r}} + b\dot{\mathbf{r}} + \nabla V(\mathbf{r}) = A \sin(\omega t)\hat{\mathbf{x}}, \tag{8}$$

where the potential well is given as [Fig. 2]

$$V(x, y) = -a \cos x + (x + \chi)y^2 - y^4, \tag{9}$$

with b, χ, A and ω being, respectively, the dissipation coefficient, a characteristic scale length, and the amplitude and frequency of the driving force.

The phase space H of this system is five-dimensional, with coordinates x, p_x, y, p_y, and $\theta \equiv \omega t$ (mod 2π), in terms of which the equation of motion can be written as the autonomous vector field

$$\begin{aligned}
\dot{x} &= p_x, && (10)\\
\dot{p}_x &= -bp_x - a \sin x - y^2 + A \sin\theta, && (11)\\
\dot{y} &= p_y, && (12)\\
\dot{p}_y &= -bp_y - 2y(x + \chi) - 4y^3, && (13)\\
\dot{\theta} &= \omega, && (14)
\end{aligned}$$

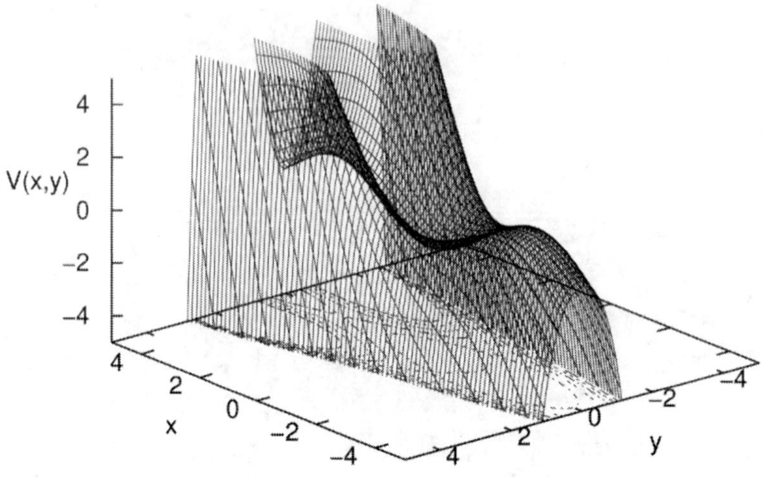

Figure 2. Potential function (9), with $a = 1.0$ and $\chi = 2.5$.

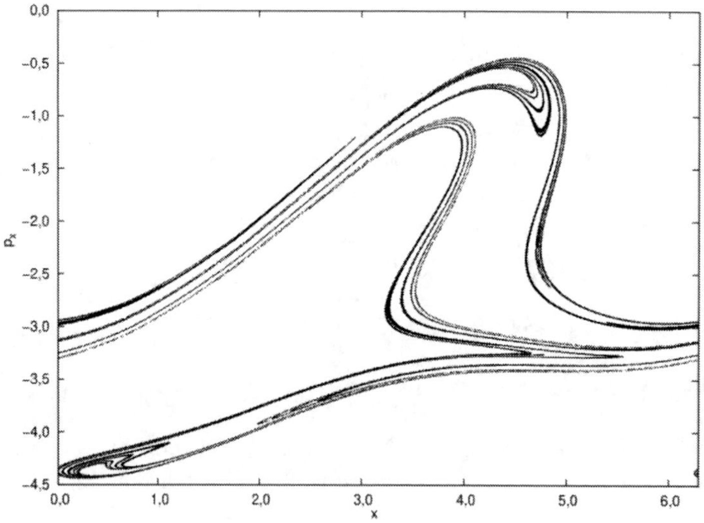

Figure 3. Chaotic attractor of the driven damped pendulum in the invariant plane $y = p_y = 0$ in the phase space, with $a = 1.0$, $b = 0.22$, $A = 2.7$, $\chi = 2.5$, and $\omega = 1.0$.

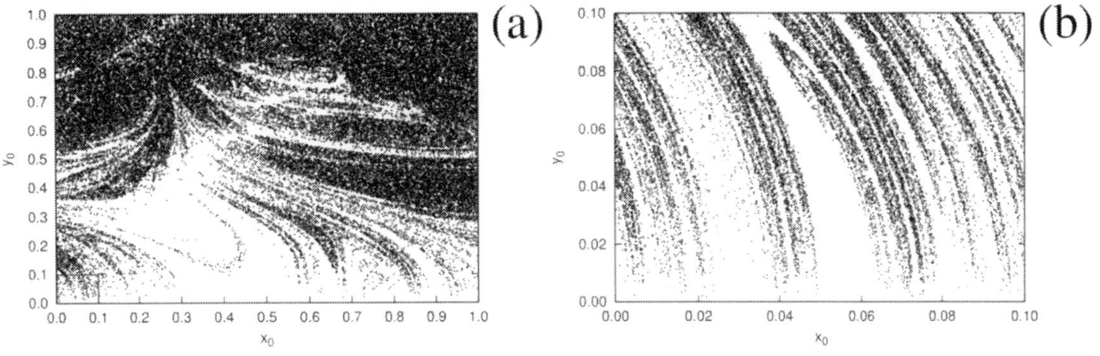

Figure 4. (a) The white and black dots represent, respectively, the basins of a chaotic attractor in the invariant subspace at $y = p_y = 0$ and the attractor at infinity. The parameters were the same as in the previous figure. (b) Magnification of a small region chosen from (a).

Due to the symmetry $V(x, y) = V(x, -y)$ the particle dynamics is invariant with respect to the $y \to -y$ and $p_y \to -p_y$ transformations in H. Thus the constraints $y = 0$ and $p_y = 0$ define a three-dimensional invariant subspace $\mathsf{M} \in \mathsf{H}$ in such a way that a trajectory with an initial condition $(x(0), y(0) = 0, p_x(0), p_y(0) = 0, \theta(0))$ belonging to M will remain there forever (condition **1**).

Restricted to this invariant subspace, the system state is described by the coordinates (x, p_x, θ), whose dynamics is governed by the equations of the forced damped pendulum [28]

$$\dot{x} = p_x, \tag{15}$$
$$\dot{p}_x = -bp_x - a \sin x + A \sin \theta, \tag{16}$$
$$\dot{\theta} = \omega. \tag{17}$$

which has, for conveniently chosen values of its parameters, like $a = 1.0$, $b = 0.22$, $\chi = 2.5$, $A = 2.7$, and $\omega = 1.0$, a chaotic attractor A in the invariant subspace M, depicted in Fig. 3, and which fulfills condition **2** for riddling. From now on we will fix all parameters but the characteristic length χ, which will be our variable parameter (however, for all considered values of ξ we have obtained chaotic attractors in the invariant subspace). There is another attractor at infinity $\mathsf{B} = (|y| = \infty, |p_y| = \infty)$ for the same set of parameters (condition **3**).

We integrated Eqs. (10)-(14) by using an adaptive stepsize fourth-order Runge-Kutta scheme. The dynamics of this time-periodic flow can be studied by means of a stroboscopic, or time-T map: we consider the values of the phase space variables at discrete time instants satisfying $\theta = 0$, mod 2π, i.e. we sample the values of $(x(t_n), p_x(t_n); y(t_n), p_y(t_n))$ where $t_n = 2n\pi/\omega$ for $n = 0, 1, 2, \ldots$. For ease of visualization we consider the two-dimensional projection of these variables in the plane (x, y).

A graphical representation of the basins is obtained by considering a unit square in this projection and sparkle randomly 10^5 over this region [Fig. 4(a)]. Each of these initial condition was integrated for 10^5 iterations. If the resulting trajectory was found to asymptote to the attractor A the corresponding initial condition was marked with a black pixel, otherwise the trajectory was supposed to go to the attractor at infinity and no dot was marked. We used $|y| > 20$ as a criterion to determine whether or not the trajectory is going to infinity (varying this number does not affect significantly our results, though). The black and white and regions in Fig. 4(a) are thus graphical approximations of the basins of A and B, respectively. Both regions appear to be densely interwoven, which is confirmed by magnifications suggesting that

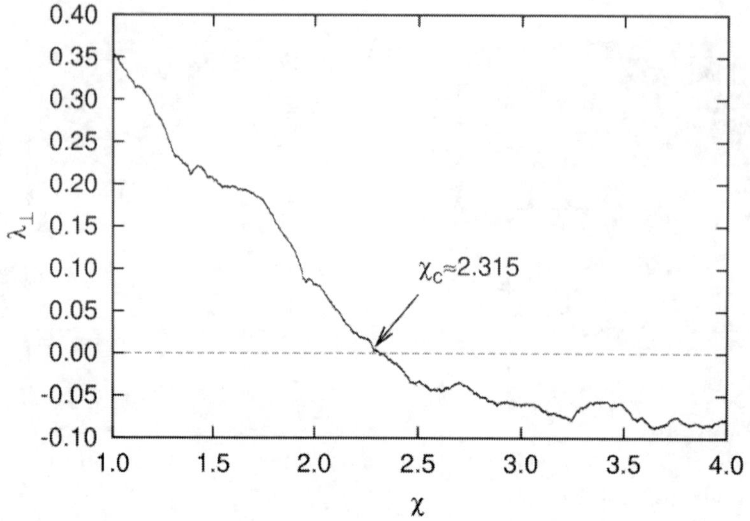

Figure 5. Transversal Lyapunov exponent as a function of the control parameter χ. The remaining parameters are the same as in the previous figures.

regions containing black dots also contain white dots [Fig. 4(b)]. As we will see later on, this will prove to be true for any region containing black dots, irrespective of how small it might be. We remark, however, that the basin of infinity *is not* riddled, since magnifications of a number of regions containing black points do not present blank areas.

Although the basins appear to be intertwined, a quantitative characterization of riddling needs a deeper analysis, where the verification of the remaining conditions for riddling demands the computation of Lyapunov exponents in the directions transversal to M. There are two such exponents, the largest one being denoted as $\lambda_\perp$, and which is computed by taking a variation of Eqs. (12) and (13) and setting $y = p_y = 0$, what results in

$$\dot{\delta y} = \delta p_y, \tag{18}$$
$$\dot{\delta p_y} = -b\delta p_y - 2\left[x(t) + \chi\right]\delta y - 4(\delta y)^3, \tag{19}$$

where $x(t)$ is computed from a trajectory in the invariant subspace M, obtained by solving Eqs. (15)-(17), and which can be regarded as a driving term for Eq. (19).

The largest finite-time transversal exponent, computed for the set of initial conditions $\mathbf{x}_0 \equiv \{x(0), y(0) = 0, v_x(0), v_y(0) = 0, \theta(0)\}$ is then given by

$$\tilde{\lambda}_\perp(\mathbf{x}_0, t) = \frac{1}{t}\ln\left[\frac{\delta(t)}{\delta(0)}\right], \tag{20}$$

where

$$\delta(t) = \sqrt{\left[\delta y(t)\right]^2 + \left[\delta v_y(t)\right]^2} \tag{21}$$

is the magnitude of a transversal displacement to the invariant subspace.

The infinite-time limit of Eq. (20), $\lambda_\perp$, determines the transversal stability of the Duffing attractor belonging to M. Figure 5 depicts its variation with the system parameter χ, showing that $\lambda_\perp$ takes on negative values for $\chi > \chi_c \approx 2.315$ (condition **4**). Moreover, if the basin of the

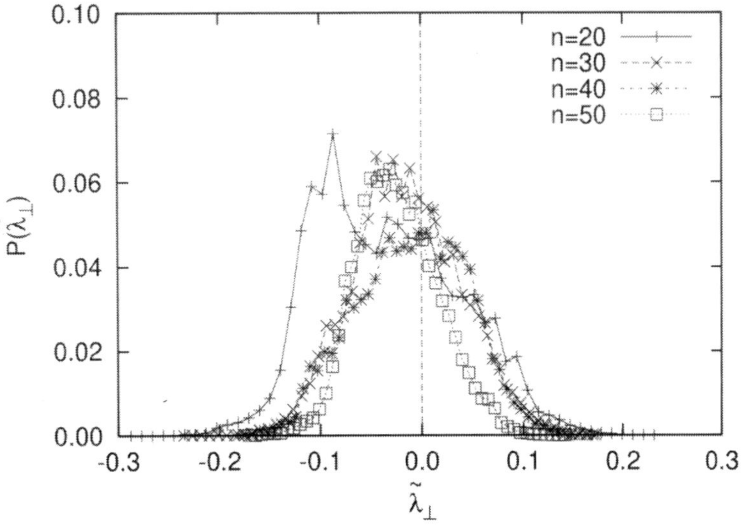

Figure 6. Probability distribution functions for the time-n transversal Lyapunov exponents for different values of t and $\chi = 2.5 < \chi_c$. The remaining parameters are the same as in the previous figures.

chaotic attractor A is riddled by the basin of the attractor at infinity, condition **5** implies that there are negative as well as positive fluctuations of $\tilde{\lambda}_\perp(t)$, in such a way that there is a nonzero fraction of positive finite-time Lyapunov exponents for $\chi > \chi_c$.

We have verified this condition by computing the probability distribution function $P(\tilde{\lambda}_\perp(\mathbf{x}_0, t))$ related to transversal time-t Lyapunov exponent. Numerical approximations of this PDF are depicted in Figure 6 for some values of t, when χ takes on a value higher than χ_c, thus in a regime where condition **4** for riddling holds. The PDF becomes increasingly more akin of a Gaussian function as the time t increases from 20 to 50, as already expected on general bases (large deviation statistics [29]). In all cases the probability distribution function includes negative as well as positive values of $\tilde{\lambda}_\perp(\mathbf{x}_0, t)$, hence there is a nonzero positive fraction $\phi(t)$, thus verifying condition **5** for riddling. Figure 7 shows the variation of the positive fraction with the control parameter χ.

4. Scaling laws for riddled basins

4.1. Fine structure of basin filaments

The onset of riddling has been shown to occur through a bifurcation suffered by some low-period unstable periodic orbit embedded in the chaotic orbit A lying on the invariant manifold M [30]. Exactly at the bifurcation, this orbit (which is already unstable along the invariant manifold, as it is embedded in a chaotic set) loses transversal stability and becomes a repeller along some direction orthogonal to M. Due to the nonlinearity of the transversal dynamics there develops, just after the bifurcation takes place, a narrow tongue anchored at the unstable orbit, as schematically shown by Fig. 8. If an initial condition starts within this narrow tongue, even if it is very close to the invariant set M, the resulting trajectory will asymptote to the infinity.

Once this bifurcation occurs for an unstable orbit in M, the same formation of tongues occurs for every preimage of this orbit, yielding a dense set of tongue-like sets anchored at the corresponding preimages [30]. These sets belong to the basin of the attractor at infinity B.

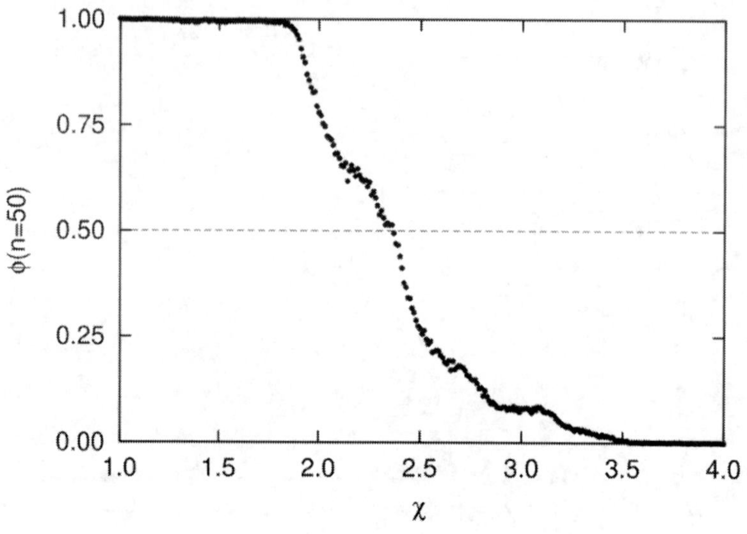

Figure 7. Positive fraction of time-50 transversal Lyapunov exponent as a function of χ.

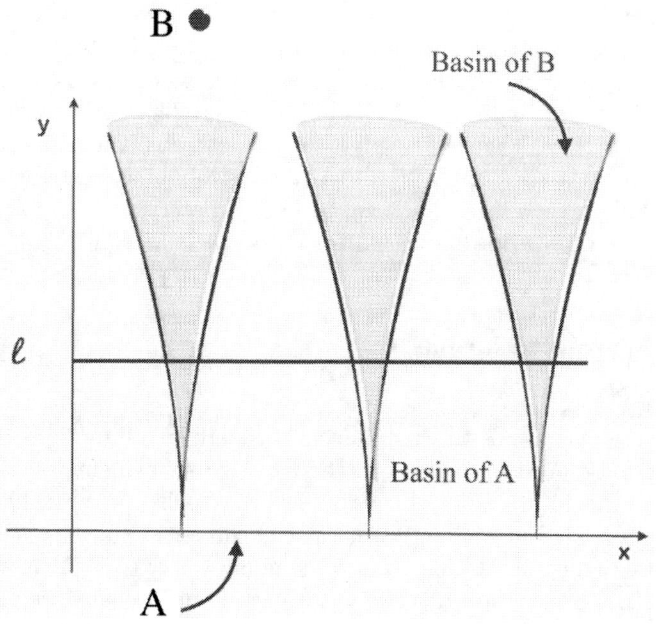

Figure 8. Schematic figure showing the structure of riddled basins near the invariant subspace that contains a chaotic attractor.

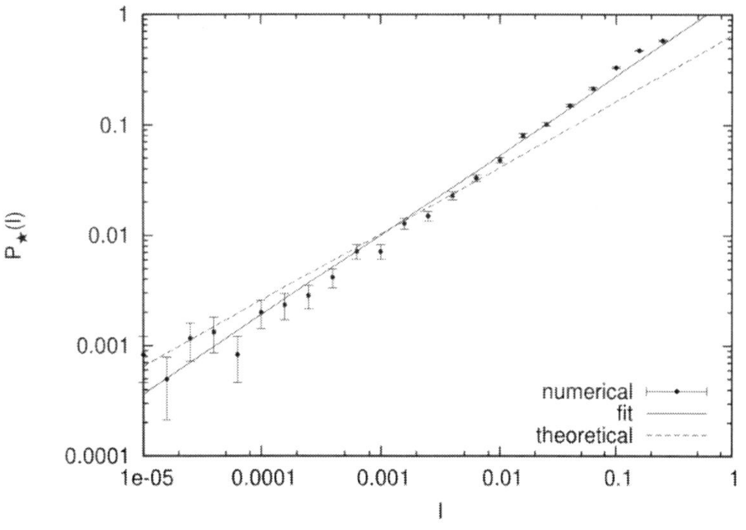

Figure 9. Fraction of length of a line at $y = \ell$ belonging to the basin of infinity, as a function of ℓ. Numerical results, least squares fit and the theoretical prediction given by Eq. (23) are given, respectively, by filled circles, thick dashed line, and thin dashed line.

The characteristic feature of riddling is that those tongues have widths that tend to zero as we approach the invariant manifold containing the chaotic attractor A. Hence the basin of A always contains pieces of basins of B, regardless the distance to the manifold M, so forming a fine structure of basin filaments.

This fine structure can be quantitatively characterized by the following numerical experiment [23, 24]: let us consider the invariant manifold at $y = 0$ and take y as the transversal distance to the manifold. We draw a horizontal line at $y = \ell$, and evaluate the fraction V_ℓ of its length that belongs to the basin of the chaotic attractor A. If the basin of A is riddled with tongues belonging to the basin of infinity, it follows that for any distance ℓ, no matter how small, there is always a nonzero value of V_ℓ. Since the tongues are anchored at the invariant subspace $y = 0$, this fraction tends to zero as $\ell \to 0$ (in the limit it is a Lebesgue measure zero set). Accordingly, the fraction of length belonging to the basin of the infinity is $P_* = 1 - V_\ell$, and it is expected to scale with ℓ as a power law

$$P_*(\ell) \sim \ell^\eta, \tag{22}$$

where $\eta > 0$ is a characteristic scaling exponent.

Figure 9 shows (filled circles) the results of such numerical experiment for the same parameter values used in plotting the basins shown in Fig. 4. The dashed line represents a least squares fit with exponent $\eta = 0.72 \pm 0.02$, where the errors were estimated from the expression $[P_*(1 - P_*)/N]^{1/2}$, following Ref. [24]. Ott and co-workers have developed a stochastic model for the time evolution of the transversal finite-time Lyapunov exponents [24]. Although they are correlated through the deterministic chaotic process governing the system dynamics [in our case the equations (10)-(14)], if the correlation time is short enough we can ignore such correlations and treat the finite-time Lyapunov exponents $\tilde{\lambda}_\perp(\mathbf{x}_0, n)$ as statistically independent innovations. This approximation is supposed to be valid if the system is near the point where the invariant manifold M loses transversal stability as a whole.

The assumption of statistically independent values of $\tilde{\lambda}_\perp(\mathbf{x}_0, n)$ enables us to model the

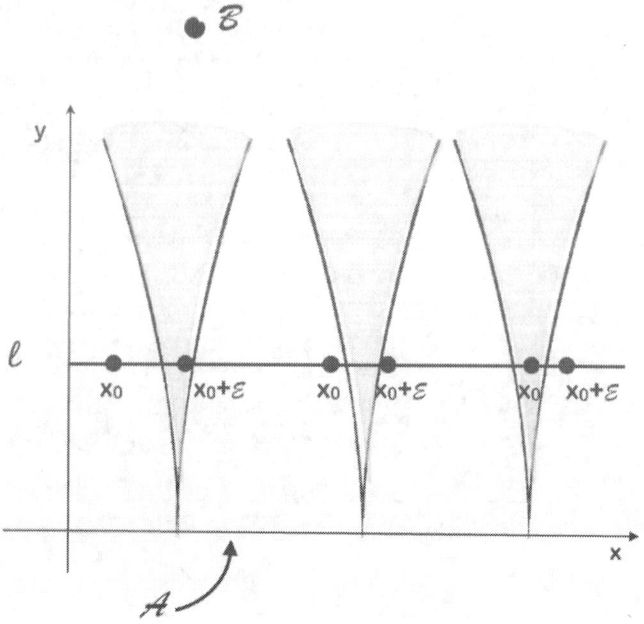

Figure 10. Schematic figure showing the numerical determination of the uncertain fraction of initial conditions.

stochastic dynamics of finite-time exponents as a biased random walk described by a Fokker-Planck equation. The stationary solution of the latter, with reflecting boundary conditions at $y = 0$, predicts a scaling law for P_* of the same form as (22), where the exponent is given by the following expression [23, 24]:

$$\eta = \frac{|\lambda_\perp|}{D}, \tag{23}$$

where $\lambda_\perp$ is the infinite-time limit of the transversal Lyapunov exponent, and D is the diffusion coefficient obtained from the Fokker-Planck equation through the variance of the statistically independent fluctuations. The thin dashed line in Fig. 9 represents the result of the theoretical model based on this stochastic dynamics assumption, giving a prediction of $\eta = 0.60$ for the scaling exponent. As we can see, the agreement is fairly good, specially when larger values of ℓ are considered.

4.2. Uncertainty exponent

Another quantitative characterization of riddled basins is based on the uncertainty exponent described in the Introduction, and which gives a measure of the extreme final-state sensitivity due to riddling. Consider again the line $y = \ell$ drawn in the phase space portrait as described earlier, and choose randomly an initial condition $x(0) = x_0$ on that line. Now choose randomly another initial condition $x'(0)$ with uniform probability within an interval of length 2ϵ and centered at x_0 [Fig. 10]. If both points belong to different basins, they can be referred to as ϵ-uncertain [2, 3].

The fraction of ϵ-uncertain points, or uncertain fraction, which will be denoted by $< p >$, is the probability of making a mistake when attempting to predict to which basin the initial condition $(x_0, y_0 = \ell)$ belongs, given a measurement uncertainty ϵ. This probability is expected, on general grounds, to scale with the latter as another power law of the form

$$< p > \sim \epsilon^{\varphi}, \tag{24}$$

where $\varphi \geq 0$ depends on both x and ℓ. The stochastic model of Ott *et al.* gives the following expression for this exponent in terms of the infinite-time Lyapunov exponents

$$\varphi = \frac{\lambda_{\perp}^2}{4D\lambda_{\|}}. \tag{25}$$

Figure 11 presents our results for this second numerical experiment, for $< p >$ *versus* ϵ for a given ℓ. The dashed line is a least squares fit of the numerical data, yielding an exponent $\varphi = 0.280 \pm 0.005$, where the errors were estimated from the formula $[< p > (1- < p >)/N]^{1/2}$, as for the previous scaling law. This value compares reasonably with the prediction of the stochastic model (25), which gives $\varphi = 0.10$, although the agreement here is not so good as for the first scaling law. Both values of this exponent are near zero, showing that indeed the fractality here is extreme. The smallness of φ is ultimately due to its quadratic dependence on $\lambda_{\perp}$ which is itself very small near the transition to a riddled basin. This explains only partially, however, the disagreement between the numerical and theoretical values for the scaling exponent. The most important factor is that the theoretical model assumes a random walk behavior for the finite-time transversal Lyapunov exponents, thus neglecting any correlation between them, what it is naturally a crude approximation, except very near the blowout bifurcation point (which, in our model, is at χ_c). Outside this rather particular case, we cannot expect but a qualitative agreement between model and numerics.

Just to give a quantitative example of how difficult is to predict the final state of the mechanical system here studied. Suppose we are to improve the accuracy with which the initial conditions are found. A great effort is then hypothetically devoted to improve this accuracy by N orders of magnitude. This means to decrease the uncertainty by about the same factor. Due to the smallness of the uncertainty exponent, the uncertain fraction is expected to decrease a mere with respect to its previous value, whatever it was estimated. Hence no significant improvement in final state predictability was gained even though a massive effort was done to decrease uncertainties. This plagues every physical system exhibiting riddling.

5. Conclusions

Riddled basins of attraction are typical in systems of physical interest. In this paper we consider a particle under a two-dimensional potential function with friction and time-periodic forcing. Thanks to a symmetry of the potential function, the system presents an invariant subspace in which there is a chaotic attractor similar to that of a driven damped pendulum. The riddled basins appear if the transversal dynamics to this subspace present a maximal negative Lyapunov exponent but also transversely unstable periodic orbits embedded in the chaotic attractor.

These two apparently contradictory requirements are possible once the natural measure of a chaotic attractor is supported by an infinite number of unstable periodic orbits. The closed set of transversely stable orbits (embedded in the chaotic attractor) which makes the Lyapunov exponent to be negative coexists with an open set of transversely unstable orbits. The latter can be detected through positive fluctuations of the finite-time Lyapunov exponent (maximal along the transversal direction).

We have used two numerical diagnostics for characterization of riddling: (i) the existence of the scaling law for the fraction of the basin of the second attractor (at infinity) when it riddles the basin of the chaotic attractor lying in the invariant subspace; (ii) the computation of the finite-time Lyapunov exponents along the transversal directions in order to verify the existence

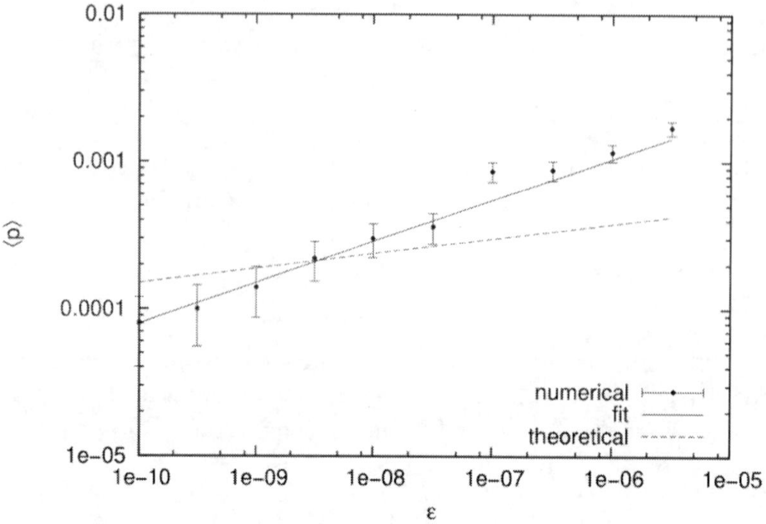

Figure 11. Fraction of uncertain initial condition as a function of the uncertainty radius. Numerical results, least squares fit and the theoretical prediction given by Eq. (23) are given, respectively, by filled circles, thick dashed line, and thin dashed line.

of positive values. In particular, we have verified that the scaling laws agree reasonably well with a theoretical model developed by Ott *et al.* [24] using a biased random walk with reflecting barrier.

Both diagnostics, however, have their shortcomings. The mere existence of a scaling law for the fraction of length of some line drawn transversely to the basin filaments does not necessarily imply riddling. The reason is that any map with an unstable fixed point belonging to an invariant subspace could display the same scaling if there are trajectories forming separatrices. The latter divide the local unstable subspace of this point into different regions according to whether points are asymptotic to an attractor inside or outside the invariant subspace. On the other hand, the mere existence of positive fluctuations of the finite-time transversal Lyapunov exponents cannot warrant the existence of riddled basins, since it may also occur in other contexts, like the proximity of a near-tangency (" glitch") of the stable and unstable subspaces of a periodic saddle orbit. We conclude that the characterization of riddling, in practice, relies on the combination of a number of numerical diagnostics as well as some knowledge of the periodic orbit structure of the system.

The agreement between our numerical results and the model of Ott *et al.* is already expected to be partial, since the stochastic model implies that the finite-time Lyapunov exponents are statistically independent variables, which obviously they are not, since the deterministic chaotic dynamics unavoidably introduces correlations which, however, decay with time. If this decay time is short enough the use of stochastic models is acceptable as an approximation. Moreover, we observed that the stochastic approximation is better if the contributions of positive and negative fluctuations are of comparable importance, a situation more likely to occur near the blowout point, where the chaotic attractor as a whole loses transversal stability.

Acknowledgements

This work was made possible with partial financial support from the CNPq, CAPES, and Fundação Araucária (Paraná).

References

[1] Wiggins S 1990 *Introduction to Applied Dynamical Systems and Chaos* (Berlin-Heidelberg-New York: Springer)
[2] Grebogi C, McDonald S W, Ott E and Yorke J A 1985 *Phys. Lett.* A **100** 1
[3] McDonald S W, Grebogi C, Ott E and Yorke J A 1985 *Physica* D **17** 125
[4] Bleher S, Grebogi C, Ott E and Brown R 1988 *Phys. Rev.* A **38** 930
[5] Kennedy J and Yorke J A 1991 *Physica* D **51** 213
[6] Bleher S, Grebogi C and Ott E 1990 *Physica* D **46** 87
[7] Smale S 1967 *Bull. Amer. Math. Soc.* **73** 747
[8] Tél T and Lai Y C 2008 *Phys. Reports* **460** 245
[9] Aguirre J, Viana R L and Sanjuán M A F 2009 *Rev. Mod. Phys.* **81** 333
[10] Alexander J C, Yorke J A, You Z and Kan I 1992 *Int. J. Bifurcat. Chaos* **2** 795
[11] Pikovsky A and Grassberger P 1991 *J. Phys.* A **24** 4587
[12] Woltering M and Markus M 2000 *Phys. Rev. Lett.* **84** 630
[13] Pikowsky A, Rosenblum M and Kurths J 2003 *Synchronization: a Universal Concept in Nonlinear Sciences* (Cambridge: Cambridge University Press)
[14] Ashwin P 2005 *Dynamics of Coupled Map Lattices And of Related Spatially Extended Systems* (Lecture Notes in Physics vol 671) eds J R Chazottes and B Fernandez (Berlin-Heidelberg: Springer)
[15] Ashwin P, Buescu J and Stewart I 1994 *Phys. Lett.* A **193** 126
[16] Heagy J F, Carroll T L and Pecora L M 1994 *Phys. Rev. Lett.* **73** 3528
[17] Cazelles B 2001 *Chaos, Solit. & Fract.* **12** 301
[18] Pereira R F, Camargo S, Pinto S E S, Lopes S R and Viana R L 2008 *Phys. Rev.* E **78** 056214
[19] Nakajima H and Ueda Y 1996 *Physica* D **99** 35
[20] Woltering M and Markus M 2000 *Chem. Phys. Lett.* **321** 473
[21] Yousefi S, Maistrenko Y and Popovich S 2000 *Discrete Dynamics in Nature and Society* **5** 161
[22] Ott E and Sommerer J C 1993 *Nature* **365** 138
[23] Ott E, Sommerer J C, Alexander J C, Kan I and Yorke J A 1993 *Phys. Rev. Lett.* **71** 4134
[24] Ott E, Alexander J C, Kan I, Sommerer J C and Yorke J A 1994 *Physica* D **76** 384
[25] Milnor J 1985 *Comm. Math. Phys.* **99** 177
[26] Hasler M and Maistrenko Y L 1997 *IEEE Trans. Circ. Syst.* I **44** 856
[27] Dorfman J R 1999 *An Introduction to Chaos in Nonequilibrium Statistical Mechanics* (Cambridge: Cambridge University Press)
[28] Grebogi C, Ott E, and Yorke J A 1987 *Science* **238** 632
[29] Ott E, 1993 *Chaos in Dynamical Systems* (Cambridge: Cambridge University Press), Sec. 9.4
[30] Lai Y C, Grebogi C, Yorke J A and Venkataramani S C 1996 *Phys. Rev. Lett.* **77** 55

Lyapunov exponent of the random frequency oscillator: cumulant expansion approach

C. Anteneodo[1] **and R. O. Vallejos**[2]

[1] Department of Physics, PUC-Rio and National Institute of Science and Technology for Complex Systems, Rua Marquês de São Vicente 225, Gávea, CEP 22453-900 RJ, Rio de Janeiro, Brazil

[2] Centro Brasileiro de Pesquisas Fisicas, Rua Dr. Xavier Sigaud 150, Rio de Janeiro, Brazil

E-mail: `celia@fis.puc-rio.br,vallejos@cbpf.br`

Abstract. We consider a one-dimensional harmonic oscillator with a random frequency, focusing on both the standard and the generalized Lyapunov exponents, λ and $\lambda^\star$ respectively. We discuss the numerical difficulties that arise in the numerical calculation of $\lambda^\star$ in the case of strong intermittency. When the frequency corresponds to a Ornstein-Uhlenbeck process, we compute analytically $\lambda^\star$ by using a cumulant expansion including up to the fourth order. Connections with the problem of finding an analytical estimate for the largest Lyapunov exponent of a many-body system with smooth interactions are discussed.

1. Introduction

The theory of Lyapunov exponents of hard-ball systems has a long history. It started with the pioneering work of Krylov [1, 2], was rigorously developed by Sinai [3] and collaborators, and completed (to some extent) by van Beijeren, Dorfman and co-workers [4, 5, 6, 7, 8]. The analytical calculation of, e.g., the largest Lyapunov exponent of a dilute rigid-sphere gas, is based on the fact that the dynamics consists of free rectilinear motions interrupted by instantaneous elastic collisions [6]; the expressions so-obtained agree quantitatively with the numerical experiments [6, 9, 10].

The case of a dilute gas with finite-range interactions can be handled in close analogy with the rigid-sphere problem: though the collisions are not trivial any more, the dynamics is still ruled by occasional pairwise encounters [6, 11, 12]. However, when one considers long-range interactions (or short-range interactions and high densities), the theoretical approach must be substantially modified.

In the general case we must deal with the full system of coupled differential equations that govern the evolution of multidimensional tangent vectors $\eta(t)$. Consider for instance a gas of N particles in three dimensions described by the Hamiltonian

$$\mathcal{H} = \sum_{i=1}^{3N} \frac{p_i^2}{2m} + \mathcal{V}(q_1, \ldots, q_{3N}), \tag{1}$$

where q_i and p_i, are conjugate position-momentum coordinates. Assuming $m = 1$, tangent vectors evolve according to

$$\dot{\eta} = \begin{pmatrix} 0 & 1 \\ -\mathbf{V}(t) & 0 \end{pmatrix} \eta \tag{2}$$

(dot meaning time derivative), where $\mathbf{V}$ is the Hessian matrix of the potential $\mathcal{V}$, namely

$$V_{ij} = \frac{\partial^2 \mathcal{V}}{\partial q_i \partial q_j} . \tag{3}$$

The Hessian depends explicitly on time because it is calculated along a reference trajectory $q(t)$. Once initial conditions $x_0 = (q_0, p_0)$ and η_0 have been specified, one can find $\eta(t)$ from Eq. (2). Then the Lyapunov exponent λ is obtained by calculating the limit [13]

$$\lambda = \lim_{t\to\infty} \frac{1}{t} \ln |\eta(t; x_0, \eta_0)| . \tag{4}$$

Assuming ergodicity on the energy-shell, λ becomes independent of initial conditions x_0, which can then be chosen randomly according to the microcanonical distribution. There will also be no dependence on initial tangent vectors, because if η_0 is also chosen randomly, it will have a non-zero component along the most expanding direction. It is this average over x_0 and η_0 that permits to treat equations (2) formally as a system of stochastic differential equations [14]. Moreover, if the dynamics can be thought of as free motion plus weak interactions, then perturbative techniques, like the cumulant expansion [14, 15, 16], can be invoked. So, the theory attempts to calculate the average

$$\lambda = \lim_{t\to\infty} \frac{1}{t} \langle \ln |\eta(t; x_0, \eta_0)| \rangle . \tag{5}$$

However, in practice, it is much simpler to develop an estimate for the generalized Lyapunov exponent [17, 18]

$$\lambda^{\star} = \lim_{t\to\infty} \frac{1}{2t} \ln \langle |\eta(t; x_0, \eta_0)|^2 \rangle . \tag{6}$$

This is essentially the approach followed by Barnett et al [19, 20, 21], Pettini et al [22, 23, 24], and the present authors [25, 26, 27]. In situations of weak intermittency both exponents are expected to be close. If one wishes to use a theoretically calculated $\lambda^{\star}$ as an approximation to λ, then a numerical check must be done first to verify that both exponents coincide. The cumulant expansion to be discussed below offers an analytical expression for λ via the replica trick. Note, however, that the difficulties involved in such a calculation are much greater than those we shall face when dealing with $\lambda^{\star}$.

Though there are some differences among the formulations of the three just-mentioned groups, it may be said that the main theoretical conclusion extracted from that body of work is: if one combines the cumulant expansion with some kind of isotropy approximation (which may be fully justified in some cases), the original problem of $6N$ differential equations can be reduced to a system of only two equations for a "representative" single degree of freedom:

$$\begin{pmatrix} \dot{\eta}_1 \\ \dot{\eta}_2 \end{pmatrix} = \begin{pmatrix} 0 & 1 \\ -\kappa(t) & 0 \end{pmatrix} \begin{pmatrix} \eta_1 \\ \eta_2 \end{pmatrix} . \tag{7}$$

In this kind of mean-field approximation, the "curvature" $\kappa(t)$ is a scalar stochastic process, whose cumulants can be related to the (operator) cumulants of the Hessian $\mathbf{V}(t)$ (see, e.g., [25]).

The comparison of theoretical results obtained with the cumulant approach versus numerical simulations has met mixed success. The agreement is very good for a many-particle system with bounded weak interactions [26, 27] and for the Fermi-Pasta-Ulam system [24]. However, the results for the 1d-XY model [24], for a dense one-component plasma [19, 28], and for a dilute Lennard-Jones gas [29] are not so satisfactory.

The purpose of this paper is to investigate the limits of validity of the cumulant approach for the Lyapunov exponent of a many particle system. We choose as a starting point the simplified mean-field setting (7) and consider two possibilities for $\kappa(t)$. It has been argued [24] that, for typical chaotic many-body systems, $\kappa(t)$ should be close to Gaussian white noise; this is the first case we shall consider. In the white-noise case the second-order expansion for $\lambda^\star$ is exact, thus this case is ideally suited for analyzing the difficulties that appear in the *numerical* calculation of $\lambda^\star$. Next, we keep the Gaussian and Markov properties but allow for finite correlation times, leading to the Ornstein-Uhlenbeck process; in this case we calculate the fourth cumulant contribution to $\lambda^\star$. Though it will not be considered here, we also mention the interesting situation of $\kappa(t)$ being a Poisson process, which appears to be the appropriate choice for modeling the tangent-vector dynamics in a dilute gas with short-range interactions.

2. Cumulant expansion for the Kubo oscillator

Formally, Eq. (7) describes a harmonic oscillator with a random frequency ω such that $\omega^2 = \kappa$ (Kubo oscillator). It is worth generalizing this model a bit to account for the possibility of damping, i.e., we shall consider an oscillator described by the dynamical equation

$$\ddot{q} + \alpha\,\dot{q} + \kappa\,q = 0\,. \tag{8}$$

Setting $\alpha = 0$, $q = \eta_1$, $p = \dot{q} = \eta_2$, recovers (7).

Some analytical results for the Lyapunov exponent of the Kubo oscillator (8) can be found in the literature (see, e.g., [30, 31]). Here we shall restrict ourselves to the analytical calculation of the generalized exponent $\lambda^\star$. For this purpose we must consider the dynamics of second moments:

$$\frac{d}{dt}\begin{pmatrix} q^2 \\ p^2 \\ qp \end{pmatrix} = \begin{pmatrix} 0 & 0 & 2 \\ 0 & -2\alpha & -2\kappa \\ -\kappa & 1 & -\alpha \end{pmatrix}\begin{pmatrix} q^2 \\ p^2 \\ qp \end{pmatrix} \equiv \mathbf{B}(t)\begin{pmatrix} q^2 \\ p^2 \\ qp \end{pmatrix}. \tag{9}$$

Let us think that, in principle, both parameters α and κ are stationary stochastic processes. If fluctuations are small enough (in a sense that will be discussed later), one can obtain the average of the second-moment vector using the first terms of the cumulant expansion, which works as follows [14]. First we split the stochastic matrix as an average plus fluctuations:

$$\mathbf{B}(t) = \mathbf{B_0} + \mathbf{B_1}(t)\,. \tag{10}$$

For long times one has:

$$\frac{d}{dt}\left\langle\begin{pmatrix} q^2 \\ p^2 \\ qp \end{pmatrix}\right\rangle = \mathbf{K}\left\langle\begin{pmatrix} q^2 \\ p^2 \\ qp \end{pmatrix}\right\rangle, \tag{11}$$

where $\mathbf{K}$ is the 3×3 matrix given by the operator cumulant expansion [14]

$$\mathbf{K} = \mathbf{B}_0 + \int_0^\infty \langle \mathbf{B}_1(\tau)\, e^{\mathbf{B}_0\tau}\, \mathbf{B}_1(0)\rangle\, e^{-\mathbf{B}_0\tau} d\tau + \ldots\,. \tag{12}$$

Dots stand for third and higher cumulants (some explicit expressions can be found in [16]). The exponent $\lambda^\star$ is related to the eigenvalue of $\mathbf{K}$ that has the largest real part:

$$\lambda^\star = \frac{1}{2}\max\,\Re\,\{k_1, k_2, k_3\}\,, \tag{13}$$

with k_i the eigenvalues of $\mathbf{K}$.

3. Gaussian white noise

When the entries of the fluctuation matrix $\mathbf{B}_1$ are Gaussian white noise (and only in this case [16]) the cumulant expansion stops at the second order, i.e., Eq. (12) is exact (without the ellipsis). This is the case we consider now.

3.1. Random frequency

Let us first study the situation where the damping α is a constant and

$$\kappa(t) = \kappa_0 + \xi(t)\,, \tag{14}$$

where $\xi(t)$ is a zero-mean Gaussian white noise. Its correlation function reads

$$\langle \xi(t)\,\xi(t')\rangle = \Delta\,\delta(t-t')\,. \tag{15}$$

With these definitions one has

$$\mathbf{B} = \begin{pmatrix} 0 & 0 & 2 \\ 0 & -2\alpha & -2\kappa_0 \\ -\kappa_0 & 1 & -\alpha \end{pmatrix} + \xi(t)\begin{pmatrix} 0 & 0 & 0 \\ 0 & 0 & 2 \\ 1 & 0 & 0 \end{pmatrix}. \tag{16}$$

After substitution into Eq. (12) we readily obtain

$$\mathbf{K} = \begin{pmatrix} 0 & 0 & 2 \\ \Delta & -2\alpha & -2\kappa_0 \\ -\kappa_0 & 1 & -\alpha \end{pmatrix}. \tag{17}$$

The generalized exponent $\lambda^\star$ can now be calculated from Eq. (13). A closed expression for the standard Lyapunov exponent can be found in the literature [30]. As an example, Fig. 1 displays analytical results for both exponents. We also show the outcomes of numerical simulations. Given that the theoretical expressions are exact, Fig. 1 constitutes a test for our numerical calculations. Numerical details, including a discussion about the difficulties found in the calculation of $\lambda^\star$, will be presented in Sec. 4.

3.2. Random damping

Now we consider an harmonic oscillator with constant frequency but in an environment with fluctuating damping coefficient

$$\alpha(t) = \alpha_0 + \xi(t)\,, \tag{18}$$

where $\xi(t)$ is also in this case a zero-mean Gaussian white noise. The corresponding stochastic differential equation (8) will be taken in Stratonovich sense. Therefore, the matrix $\mathbf{B}$ in Eq. (9) can be decomposed as

$$\mathbf{B} = \begin{pmatrix} 0 & 0 & 2 \\ 0 & -2\alpha_0 & -2\kappa \\ -\kappa & 1 & -\alpha_0 \end{pmatrix} + \xi(t)\begin{pmatrix} 0 & 0 & 0 \\ 0 & 2 & 0 \\ 0 & 0 & 1 \end{pmatrix}. \tag{19}$$

Hence, substitution into (12) yields

$$\mathbf{K} = \begin{pmatrix} 0 & 0 & 2 \\ \Delta & -2\alpha_0 + 2\Delta & -2\kappa \\ -\kappa & 1 & -\alpha_0 + \Delta/2 \end{pmatrix}. \tag{20}$$

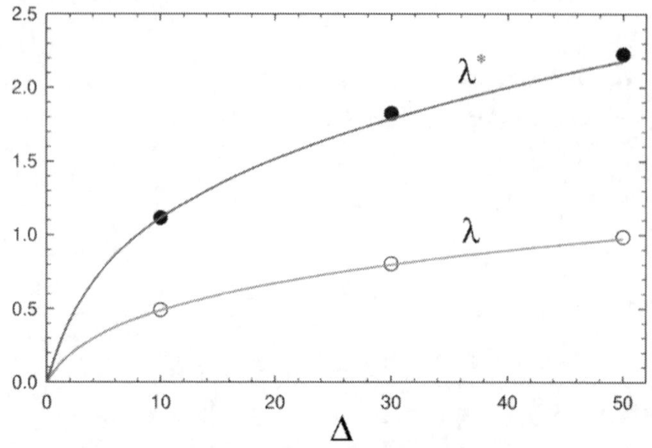

Figure 1. Harmonic oscillator with random frequency. Shown are the Lyapunov exponents λ and $\lambda^\star$ as a function of the noise strength Δ. Solid lines correspond to the theoretical expressions given by inserting (17) into (13), and in Ref. [30] for $\lambda^\star$ and λ, respectively. Symbols are the results of numerical simulations (averaged over 10^4 trajectories). We chose $\alpha = 0$ and $\kappa_0 = 1$.

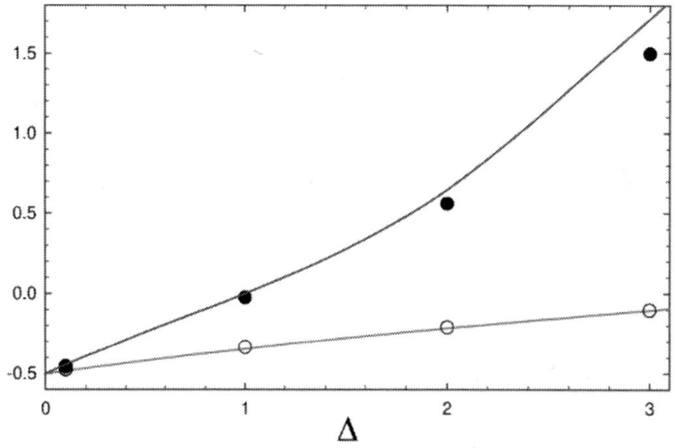

Figure 2. Harmonic oscillator with random damping. Shown are the Lyapunov exponents λ and $\lambda^\star$ as a function of the noise strength Δ. Solid lines correspond to the theoretical expressions given by (13,20) and Ref. [31], for $\lambda^\star$ and λ, respectively. Symbols are the results of numerical simulations (averaged over 10^4 trajectories). We chose $\kappa = 1$ and $\alpha_0 = 1$.

From the eigenvalues of **K** we obtain $\lambda^\star$ following Eq. (13). A theoretical expression for λ can be found in Ref. [31]. Fig. 2 exhibits numerical and analytical results for both exponents, as a function of the noise intensity.

Note that in the cases considered above λ and $\lambda^\star$ do not coincide. We have checked that the difference between them (which is a quantifier of the degree of intermittency of the dynamics) may be controlled by suitable choice of the parameters of the oscillator. We preferred to consider intermittent cases, because it is in these regimes that the numerical difficulties arise, as we discuss in the following section. We remark that there are situations of interest where both exponents practically coincide, e.g., for a dilute Lennard-Jones gas [29]. In such cases a theory capable of estimating $\lambda^\star$ will also produce a good estimate for the standard Lyapunov exponent λ.

4. Numerical treatment

Numerical simulations were performed by means of the Euler algorithm with time step $dt = 10^{-3}$. For each trajectory we computed the norm $|\eta(t)|^2 = q^2 + p^2$ as a function of time t. The Lyapunov exponent is approximated by the average over initial conditions of the finite-time exponents:

$$\lambda \approx \frac{1}{t}\langle \ln|\eta(t; x_0, \eta_0)|\rangle \equiv \langle \lambda(t; x_0, \eta_0)\rangle \,, \tag{21}$$

where t is large enough to guarantee the convergence of the average to the desired precision. In order to obtain the asymptotic value of the generalized exponent (6), in principle, one must calculate the squared-norm averaged over several realizations at a given large time. However, we must keep in mind that such an average is dominated by the extreme positive values of the local exponent $\lambda(t)$. Hence, direct averaging over $|\eta(t)|^2$ may yield spurious results whenever the variance of $\lambda(t)$ fails to vanish with time fast enough. To avoid this problem, instead of the straightforward averaging, we preferred to estimate the local generalized exponent from the cumulants of the distribution of $\lambda(t)$:

$$\lambda^\star(t) = \frac{\ln\langle|\eta(t)|^2\rangle}{2t} = \frac{\ln\langle e^{2\lambda(t)t}\rangle}{2t} = \sum_{n\geq 1}\frac{(2t)^{n-1}}{n!}\kappa_n(t)\,, \tag{22}$$

where κ_n are the nth-order cumulants of the distribution of local exponents $\lambda(t)$. Fig. 3 illustrates, for the white-noise random frequency oscillator, the $\langle\lambda(t)\rangle$ as a function of time (first-order truncation of (22)), as well as the expansion (22) truncated at the second and third orders. For comparison, also plotted is the crude estimate (6). Clearly, the expansion (22) has to be considered in order to properly estimate $\lambda^\star$. In Figs. 1 and 2, $\lambda^\star$ was numerically computed from the third-order truncation, because the next (noisier) terms do not contribute significantly.

5. Correlated noise

For white noise fluctuations, either in the frequency or in the damping, we have verified in Sec. 3 (see Figs. 1 and 2), that our theory for $\lambda^\star$ is in agreement with numerical results, provided the later are obtained by means of the procedure described in the preceding section. The analysis in Sec. 3 also allows to quantify the discrepancy between λ and $\lambda^\star$, which typically increases with increasing amplitude of the fluctuations.

Now we shall analyze the effect of introducing noise correlations. We consider again the case of a random frequency, as in Eq. (14), but now the noise is a zero-mean Ornstein-Ulhenbeck process, i.e., with correlation function

$$\langle\xi(t)\xi(t')\rangle = \frac{\Delta}{2\tau}\exp(-|t-t'|/\tau) \equiv \sigma^2\exp(-|t-t'|/\tau)\,. \tag{23}$$

For simplicity we set $\alpha = 0$ and $\kappa_0 = 0$. By inserting Eq. (16) into Eq. (12), the second-cumulant matrix $\mathbf{K}^{(2)}$ becomes

$$\mathbf{K}^{(2)} = \begin{pmatrix} 0 & 0 & 2 \\ \Delta & -2\Delta\,\tau^2 & 0 \\ \Delta\,\tau & 1 & -2\Delta\,\tau^2 \end{pmatrix}. \tag{24}$$

Notice that in the limit $\tau \to 0$ the white-noise case is recovered.

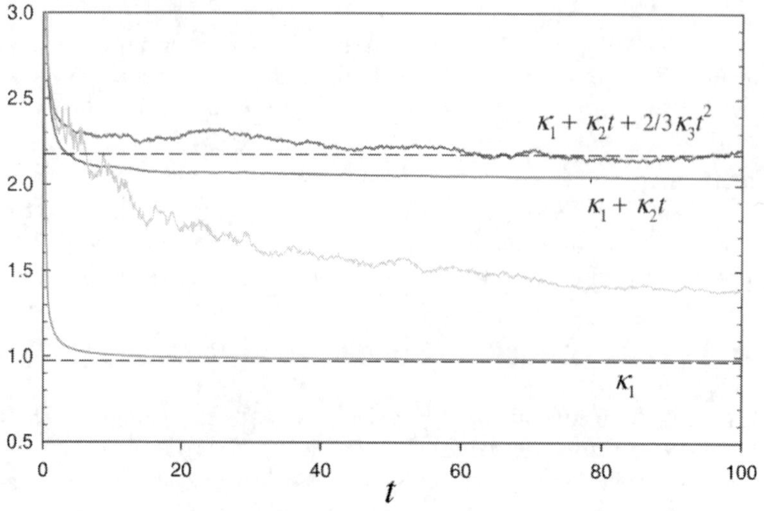

Figure 3. Numerical difficulties in the calculation of $\lambda^\star$. We plot the finite-time exponent $\lambda(t) \equiv \kappa_1(t)$ as a function of time (red line) for the random frequency oscillator with $\alpha = 0$ and $\Delta = 50$. Averages were computed over 10^5 realizations. Also plotted are the corrections arising from the second and third cumulants of the distribution of finite-time Lyapunov exponents Eq. (22) (blue). For comparison we also show the straightforward average $\ln\langle|\eta(t; x_0, \eta_0)|^2\rangle/2t$ (light gray). Dashed lines correspond to the theoretical asymptotic values.

In the presence of correlations the second-order truncation of the cumulant expansion (12) is not exact. In order to improve the theory one must calculate higher cumulants. For the present case the third cumulant is null. Explicit expressions for the fourth cumulant were given by Fox [16] and by Breuer et al [32]. So, the fourth cumulant can be calculated without great effort (with the aid of algebraic manipulation programs). The fourth order approximation to $\mathbf{K}$ reads

$$\mathbf{K}^{(4)} = \mathbf{K}^{(2)} + \frac{1}{2}\Delta^2\tau^3 \begin{pmatrix} 0 & 0 & 0 \\ 13 & 74\tau^2 & -57\tau \\ 17\tau & 173\tau^3 & -99\tau^2 \end{pmatrix}. \tag{25}$$

The comparison between the theoretical results for $\lambda^\star$ (with the second (blue) and fourth (dark blue) order corrections) and numerical outcomes is shown in Fig. 4. Notice that in numerical estimates the fourth-order correction is very small in comparison with the third-order one, suggesting that the cumulant expansion is rapidly converging.

5.1. Kubo number

The perturbation parameter controlling the convergence of the cumulant expansion is the so-called Kubo number ε. General considerations led van Kampen [14] to conclude that the Kubo number is the product of the amplitude of the fluctuations and the correlation time, that is $\sigma\tau$. However, in the present case it is clear that such a combination is not adimensional. The correct Kubo number is instead

$$\varepsilon = \sigma\tau^2 = \sqrt{\frac{\Delta\tau^3}{2}}. \tag{26}$$

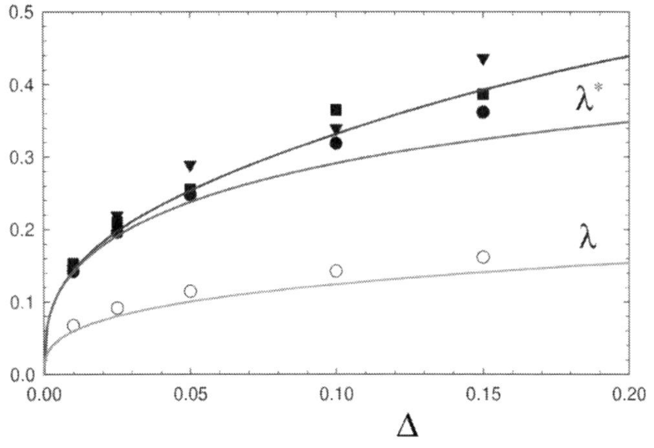

Figure 4. Harmonic oscillator with correlated random frequency. We show λ and $\lambda^\star$ as a function of the noise amplitude Δ. Solid lines correspond to the theoretical results obtained from (13,24) (blue) and (13,25) (dark blue) for $\lambda^\star$ and to the approximate expression (using the decoupling ansatz) following Ref. [30] for λ. Symbols are the results of numerical simulations (averaged over 10^5 trajectories), corresponding to the second (circles), third (squares) and fourth (triangles) order corrections of (22). Parameters are $\alpha = 0$, $\kappa_0 = 0$ and $\tau = 1$.

This can be checked explicitly from the second and fourth cumulants above. Consider, for instance, the element $\mathbf{K}_{21}$, which dominates the Lyapunov exponent for small correlation times:

$$\mathbf{K}_{21} = \Delta + \frac{13}{2}\Delta^2\tau^3 + \cdots = \Delta\left(1 + \frac{13}{2}\Delta\tau^3 + \ldots\right). \tag{27}$$

In the white-noise limit, i.e., $\tau \to 0$ with Δ fixed, the Kubo number tends to zero –as it should be.

6. Final remarks

We have taken the first step towards the application of the cumulant expansion to calculate the largest Lyapunov exponent of a dilute gas.

The case of white-noise fluctuations (either in the frequency or in the damping) was considered first. This study was very useful to understand the difficulties behind the numerical calculation of the generalized exponent $\lambda^\star$. It was verified that $\lambda^\star$ can be obtained with a satisfactory precision by using the cumulant expansion for the distribution of the finite-time Lyapunov exponent.

We also analyzed briefly the case of correlated noise. For the Ornstein-Uhlenbeck noise we were able to obtain the fourth cumulant contribution to the analytical $\lambda^\star$, which showed an improvement with respect to the second order truncation, when compared with numerical outcomes. Moreover, we showed that the correct perturbative parameter for the present problem is the product $\sigma\tau^2$, and not $\sigma\tau$, as a literal reading of van Kampen's discussion [14] would suggest.

It is expected that the present results will be helpful for the correct application of the cumulant approach in higher dimensionality systems, as well as for the numerical checking of its validity.

Acknowledgements:
We acknowledge Brazilian agencies Faperj and CNPq for partial financial support.

[1] Krylov NS 1979 *Works on the Foundations of Statistical Physics* (Princeton University Press, Princeton)
[2] Ma S-K 1985 *Statistical Mechanics* (World Scientific, Singapore)
[3] Sinai YaG 1970 *Russ. Math. Surv.* **25** 137
[4] van Beijeren H, Dorfman JR 1995 *Phys. Rev. Lett.* **74** 4412
[5] van Beijeren H, Latz A, Dorfman JR 1998 *Phys. Rev. E* **57** 4077
[6] van Zon R, van Beijeren H, Dellago Ch 1998 *Phys. Rev. Lett.* **80** 2035
[7] Kruis HV, Panja D, van Beijeren H 2006 *J. Stat. Phys.* **124** 823
[8] Dorfman JR 1999 *An Introduction to Chaos in Nonequilibrium Statistical Mechanics* (Cambridge University Press, Cambridge, UK)
[9] Dellago Ch, Posch HA, Hoover WG 1996 *Phys. Rev. E* **53** 1485
[10] Dellago Ch, Posch HA 1997 *Physica A* **240** 68
[11] Kimball JC 2001 *Phys. Rev. E* **63** 066216
[12] Elyutin PV 2004 *Phys. Lett. A* **331** 153
[13] Benettin G, Galgani L, Strelcyn J-M 1976 *Phys. Rev. A* **14** 2338
[14] van Kampen NG 1981 *Stochastic Processes in Physics and Chemistry* (North-Holland, Amsterdam)
[15] Kubo R 1962 *J. Phys. Soc. Japan* **17** 1100
[16] Fox RF 1974 *J. Math. Phys* **15** 1479
[17] Benzi R, Paladin G, Parisi G, Vulpiani A 1985 *J. Phys. A* **18** 2157
[18] Castiglione P, Falcioni M, Lesne A, Vulpiani A 2008 *Chaos and Coarse Graining in Statistical Mechanics* (Cambridge University Press, New York, 2008)
[19] Barnett DM, Tajima T, Nishihara K, Ueshima Y, Furukawa H 1996 *Phys. Rev. Lett.* **76** 1812
[20] Barnett DM, Tajima T, Nishihara K, Ueshima Y, Furukawa H 1997 *Phys. Rev. E* **55** 3439
[21] Barnett DM, Tajima T 1996 *Phys. Rev. E* **54** 6084
[22] Casetti L, Livi R, Pettini M 1995 *Phys. Rev. Lett.* **74** 375
[23] Casetti L, Clementi C, Pettini M, 1996 *Phys. Rev. E* **54** 5969
[24] Casetti L, Pettini M, Cohen EGD. 2000 *Phys. Rep.* **337** 238
[25] Vallejos RO, Anteneodo C 2002 *Phys. Rev. E* **66** 021110
[26] Anteneodo C, Maia RNP, Vallejos RO, 2003 *Phys. Rev. E* **68** 036120
[27] Vallejos RO, Anteneodo C 2004 *Physica A* **340** 178
[28] Torcini A, Dellago Ch, Posch HA 1999 *Phys. Rev. Lett.* **83** 2676
Barnett DM, Tajima T, Ueshima Y 1999 *ibid.* **83** 2677
[29] Anteneodo C, Cirto L, Vallejos RO (unpublished)
[30] Mallick K, Peyneau PE 2006 *Physica D* **221** 72
[31] Leprovost N, Aumaître S, Mallick K 2006 *Eur. Phys. J. B* **49** 453
[32] Breuer H-P, Ma A, Petruccione F 2002 *preprint* arXiv:quant-ph/0209153v1
[33] Risken H 1984 *The Fokker-Planck Equation: Methods of Solution and Applications* (Springer-Verlag, Berlin)
[34] Gardiner C W 1985 *Handbook of stochastic methods for Physics, Chemistry and Natural Sciences* (Springer-Verlag, Berlin)

Bouncing Ball Problem: Numerical Behavior Characterization

E. E. N. Macau[1], M. V. Carneiro[2], J. J. Barroso[3]

[1,2]Computing and Applying Mathematics Laboratory
[3]Associated Plasma Laboratory
National Institute for Space Research
12227-010 – São José dos Campos – SP
Brazil

[1]elbert@lac.inpe.br, [2]mvcmartins@gmail.com, [3]barroso@plasma.inpe.br

Abstract. This paper gives an overview of the simple yet fundamental bouncing ball problem, which consists of a ball bouncing vertically on a sinusoidally vibrating table under the action of gravity. The dynamics is modeled on the basis of a discrete map of difference equations, which numerically solved fully reveals a rich variety of nonlinear behaviors, encompassing irregular non-periodic orbits, subharmonic and chaotic motions, chattering mechanisms, and also unbounded non-periodic orbits.

1. Introduction

Consisting of a ball bouncing vertically under the action of gravity on a massive sinusoidally vibrating platform, such a deterministic system exhibits large families of irregular non-periodic solutions, fully developed chaos in addition to harmonic and subharmonic motions [1] depending upon the amplitude and frequency of the driving platform and also on the coefficient of restituition $0 \le \varepsilon \le 1$, which accounts for the amount of energy dissipated during each collision, with the elastic limit $\varepsilon = 1$ giving rise to unbounded non-periodic or stochastic motions [2]. This system shows to be meaningful in modeling relevant nonlinear systems subject to periodic excitation [3-7].

In past studies, the bouncing-ball problem has been examined in many of its ramifications. The first systematic study is attributed to Holmes [8], upon constructing a discrete map whereby the conditions for stability and bifurcation of periodic trajectories are determined on the assumption the jumps of the ball are larger compared to the overall displacement of the table. Other studies based on the differential equation of motion of the ball [9], or else using a mapping approach similar to Holmes's investigated chaotic response and manifold collisions [10], period doubling regime [11], noise-induced chaotic motion [12], the completely inelastic case [13], rate of energy input into the system [14, 15], and chattering mechanisms through which the ball gets locked on the vibrating table [16,17].

However, there still remains a lack of information on how to access the appropriate driving parameters and also the starting conditions so as to drive the ball into a prescribed oscillation mode at a given coefficient of restitution. To supplement this kind of information and extend past studies, the present paper embraces numerical methods based on computer simulation to examine in a unified way the rich phase behavior of the bouncing ball with emphasis on the driving and launching parameters so that the ball dropped at zero initial velocity might evolve to a desired periodic orbit and keep bouncing there.

2. Bouncing-ball behavior under various starting conditions

Let us consider an elastic ball, with coefficient of restitution ε, which is kept continually bouncing off a vertically oscillating base. Infinitely massive, the platform is fixed to a rigid frame that vibrates sinusoidally as $S(t) = A\sin\omega t$ so as to maintain the motion of the ball, whose dynamics is governed by a gravitational field g and the impacts with the base (Fig. 1). The next collision time after the departure time t_i from the platform is the smallest solution $t_{i+1}>t_i$ of the discrete-time dynamics map

$$A(\sin\omega t_{i+1} - \sin\omega t_i) = V_i(t_{i+1} - t_i) - \frac{1}{2}g(t_{i+1} - t_i)^2 \tag{1}$$

where V_i is the post impact velocity (Fig. 1), which relates to the pre-impact velocity U_{i+1} at time t_{i+1} through

$$U_{i+1} = V_i - g(t_{i+1} - t_i) \tag{2}.$$

As far as the collision is partially elastic, the ball bounces back instantanteously at t_{i+1} with a relative velocity

$$V_{i+1} - \dot{S}(t_{i+1}) = -\varepsilon[U_{i+1} - \dot{S}(t_{i+1})] \tag{3}$$

where the relative landing velocity $U_{i+1} - \dot{S}(t_{i+1})$ is always negative. Physically, the coefficient ε (defined as the ratio of the relative velocities before and after the collision) gives a measure through the quantity $(1-\varepsilon^2)$ of the energy lost in the collision. Combining (1)-(3) and non-dimensionalizing the time and velocity variables according to $t_i \to \omega t_i \equiv \phi_i$, and $v_i \to V_i(\omega / g)$ gives the phase and velocity maps

$$\phi_{i+1} = \phi_i + \tau_i \tag{4a}$$

$$\Gamma[\sin(\phi_i + \tau_i) - \sin\phi_i] = v_i\tau_i - \frac{1}{2}\tau_i^2 \tag{4b}$$

$$v_{i+1} = -\varepsilon(v_i - \tau_i) + (1+\varepsilon)\Gamma\cos(\phi_i + \tau_i) \tag{4c}$$

where $\Gamma = A\omega^2 / g$ is the dimensionless shaking acceleration. With state characterized by the phase ϕ_i and the post-velocity v_i, the above discrete map describes the complete bouncing ball dynamics, which is controlled by two parameters, namely, Γ and ε.

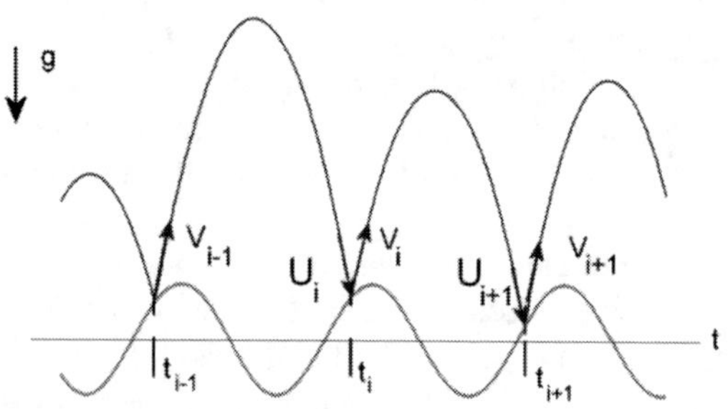

Figure 1 One-dimensional ball bouncing off a sinusoidally vibrating table. The table and ball trajectories are depicted by solid lines over time

We now proceeding to derive the initial starting conditions for the ball to execute a periodic motion motion upon collision with the vibrating platform. Then we discuss the various characters of the bouncing-ball trajectories determined numerically from the full system (4). The numerical solutions are obtained by using an event driven procedure [18] that consists in monitoring a sequence of events for which the force and trajectory equations (4) are solved without resorting to any approximation.

Let us consider a ball that drops at zero initial velocity form the height H_0 at initial phase ϕ_0. Changing to the nondimensionalized coordinages, $h_0 = H_0 w^2/g$, the free flight is described by $h = h_0 - (\phi - \phi_0)^2/2$, such that at the collision phase ϕ, the height is $h = \Gamma \sin\phi$, and since for a periodic motion of subharmonic index n the initial phase ϕ_0 is symmetrically spaced from the phases for the n-1 and nth impacts, the following relation

$$\phi = \phi_0 + n\pi \tag{5}$$

holds, and therefore

$$h_0 = \frac{1}{2}(n\pi)^2 + \Gamma \sin\phi \tag{6}.$$

The exact phase at which impact occurs is such that the upward movement of the platform compensates for the energy loss from the inelastic collisions so as the ball lands and departs from the platform at the same speed $v = n\pi$ (relative to the laboratory frame), which is consistent with the final velocity the ball reaches after the time interval $\phi - \phi_0 = n\pi$ given in (9). Recalling that the coefficient of restitution ε relates to Γ and ϕ through

$$\Gamma \cos\phi = n\pi \frac{1-\varepsilon}{1+\varepsilon} \tag{7}$$

then for a given ε and period index n the initial height h_0 and phase ϕ_0 are calculated from (5)-(7) provided the constraint $0 < \Gamma \sin\phi < 1$ is fulfilled to ensure stability of the periodic orbits. Assuring the existence of periodic orbits, Eq. (11) has two solutions, one of which is unstable [19]. To this end, we set $\Gamma \sin\phi = 1/2$ at $\varepsilon = 0.85$ and n=1, this resulting in a complex pair of eigenvalues $\lambda_{1,2} = \pm i\sqrt{\varepsilon}$. Using(11) gives $\Gamma = 0.5611$ and $\phi = \pm 0.3501\pi$, which combined with Eqs (5)-(7) yields two solutions with the corresponding starting conditions $(\phi_0 = -0.6449\pi, h_0 = (\pi^2+1)/2)$ and $(\phi_0 = 0.6449\pi, h_0 = (\pi^2-1)/2)$. If dropped at such consistent starting parameters the ball immediately enters the 1-periodic mode, i. e., without overshoot or transient as shown in Fig. 2. Similar behavior is exhibited by the second (and unstable) solution indicated by the dotted line. Concerning the first solution, it shows a robust stability with respect to initial heights ranging in the interval as wide as [1.823, 11.510].

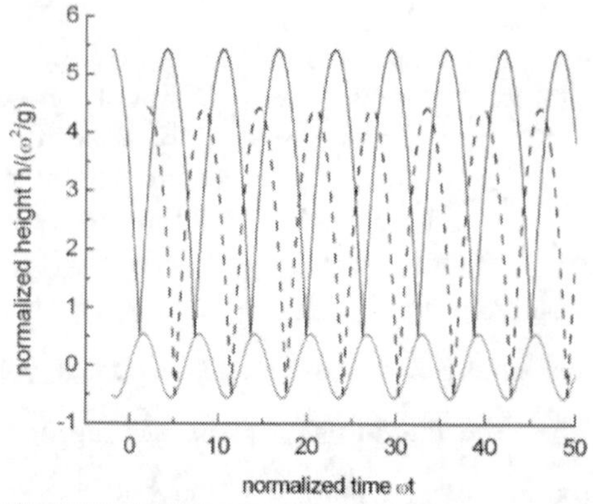

Figure 2 For ε =0.85, period-1 modes driven at Γ = 0.5611 with collision phases ϕ = 0.3501 π (h0= (π^2 +1)/2, solid line) and $\phi = -$ 0.3501 π (h0= (π^2 -1)/2, dashed line).

At $\phi_0 = 65001\pi$, if we now drop the ball from initial height h_0=*1.822*, which is located just on the left limit of the stability range for h_0, now the the ball is unable to sustain its motion and then comes to a permanent contact with the platform by executing a convergent sequence of decaying jumps. As shown in Fig. 3, after the 10th collision the impact position commutes to a descending phase when the base, moving downward, has a negative velocity. The ball loses energy and in the next collision the ball, upon rising to a lower height, arrives further delayed; the lost synchronism cannot be restored and the ball rests immobile on the platform.

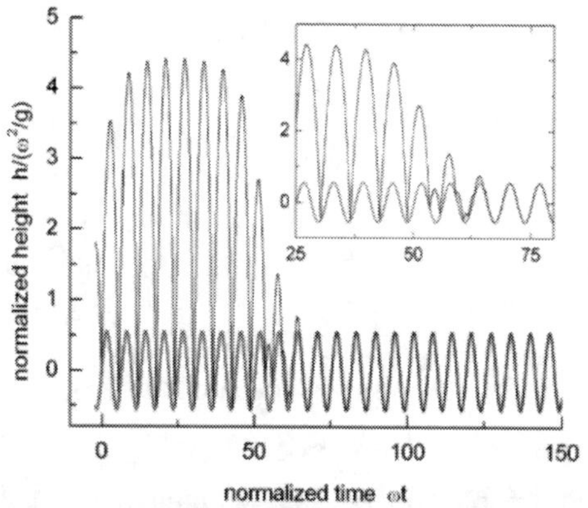

Figure 3 , the ball dropped from h_0 =1.822 is unable to enter the period-1 mode. Transient motion is zoomed in the inset.

To drive a higher-order nth-subharmonic periodic-1 mode from required specifications, for instance n=5 at $\varepsilon = 0.85$ and $\Gamma\sin\phi = 0.8$ we use Eqs. (5)-(7) to consistently obtain the starting parameters $\phi_0 = 4.821\pi,\ h_0 = \left(\Gamma\sin\phi + (5\pi)^2/2\right) = 124.17$ and the drive amplitude $\Gamma = 1.504$, by noting that the collision phase $\phi = 0.1785\pi$, relates with the initial phase by $\phi = \phi_0 + 5\pi$. Instead of dropping the ball from the calculated height (h_0=124.17), at which the ball would enter the n=5, period-*1* motion without transient, let us drop the ball from a larger height h_0=132.0. Preceded by a persistent sequence of somehow period-tripling oscillations, as shown in Fig. 4, the ball ultimately reaches its steady state of motion past the time span of nearly 3500/2 π oscillation periods through exponentially damped oscillations, as in this case the resulting eigenvalues $\lambda_{1,2} = -0{,}5500 \pm 0.7362i$

typify a stable focus. After the transient is finished, the ball reaches the collision velocity $v = 5\pi$ at the phase $\phi = 0.1785\pi$ as portrayed in the phase-space plot in Fig. 5. Detailed in Fig. 6, the steady-state oscillations consist of equal jumps separated by a periodic time span of 5π; each jump in 25 times as high as the single jump in the n=1 periodic mode discussed in Fig. 2. Here we note that for a *nth*-subharmonic periodic mode the maximum height relative to the impact point is just $(n\pi)^2/2$, irrespective of the drive amplitude Γ. In spite of increasing Γ the relative height remains constant, otherwise the ball would start jumping higher, thus leading to longer flights which would not be synchronized with the oscillation period of the platform. In preserving its relative height to the collision point, the wavetrain of parabolic jumps shifts as a whole by searching a new footpoint so as to keep both the landing and departure velocities matched at the synchronous value $v = n\pi$ [19].

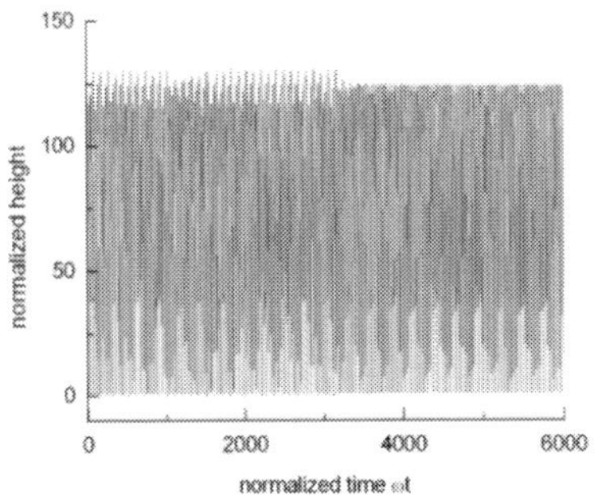

Figure 4 Dropped from h_0=132.0, at ε = 0.85 and with Γ =1.504, the ball reaches the 5-period mode

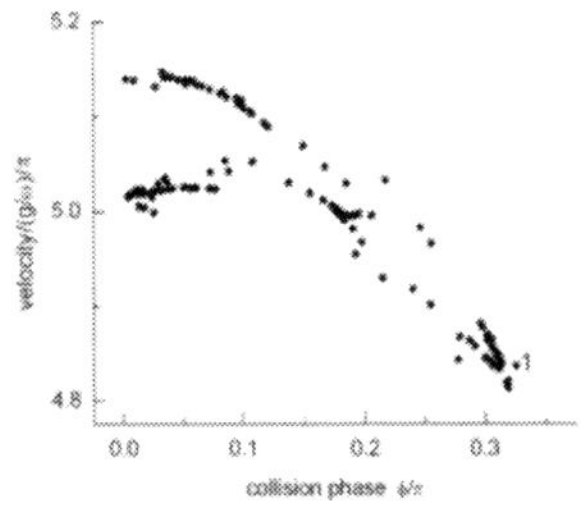

Figure 5 Phase-space plot of the ball motion shown in Fig. 4. The system evolves to the converging point (ϕ/π, v/π)=(0.1775, 5) identifying a n=5 periodic mode; the three clusters of points displays the period-tripling transient time series displayed in Fig. 4

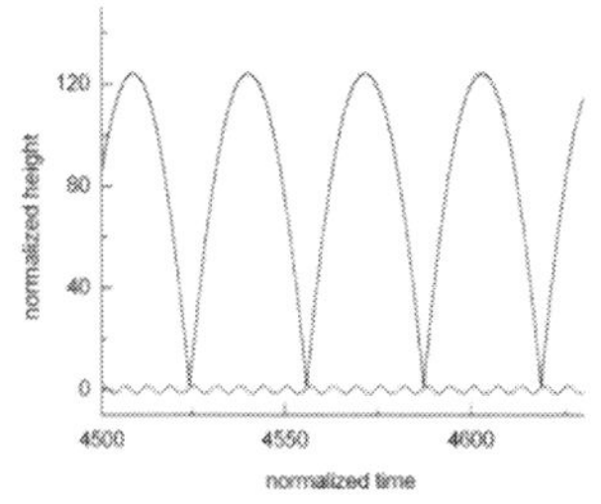

Figure 6 . Periodic mode with subharmonic index n=5: both the landing and departure velocities are 5 π and two consecutive collision phases are spaced by 5 π .

Without changing the previous control parameters Γ and ε, if now the ball is dropping from larger height, namely h_0=135.0, chaotic oscillations are generated (Fig. 7). Chaotic trajectories do not fill the phase space in a random like manner. Instead, as seen in the map map (Fig. 8) encompassing many unstable orbits which remain in the system, the trajectories fall onto a complex but well defined and bounded object (chaotic attractor) which is cut at the bottom by a cosine-shaped boundary rendered by the velocity time profile ($\Gamma\cos\omega t_i$ with $\omega t_i=\phi_i$) of the table oscillation. At Γ=1.504 and ε=0.85 one finds $|v|$=10.027, which is in good agreement with the velocity range [-1.504, 9.291] portrayed in Fig. 8 [19].

Journal of Physics: Conference Series **246** (2010) 012003 doi:10.1088/1742-6596/246/1/012003

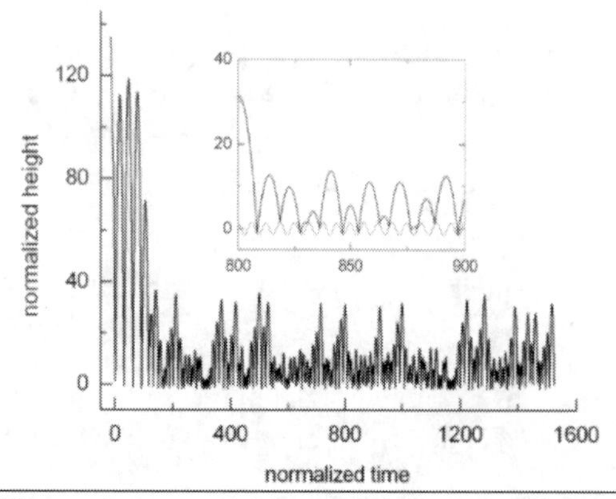

Figure 7 At $\varepsilon = 0.85$ and $\Gamma\sin\phi = 0.8$ with $\Gamma = 1.504$, but dropped from $h_0 = 135.0$ the ball enters a chaotic motion.

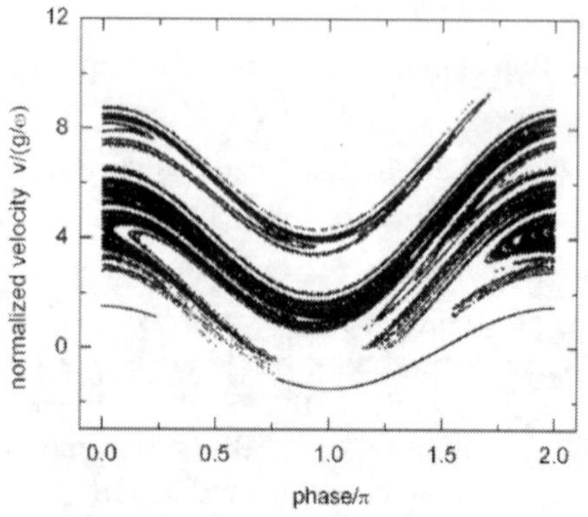

Figure 8 Phase-space plot of the chaotic motion in Fig. 7.

Discussing now the elastic case $\varepsilon = 1$, by considering first the equation $v_2 = \varepsilon v_1 + (1+\varepsilon)\Gamma\cos\phi$ that relates the pre- and post-collisional velocities of the ball upon impact against the moving platform, where $v_1 > 0$ denotes the incoming velocity. Then, at $\varepsilon = 1$, and for head-on collisions, in which the base moves upwards, the post-collisional reduces to $v_2 = v_1 + 2\Gamma\cos\phi$, while for overtaking collisions (with the base moving downwards) the post velocity turns into $v_2 = v_1 - 2\Gamma\cos\phi$, where ϕ is the collision phase. If the post-collision velocity is negative (particle still moving down) then a second impact will occur, provided that $V < v_1 < 2V$ where $V = |2\Gamma\cos\phi|$. But for both types of collisions, when $\phi = n\pi/2$, n integer, the ball bounces back after collision with a departure velocity which is simply the reverse of the velocity before the bounce, neither gaining nor losing energy on the collision. For the 1-periodic mode, at $\phi_0 = -\pi/2$, $\Gamma = 0.5$, and $h_0 = \pi^2/2 + \Gamma$, this situation is shown in Fig. 9. At such starting conditions, the time length between collisions is matched with the period of the platform motion, with the ball performing a sequence of perfectly equal jumps and seeing the base as it were static. Accordingly, by dropping the ball from $h_0 = (6\pi)^2/2 + \Gamma$ at $\phi_0 = -11\pi/2$, the ball will execute a n=6 subharmonic motion (Fig. 10), in which two consecutive collisions are spaced by six periods of the base oscillation.

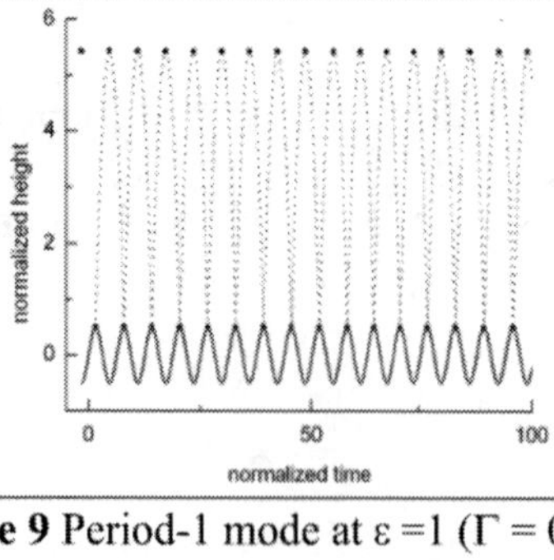

Figure 9 Period-1 mode at ε =1 (Γ = 0.5, ϕ = π/2).

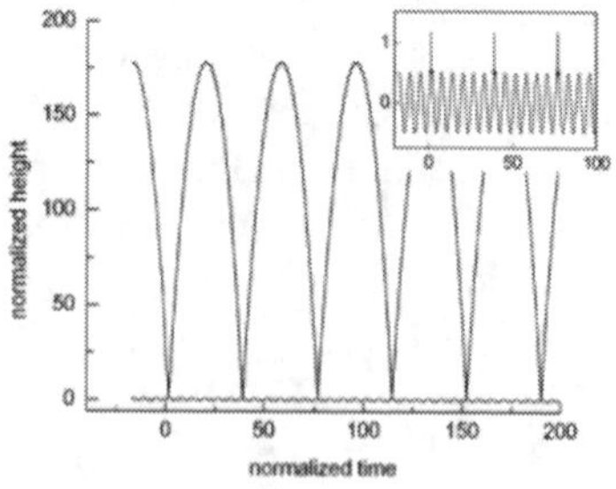

Figure 10 6th subharmonic mode at $\varepsilon = 1$. The inset shows that two consecutive collisions are spaced by six periods of the base oscillation

Releasing now the ball from h_0=0.5 at $\Gamma = 0.5$ and $\phi_0 = -\pi/2$, we see in Fig. 11 that the jumps starts increasing without limit, an example that exhibits Fermi acceleration—a process in which the particle gains energy by collision against a moving scatterer. Gain or loss of energy occurs on head-on (base moves towards the incoming particle) and overtaking collisions (base moves away from the particle), but the net result will be an average gain by the reason that increasing velocities make head-on collisions more frequent. In fact, averaging from $v_2=v_1+2\Gamma\cos\phi$ the square velocity v_2 over an oscillation period leads to $\langle v_2^2\rangle = \langle v_1^2\rangle + 2\Gamma^2$, which describes a constant net energy gain per collision, thus meaning that the particle's energy (or height) tends to increases linearly with time. It means that there is an increase of the particle energy roughly follows a linear growth, thus characterizing a random walking particle, for which the average velocity scales with the collision number as $\langle v\rangle \sim \sqrt{N}$ [5]. In the phase-space plot (Fig. 12) of this motion, there appear stochastic layers which, separated from each other, seem to be limited in their width.

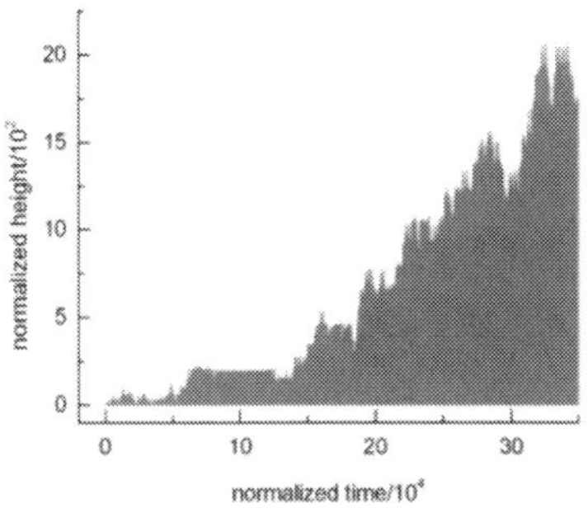

Figure 11 At the start conditions h_0=0.5, ϕ_0=-π/2, and Γ=0.5 the amplitude of the period-1 mode shows a boundless increase.

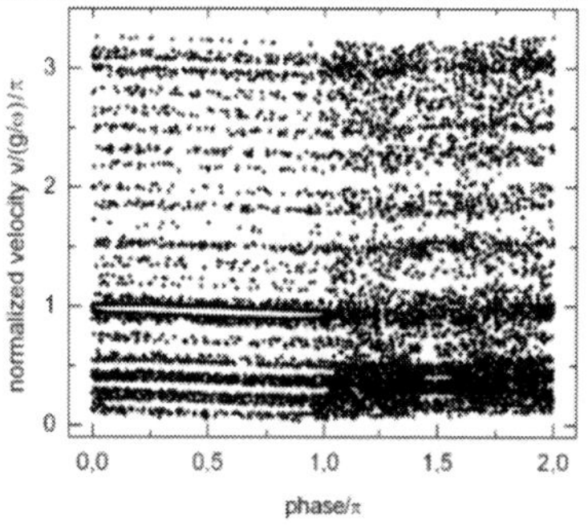

Figure 12 Phase-space plot of the motion in Fig. 22

3. Conclusion

Through numerical examples from computer simulations guided by analytical considerations, this paper has presented a quantitative description of the bouncing ball problem. The dynamics is modeled on the basis of a discrete map of difference equations for the trajectory (describing the ball free fall under gravitational acceleration), velocity, and phase of the ball's motion on assuming a constant restituition coefficient and the collisions to be instantaneous.

Once iterated numerically the equations fully reveal a rich variety of nonlinear behaviors, including non periodic motions as well as chaotic and stochastic phenomena. Numerical examples have demonstrated that the bouncing-ball behavior is strongly dependent on the control parameters (Γ and ε) and also on the initial conditions.

This work has been supported by the National Council for Scientific and Technological Development (CNPq) and Sao Paulo State Council for Scientific Development (FAPESP).

Journal of Physics: Conference Series **246** (2010) 012003 doi:10.1088/1742-6596/246/1/012003

References

[1] Pippard A B 2006 *The Physics of Vibration* (Cambridge: Cambridge University Press)

[2] Zaslavski G M and Chirikov B.V 1972 *Sov. Phys. Usp.* **14** 549

[3] Pastor J M, Maza D, Zuriguel I, Garciamartín and. Boudet J –F 2007 *Physica D* **232** 128

[4] Fermi E 1949 *Phys. Rev.* **75** 1169

[5] Leonel E D and Livorati A L P 2008 *Physica A* **387** 1155

[6] Thompson J M T and Stewart H B 1991 *Nonlinear Dynamics and Chaos* (John Wiley)

[7] Burnham N A, Kulik A J and Gremaud G 1995 *Phys. Rev. Lett.* **25** 5092

[8] Holmes P J 1982 *J. Sound Vib.* **84**(2) 173

[9] Luo A C J and Han R P S 1996 *Nonlinear Dynamics* **10** 1

[10] Everson R M 1986 *Physica D* **19** 355

[11] Tufillaro N B, Mello T M, Choi Y M and Albano A M 1986 *J. Physique* **47** 1477

[12] Kowalik Z J, Franaszek M and Pieranski P 1988 *Phys. Rev. A* **37**, 4016

[13] Mehta A and Luck J M 1990 *Phys. Rev. Lett.* **65** 393

[14] Warr S and Huntley J M 1995 *Phy. Rev. E* **52** 5596

[15] Géminard J -C and Laroche C 2003 *Phys. Rev. E* **68** 031305

[16] Giusepponi S, Marchesoni F and Borromeo M 2005 *Physica A* **351** 142

[17] Bapat C N and Sankar S 1986 *J. Sound Vib.* **108**(1) 99

[18] Luding S, Clément E, Blumen A, Rajchenbach J and Duran J 1994 *Phys. Rev. E* **49** 1634

[19] Barroso J J, Carneiro M V, Macau E E N 2009 *Phys. Rev. E* **79** 026206

Journal of Physics: Conference Series **246** (2010) 012004 doi:10.1088/1742-6596/246/1/012004

Edge effects in billiards: stickiness and long-lived states

M. S. Custódio and M. W. Beims

Departamento de Física, Universidade Federal do Paraná, 81531-990 Curitiba, PR, Brazil
Email: mbeims@fisica.ufpr.br

Abstract. The effect of rounded edges in the rectangular billiard is analyzed via the escape times statistic. Such realistic and sometimes intrinsic edges are shown to generate autosimilar structures and power law decays for the escape times statistic. As the rounding effects increase, stickiness and long lived trajectories are observed. Rounding edges of around $0.01\% \rightarrow 1\%$ from the whole billiard size generate the sticky motion.

1. Introduction

The boundary of realistic physical devices always present irregularities like roughness, softness, defects, edges and border effects. Such sometimes intrinsic irregularities affect the dynamics of particles which collide with the boundary. From the theoretical point of view it is very difficult to describe, in general, the dynamics of colliding particles with irregular boundaries. Therefore in recent years more and more attention has been given to the description of particles confined inside boundaries (or billiards) which present some *specific* edges, softness etc. To mention some examples we have the edge roughness in quantum dots [1], unusual boundary conditions in two-dimensional billiards [2, 3], effects of soft walls [4] and edge collisions [5] of interacting particles in a 1D billiard, edge diffractions and the corresponding semiclassical quantization [6, 7], rounding edge [8] and edge corrections [9] in a resonator, deformation of dielectric cavities [10], among others. Although such irregularities are sometimes very small, they can change significantly the dynamics of quantum and classical systems.

From the nonlinear dynamics point of view, it has been shown that soft walls and soft edges do not destroy trajectories found in the hard-wall limit [11] and may induce the appearance of regular islands in phase space [4, 12, 13, 14]. Such regular islands inside the chaotic sea induce a "sticky"(or trapped) motion, which is a common phenomenum in conservative systems [15]. Such motion usually arises from broken Kolmogorov-Arnold-Moser (KAM) curves and generates a rich dynamics in quasi-integrable systems [15]. Stickiness may generate an anomalous behavior of the transport [15, 16, 17] and is therefore very important to be characterized in realistic physical systems. In the context of soft walls, the sticky motion has been observed theoretically and experimentally in the atom-optic billiard [18, 19] and has shown to affect the quantum conductance in the soft wall microwave billiard [20].

In this work we show that tiny rounding edges are able to change the escape times statistics, generating long lived trajectories and sticky motion. As a model we use the open rectangular billiard, but results should be valid in general for open billiards with edges.

2. Escape times statistics and the model

The escape times statistic is defined by [21]

$$Q(\tau) = \lim_{N\to\infty} \frac{N_\tau}{N}, \tag{1}$$

where N is the total number of trajectories which escape the billiard and N_τ is the number of trajectories which escape the billiard after the time τ. For systems with stickiness the escape times statistic decays as a power law [15] $Q(\tau) \propto \tau^{-\gamma_{esc}}$, where $\gamma_{esc} > 1$ is the scaling exponent. For hyperbolic chaotic systems and long times the escape times statistic decays exponentially. It is known that the diffusion exponent μ from the MSD of the position $\langle x^2 \rangle \sim t^\mu$, is related to γ_{esc} via $\mu = 3 - \gamma_{esc}$ [16, 17].

The model used here is the open rectangular billiard with rounded edges (See Fig. 1). We consider a point-like collision of particles with the rounded edge. The edge is modelled by a quarter of a circle. The escape times is studied as a function of the initial incoming angles θ_0 and the ratio R/L. L is the billiard height, D its width and R is the radius of the semicircle which models the rounded edge. In all simulations $L = 4, D = 10$ and R changes. The escape point with constant aperture $a = 1.0 \times 10^{-1}$ is located at $D/2$.

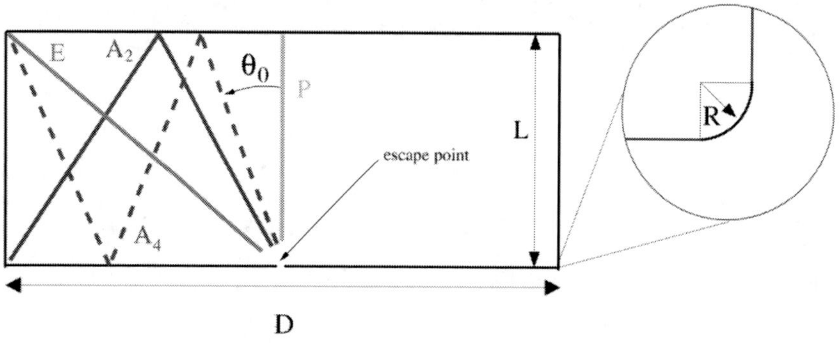

Figure 1. (Color online) The rectangular billiard with dimension $L \times D$. Schematically the rounded edge with radius R is shown for just one edge. In all simulations the rounded edge is applied for all four edges. The escape point with aperture $a = 1.0 \times 10^{-1}$ lies at $D/2$. Initial angle θ_0 and, schematically, the shortest escape trajectories (P, E, A_2 and A_4) are shown.

Without rounding edges ($R/L = 0$) the rectangular billiard is integrable and has zero Lyapunov exponents. It has marginally unstable periodic orbits, which are orbits bouncing perpendicularly between the parallel walls. They are obtained from the initial conditions $\theta = 0, \pi/2$. As we will see, for small values of the ratio R/L the rounded edges generate the sticky motion. Since marginally unstable periodic orbits in billiards are known [25, 26] to generate exponent $\gamma_{esc} = 2$ for the escape times, we also expect to see $\gamma_{esc} \to 2$ as R/L increases. For $R/L = 1/2$ our model is the Bunimovich stadium [22], which is known to be chaotic having an almost exponential escape time decay [23].

We start discussing the quantity $Q(\tau)$ for different values of the ratio R/L, where L is kept fixed. At times $t = 0$ particles always start at the escape point, towards the inner region of the billiard with an initial angle θ_0. We use 10^5 initial conditions uniformly distributed in the interval $0.1 \lesssim \theta_0 \lesssim 1.5$. For each initial condition the simulation is done until the particle escapes through the hole. The dynamics for $-\pi/2 \leq \theta_0 \leq 0.0$ is symmetric. Results for $Q(\tau)$ are shown in the log-log plot of Fig. 2 for different values of $R/L = 0.0, 1.0\times10^{-4}, 1.0\times10^{-3}, 1.0\times10^{-2}, 1.0\times10^{-1}, 0.5$. For comparison the straight dashed line shows the power law decay with $\gamma_{esc} = 2$ (Straight lines

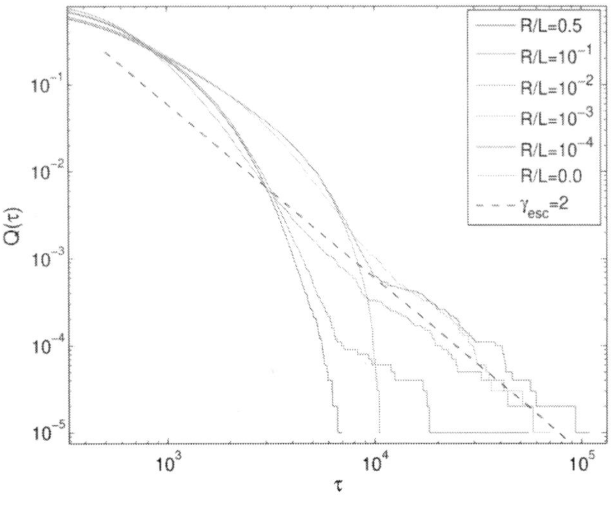

Figure 2. (Color online) Log-log plot of $Q(\tau)$ for different values of the ratio R/L.

in this plot mean a power law). First observation is that for $R/L = 0.0$ (no edge effects) $Q(\tau)$ did not has a power law. It should [24] obey a quasi-algebraic decay for short and intermediate times and a power law decay for large times. In fact, we did some simulations (not shown) using only initial conditions very close to the marginally unstable periodic orbit $\theta_0 \sim \pi/2$. We found in this case that $Q(\tau)$ has a power law tail for very long times with $\gamma_{esc} \sim 1.0$. This shows that marginally unstable periodic orbits without the focusing component from the edge generates the exponent $\gamma_{esc} \sim 1.0$. This agrees with results obtained in [25] for some specific initial conditions.

Increasing the rounding effects to $R/L = 1.0 \times 10^{-4}$, the qualitative behaviour of $Q(\tau)$ starts to change (when compared to the case $R/L = 0.0$) for times $\tau \gtrsim 1.0 \times 10^4$, i. e. for those trajectories which stay longer inside the billiard. This means that the very few trajectories which collide with the rounded edges tend to stay longer inside the billiard and also change the qualitative behavior of $Q(\tau)$. Since the probability to collide with the edges increases in time, just very long-lived trajectories essentially affect $Q(\tau)$. For rounded edges of size $R/L = 1.0 \times 10^{-3}$ and $R/L = 1.0 \times 10^{-2}$ we observe a power law decay for times $\tau \gtrsim 2 \times 10^3$. In all cases we found that the long-lived trajectories can be fitted by an escape exponent of $\gamma_{esc} \sim 2.0$, as can be seen by comparing these decays with the straight dashed line for $\gamma_{esc} = 2.0$ (see Fig. 2). For $R/L = 1.0 \times 10^{-1}$ the power law behavior is not relevant anymore and the decay is better fitted by an exponential. For $R/L = 0.5$ no power law decay is obtained (for this time and hole aperture) since this is the stadium billiard with totally chaotic behavior. This agrees with the almost exponential escape times decay observed [23] for the Bunimovich stadium when the aperture is small enough.

The escape times from the long-lived trajectories present relevant characteristic of sticky motion for $R/L = 10^{-4}, 10^{-3}, 10^{-2}$. In other words, sticky motion and long living trajectories start to occur for *very* small rounding edges. Therefore edges of around $0.01\% \rightarrow 1\%$ of the whole billiard size are sufficient for this purpose. Visually such rounded edges are almost negligible.

3. Self-similar escape times

Figure 3 shows the behavior of the escape times as a function of the initial angles θ_0 and for values (a) $R/L = 0.0$ and (b)-(c) $R/L = 1.0 \times 10^{-3}$. For $R/L = 0.0$ we observe in Fig. 3(a) that the escape times alternate between minima (with values around $\tau \sim 10$) and maxima (with values around $\tau \sim 10^3$). For angles $\theta_0 \to 1.5$ (almost horizontal trajectories) the escape times increase since trajectories need more time to reach the escape point which lies in the horizontal wall. The minima in the escape times are related to the shortest trajectories. Some of these

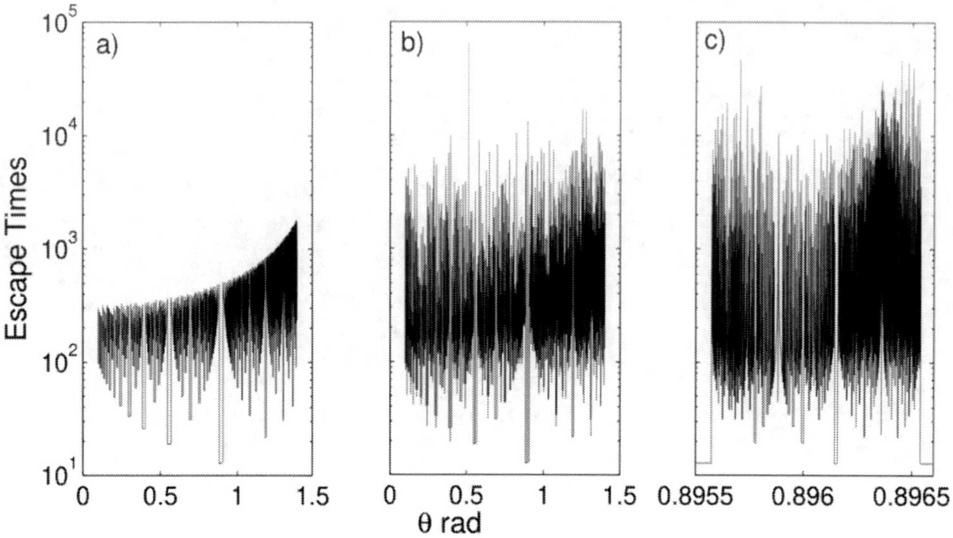

Figure 3. Semi-log plot of escape times as a function of initial angle θ_0 for (a) $R/L = 0.0$, (b) $R/L = 1.0 \times 10^{-3}$ (c) magnification of (b).

trajectories are shown schematically in Fig. 1, where A_n refers to the trajectory which collides once with the rounded edge and n times with the horizontal boundary. Besides trajectory P ($\theta_0 \sim 0.0$), which collides directly with the wall in front of the escape point and leaves the billiard, the shortest trajectory is labelled by letter E (Fig. 1) and has $\theta_0 = \arctan(L/D) \sim 0.89$. This agrees with the smaller minimum shown in Fig. 3(a). The width of the minima is related to the aperture $a \sim 0.4$. Close to the border of these widths, larger escape times are observed and are related to trajectories which jump over the escape point, staying longer inside the billiard. As the rounded edge increases to $R/L = 10^{-3}$, the escape times increase very much, as shown in Fig. 3(b). Trajectories with escape times close to 2×10^4 are observed. It means that edge effects strongly increase the time that trajectories will stay inside the billiard. More interesting is that for small edges the escape times show a self-similar structure as a function of θ_0, as can be seen in the magnification shown in Fig. 3(c), where the escape times are plotted for initial angles in the small interval $[0.8956, 0.8966]$. This means that no matter how small the interval of initial angles, small rounded edges always induce long-lived trajectories and stickiness.

4. Summary

Rounded edges effects in a rectangular billiard are shown to strongly influence the dynamics. We show that rounded edges of about $0.01\% \to 1\%$ are sufficient to generate the sticky motion and long lived trajectories inside the billiard. The sticky motion is characterized here by the escape times statistics decay. Self-similar escape times are also observed, showing that no matter how small is the interval of initial angles, tiny rounded edges always induce long-lived trajectories and stickiness.

Very recently [27] it was shown that edge effects are essential for the directional emission in optical mushroom-shaped billiards, where marginally unstable periodic orbits are generated close to the edge. From results of the present work we suggest that, due to stickiness and long-lived trajectories, musrhroom-shaped billiards with *rounded* edges could be more efficient to generate the directional emission. In addition, in the periodic orbit quantization of the stadium billiard it was shown [7] that important contributions are due to edge orbits. Such contributions should also increase with rounded edges. Stickiness are also expected to strongly influence the transport of particles along billiard systems or periodic potentials in classical [28, 29, 30] and quantum mechanics [31, 32, 33].

Acknowledgments

The authors thank CNPq and CAPES for financial support.

[1] Libisch F, Stampfer C and Burgdörfer J 2009 *Phys. Rev. B* **79** 115423.
[2] Bogomolny E, Dennis M R and Dubertrand D 2009 *J. Phys. A* **41** 335102.
[3] Berry M V and Dennis M R 2008 *J. Phys. A* **41** 135203.
[4] Oliveira H A, Manchein C and Beims M W 2008 *Phys. Rev. E* **78** 046208.
[5] Beims M W, Manchein C and Rost J M 2007 *Phys. Rev. E* **76** 056203.
[6] Bruus H and Whelan N D 1996 *Nonlinearity* **9** 1023.
[7] Alonso D and Gaspard P 1994 *J. Phys. A* **27** 1599.
[8] Wiersig J 2003 *Phys. Rev. A* **67** 023807.
[9] Vaa C, Koch P M and Blümel R 2005 *Phys. Rev. E* **72** 056211.
[10] Dubertrand R, Bogomolny E, Djellali N, Lebental M and Schmit C 2008 *Phys. Rev. A* **77** 013804.
[11] Rapoport A, Rom-Kedar V and Turaev D 2007 *Commun. Math. Phys.* **272** 567.
[12] Turaev D and Rom-Kedar V 1998 *Nonlinearity* **11** 575.
[13] Rom-Kedar V and Turaev D 1999 *Physica* **130D** 187.
[14] Turaev D and Rom-Kedar V 2003 *J. Stat. of Phys.* **112** 765.
[15] Zaslavski G M 2002 *Phys. Rep.* **371** 461.
[16] Meiss J D 1997 *Chaos* **7** 139.
[17] Zumofen G and Klafter J 1999 *Phys. Rev. E* **59** 3756.
[18] Kaplan A, Friedman N, Andersen M and Davidson N 2001 *Phys. Rev. Lett.* **87** 274101.
[19] Kaplan A, Friedman N, Andersen M and Davidson N 2004 *Physica D* **187** 136.
[20] Weingartner B, Rotter S and Burgdörfer J 2005 *Phys. Rev. B* **72** 036223.
[21] Altmann E G, Motter A and Kantz H 2005 *Chaos* **15** 033105.
[22] Bunimovich L A 1979 *Math. Phys.* **65** 295.
[23] Alt H, Gräf H-D, Harney H L, Hofferbert R, Richter A and Schardt P 1996 *Phys. Rev. E*, **53** 2217.
[24] Kokshenev V B and Nemes M C 2000 *Physica A* **275** 70.
[25] Armstead D N, Hunt B R and Ott E 2004 *Physica D* **193** 96.
[26] Gaspard P and Dorfman J R 1995 *Phys. Rev. E* **52** 3525.
[27] Andreasen J, Cao H, Wiersig J and Motter A E 2009 *Phys. Rev. Lett.* **103** 154101.
[28] Rosa J and Beims M W 2008 *Phys. Rev. E* **78** 031126.
[29] Rosa J and Beims M W 2007 *Physica A* **386** 54.
[30] Rosa J and Beims M W 2004 *Physica A* **342** 29.
[31] Andrade F M, Cheng B K, Beims M W and da Luz M G E 2003 *J. Phys. A* **36** 227.
[32] da Luz M G E and Cheng B K 1998 *J. Phys. A* **31** 2975.
[33] da Luz M G E, Cheng B K and Beims M W 2001 *J. Phys.A.* **34** 5041.

Effect of robust torus on the dynamical transport

Caroline G. L. Martins[1], **R. Egydio de Carvalho**[1], **I. L. Caldas**[2] **and M. Roberto**[3]

[1] UNESP-Univ Estadual Paulista; Instituto de Geociências e Ciências Exatas; Departamento de Estatística, Matemática Aplicada e Computação, Av. 24A, 1515, 13506-900 Rio Claro , SP, Brazil.
[2] Universidade de São Paulo; Instituto de Física 05315-970 São Paulo, SP, Brazil.
[3] Instituto Tecnológico da Aeronáutica; Departamento de Física 12228-900 São José dos Campos, SP, Brazil.

E-mails: carolinegameiro@gmail.com, regydio@rc.unesp.br, ibere@if.usp.br,

marisar@ita.br

Abstract In the present work, we quantify the fraction of trajectories that reach a specific region of the phase space when we vary a control parameter using two symplectic maps: one non-twist and another one twist. The two maps were studied with and without a robust torus. We compare the obtained patterns and we identify the effect of the robust torus on the dynamical transport. We show that the effect of meandering-like barriers loses importance in blocking the radial transport when the robust torus is present.

1. Introduction

In a previous work [1] we introduced a non-twist map with a new transport barrier called robust torus (RT), through a known Hamiltonian derivation in which the RT had been introduced in a systematic way [2,3,4]. The motivation to create this new map was the relevance of transport barrier to plasma confinement in tokamaks [5-7] and also because our map could be a natural extension of usual maps [8-13] applied to study theoretically transport in tokamaks. In another work [14] we adapted our map with a single robust torus and adjusted the parameters according to the experimental ones used in the tokamak TCABR. We have verified that the RT regularizes the orbits close to its position in the phase space and blocks the radial transport inside the tokamak plasma. The main idea to create a RT is to introduce an appropriate polynomial pre-factor multiplying the resonant perturbations which act on the system in such way that their effects vanish on particular positions of the phase space even if the control parameter is turned on. This vanishing will occur in the values of the roots of the pre-factor. The RT in fact corresponds to an invariant curve which remains intact under the action of any generic perturbation.

In this work, we quantify the portion of trajectories that reach a specific region of the phase space when we vary a control parameter using two maps: one non-twist and another twist, both of them with or without a robust torus. The non-twist maps present isochronous resonances [15], which can reconnect themselves and generate the well known meandering barriers [16].

We compare the obtained patterns for the non-twist and twist maps, with and without RT, and we identify the effect of the RT on the dynamical transport. We create a RT near a secondary resonant mode and we show that the robustness of the RT is transferred to its neighborhood. Because of that, the effect of the other meandering-like barrier, present in the non-twist map, loses importance in blocking the radial transport when the RT is present.

The paper is organized as follows; in Section 2 we present a brief Hamiltonian approach relevant for plasma confinement in tokamak. In Section 3 we show the corresponding non-linear maps, twist and non-twist, which we are going to use. In Section 4 we discuss the results on transport in the phase space and in last section we present final remarks.

2. Hamiltonian Approach

As a background motivation, in this section we review the usual Hamiltonian approach developed for plasma confinement in tokamaks in order to introduce the RT in this formalism and also to link it with our map. We propose an equilibrium Hamiltonian, H_0, with toroidal symmetry and we define the toroidal angle, φ, as the canonical time t. This non-perturbed Hamiltonian corresponds to an integrable system and it is described by an analytical solution of the non-linear Grad-Shafranov equation [17]. H_0 is given in terms of the action J, corresponding to the toroidal normalized flux, and ϑ the poloidal angle canonically associated to J [18]. The perturbations are generated by a helical electric current applied from thins ergodic magnetic limiters (EML) distributed along the poloidal direction in toroidal sections of the tokamak [19]. The perturbing Hamiltonian, H_1, is a function of J, ϑ and t, and it is represented by a Fourier expansion of delta-kicks due to the EML rings [10]. The perturbations create a region with chaotic magnetic field lines at the plasma edge.

We expand H_0 around a magnetic surface with action J_0 and frequency Ω_0=n/m where Ω_0 is equal to the inverse of the safety factor q and *n* and *m* are integers. In the Fourier perturbation expansion, we keep only the two main resonances in such way the Hamiltonian has only the dominant resonant modes with poloidal wave numbers *m* and *(m+1)* [10,19],

$$H(\Delta J,\vartheta,t) = H_0(\Delta J) + P(\Delta J)[\beta\cos(m\vartheta)+\eta\cos((m+1)\vartheta+nt)]\sum_{k=-\infty}^{+\infty}\delta\left(t-\frac{2\pi}{N_r}k\right) \tag{1}$$

where $\Delta J = J - J_0$ and N_r=4 is the quantity of EML rings equally spaced around the tokamak.

We also consider two equilibrium configurations for the frequency profiles $\Omega_0 \equiv dH_0/d\Delta J$ given by:

$$H_0(\Delta J)=\frac{1}{2}\Delta J^2 \tag{2}$$

for the twist case and

$$H_0(\Delta J)=\frac{1}{2}\Delta J^2-\frac{\alpha}{3}\Delta J^3 \tag{3}$$

for the non-twist case.

The parameters β and η in Eq. (1) represent the perturbing electric current applied in EML rings and the parameters α in Eq. (3) is determined by the safety factor profile defined by H_0. The pre-factor $P(\Delta J)$ in Eq. (1) is a polynomial function and it allows us to introduce the RT.

We emphasize that, for a generic polynomial function, it is possible to have a number of RT equal to the number of real roots of this polynomial. Thus, this procedure enables us to construct a Hamiltonian to study the alterations of the magnetic field lines topology due to the presence of an infinity barrier, a RT.

Journal of Physics: Conference Series **246** (2010) 012005 doi:10.1088/1742-6596/246/1/012005

3. Non-linear maps

In order to investigate the characteristics of the phase space concerning transport, due to the presence of a RT, we consider the Hamiltonian of Eq. (1) with the non-perturbed Hamiltonian, H_0, of Eq. (2) and (3), twist and non-twist cases respectively. We also choose two configurations for the pre-factor, one with $P(\Delta J)=1$ for the usual Hamiltonian without RT and another one with $P(\Delta J)=(\Delta J-a)$ introducing one RT in the position $\Delta J=a$. However, instead of working with the Hamiltonian formulation we will take the corresponding non-linear map associated to the Hamiltonian of Eq. (1) that was derived in details by us in [1]. Thus, the non-twist map with two resonant modes and without RT is,

$$\begin{aligned}
\Delta J_{k+1} &= \Delta J_k + \beta m \sin(m\vartheta_k) + \eta(m+1)\sin[(m+1)\vartheta_k + nt_{k+1}] \\
\vartheta_{k+1} &= \vartheta_k + \frac{2\pi}{N_r}(\Delta J_{k+1} - \alpha\,\Delta J_{k+1}^2)\ , \quad \mathrm{mod}(2\pi) \\
t_{k+1} &= t_k + \frac{2\pi}{N_r m}
\end{aligned} \tag{4}$$

The non-twist map with two resonant modes and with one RT is,

$$\begin{aligned}
\Delta J_{k+1} &= \Delta J_k + (\Delta J_{k+1} - a)\{\beta\, m \sin(m\vartheta_k) + \eta\,(m+1)\sin[(m+1)\vartheta_k + nt_{k+1}]\} \\
\vartheta_{k+1} &= \vartheta_k + \frac{2\pi}{N_r}(\Delta J_{k+1} - \alpha\,\Delta J_{k+1}^2) + \beta\cos(m\vartheta_k) + \eta\cos[(m+1)\vartheta_k + nt_{k+1}]\ , \quad \mathrm{mod}(2\pi) \\
t_{k+1} &= t_k + \frac{2\pi}{N_r m}
\end{aligned} \tag{5}$$

In all numerical calculations we chose $m=3$ and $n=1$ introducing the resonant modes (1:3) and (1:4). The constant α that appears in Eq. (3) is adjusted for both maps, in the twist case $\alpha=0$ and in the non-twist case $\alpha=160.15$. We will split our study in four analyses:

i) Case Non-twist without RT, Eq. (4) with $\alpha=160.15$;
ii) Case Non-twist with one RT, Eq. (5) with $\alpha=160.15$;
iii) Case Twist without RT, Eq. (4) with $\alpha=0$;
iv) Case Twist with one RT, Eq. (5) with $\alpha=0$.

For the non-twist equilibrium, there are isochronous resonances in the phase space. Figure 1(a) corresponds to the case **i)**, where two isochronous resonances (1:3) has already reconnected and they are dimerized and separated by invariant meandering curves. Such invariant curves encircle both sets of surviving islands (1:3) and they exist only in non-twist maps. In plasma confinement approaches, the meandering curves are located in the shearless region, which traps the magnetic field lines for a long time hampering the radial diffusion, in such way that an internal transport barrier (ITB) takes place due to a strong stickiness effect [20].

Figure 1(b) corresponds to the case **ii)**, the RT is indicated in blue color at the position $\Delta J=3=a$, and the alterations introduced in the dynamics are noticeable. We still have the two *(1:3)* and *(1:4)* island chains, but now an interesting topological rearrangement occurred, the chaotic sea that was close to the upper island chain *(1:4)*, in Figure 1(a), had been suppressed due to the presence of the RT. The motion near the RT is kept integrable while in the other side of the meandering curves there is still a significant destruction of the invariant curves around the low resonance *(1:4)* showing a visible chaotic sea.

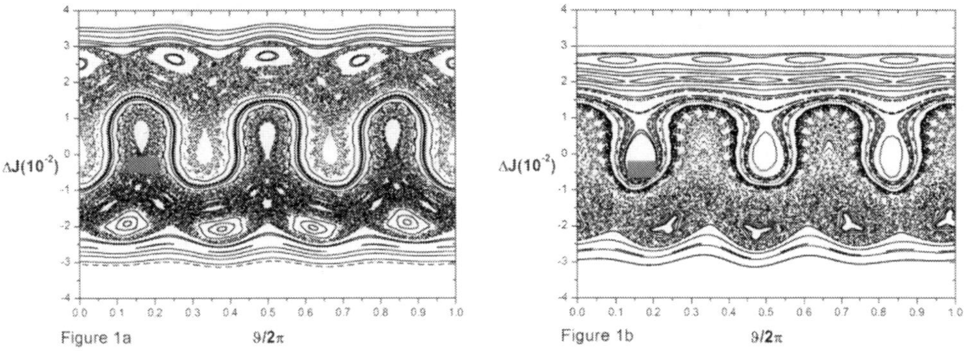

Figure.1. Poincaré sections for the resonant modes (1:3) and (1:4) for α=160.15: **(a)** Non-twist map, Eq. (4), without RT with $\beta=-1.3\text{x}10^{-4}$ and $\eta=1.15x10^{-4}$, the blue line at ΔJ=3 is only to guide the eyes, it is not a RT; **(b)** Non-twist map, Eq. (5), with RT (in red) at ΔJ=3, with $\beta=-4.0\text{x}10^{-3}$ and $\eta=9.2\text{x}10^{-3}$.

Figure 2(a) corresponds to the case **iii)** and we see the dominant island chain (1:3) and the main secondary resonance (1:4) embedded in a chaotic sea with several other secondary resonances. In this configuration, the barriers are in fact the island structures, which will be destroyed as the parameters are varied. On the other hand, in Figure 2(b), which corresponds to the case **iv)**, we can see the stabilizing alterations introduced by the RT, which is indicated in red colour in $\Delta J=a=3$. Note that, for the parameters used to plot Figure 2(b), the chaotic sea near the RT, around the island (1:3), is suppressed and the neighbourhood around $\Delta J=3$ is more stable than the one of the Figure 2(a) evidencing the local effect of the RT.

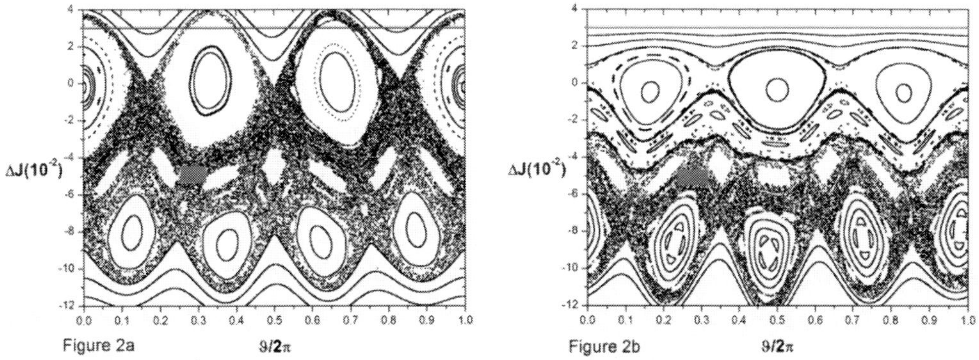

Figure.2. Poincaré sections for the resonant modes (1:3) and (1:4) with α=0 for: **(a)** Twist map, Eq. (4), without RT with $\beta=-6.0\text{x}10^{-4}$ and $\eta=2.99\text{x}10^{-4}$, the blue line at ΔJ=3 is only to guide the eyes, it is not a RT. **(b)** Twist map, Eq.(5), with RT (in red) at ΔJ=3, with $\beta=-7.5\text{x}10^{-3}$ and $\eta=8.2\text{x}10^{-3}$.

Journal of Physics: Conference Series **246** (2010) 012005 doi:10.1088/1742-6596/246/1/012005

We should emphasize that different Hamiltonians (maps) govern the system without RT (Figures 1(a) and 2(a)) and with RT (Figures 1(b) and 2(b)) and hence the scale of the parameters are different. In addition, the ranges of β are different because the non-twist system has one term that does not appear in the twist system.

4. Dynamical transport on the phase space

In this section we present the transport of an ensemble of initial conditions given inside the small red boxes in Figures 1 and 2, with and without RT, as function of the parameter β. We iterated 110 initial conditions inside the red squares and we varied the perturbation parameter β in order to investigate how many initial conditions reach the position $\Delta J=3$ (in blue), after 2500 iterations of the maps of equations (4) and (5). This small quantity of collected iterations does not lead to any loss of generality because we intend to investigate the effect of barriers or stickiness in the transport, which could be neglected if we had taken a bigger quantity of iterations than the one considered here.

For the non-twist cases, the meandering curves play the role of partial barrier and they interfere differently on the transport depending on the presence of the RT. When the system does not have a RT, case **i)**, the trajectories reach the reference line after the destruction of the meandering curves. However, as showed in [21] this kind of curves has a discontinuous behaviour, they disappear for some values of β but they exist for many other intercalated values of β. So, for the case i) without RT, the fluctuations seen in Figure 3(a) show exactly the sensibility of the meandering curves for different values of the perturbation, while for the case ii) with a RT, in Figure 3(b) the trajectories reach the reference line, which is in fact the RT, only after the destruction of the resonance mode (1:4). Hence, in this case, the destruction or existence of the meandering curves is not a sufficient condition for the considered orbits reach the reference position. For the twist case, the small boxes in Figures 1 and 2 are above the resonance (1:4), thus we are going to investigate the effect of the RT in the neighborhood of the resonance mode (1:3). Figure 3(c) and 3(d) correspond to the case **iii)** and **iv)** respectively, and we observe a similar behaviour in the transport rates but with a small nuance. In the case without RT, Figure 3(c), as the structures are being destroyed the transport increases but the trajectories still find some regions with a soft stickiness. In the case with RT, Figure 3(d), as the resonance structures are destroyed the transport increases abruptly and we observe a soft stickiness only around the RT.

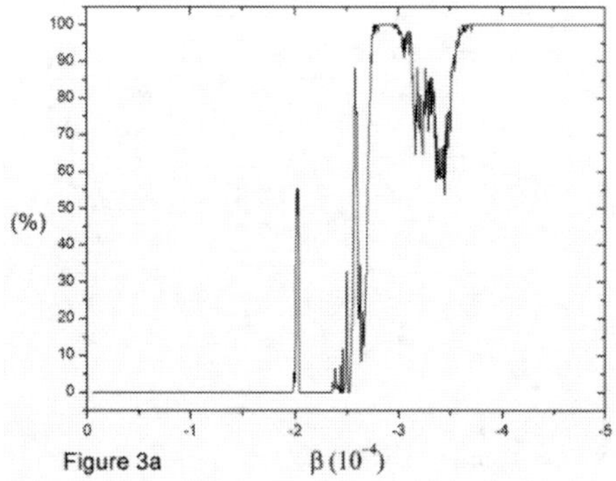

Figure 3a

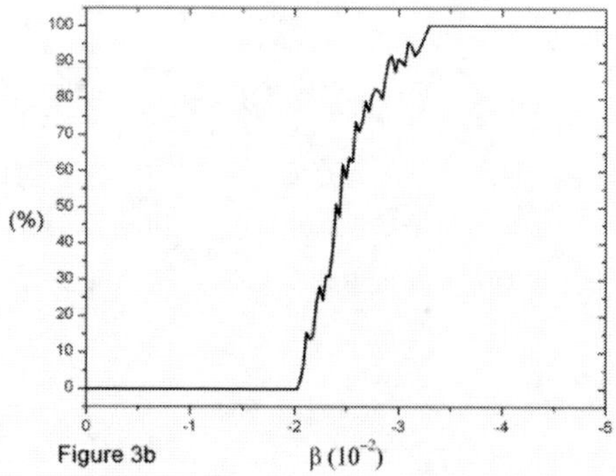

Figure 3b

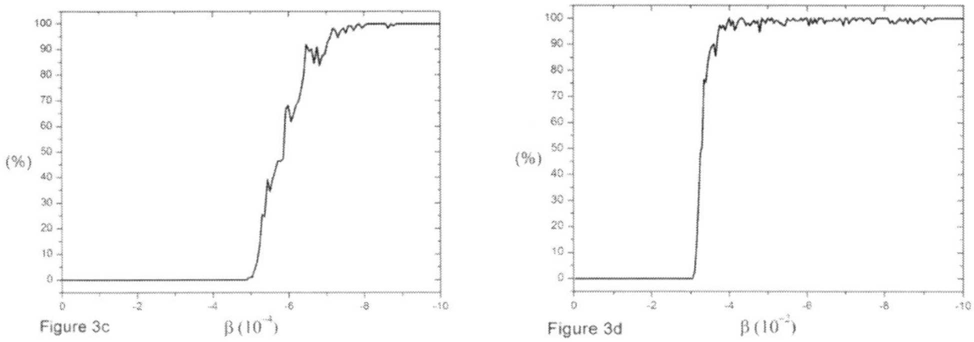

Figure.3. Percentage of initial conditions that arrived $\Delta J=3$ up 2500 iterations: **(a)** Non-twist map without RT, Eq. (4); **(b)** Non-twist map with RT, Eq. (5); **(c)** Twist map without RT, Eq. (4); **(d)** Twist map with RT, Eq. (5).

5. Conclusions

The non-liner maps that we use here may describe the equilibrium magnetic field lines perturbed by resonances created by rings of ergodic magnetic limiters. In previous works [1,14], robust torus had proved to be, theoretically, an efficient transport barrier that prevented the magnetic field lines to reach the tokamaks wall, avoiding plasma-wall interactions.

We showed here the behavior of two different transport barriers, one formed by the reconnection process of isochronous resonances, the meandering curves, and another formed by the vanishing of the perturbation in a particular region, through the introduction of a RT. In our approach the RT is stronger than the meandering barriers due to the stabilizing effect it induces in its neighborhood. We checked the behaviour of dynamical transport from our maps for different resonant modes with rotation number close to the resonant mode (1:3) used here, and we have observed that the qualitative behavior of the transport is similar with the plots showed in figure (3). The RT plays a remarkable influence on the dynamical transport; this is the reason that this kind of robust barrier is relevant for Hamiltonian plasma approaches.

Acknowledgments
The authors thank the scientific Brazilian agencies FAPESP and CNPQ for financial support.

References

[1] Martins C G L, Egydio de Carvalho R, Caldas I L, Roberto M 2010 *J. Phys. A* to appear
[2] Voyatzis G and Ichtiaroglou S 1999 *Int. J. Bif. and Chaos* **9**(5) 849
[3] Egydio de Carvalho R and Favaro G M 2005 *Physica A* **350** 173
[4] Egydio de Carvalho R, Martins C G L and Favaro G M 2009 *Braz. J. Phys.* **39**(3) 606
[5] Horton W 1999 *Rev. Mod. Phys.* **71** 735
[6] Wolf R C 2003 *Plasma Phys. Control. Fusion* **45** R1
[7] Connor J W et al 2004 *Nucl. Fusion* **44** R1
[8] Portela J S E, Caldas I L, Viana R L 2008 *Eur. Phys. J. Special Topics* **165** 195
[9] Roberto M, da Silva E C, Caldas I L, Viana R L 2004 *Braz. J. Phys.* **34(4B)** 1769
[10] Roberto M, da Silva E C, Caldas I L, Viana R L 2004 *Physica A* **342** 363
[11] Roberto M, da Silva E C, Caldas I L and Viana R L 2004 *Phys. Plasma* **11(1)** 214.
[12] Wurm A, Apte A and Morrison P J 2004 *Braz. J. Phys.* **34(4B)** 1700.
[13] Caldas I L, Viana R L, Araújo M S T, Vannucci A, da Silva E C, Ullmann K and Heller M V A P 2002 *Braz. J. Phys.* **32(4)** 980.
[14] Martins C G L, Egydio de Carvalho R, Caldas I L, Roberto M 2010 submitted

Journal of Physics: Conference Series **246** (2010) 012005 doi:10.1088/1742-6596/246/1/012005

[15] Egydio de Carvalho R and Ozório de Almeida A M 1992 *Phys. Lett. A* **162** 457
[16] Wurm A, Apte A, Fuchss K and Morrison P J 2005 *Chaos* **15** 023108
[17] Kucinski M Y and Caldas I L 1987 *Z. Naturforsch A: Phys. Sci* **42** 1124.
[18] Roberto M, Silva E C, Caldas I L and Viana R L 2005 *J. Phys.:Conf. Series* **7** 163
[19] Karger F and Lackner F 1975 *Phys. Lett. A* **61(6)** 385
[20] Szezech J D, Caldas I L, Lima G Z S, Viana R L, Morrison P J 2009 *Chaos* **19** 043108
[21] Morrison P J 2000 *Phys. Plasma* **7(6)** 2279

Journal of Physics: Conference Series **246** (2010) 012006 doi:10.1088/1742-6596/246/1/012006

Stability analysis under dilatation of coordinates

Paulo C. Rech

Departamento de Física, Universidade do Estado de Santa Catarina, 89223-100 Joinville, Brazil

Marcus W. Beims

Departamento de Física, Universidade Federal do Paraná, 81531-990 Curitiba, Brazil

E-mail: mbeims@fisica.ufpr.br

Abstract. The stability of periodic orbits under *translation* and also *dilatation* (or *contraction*) of coordinates is analysed using a q-Jacobian. Specifically we analyse the stability of period-1 and period-2 orbits of the quadratic map and show that the dilatation changes the parameter values for which bifurcations take place. The dilatation ($|q| > 1$) or contraction ($|q| < 1$) can be used to stabilize (destabilize) the unstable (stable) periodic orbits. In addition, dilatation effects on the stability depend only on the period of the orbit, and not on the location of orbital points in phase-space. This is shown to be true for any period-n orbit of the quadratic map. As a possible practical application, we suggest that the dilatation (contraction) considered here could represent temperature variations in physical systems.

1. Introduction

The study of basic hypergeometric series (also called q-hypergeometric series, where q is a parameter) has essentially started in 1748 by Euler [1]. These series have the additional parameter q which deforms the usual hypergeometric functions. The connection between the basic hypergeometric series and the usual hypergeometric functions is nicely shown in Ref. [2]. Although in the last decades many works considered the subject of q-deformations, ranging from mathematical physics, statistical physics, quantum physics to relativistic mechanics, it is still not clear what is the physical interpretation of the parameter q. The related literature is quite extensive and in this introduction we attain us to mention just some classical q-deformations in the context of the present work. From the point of view of nonlinear dynamical systems it was shown that the q-deformed analysis constitutes a natural way to treat the fractal properties of functions [3]. The q-deformed exponential function was also used to show a relation between the fractal dimension of an attractor and the parameter q [4], and to show the power-law sensitivity to the initial conditions at the edge of chaotic motion [5, 6]. More recently, q-deformation has been proposed to study the properties of the q-deformed quadratic map [7] and also the q-deformed Gaussian map [8]. In both cases the q deformation enters in the map itself and they observed the co-existence of attractors in the one-dimensional maps. In this paper we use the q-deformed derivative [3]

$$\partial_x^{(q)} f(x; y; ...) \equiv \frac{f(qx; y; ...) - f(x; y; ...)}{(q-1)x}, \tag{1}$$

to analyse stability in the quadratic map. Our quadratic map is the usual one but the *derivative*, used to determine the stability via the Jacobian of the map, is deformed. The derivative (1) measures the rate of changes of the function $f(x; y; ...)$ with respect to a *dilatation* of its argument by a factor of q. In the context of the linear stability analysis in dynamical systems, we can use Eq. (1) to study the stability of periodic points under a dilatation of the dynamical variable x. In the limit of $q \to 1$, the usual derivative is recovered.

The usual way [9] to study the stability of periodic orbits in one-dimensional maps is to analyse the eigenvalues of the Jacobian given by

$$J = \frac{\partial f(x)}{\partial x}, \tag{2}$$

where $f(x)$ is the function which describes the dynamics of the map. The stability index m from an n-periodic orbit can be calculated from the equation [10, 11, 12]

$$W_n^m = J_n \times J_{n-1} \times ... \times J_1 - m = 0. \tag{3}$$

For $|m| < 1$ the orbits are stable, for $|m| > 1$ the orbits are unstable, and for $|m| = 1$ there is a bifurcation point. In this paper we use the q-derivative (1) to determine the q-Jacobian and to analyse the stability of the periodic points under *dilatation* ($|q| > 1$) or *contraction* ($|q| < 1$) of the argument of the map. Using the q-derivative defined in [3], the q-Jacobian is calculated from

$$J^q = \partial_x^{(q)} f(x) = [\psi]_q \frac{f(x)}{x}, \tag{4}$$

where

$$[\psi]_q \equiv \frac{q^\psi - 1}{q - 1}, \tag{5}$$

and ψ is the degree of the homogeneous function $f(x)$. The deformed stability index m_q is determined from

$$W_n^{m_q} = J_n^q \times J_{n-1}^q \times ... \times J_1^q - m_q = 0. \tag{6}$$

In the limit $q \to 1$ we expect to obtain $J^q \to J$. The deformed stability index m_q recovers the usual stability index m when $q \to 1$. Also in this q-analysis we expect to have $|m_q| < 1$ for q-stable orbits, $|m_q| > 1$ for q-unstable orbits and $|m_q| = 1$ for q-bifurcating orbits. It is important to mention that q-stable and q-unstable means that the periodic orbits are respectively stable and unstable under simultaneous translation and dilatation. Each time we just mention the word stable or unstable, we are referring to the stability of the periodic orbits without dilatation. We also use the expression q-bifurcation when $q \neq 1$, otherwise we use bifurcation.

This work analyses the implication of dilatation or contraction of coordinates on the stability of a physical model, namely the quadratic map. Since such simple map presents the main features of more complicated dynamical systems, its dynamics under dilatation (contraction) of coordinates should also present the main features observed in more complicated dilatated (contracted) systems. We believe that our method can be very useful to describe the stability of systems when the dynamical variable experiments dilatation. It may also help us to elucidate the physical relevance of the parameter q in dynamical systems.

2. Quadratic Map

As a prototype model we use the quadratic map

$$f(x_n) = x_{n+1} = a - x_n^2, \tag{7}$$

which experiences bifurcations at the parameter values $a = -1/4$ (born of period-1 orbits), $a = 3/4$ (bifurcation $1 \to 2$), $a = 5/4$ (bifurcation $2 \to 4$), $a = 7/4$ (bifurcation $4 \to 8$), and so on. The usual Jacobian from Eq. (7) is given by

$$J_n = \frac{\partial f}{\partial x_n} = -2x_n, \tag{8}$$

while the q-Jacobian is

$$J_n^q = \partial_x^{(q)} f = -\frac{(q^2-1)}{(q-1)} x_n. \tag{9}$$

In the limit $q \to 1$ we obtain $J_n^q \to J_n$. To understand in details the effects of the dilatation on the stability analysis, we determine now the (a, q) dependence of the deformed stability index m_q for the simplest cases of period-1 and period-2 orbits.

2.1. Fixed points

The two fixed points are obtained from Eq. (7) using $n = 0$ and the condition $x_0 = x_1$. The results are

$$x_1^{\pm} = -\frac{1}{2}(1 \pm \eta), \tag{10}$$

where $\eta = \sqrt{1+4a}$. Using the deformed Jacobian (9) the condition for the stability index m_q is obtained from Eq. (6) and given by

$$W_1^{m_q} = m_q - \frac{1}{2}(q+1)(1 \pm \eta) = 0, \tag{11}$$

and the solutions are

$$m_q^{\pm} = \frac{(q+1)}{2}(1 \pm \eta). \tag{12}$$

When $q \to 1$, Eq. (12) reduces to

$$m_1^{\pm} = m^{\pm} = (1 \pm \eta), \tag{13}$$

which is the usual stability index under translation. Using $m = 1$ in this equation we obtain the condition $a = -1/4$, which is the value of the parameter where both fixed points are born. When $m = -1$ we recover the $1 \to 2$ bifurcation point which occurs at $a = 3/4$. For $-1/4 < a < 3/4$ only the fixed point x_1^- is stable. The fixed point x_1^+ is unstable. The relevant point here is that Eq. (12) imposes a relation between the parameter a and the deformation q. It means that when we consider $q \neq 1$ the dilatation *changes* the values of a where the q-bifurcation occurs. We can also vary q so that the fixed points change their stability. We will analyse this explicitly next.

2.1.1. Changing the stability of the fixed points using dilatation (contraction) Looking at Eq. (12) it is easy to observe that the stability for a given fixed point may change by varying the parameter q. As an example, let us discuss the stable fixed point x_1^- for $a = 1/2$. Using this fixed point, the deformed stability index is $m_q^- = (q+1)(1-\sqrt{3})/2$. We can now "destabilize" the point x_1^- by making an appropriate dilatation (contraction). Imposing the instability condition $|m_q^-| > 1$ we find the values $q < -3.732$ and $q > 1.732$. For these values of q the stable point x_1^- becomes q-unstable. It is also possible to "stabilize" the unstable fixed point x_1^+. Using $m_q^+ = (q+1)(1+\sqrt{3})/2$ and forcing $|m_q^+| < 1$ we obtain the interval $-1.732 < q < -0.268$.

Such (de)stabilization condition can be generalized for any value of a. Forcing $|m_q^-| > 1$ and $|m_q^+| < 1$ in Eq. (12) we obtain the "destabilization" condition for the stable point x_1^-

$$q < -\frac{1+\eta}{-1+\eta}, \qquad q > \frac{3-\eta}{-1+\eta}, \tag{14}$$

and the "stabilization" condition for the unstable point x_1^+

$$\frac{-3-\eta}{1+\eta} < q < \frac{1-\eta}{1+\eta}. \tag{15}$$

2.1.2. Case $m_q = 1$ (Born of q-fixed points) Here and in Sec. 2.1.3 we analyse the effect of the dilatation and contraction from another perspective. The question is what happens *at the q-bifurcation* points defined through $|m_q| = 1$?

We start with $m_q^\pm = 1$, where the q-fixed points $x_1^\pm$ are born. Solving Eq. (12) for q we obtain

$$q^\pm = -\frac{1}{2a}(1 + 2a \mp \eta). \tag{16}$$

For each value of a there is a given dilatation (or contraction) $q^\pm$ for which the q-stable and q-unstable fixed points are born. From Eq. (16) we determine

$$q^+ - q^- = 2\sqrt{1+4a}, \tag{17}$$

which says that the dilatation for both fixed points is *different at the q-bifurcation points.* The only exception occurs at the bifurcation point $a = -1/4$, where $x_1^+ = x_1^-$ and $q^+ = q^- = 1$.

It is also easy to verify that

$$q^- x_1^- = x_1^+ \qquad \text{and} \qquad q^+ x_1^+ = x_1^-, \tag{18}$$

which is independent of a. This means that the stable (unstable) point x_1^- (x_1^+) when multiplied by the dilatation factor q^- (q^+), will be located at the unstable (stable) fixed point x_1^+ (x_1^-). There is an interchange between the fixed points. This is a more or less obvious result since the stable point x_1^- becomes unstable when it is located at the unstable point x_1^+, and vice-versa. For later purposes we call to attention that in this section we found a relation between stable and unstable fixed points.

2.1.3. Case $m_q = -1$ (q-Bifurcation $1 \to 2$) Now imposing $m_q^\pm = -1$ in Eq. (12) and solving in q we obtain

$$q^\pm = -\frac{1}{2a}(-1 + 2a \mp \eta), \tag{19}$$

which again connects the parameter a and the dilatation q. It is easy to verify that, independent of a, we have

$$q^- x_1^- = 2 + x_1^+, \qquad \text{and} \qquad q^+ x_1^+ = 2 + x_1^-. \tag{20}$$

Different from Eq. (18) here the dilatation factors do not transform the stable (unstable) fixed point to the unstable (stable) fixed point. There is a factor 2 in between. Equation (20) gives a relation between a q-bifurcating point and an unstable fixed point, and also says that by a convenient dilatation (or contraction), the dilatated fixed point $q^- x_1^-$ bifurcates at $2 + x_1^+$, while the dilatated fixed point $q^+ x_1^+$ bifurcates at $2 + x_1^-$.

2.2. *Period-2 orbits*

Iterating twice the map (7) we find the points of the period-2 orbit, namely

$$x_2^{(1)} = \frac{1}{2}(1+\beta), \qquad \text{and} \qquad x_2^{(2)} = \frac{1}{2}(1-\beta), \tag{21}$$

where $\beta = \sqrt{-3+4a}$. This orbit bifurcates from the stable fixed point x_1^- at $a = 3/4$. The deformed stability index is calculated from Eq. (6), and gives

$$m_q = J_2^q \times J_1^q = \left[-\frac{q^2-1}{q-1} x_2^{(1)} \right] \left[-\frac{q^2-1}{q-1} x_2^{(2)} \right]. \tag{22}$$

Using the expression for $x_2^{(1)}$ and $x_2^{(2)}$ from Eq. (21) we obtain

$$m_q = (q+1)^2(1-a), \tag{23}$$

which is the deformed stability index for the period-2 orbit. In the limit $q \to 1$ we obtain from Eq. (23) the condition $m_1 = m = 4 - 4a$. Using $m = 1$ we get $a = 3/4$, which is the value where the bifurcation $1 \to 2$ occurs. For $m = -1$ we obtain $a = 5/4$, where the bifurcation $2 \to 4$ occurs. In the interval $3/4 < a < 5/4$ both points $x_2^{(1)}$ and $x_2^{(2)}$ are stable. For $a > 5/4$ both points are unstable.

2.2.1. Changing the stability by dilatation (contraction) As discussed in Sec. 2.1.1, also here the stability of the period-2 orbit can be changed by varying q, as can be observed from Eq. (23). Since the orbit is stable, we can only destabilize it. The "destabilization" condition for the period-2 orbits in the interval $3/4 < a < 5/4$ is obtained by imposing $|m_q| > 1$ to Eq. (23), to obtain

$$\frac{\sqrt{1-a}}{(1-a)} - 1 < q < \frac{\sqrt{a-1}}{(a-1)} - 1. \tag{24}$$

2.2.2. Case $m_q = 1$ (q-Bifurcation $1 \to 2$) Making $m_q = 1$ in Eq. (23) we obtain

$$q = \frac{\sqrt{1-a}}{1-a} - 1, \tag{25}$$

which is real for $a < 1$. At the original bifurcation point $a = 3/4$ we obtain $q = 1$, which is equal to q^- obtained in last section for the bifurcation $1 \to 2$ ($m_q^- = -1$). Therefore, the points q and q^- give correctly the same value at the bifurcation $1 \to 2$.

In order to observe the effect of the dilatation q we determine

$$q\, x_2^{(1)} = x_2^{(2)} + \frac{(1+\beta)\sqrt{1-a}}{2(1-a)} - 1, \tag{26}$$

and

$$q\, x_2^{(2)} = x_2^{(1)} + \frac{(1-\beta)\sqrt{1-a}}{2(1-a)} - 1, \tag{27}$$

which gives a relation between the two originally stable points. At the bifurcation point $a = 3/4$ we obtain $x_2^{(1)} = x_2^{(2)}$ as expected.

2.2.3. Case $m_q = -1$ (q-Bifurcation $2 \to 4$) Now we use $m_q = -1$ in (23) to obtain

$$q = \frac{\sqrt{a-1}}{a-1} - 1, \tag{28}$$

which is real for $a > 1$. It is also straightforward to obtain

$$q\, x_2^{(1)} = x_2^{(2)} + \frac{(1+\beta)\sqrt{a-1}}{2(a-1)} - 1, \tag{29}$$

and

$$q\, x_2^{(2)} = x_2^{(1)} + \frac{(1-\beta)\sqrt{a-1}}{2(a-1)} - 1. \tag{30}$$

Observe that these relations between the points are not valid for $a < 1$. For the point $a = 5/4$ (bifurcation $2 \to 4$) we obtain from the above equations $x_2^{(1)} = x_2^{(2)} + \sqrt{2}$ and $x_2^{(2)} = x_2^{(1)} - \sqrt{2}$, respectively.

2.3. Period-n orbits
From Eq. (22) we can observe that the dilatation effect on the deformed stability index m_q is a global factor. For a period-n orbit we obtain from Eqs. (6) and (9)

$$m_q = J_n^q \times \ldots J_2^q \times J_1^q = [-(q+1)]^n \, x^{(n)} x^{(n-1)} \times \ldots \times x^{(2)} x^{(1)}, \tag{31}$$

where $x^{(i)}, i = 1, \ldots n$ are the orbital points. The term $[-(q+1)]^n$ is a global factor, independent of the orbital points. Taking the limit $q \to 1$ in the above equation we observe that the stability index m can be written as

$$m = J_n \times \ldots J_2 \times J_1 = [-2]^n \, x^{(n)} x^{(n-1)} \times \ldots \times x^{(2)} x^{(1)}. \tag{32}$$

From Eqs. (31) and (32) we obtain the interesting relation

$$\frac{m_q}{m} = \left[\frac{(q+1)}{2}\right]^n = e^{n \ln\left[\frac{q}{2} + \frac{1}{2}\right]}. \tag{33}$$

This shows us the connection between the stability index m and the q-deformed stability index m_q for *any* period-n orbit. Some interesting conclusions can be obtained from Eq. (33):

- For $q = 1$ we have $m_q = m$, as expected. For $q = -1$ we have $m_q = 0$, a superstable orbit.
- For $-1 < q < 1$ (contraction) we obtain $|m_q| < |m|$. Therefore the contraction tends to stabilize the periodic orbits. For $n \to \infty$ we obtain $m_q \to 0$ (superstability).
- For $|q| > 1$ (dilatation) we obtain $|m_q| > |m|$. The dilatation tends to destabilize the periodic orbits. For $n \to \infty$ we obtain $m_q \to \infty$ (a very large instability).
- From the last two conclusions we can conjecture that at the border line between periodic and chaotic motion (assuming n is very large), while contractions leave to superstable orbits, dilatation leaves to very unstable orbits.
- At the bifurcation points $m_q = m = \pm 1$ (or more generally at $m_q = m$) we always obtain $q = 1$. This means that, keeping a fixed, changes in the stability of the periodic orbits will only occur when $q \neq 1$.
- Changes in the stability ($m_q \neq m$) of periodic orbits via dilatation depend *only on the period* of the orbit, and not on their location in phase-space.

3. Summary

This paper proposes to use a q-deformed derivative [3] to analyse the stability of periodic points under simultaneous *dilatation* and *translation* of coordinates. Different from recent works [7, 8], where q-deformed maps were proposed, here the map is the usual one but the *derivative* utilized to determine the Jacobian of the map is deformed. Specifically we analyse the stability of period-1 and period-2 orbits of the quadratic map. We show that the dilatation can change the parameter values $a = -1/4, 3/4$, and $5/4$, where bifurcations of the low periodic orbits from the quadratic map occur. The q-deformation can be used to stabilize (destabilize) the unstable (stable) periodic orbits under dilatation ($|q| > 1$) or contraction ($|q| < 1$). We also show that the effect of dilatation and contraction on the stability depends only on the period of the orbit, and not on the location of the orbital points in phase-space. This occurs for any period-n orbit of the quadratic map and suggests that at the border line between periodic and chaotic motion (assuming n very large), contractions leave to superstable orbits while dilatation to very unstable orbits. More generally, the stability analysis proposed here, via the q-derivative, can be applied to any dynamical system, discrete or continuous, and explores a new approach to analyse the stability of physical systems. To finish we would like to make an attempt to give a physical interpretation for the dilatation and contraction: consider a one-dimensional monoatomic ideal gas with temperature defined by $\frac{1}{2}kT = \frac{1}{2}\langle mv^2\rangle$. By making the dilatation (contraction) $x \to qx$, and therefore $v \to qv$, we obtain $\frac{1}{2}kT = \frac{1}{2}q^2\langle mv^2\rangle$, which shows that dilatation ($q^2 > 1$) are related to a temperature increase and contractions ($q^2 < 1$) are related to a temperature decrease. Obviously that our quadratic map did not represent the dynamics of an ideal gas, but the above arguments suggest that dilatation (contraction) of coordinates could be related to a temperature increase (decrease). The dynamics of a particle moving on a metallic plate which changes its temperature, or a particle moving on the surface of a balloon which inflates are just another examples of the dynamics on dilatated coordinates. The key point is that since the quadratic map presents the main features of more complicated dynamical systems, its stability under dilatation (contraction) of coordinates should also present the main features observed in more complicated dilatated (contracted) systems, whatever induces the dilatation (contraction).

Acknowledgments

The authors thank CNPq for financial support.

[1] Euler L 1748 *Introduction in Analysis Infinitorum* (Bousquet: Lausanne)
[2] Gasper G and Rahman M Basic 1997 *Hypergeometric Series* (University Press: Cambridge)
[3] Ayse E 1997 *Phys. Rev. Lett.* **78** 3245
[4] Costa U M S, Lyra M L, Plastino A R and Tsallis C (1997) *Phys. Rev. E* **56** 245
[5] Tsallis T, Plastino A R and Zheng W M 1997 *Chaos Solitons Fractals* **8** 885
[6] Lyra M L and Tsallis C 1998 *Phys. Rev. Lett.* **80** 53
[7] Jaganathan R and Sinha S 2005 *Phys. Lett. A* **338** 277
[8] Patidar V and Sud K K 2009 *Commun. Nonlinear Sci. Numer. Simulat.* **14** 827
[9] Lichtenberg A J and Lieberman M A 1992 *Regular and Chaotic Dynamics* (Springer-Verlag: New York)
[10] Rech P C, Beims M W and Gallas J A C 2005 *Phys. Rev. E* **71** 17202
[11] Beims M W and Gallas J A C 1997 *Physica A* **238** 225
[12] Rech P C, Beims M W and Gallas J A C 2000 *Europhys. Lett.* **49** 702

Journal of Physics: Conference Series **246** (2010) 012007 doi:10.1088/1742-6596/246/1/012007

Dynamical properties of the weak stability boundary and associated sets

Priscilla A Sousa Silva and Maisa O Terra

Instituto Tecnológico de Aeronáutica, Departamento de Matemática, Praça Marechal Eduardo Gomes, 50, 12.228-900, São José dos Campos, SP, Brazil

E-mail: `priandss@ita.br, maisa@ita.br`

Abstract. In this contribution the weak stability boundary algorithmic definition was numerically accomplished with the inclusion of lunar and earth collisional sets and a subclassification of the unstable set. Then, the associated sets to WSB definition were analyzed and characterized according to relevant dynamical properties in order to clarify their applicability in earth-moon transfer orbit design. The obtained stable, unstable, and collisional sets are defined as a function of the osculating ellipse eccentricity for prograde and retrogade initial conditions. The stable sets, candidates to ballistic capture transfers, are subclassified according to chosen specific criteria, namely, the Jacobi constant intervals defined by distinct classes of Hill regions, the location of the final state after a complete cycle with respect to the Hill sphere, the permanence in the lunar sphere of influence in a full cycle around the moon, and exit basins for retrograde evolution. By the first time, with this investigation, elucidative criteria based on three-body problem elements are employed to identify initial condition subsets with required properties to design ballistic capture transfers.

1. Introduction

The Weak Stability Boundary (WSB) concept was heuristically proposed by E. Belbruno (1987 and beyond) [1] in the context of ballistic lunar capture in low energy earth-moon transfer orbit design. Traditional techniques usually rely on two-body dynamics, building transfer trajectory approximations by patching several segments of conic solutions obtained by considering successive two-body interactions, with one body always being the spacecraft. As space missions became more demanding, alternative design methods have been proposed aiming to employ the nonlinear dynamical characteristics of three and four body systems in a more natural way. By doing so, total fuel consumption may be reduced, allowing a higher diversity of mission profiles.

Particularly, a class of low energy lunar transfers associated with the WSB concept was introduced, including the gravitational effects of the earth, the moon and the sun in modeling the motion of the spacecraft [2]. In the final portion of the transfer trajectory, the spacecraft approaches the moon in a state defined as lunar ballistically captured, thus reducing the propellant mass needed to stabilize its motion into the final prescribed selenocentric orbit. Some expressions found in literature to describe the WSB concept are "*regions in the phase space where the perturbative effects of the earth-moon-sun acting on the spacecraft tend to balance* " [3] and "*a location near the moon where the spacecraft lies in the transition between ballistic capture and ejection* " [4].

The WSB concept was successfully employed for the first time in the Japan's Hiten spacecraft rescue to the moon in 1991 [3]. Originally, the Hiten mission, known before launch as Muses-A, had been designed to orbit the earth, serving as a communication relay for Hagoromo, a second spacecraft, that would be send to the moon. However, Hagoromo failed due to loss of communication, so it was proposed to send Hiten to the moon. The low amount of available on board propellant (10% of necessary fuel for a classical transfer design) required an alternative transfer orbit design technique and WSB strategy was able to carry Hiten from an earth orbit to a lunar orbit. This successful application motivated a number of papers regarding WSB earth-moon transfer projects [5–8].

The WSB algorithmic definition [1, 9] was recently refined by F. García and G. Gómez [6]. In this approach, the WSB concept is established in the framework of the Planar Circular Restricted Three-Body Problem (PCRTBP). Departing from an initial condition set, trajectories are generated and classified according to a prescribed stability criterion, in such a way that the WSB set is defined as the boundary set between stable and unstable sets.

This contribution reports a modified numerical implementation of the usual algorithmic definition and a series of dynamical analyses performed in order to investigate the applicability aspects of the generated sets in ballistic capture transfer projects. In this work, the collision of the third body with the surfaces of the primaries, namely, the earth and the moon, is taken into account with the inclusion of the mean radius of each primary since the beginning of the implementation. The algorithmic definition numerical implementation is performed for distinct osculating ellipse eccentricity values and for prograde and retrograde initial conditions. Moreover, the unstable set is explicitly generated and subclassified according to the associated unstable behavior of each initial condition, revealing the existence of subcases for the stable-unstable boundary.

The preliminary analysis classifies the full initial condition set according to the integral of motion of the PCRTBP, the Jacobi constant, in order to verify which possible transport channels are available for the set. Then, we focus our attention on the stable set analysis, since its elements are natural candidates to be employed in ballistic capture transfers. The stable set is subclassified according to convenient specific criteria, namely, the Jacobi constant intervals defined by distinct classes of Hill regions, the location of the final state after a complete cycle in relation to the Hill sphere, the permanence in the lunar sphere of influence in a full cycle around the moon, and exit basins for retrograde evolution. The purpose of this investigation is to elucidate the applicability characteristics of the trajectories associated to the defined set of initial conditions in ballistic capture transfers.

The present paper is organized as follows: in Section 2, we present the theoretical framework upon which the work is based on, namely, we describe the dynamical model, the PCRTBP, several capture definitions, and the revised version of the WSB algorithmic definition, as in [6]. Our particular implementation, analyses and results are delineated in Section 3. The last section is devoted to the final remarks and conclusions.

2. Theoretical Framework

We proceed with the description of the Planar Circular Restricted Three-Body Problem (PCRTBP), since it provides the dynamical framework in which the WSB has been defined. Then some definitions related to capture are stated in order to set ground to define a special kind of capture in the PCRTBP, called ballistic capture, related to the WSB [9]. Finally, definitions for the WSB, proposed in the literature, are stated.

2.1. Dynamical Model

The Restricted Three-Body Problem (RTBP) describes the motion of a particle P_3 of negligible mass moving under the gravitational influence of two bodies P_1 and P_2, called the primaries, of masses m_1 and m_2, respectively. The motion of P_1 and P_2 is not perturbed by P_3, being a solution of a Kepler Problem, with two bodies moving under mutual gravitational influence. In the planar circular version of the RTBP, the PCRTBP, P_1 and P_2 describe circular coplanar orbits around the barycenter of this two-body system and the motion of the third body is restricted to the orbital plane of the primaries [10]. This model is suited to describe the earth-moon-spacecraft system (*earth* $\Leftrightarrow P_1$, *moon* $\Leftrightarrow P_2$, *spacecraft* $\Leftrightarrow P_3$), given that the moon's orbit eccentricity around the earth is 0.05490 and its inclination with respect to the orbit of the planet is $5°09'$. This dynamical model is expressed in non-dimensional variables, in such a way that the distance between P_1 and P_2, the sum of their masses and their angular velocity around the barycenter are normalized to one. The normalized masses of P_1 and P_2 are given, respectively, by $1-\mu$ and μ, with $\mu = m_2/(m_1+m_2)$, $m_1 > m_2$, being the only parameter of the model. For the earth-moon system, $\mu = 0.0121506683$.

In the synodic coordinate system x-y, with origin in the barycenter of P_1-P_2 and that rotates with the primaries with respect to an inertial frame X-Y, P_1 and P_2 are located at $(\mu, 0)$ and $(\mu-1, 0)$, respectively, as shown in Figure 1.

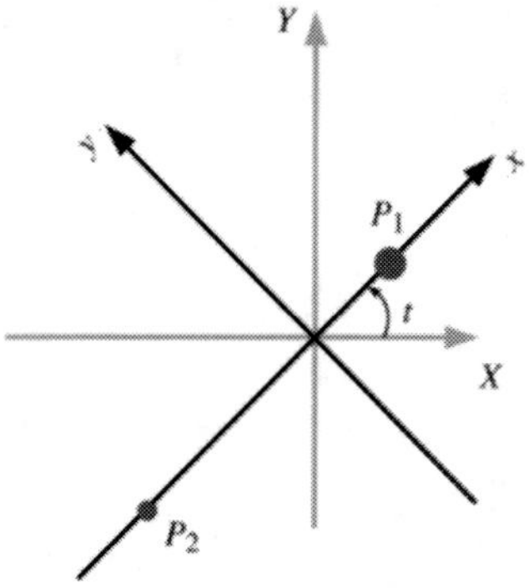

Figure 1. The x-y synodic coordinate frame rotates with unitary angular velocity with respect to the X-Y inertial reference frame.

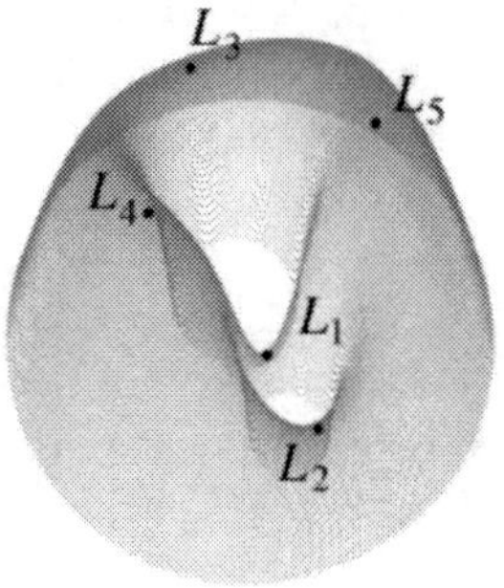

Figure 2. Effective potential Ω for $\mu = 0.3$.

Let $(x, y, \dot{x}, \dot{y})$ represent the state of P_3 in the synodic reference system, the particle's equations of motion are given by

$$\begin{aligned} \ddot{x} - 2\dot{y} &= \Omega_x, \\ \ddot{y} + 2\dot{x} &= \Omega_y, \end{aligned} \tag{1}$$

where

$$\Omega(x,y) = \frac{1}{2}(x^2+y^2) + \frac{1-\mu}{r_1} + \frac{\mu}{r_2} + \frac{\mu(1-\mu)}{2}, \tag{2}$$

is the effective potential, shown in Figure 2, for $\mu = 0.3$, and $r_1 = [(x-\mu)^2 + y^2]^{1/2}$ and $r_2 = [(x+1-\mu)^2 + y^2]^{1/2}$ are the distances from P_1 and P_2, respectively. Both in Figures 2 and 3, the value of $\mu = 0.3$ was adopted for the sake of visualizing clarity.

This system has an integral of motion given by

$$J(x, y, \dot{x}, \dot{y}) = 2\Omega(x,y) - (\dot{x}^2 + \dot{y}^2) = C, \tag{3}$$

where C is the Jacobi constant. The conservation associated to J defines a three-dimensional invariant manifold immersed in the four-dimensional phase space by

$$\mathcal{M}(\mu, C) = \left\{(x, y, \dot{x}, \dot{y}) \in \mathbb{R}^4 | J(x, y, \dot{x}, \dot{y}) = \text{constant}\right\}. \tag{4}$$

The dynamical model has five equilibria, called libration points or Lagrangian points, defined by

$$\frac{\partial J}{\partial x} = \Omega_x = 0, \quad \frac{\partial J}{\partial y} = \Omega_y = 0, \quad \frac{\partial J}{\partial \dot{x}} = 0 \Rightarrow \dot{x} = 0, \quad \frac{\partial J}{\partial \dot{y}} = 0 \Rightarrow \dot{y} = 0. \tag{5}$$

Three of them, the collinear equilibria $L_k, k = 1, 2, 3$, are located on the x-axis and are saddle-center points. The other two, the triangular points $L_k, k = 4, 5$, are located at the vertices of equilateral triangles formed with the positions of primaries and are stable if $m_1/m_2 > 24.96$, as it is the case for the earth-moon system. The Jacobi constant values evaluated in these static solutions are denoted by $C_k, k = 1, 2, 3, 4, 5$.

The regions obtained by the projection of the $\mathcal{M}$ surface onto the position space x-y are called the Hill regions, M, defined by

$$M(\mu, C) = \{(x, y) | \Omega(x, y) \geqslant C/2\} \tag{6}$$

and bounded by the *zero-velocity curves*, the locus of points in the x-y plane where the kinetic energy vanishes. The Hill regions constitute the accessible areas to the trajectories for each given C, or analogously, energy value. For a given μ, there are five basic configurations for these regions, defined by the C_k values. The first four cases are shown in Figure 3. In the fifth case, motion over the entire x-y plane is possible.

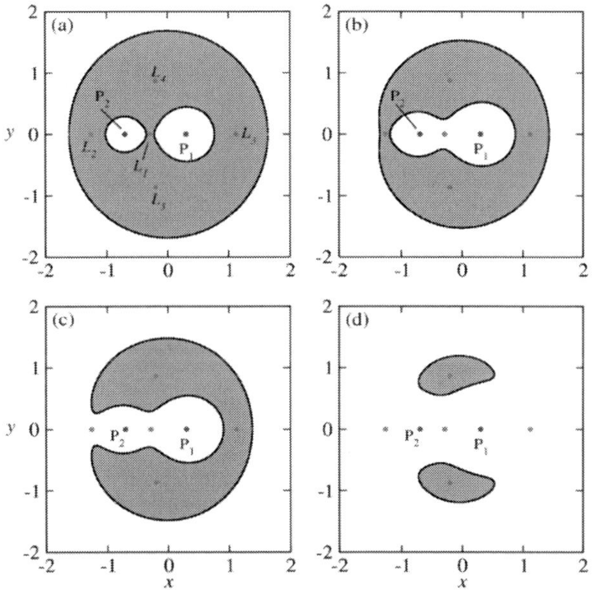

Figure 3. Four possible Hill regions cases for $\mu = 0.3$: the white areas correspond to the Hill regions. The inaccessible areas and the zero velocity curves are shown in grey and black, respectively. The blue points correspond to the primaries and the red ones to the equilibrium points $L_k, k = 1, 2, 3, 4, 5$. (a) Case 1 ($C > C_1$): no transit orbits between the primaries are possible. (b) Case 2 ($C_1 > C > C_2$): the realms around P_1 and P_2 are connected by transit orbits through the neck region around L_1. (c) Case 3 ($C_2 > C > C_3$): the same as in Case 2, besides the possibility of transit to the external realm of motion through the neck region around L_2. (d) Case 4 ($C_3 > C > C_4 = C_5$): there is a neck region also around L_3. In the fifth case ($C < C_4 = C_5$), motion over the entire x-y plane is possible.

There are periodic orbits, called Lyapunov orbits, around each collinear Lagrangian point [11]. The unstable and stable invariant manifolds associated to a Lyapunov orbit are locally homeomorphic to two-dimensional cylinders and act as separatrices of the phase space, defining four categories of trajectories in the neck region near each equilibrium [12], namely transit, nontransit and asymptotic orbits, besides the periodic orbit itself. The x-y projection R of the region $\mathcal{R}$ around a generic equilibrium $L_k, k = 1, 2, 3$ is shown in Figure 4.

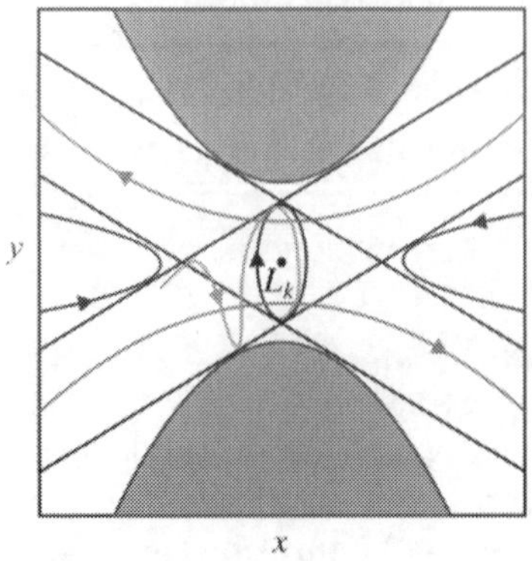

Figure 4. The four categories of orbits are shown in the x-y projection R of the neck region $\mathcal{R}$ around an equilibrium $L_k, k = 1, 2, 3$. The periodic orbit is the black curve around L_k and the transit, nontransit and asymptotic orbits are represented by the red, blue and green curves, respectively.

2.2. Capture

Let $\mathbf{q} = (x, y) \in \mathbb{R}^2$ be the position vector of P_3. In general, a capture state of P_3 is defined when P_3 is somehow geometrically *bounded* to P_1 and (or) P_2. Following [9], we state the definitions of some relevant types of capture.

Definition 1 (Permanent capture) *P_3 is permanently captured into the P_1-P_2 system in forward time if $|\mathbf{q}|$ is bounded as $t \to \infty$, and $|\mathbf{q}| \to \infty$ as $t \to -\infty$. P_3 is permanently captured in backward time if $|\mathbf{q}|$ is bounded as $t \to -\infty$, and $|\mathbf{q}| \to \infty$ as $t \to \infty$.*

For the General Problem of Three Bodies, is was proved by Chazy that the set of initial values leading to permanent capture comprises a set of measure zero. The existence of such orbits was later shown by Alekseev. For further discussion we refer the reader to [13].

Definition 2 (Temporary capture) *P_3 has temporary capture at $t = t^*, |t^*| < \infty$, if $|q(t^*)| < \infty$ and*

$$\lim_{t \to \pm\infty} |q(t)| = \infty.$$

The temporary capture state implies the ejection of the particle as $t \to \pm\infty$, that is, P_3 is neither permanently captured nor describes an unbounded oscillatory orbit.

Definition 2 can be modified to *finite temporary capture* considering that P_3 achieves a prescribed finite distance $|q(t)| = d > 0$ from P_2, for instance, at time t^*. In this case, the time interval $\Delta t = |t^* - T^*| > 0$ represents the duration of the capture relative to a reference time T^*. This definition measures the time that P_3 remains within a given neighborhood of P_2.

An alternative type of capture, called ballistic capture, said to be defined analytically as to be distinguished from the previously geometrically defined types of capture, is presented in [9].

Let $(\tilde{x}, \tilde{y}, \dot{\tilde{x}}, \dot{\tilde{y}})$ be the state of P_3 in an inertial reference frame with origin in P_2. The Keplerian (two-body) energy of P_3 w.r.t. P_2 is

$$h_K = \frac{1}{2}\left(\dot{\tilde{x}}^2 + \dot{\tilde{y}}^2\right) - \frac{Gm_2}{\tilde{r}_2}, \tag{7}$$

where $\tilde{r}_2 = \sqrt{\tilde{x}^2 + \tilde{y}^2}$ is the distance between P_3 and P_2, m_2 is the mass of the primary and G stands for the universal gravitational constant [14].

Definition 3 (Ballistic capture) *P_3 is ballistically captured by P_2 at time $t = t_c$ if*

$$h_K(\varphi(t_c)) \leq 0$$

for a solution $\varphi(t) = (x(t), y(t), \dot{x}(t), \dot{y}(t))$ of the RTBP.

We note that this definition is not unique. In [15], for example, ballistic capture by the moon refers to an orbit which under natural dynamics of the RTBP gets within a 20,000 km radius region around the moon and performs at least one revolution around that primary.

2.3. The Weak Stability Boundary

Let $(x, y, \dot{x}, \dot{y})$ be the state of P_3 in the synodic reference frame and $J(x, y, \dot{x}, \dot{y})$ be the integral of Jacobi (Eq.(3)). Consider the sets

$$\begin{aligned} J^{-1}(C) &= \{(x, y, \dot{x}, \dot{y}) | J(x, y, \dot{x}, \dot{y}) = C\}, \\ \Sigma &= \{(x, y, \dot{x}, \dot{y}) | h_K(x, y, \dot{x}, \dot{y}) \leq 0\}, \quad \sigma = \{(x, y, \dot{x}, \dot{y}) | \dot{r}_2 = 0\}, \end{aligned} \tag{8}$$

where $\dot{r}_2$ is the radial velocity of P_3 relative to P_2, so that σ is the set of points corresponding to local periapsis or apoapsis of osculating ellipses, and h_K is the Keplerian energy of P_3 w.r.t. P_2.

Definition 4 (WSB analytically defined) *The set W, associated to a certain value of the Jacobi constant C, given by*

$$W = J^{-1}(C) \cap \Sigma \cap \sigma, \tag{9}$$

is called the Weak Stability Boundary, a special set where ballistic capture occurs in the RTBP.

There is also an algorithmic definition of the WSB [3,9], discussed and revised by F. García and G. Gómez [6]. Following this reference, consider a radial segment $l(\theta)$ departing from P_2 and making an angle θ with the x-axis. Take trajectories for P_3, starting on $l(\theta)$ which satisfy:

(i) P_3 starts its motion at the periapsis of an osculating ellipse around P_2, at a certain point on $l(\theta)$. If r_2 denotes the distance between P_3 and P_2 at this point, then $r_2 = a(1 - e)$, where a and e are, respectively, the semi-major axis and the eccentricity of the ellipse.

(ii) The initial velocity vector of the trajectory is perpendicular to the radial segment $l(\theta)$.

(iii) The initial Kepler energy h_K of P_3 w.r.t. P_2 is negative, which occurs for $e \in [0, 1)$, since the two-body energy computed at the periapsis is

$$h_K = \frac{\mu(e-1)}{2r_2}. \tag{10}$$

(iv) The eccentricity of the initial Keplerian motion is held fixed along $l(\theta)$. According to Eq.(10), this condition is equivalent to fix the value of h_K for each value of r_2 and $\forall$ $\theta \in [0, 2\pi]$.

The following definition of stability and instability is given for the trajectories with the initial conditions specified above.

Definition 5 *The motion of a particle is said to be* ***stable*** *about P_2 if after leaving $l(\theta)$ it makes a full cycle about P_2 without going around P_1 and returns to $l(\theta)$ at a point with negative Kepler energy w.r.t. P_2. Otherwise, the motion will be* ***unstable****. (See Figure 5.)*

As the initial conditions vary along $l(\theta)$, there is a distance $r^*(\theta, e)$ such that:

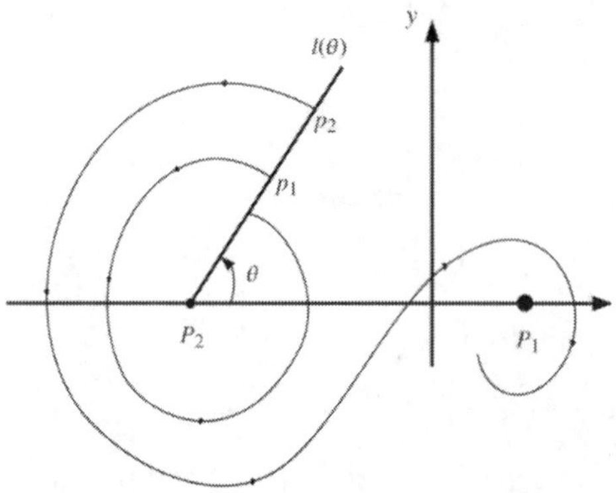

Figure 5. Schematic representation of the definition of stable and unstable trajectories, represented by p_1 and p_2, respectively.

- If $r_2 < r^*$, the motion is stable.
- If $r_2 > r^*$, the motion is unstable.

Furthermore, $r^*(\theta, e)$ is a smooth function of θ and e which defines the WSB as the locus of all points $r^*(\theta, e)$ along $l(\theta)$, $\theta \in [0, 2\pi]$, for which there is a change of stability [4,9].

Definition 6 (WSB algorithmic definition) *The Weak Stability Boundary is given by the set*

$$\mathcal{W} = \{r^*(\theta, e) | \theta \in [0, 2\pi], e \in [0, 1)\}, \tag{11}$$

with smooth $r^*(\theta, e)$ determined by Definition 5. The subset obtained by fixing the eccentricity e of the osculating ellipse is

$$\mathcal{W}^e = \{r^*(\theta, e) | \theta \in [0, 2\pi], e = constant\}. \tag{12}$$

As pointed in [6] this definition is not accurate in the following points:

- The requirements on the initial conditions fix the modulus and direction of the velocity but not its sense. So, for a fixed position on $l(\theta)$ there are two different initial velocities, fulfilling the four restrictions, which can produce orbits with different stability behavior. The modulus of the initial velocity of P_3, is given by

$$\nu^2 = \mu \left(\frac{2}{r_2} - \frac{1}{a} \right) = \frac{\mu(1+e)}{r_2}. \tag{13}$$

  The initial conditions with positive velocity (prograde osculating motions about P_2) are

$$\begin{aligned} x &= -1 + \mu + r_2 \cos\theta, & y &= r_2 \sin\theta, \\ \dot{x} &= r_2 \sin\theta - \nu \sin\theta, & \dot{y} &= -r_2 \cos\theta + \nu \cos\theta, \end{aligned} \tag{14}$$

  and the initial conditions with negative velocity (retrograde osculating motions about P_2) are

$$\begin{aligned} x &= -1 + \mu + r_2 \cos\theta, & y &= r_2 \sin\theta, \\ \dot{x} &= r_2 \sin\theta + \nu \sin\theta, & \dot{y} &= -r_2 \cos\theta - \nu \cos\theta. \end{aligned} \tag{15}$$

- Contrary to what is suggested by the definition of $\mathcal{W}$, for fixed values of θ and e, there may be several transitions from stability to instability.
- Some maximum time interval must be fixed for the numerical integration. In [6], T_{max} was taken as 80 non-dimensional time units. If after this time interval, a trajectory does not cross again the segment $l(\theta)$, the orbit is considered unstable.

The algorithmic definition is extended in [6], considering the stability criterion after n revolutions of P_3 about P_2 without going around P_1. Also some slight modification in the definition is introduced in [8] to study the behavior in the first stable-unstable transition.

3. Results and dynamical analysis

3.1. Results of the considered WSB algorithmic definition implementation

Once that this work deals with the applicability of the WSB and associated sets in realistic mission design involving ballistic capture transfers, the punctual mass idealization of the mathematical model is substituted by finite bodies with the explicit inclusion of the mean radii of the earth and the moon. Being so, this WSB algorithmic definition implementation generates, besides stable and unstable sets, the moon and earth collisional sets, i.e., sets of initial conditions of trajectories which collide with each primary. Moreover, the unstable set was subclassified according to five proposed instability criteria: (i) unstable due to negative Keplerian energy, when trajectories return to $l(\theta)$ after one turn around the moon, (ii) primary interchange through L_1 Lagrangian point with $C > C_3$, (iii) primary interchange through L_2 Lagrangian point with $C > C_3$, (iv) geometric escape with $C < C_3$, and (v) unstable due to exceeding the maximum integration time, without returning to $l(\theta)$. With the unstable set subclassification, the WSB can be decomposed in distinct subsets of stable-unstable transitions.

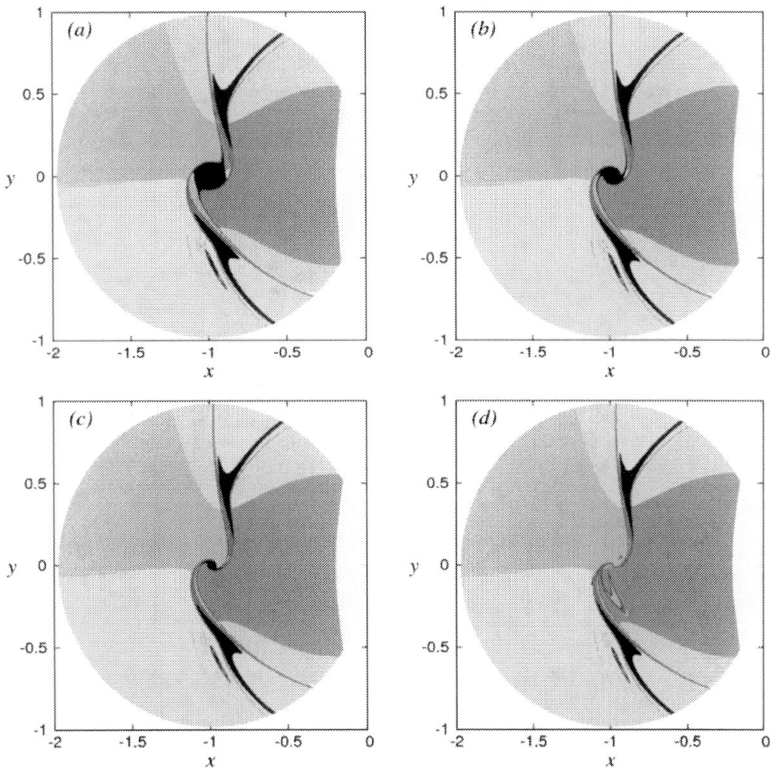

Figure 6. Projection onto the x-y plane of the sets generated by the WSB algorithmic definition with positive initial velocity (osculating prograde motion about P_2) for: (a) $e = 0.0$, (b) $e = 0.3$, (c) $e = 0.6$, and (d) $e = 0.9$. **Black**: $\mathcal{S}$, stable; **Red**: C^M, collision with the moon; **Yellow**: C^E, collision with the earth; **Green**: $\mathcal{E}$, unstable due to negative Kepler energy; **Dark grey**: $\mathcal{G}^1$, primary interchange through L_1; **Grey**: $\mathcal{G}^2$, primary interchange through L_2; **Light grey**: $\mathcal{G}^3$, geometric escape with $C > C_3$.

The implementation was performed holding fixed the osculating ellipse eccentricity value, in such a way that the investigated sets are given as functions of r and θ.

Figure 6 presents the projection onto the x-y plane of the $\mathcal{W}^e$ associated sets generated by the prescribed WSB algorithmic definition with initial positive velocity (osculating prograde motion about P_2) and for the eccentricity values (a) $e = 0.0$, (b) $e = 0.3$, (c) $e = 0.6$, and (d) $e = 0.9$. The same is shown by Figure 7 for initial conditions with negative velocities (osculating retrograde motion about P_2). The ranges of the spatial coordinates were chosen following [6]. The generated sets of initial conditions correspond to arrival states of the complete transfer orbits, and it is desirable that these states are in a near enough neighborhood of the moon. So, the presence of initial conditions outside the lunar Hill region seems not to be appropriate for transfer orbit design from the beginning.

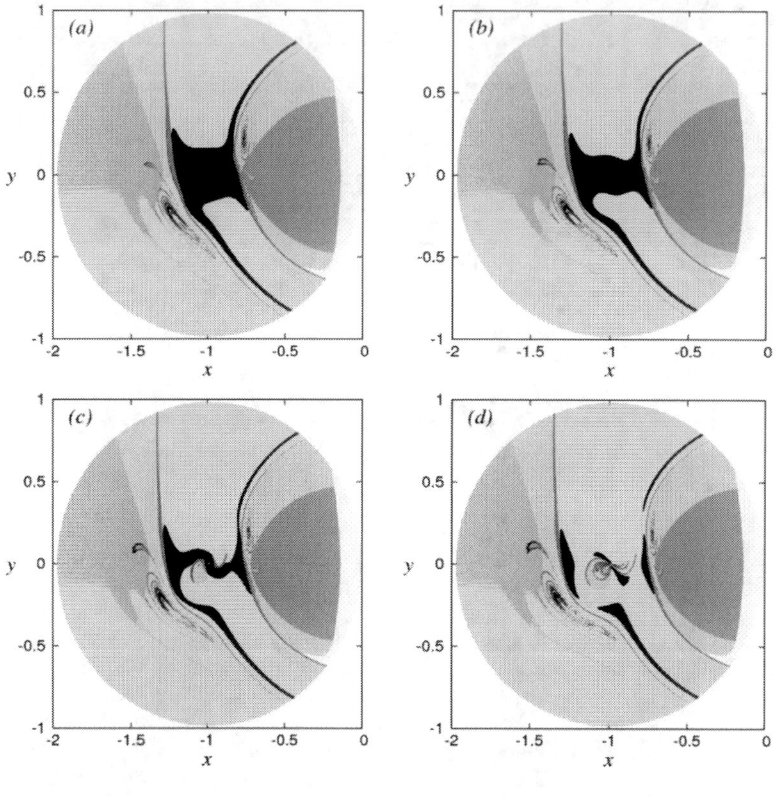

Figure 7. Projection onto the x-y plane of the sets generated by the WSB algorithmic definition with negative initial velocity (osculating retrograde motion about P_2) for: (a) $e = 0.0$, (b) $e = 0.3$, (c) $e = 0.6$, and (d) $e = 0.9$. **Black**: $\mathcal{S}$, stable; **Red**: $\mathcal{C}^M$, collision with the moon; **Yellow**: $\mathcal{C}^E$, collision with the earth; **Green**: $\mathcal{E}$, unstable due to Kepler energy; **Dark grey**: $\mathcal{G}^1$, primary interchange through L_1; **Grey**: $\mathcal{G}^2$, primary interchange through L_2; **Light grey**: $\mathcal{G}^3$, geometric escape with $C > C_3$.

The prograde and retrograde initial conditions generate completely different resulting sets. The stable set for low eccentricity usually is localized in a large region around the moon. From this central region, thin structures spread in the vertical direction. As the eccentricity increases, this core region diminishes and for negative velocities, its shape becomes irregular. The lunar collisional set, in general, spreads along the longitudinal direction of the stable set boundary.

The ballistic capture definition is quantitatively based on a two-body scalar, the Kepler energy, whose validity of application in a many-body context is restricted to the inner region of the moon sphere of influence (SOI) [16], here approximated by the region delimited by the Hill radius (about 64,483 km around the moon). The details of the central regions of Figure 6(a) and (d) and Figure 7(a) and (d) are presented, respectively, in Figures 8 and 9, at which the lunar Hill region boundary is depicted by a white curve. The portion of the stable set contained in the lunar SOI diminishes drastically with increasing eccentricity values for both signs of initial velocity.

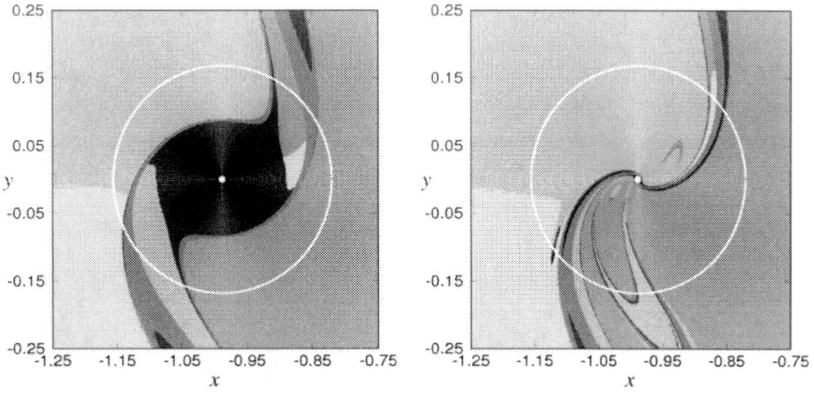

Figure 8. Detail of the central region of Figure 6(a) and (d), exhibiting the lunar Hill radius depicted by the white curve.

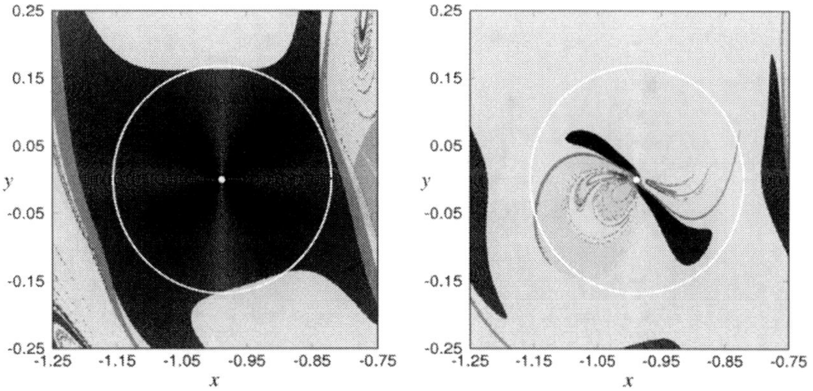

Figure 9. Detail of the central region of Figure 7(a) and (d), exhibiting the lunar Hill radius depicted by the white curve.

There are important remarks to be pointed out here: (i) It is claimed by [9] that the ideal transfer trajectories lie on the WSB set, i.e., at the boundary between sets with different stability behavior. These figures illustrate that just a reduced fraction of the full WSB set is interior to the lunar SOI for the considered full initial condition sets, as defined by [6]. (ii) Due to the restricted adequacy of the two-body quantifier, the exterior lunar SOI region of initial conditions can produce highly nonlinear trajectories for which many-problem effects are not negligible. (iii) Our new approach for the numerical implementation reveals that a large portion of the WSB set, mathematically defined, is contained in the moon collisional set, thus restricting the practical applicability of these trajectories to the circularization at the initial condition.

3.2. Dynamical investigation of the generated sets

This subsection presents a series of dynamical analyses of the obtained associated sets. The first analysis is applied for sets of both senses of initial velocity aiming their general characterization. However the other analyses were applied just to the prograde initial condition sets due to their great interest for practical applicability.

It is worthy of notice that Belbruno's original WSB algorithmic definition $\mathcal{W}^e$ holds fixed a two-body problem quantity, the osculating ellipse eccentricity, and maintains free a relevant three-body problem quantity: the invariant of the PCRTBP, the Jacobi constant. A very elucidative clarifying preliminary analysis is to evaluate the Jacobi constant for each initial condition of the algorithmic definition, once that transport properties in the PCRTBP is intimately related with the five possible Hill region profiles which are delimited by the critical Jacobi constants, C_k. Different possibilities of transport channels may occur between the realms of each primary, and the external region (lunar, earth, and exterior realms, respectively) as a function of C.

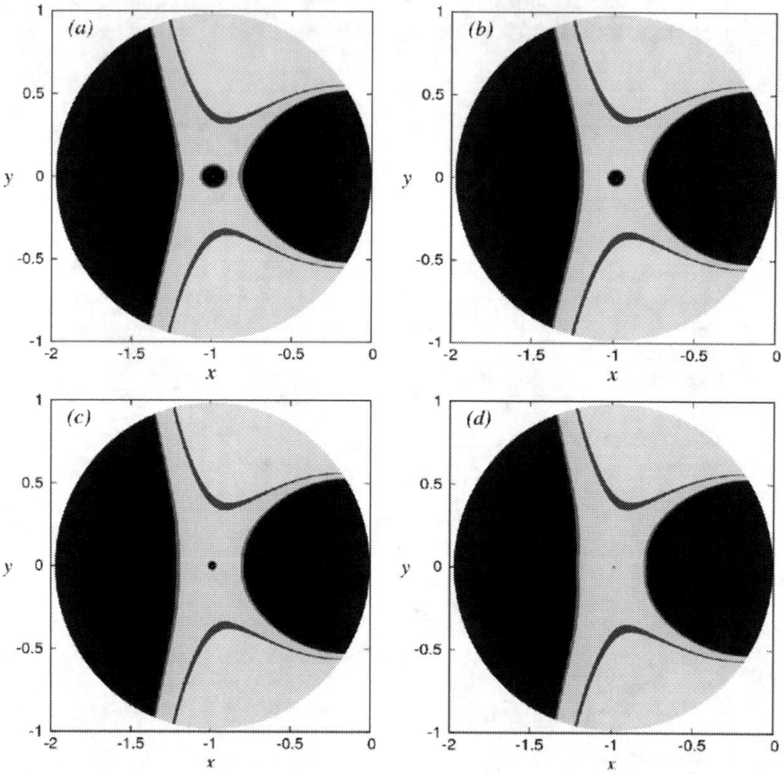

Figure 10. Classification of the prograde initial condition set according to the Jacobi constant for (a) $e = 0.0$, (b) $e = 0.3$, (c) $e = 0.6$, and (d) $e = 0.9$. **Black**: Case 1, $C > C_1$; Red: Case 2, $C_1 > C > C_2$; Green: Case 3, $C_2 > C > C_3$; **Blue**: Case 4, $C_3 > C > C_4 = C_5$; Cyan: Case 5, $C < C_4 = C_5$.

In the first case, $C > C_1$, no transit orbits between the primaries are possible. In the second case, $C_1 > C > C_2$, the realms around P_1 and P_2 are connected by transit orbits through the neck region around L_1, while for the third case, $C_2 > C > C_3$, the same as in the second case occurs plus the possibility of transit to the external realm of motion through the neck region around L_2. In the fourth case, $C_3 > C > C_4 = C_5$, there is a neck region also around L_3. In the fifth case ($C < C_4 = C_5$), motion over the entire x-y plane is possible. The availability of these transport channels is essential in the capture process.

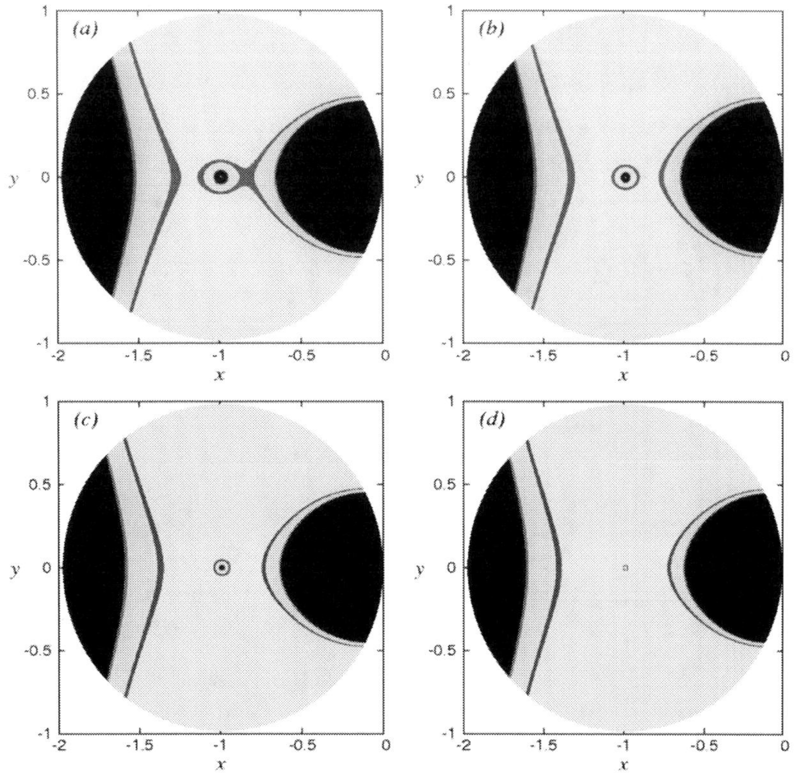

Figure 11. Classification of retrograde initial condition set according to the Jacobi constant for (a) $e = 0.0$, (b) $e = 0.3$, (c) $e = 0.6$, and (d) $e = 0.9$. **Black**: Case 1, $C > C_1$; Red: Case 2, $C_1 > C > C_2$; Green: Case 3, $C_2 > C > C_3$; **Blue**: Case 4, $C_3 > C > C_4 = C_5$; Cyan: Case 5, $C < C_4 = C_5$.

Figures 10 and 11 exhibit this Hill region classification of the full initial condition set, respectively, for prograde and retrograde initial velocities, for four eccentricity values, revealing that the distinct five classes of Hill regions occur for these initial condition sets. As it can be seen, a large amount of initial conditions belongs to the first case at which direct transfer orbits do not exist. In this case, the three distinct accessible isolated Hill regions, namely, the moon (in the central region), earth (at the right side region), and exterior (at the left side region) realms, are present in the full set. This observation corroborates that there is no reason or convenience to consider such extended initial condition ranges.

The same analysis was performed for the stable set with prograde and retrograde initial conditions as shown in Figures 12 and 13. The stable sets were subclassified in subsets S_j^k, $k = 1, 2, 3, 4, 5$ according to the Hill region cases, and $j = 1, 2$, respectively, for prograde and retrograde initial velocities. The first subset, S_1^1, is predominant in the prograde stable low altitude initial conditions for low eccentricity values. As the eccentricity increases, S_1^1 shrinks, being almost inexistent for higher eccentricities. For retrograde stable set, most of the initial conditions belongs to the high energy cases. In particular, for $e = 0.9$ the first two subsets, S_2^1 and S_2^2, are completely absent in the stable set, and S_2^5 is prevalent. For usual altitudes of selenocentric final orbits considered in realistic projects, initial conditions located far away from

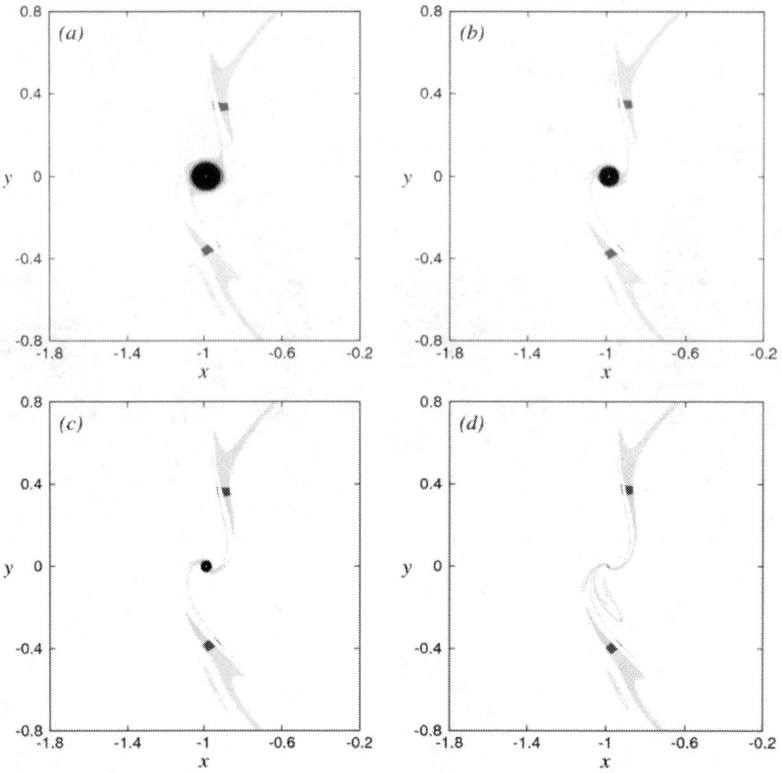

Figure 12. Classification of the prograde stable set of initial conditions according to the Jacobi constant for (a) $e = 0.0$, (b) $e = 0.3$, (c) $e = 0.6$, and (d) $e = 0.9$. **Black**: S_1^1 subset ($C > C_1$); **Red**: S_1^2 subset ($C_1 > C > C_2$); **Green**: S_1^3 subset ($C_2 > C > C_3$); **Blue**: S_1^4 subset ($C_3 > C > C_4 = C_5$); and **Cyan**: S_1^5 subset ($C < C_4 = C_5$).

the moon, such as those of S_1^4 and S_1^5, correspond to solutions of high energetic cost.

In the second and third analyses of the stable set, the permanence in the lunar sphere of influence in a full cycle around the moon and the location of the final state after a complete cycle around the moon inside or outside the lunar sphere of influence (or Hill sphere) are investigated. Second and third analyses for prograde initial condition sets are illustrated by Figures 14 and 15 for various eccentricities.

The purpose of the second analysis is to identify orbits which during a full cycle remain in a near enough vicinity of the moon. As expected, the remnant subset prevails on the stable core region for all the considered cases, but most of them are contained in the first Hill region case, that is useless for direct transfer feasibility. For $e = 0.0$ the remnant subset is about 80% of the stable set. This value decreases to about 74%, 57% and 5%, respectively, when the eccentricity assumes 0.3, 0.6 and 0.9 values.

The third investigation restricts the previous one to the analysis of the final state after a full cycle. Its aim is to verify the adequacy of the stability criterion based on the Kepler energy computation when the trajectory returns to $l(\theta)$. The quantity of final states inside the lunar sphere of influence is very high for all the eccentricities. The combination of these two analyses reveals that only the observation of the final state does not guarantee the fulfillment of qualitative aspects of the full cycle orbits which can be required in a ballistic capture transfer.

In the next analyses, the retrograde time evolution of each initial condition of the stable

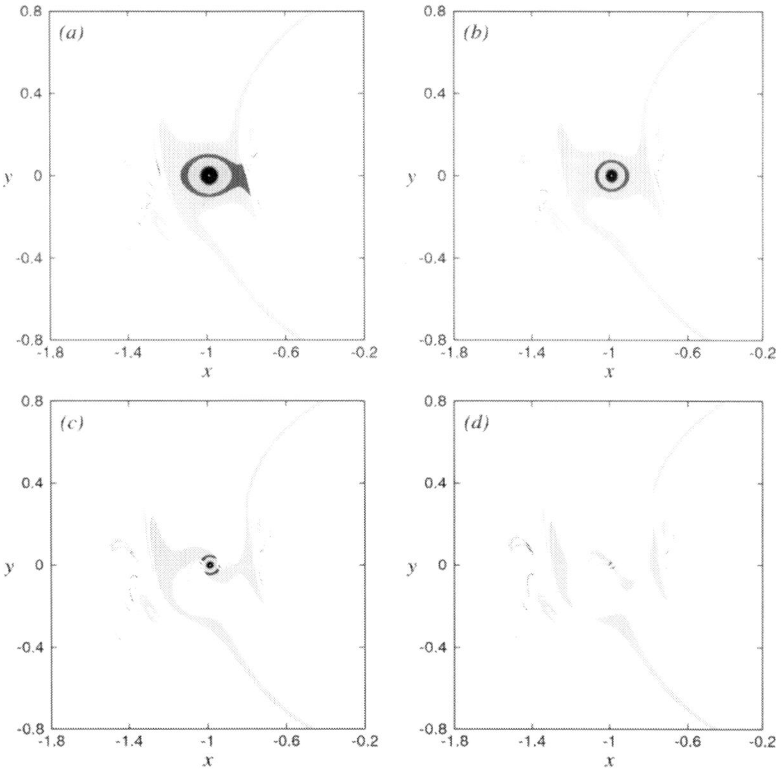

Figure 13. Classification of the retrograde stable set of initial conditions according to the Jacobi constant for (a) $e = 0.0$, (b) $e = 0.3$, (c) $e = 0.6$, and (d) $e = 0.9$. **Black**: S_2^1 subset ($C > C_1$); Red: S_2^2 subset ($C_1 > C > C_2$); Green: S_2^3 subset ($C_2 > C > C_3$); **Blue**: S_2^4 subset ($C_3 > C > C_4 = C_5$); and Cyan: S_2^5 subset ($C < C_4 = C_5$).

S_1^2 and S_1^3 subsets is performed in order to classify it according to its exit basin [17] (entrance basin for direct time evolution). Aiming to build lunar capture solutions, this dynamical aspect investigation is an essential requirement to define which kind of transfer possibility is available for the trajectories of each set.

For the S_1^2 subset, three possible retrograde exit basins are defined, namely, the earth exit basin of trajectories which escape through the neck around L_1 stable solution; the lunar collisional set, and the bounded set of trajectories which do not escape from the lunar realm in the considered time. The retrograde time evolution of the population percentage of each of these basins for the S_1^2 subset is shown in Figure 16 from one month to one year, for eccentricity values (a) $e = 0.0$, (b) $e = 0.3$, (c) $e = 0.6$, and (d) $e = 0.9$. For low eccentricity values, the convergence of the populations occurs slowly with time, while it becomes fast for higher e values. The earth exit basin population assumes higher asymptotic values for low eccentricities. This limit value diminishes with increasing e, due to the growth of the asymptotic value of the lunar collisional basin population (from 10% for $e = 0.0$ to 60% for $e = 0.9$). For the considered time scale, the bounded set converges to a population of about 10% for low e and goes to zero for higher eccentricities, as it occurs for $e = 0.9$. Among the considered cases, the numerical experiment for $e = 0.9$ has an outstanding behavior since the asymptotic earth exit basin population value is practically reached for one month.

A new basin must be introduced for the S_1^3 subset, i.e., the exterior basin of trajectories which

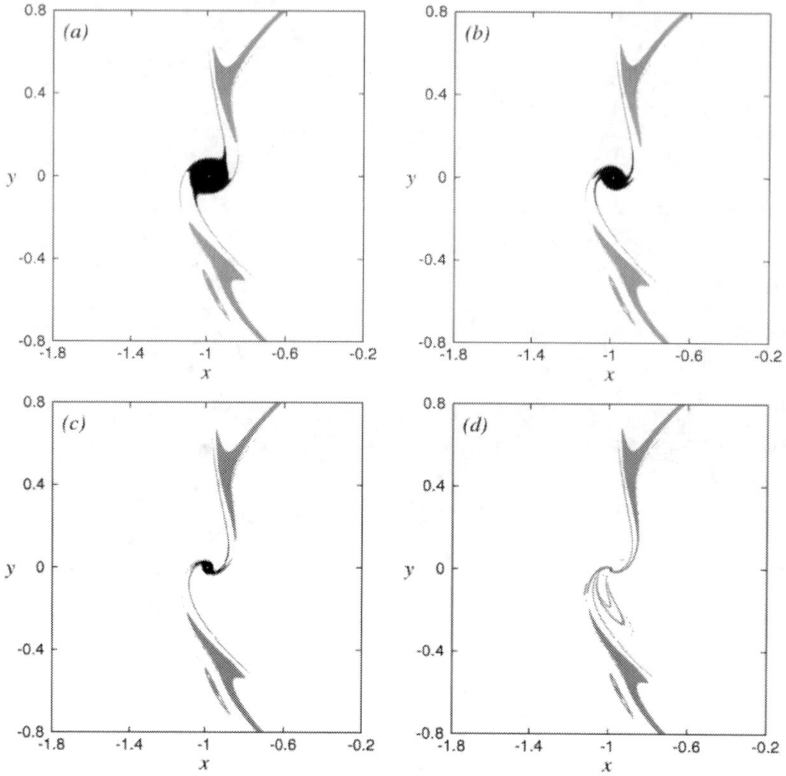

Figure 14. Classification of permanence within the moon's sphere of influence during a full cycle for the prograde stable set, for (a) $e = 0.0$, (b) $e = 0.3$, (c) $e = 0.6$, and (d) $e = 0.9$. The black points correspond to the remnant subset, and the orange one to the complementary subset.

pass through the neck around L_2 fixed point. Figure 17 presents the retrograde time evolution of the population percentage of each of the $S_{\bar{1}}^3$ subset escape basins. The main dynamical process observed in this subset is the population interchange between the exterior basin and the bounded set for all the considered eccentricities, while the earth basin and the collisional set reach their limit values in a short span of time, one or two months. In general, for each eccentricity, the bounded basin set is already small at the first month and vanishes very quickly. Unlike the $S_{\bar{1}}^2$ case, the asymptotic collisional set population does not grow with the eccentricity increasing.

Specifically, for direct interior transfers, it is important to investigate the spacecraft's maximum approximation to the earth. With this goal, we computed the minimum distance $r_{\bar{1}}^*$ of the spacecraft from the earth for all the obtained trajectories generated with the earth exit basin for both $S_{\bar{1}}^2$ and $S_{\bar{1}}^3$ subsets as a function of the time of the retrograde evolution. This analysis results is presented by Figure 18. The minimum approximation occurs for the $S_{\bar{1}}^3$ subset and seems not to depend strongly on the eccentricity value. Typical high and medium altitudes of usual earth parking orbits, HEO and MEO, are achieved in four months in the $S_{\bar{1}}^3$ subset.

This retrograde investigation was restricted to $S_{\bar{1}}^2$ and $S_{\bar{1}}^3$ stable subsets once that $S_{\bar{1}}^1$ can not produce any transfer and $S_{\bar{1}}^4$ and $S_{\bar{1}}^5$ are typically associated to high energy demands for the stabilization of the final usual altitude selenocentric orbit in ballistic capture transfer context.

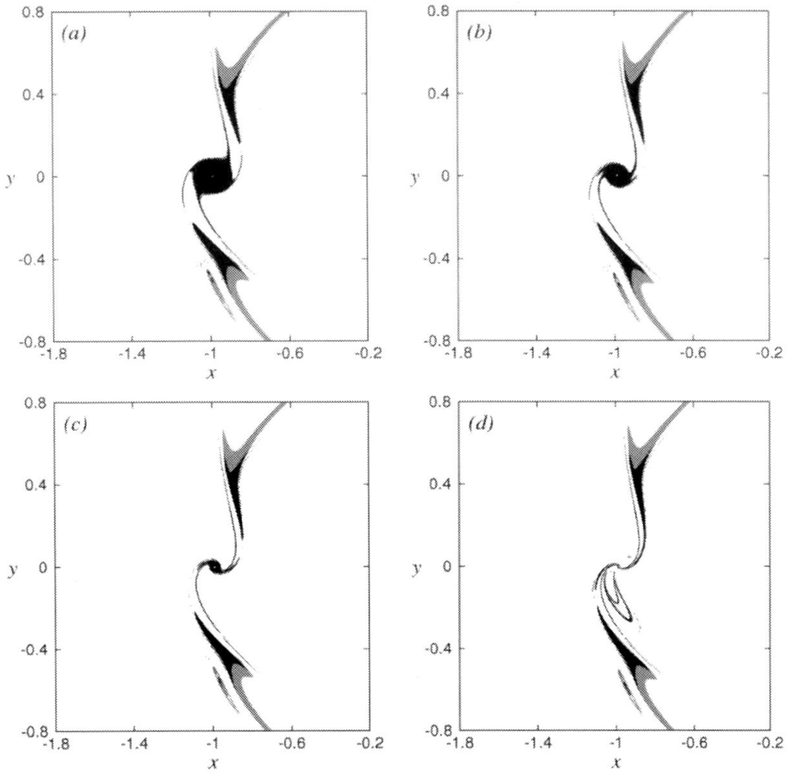

Figure 15. Classification of the final state after a full cycle around the moon with respect to the moon's sphere of influence for the prograde stable set, for (a) $e = 0.0$, (b) $e = 0.3$, (c) $e = 0.6$, and (d) $e = 0.9$. The black points correspond to interior subset, and the orange one to the external final state subset.

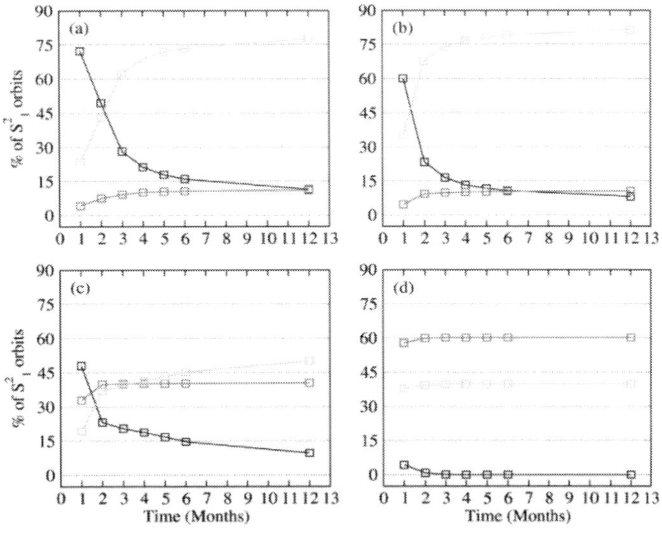

Figure 16. Retrograde time evolution of the population percentage of each of the S_1^2 subset escape basins for (a) $e = 0.0$, (b) $e = 0.3$, (c) $e = 0.6$, and (d) $e = 0.9$. The black, red, and green curves stand for bounded, lunar collisional, and earth basins, respectively.

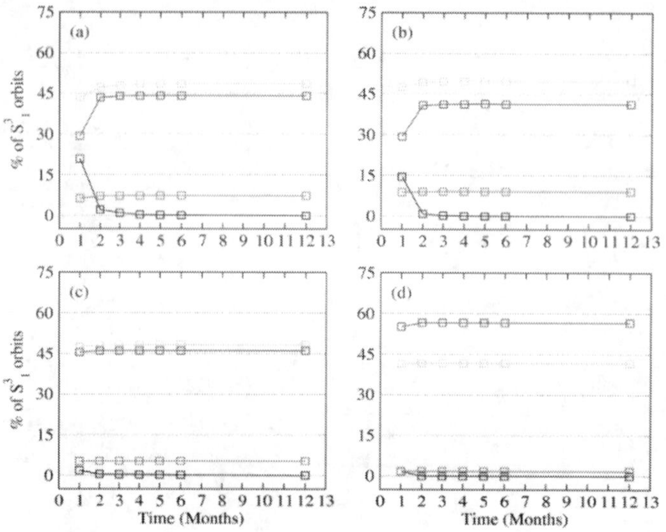

Figure 17. Retrograde time evolution of the population percentage of each of the S_1^3 subset escape basins for (a) $e = 0.0$, (b) $e = 0.3$, (c) $e = 0.6$, and (d) $e = 0.9$. The black, red, green, and blue curves stand for bounded, lunar collisional, earth, and exterior basins, respectively.

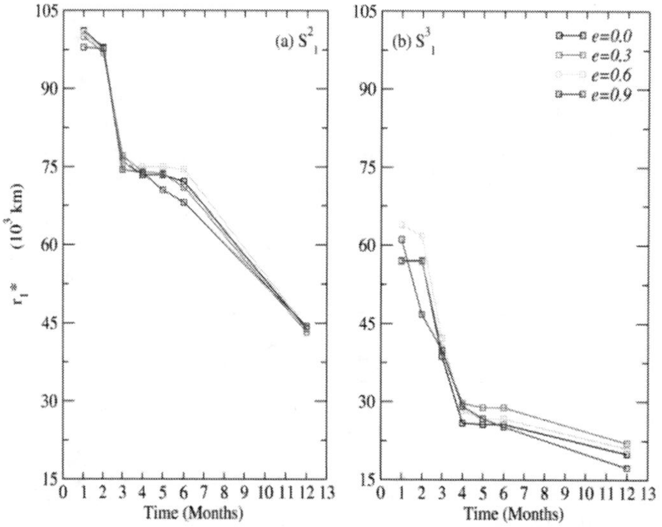

Figure 18. Retrograde time evolution of the maximum earth approximation of (a) S_1^2 and (b) S_1^3 subsets for four eccentricities values, namely, $e = 0.0$, $e = 0.3$, $e = 0.6$, and $e = 0.9$, calculated for all the trajectories belonging to the earth exit basin.

4. Conclusions and final remarks

In this contribution the numerical implementation of the WSB algorithmic definition was rebuilt with the inclusion of lunar and earth collisional sets and the classification of the unstable set according to the instability criteria based on the PCRTBP dynamical behavior. Several analyses were performed in order to clarify the applicability of the WSB associated sets in the context of low energy earth-moon ballistic capture transfer design.

The Jacobi constant interval criterion classifies the associated sets according to possible available transport channels. As expected, the initial conditions external to the lunar Hill region are not convenient. Besides of the fact they require a high energetic cost for stabilization on usual low lunar altitudes, a large portion of them belongs to the earth or external isolated Hill region for $C > C_1$.

The verification of the validity of the use of the stability two-body quantifier at the final state after a full cycle is performed for positive initial velocities, showing that a large amount of the initial conditions satisfy this criterion. The stable subset which remains in the lunar sphere of influence for a full cycle around the moon is detected in order to guarantee the permanence in a near enough neighborhood of the moon. The observation of orbits which evolve outside the nearby region of the central body, P_2, reveals that highly nonlinear behavior due to the third-body gravitational effect can be produced, what is improper for a selenocentric final orbit if no control is applied at the initial condition.

The dynamical transport investigation is complemented by examining the exit basins for retrograde temporal evolution of two suitable stable subsets which are the best candidates for low energy projects with prograde initial conditions, allowing the verification of the existence of initial conditions which can provide interior and external transfers.

In particular, the numerical implementation of the associated WSB sets of this contribution was able to reveal that a large portion of the stable-unstable boundary set is contained in the lunar collisional set, restricting its practical applicability under natural dynamics.

These original analyses exhibit the dynamical characteristics of trajectories associated to the WSB algorithmic definition, establishing criteria to determine possible adequate final portions of ballistic capture transfer trajectories, and setting ground to a future refined investigation of the boundary set between the several possible stable-unstable transitions. A detailed investigation of the retrograde initial condition sets still deserves attention in order to verify their adequacy in transfer orbit design. This approach is valuable even when four or more body models are considered to design a complete transfer mission.

Acknowledgments

This work was partially supported by Capes.

References

[1] Belbruno E 1987 *Proceedings of AIAA/DGLR/JSASS International Eletric Propulsion Conference, Paper No. AIAA 87-1054*
[2] Belbruno E 1994 *Journal of The British Interplanetary Society* **47** 73–80
[3] Belbruno E and Miller J 1993 *Journal of Guidance, Control and Dynamics* **16** 770–775
[4] Belbruno E, Topputo F and Gidea M 2008 *Advances in Space Research* **42** 1330–1351
[5] Circi C and Teofilatto P 2001 *Celestial Mechanics and Dynamical Astronomy* **79** 41–72
[6] García F and Gómez G 2007 *Celestial Mechanics and Dynamical Astronomy* **97** 87–100
[7] Topputo F and Belbruno E 2009 *Celestial Mechanics and Dynamical Astronomy* **105** 3–17
[8] Romagnoli D and Circi C 2009 *Celestial Mechanics and Dynamical Astronomy* **103** 79–103
[9] Belbruno E 2004 *Capture Dynamics and Chaotic Motions in Celestial Mechanics* (Princeton University Press)
[10] Szebehely V 1967 *Theory of Orbits* (Academic Press)
[11] Moser J 1958 *Communications on Pure and Applied Mathematics* **11** 257–271
[12] Conley C 1969 *Journal of diffrential Equations* **5** 136–158
[13] Siegel C and Moser J 1971 *Lectures on Celestial Mechanics* vol 187 (Springer-Verlag)
[14] Bate R, Mueller D and White J 1971 *Fundamentals of Astrodynamics* (Dover Publications)
[15] Koon W, Lo M, Marsden J and Ross S 2001 *Celestial Mechanics and Dynamical Astronomy* **81** 63–73
[16] Roy A 2005 *Orbital Motion* 4th ed (Institute of Physics Publishing)
[17] Aguirre J, Viana R and Sanjuan M 2009 *Reviews of Modern Physics* **81** 333–386

XI Latin American Workshop on Nonlinear Phenomena IOP Publishing
Journal of Physics: Conference Series **246** (2010) 012008 doi:10.1088/1742-6596/246/1/012008

Kink propagation in inhomogeneous systems driven by spatiotemporal perturbations

M. A. García-Ñustes[1], **I. Rondón**[1], **Jorge A. González**[1], **and R. Chacón**[2]

[1] Centro de Física, Instituto Venezolano de Investigaciones Científicas, Apartado 21827, Caracas 1020-A, Venezuela

[2] Departamento de Física Aplicada, Escuela de Ingenierías Industriales, Universidad de Extremadura, Apartado Postal 382, E-06071 Badajoz, Spain

E-mail: mogarcia@ivic.gob.ve

Abstract.
We investigate the propagation of kinks in inhomogeneous media. We show that the extended character of the kink, the internal mode instabilities and the phenomenon of disappearance of the translational mode can affect the kink motion in the presence of space-dependent external perturbations. We apply the results to the analysis of kink ratchets and the propagation of kinks driven by wave fields.

1. Introduction

Unlike point-like particles kink-solitons solutions are extended objects. The differences in their stability were theoretically demonstrated in several studies [1–3]. Moreover, effects such as length scale competition, i.e., the non-propagation of solitons under the action of external forces when the perturbation wavelength is of the order of the soliton width [4], can be explained from the point of view of instabilities of the internal modes of the soliton induced by external perturbations [5, 6]. In other words, the extended character of the soliton plays a crucial role in its propagation properties in inhomogeneous media. Besides, propagation of particles in inhomogeneous potentials have been studied on the context of many physical phenomena as spatiotemporal pattern dynamics [7–9], wave fields [10, 11], stochastic resonances [12, 13] and ratchet motion [14–17]. These studies have lead to multiple applications in models of molecular motors [14, 18–20], vortex dynamics in superconductors [21]and fluxons motion in Josephson junctions [22, 23].

In the present work, we show that under certain conditions of the parameters of the external force different phenomena, linked to the extended character of the soliton, could exist: the length scale competition, the non-existence of the translational mode, and the kink-soliton break-up. We discuss how all those phenomena can influence the propagation of kink-solitons in inhomogeneous media as periodic and ratchet-like potentials. We also investigate an optimization process of soliton transport when they are forced by the presence of wave fields.

In the following section, we discuss the main differences in stability conditions if we consider a point-like particle and a extended object, the length scale competition and the phenomenon of non-existence of translational mode and kink break-up by the presence of internal instabilities.

In section 3.1, we show how the kink motion in a periodic medium depends of the distance between the zeros of the external force. Next, in section 3.2 and 3.3, we analyze kink ratchets and propagation driven by a wave field, based on the results mentioned in previous sections. Finally, in section 4 we discuss our results and present some concluding remarks.

2. The kink as an extended object

2.1. Differences in stability conditions

The perturbed Klein-Gordon equation with dissipation is

$$\phi_{tt} - \phi_{xx} + \gamma\phi_t + \frac{\partial U}{\partial \phi} = F(x), \tag{1}$$

where where $U(\phi)$ is a potential that possesses at least two minima and γ is the damped coefficient. The sine-Gordon and ϕ^4 equations are particular examples of (1).

It is known that the external force $F(x)$ creates an effective potential $V(x)$ for the kink-soliton solution. When the soliton is treated as a pointlike particle, the zeroes of $F(x)$ are generally considered equilibrium positions for the motion of the kink [1, 24]. However, the stability conditions for these equilibrium positions are quite different if we analyze a point-like particle and a extended object. In fact, the soliton behaves as a particle only when the width of the potential makes the soliton appear pointlike. Otherwise, the soliton can have wavelike extended character [1,3,25].

Suppose we have a static kink whose center of mass is placed at an equilibrium point created by a zero of $F(x)$. When $F(x)$ are suitably chosen functions, the stability problem for the kink [1,2] can be solved exactly leading to the following eigenvalue problem:

$$\hat{L}f = \Gamma f, \tag{2}$$

where $\phi(x,t) = \phi_k(x) + f(x)e^{\lambda t}$, $\hat{L} = -\partial_x^2 + \left\{ \frac{\partial^2 U}{\partial u^2}\Big|_{\phi=\phi_k(x)} \right\}$, $\Gamma = -\lambda^2 - \gamma\lambda$.

For instance, if the distance between the zeros of $F(x)$ is much larger than l_k, where l_k is the kink's width, the kink feels the zeros of $F(x)$ as equilibrium positions. Let x_* be a zero of $F(x)$ ($F(x_*) = 0$). If $\frac{\partial F(x)}{\partial x}\Big|_{x=x_*} > 0$, the kink's equilibrium position is stable. If $\frac{\partial F(x)}{\partial x}\Big|_{x=x_*} < 0$, the equilibrium position is unstable for the kink. Nevertheless, if the distance between the zeros of $F(x)$ is comparable with or less than l_k, then the situation can change drastically. Let us say that $F(x)$ has three zeros: x_1, x_2 and x_3 such that $\frac{\partial F(x)}{\partial x}\Big|_{x=x_{1,3}} > 0$, $\frac{\partial F(x)}{\partial x}\Big|_{x=x_2} < 0$, from the point of view of the effective potential, this would imply the existence of a barrier between two stable potential wells for the kink. But when $|x_2 - x_1| < l_k$, $|x_3 - x_2| < l_k$, the kink does not feel the point x_2 as an unstable equilibrium position, i.e., the kink does not feel the barrier at point x_2. Conversely, if $\frac{\partial F}{\partial x}\Big|_{x_{1,3}} < 0$, $\frac{\partial F}{\partial x}\Big|_{x_2} > 0$, the effective potential has two unstable points and one stable between them. For the "point-like" kink placed in x_2 exists a potential well between two barriers. But, when l_k is larger than the distance between zeros, the kink will not be trapped by the potential well.

2.2. Length scale competition

Let us consider the following equation,

$$\phi_{tt} - \phi_{xx} + [1 + g(x)]\frac{\partial U}{\partial \phi} = 0. \tag{3}$$

In this case, the extrema of $g(x)$ are generally considered equilibrium positions for the movement of the kink. In Ref. [4], the authors considered a periodic function with zero average of the form

$g(x) = \varepsilon \cos(kx)$ and they observed that when the distance between the extrema of $g(x)$ is very close to the kink's width, the kink propagation in a inhomogeneous medium is impossible. This phenomenon was called length-scale competition [4–6].

The length scale competition phenomenon has been analyzed from the point of view of collective coordinate approach and stability conditions [4–6]. Next, we show the main ideas of both approaches in order to explain the phenomenon and how it can affect the kink dynamics in the presence of perturbations. The starting point of the collective coordinate method is to derive expresions for the two functions $x_{cm}(t)$ and $l(t)$ that represent, respectively, the center and the width of the kink. For this, it is proposed the following ansatz:

$$\phi(x,t) = \phi_k \left\{[x - x_{cm}(t)]/l(t)\right\}, \tag{4}$$

where ϕ_k is the kink solution.

It is possible to obtain dynamical equations for the variables $x_{cm}(t)$ and $l(t)$ by using the method based on the Lagrangian [4–6] The soliton dynamics obtain is equivalent to the particle motion in an effective potential [4]:

$$V(x_{cm}, l) = \frac{M_0}{2}\left(\frac{l_0}{l} + \frac{l}{l_0}\right) + \frac{\varepsilon}{k}\cos(kx)W_{(kl)}, \tag{5}$$

where M_0 and l_0 are the kink "mass" and natural width, respectively, without perturbation [5,6]. Their specific values depend on the function $U(\phi)$ considered. For example for the sine-Gordon equation $M_0 = 8$, $l_0 = 1$ and $W(y) = \frac{2\pi y^2}{\sinh(\frac{\pi y}{2})}$.

It can be proved that $F(x_{cm}) = -\frac{\partial V_{cm}}{\partial x_{cm}}$ and the effective force that acts on the kink is

$$F(x_{cm}) = -2\int_{-\infty}^{\infty} F(x) f_0(x - x_{cm})dx,$$

where $f_0(x)$ is the translational or Goldstone mode (see [6] and references therein for more details). We should remark that other approaches as the Fredholm-alternative based technique (solvability conditions) reports similar results where the kink dynamics is obtained through the Goldstone modes around the center of mass of the kink [9, 26, 27].

Carrying out the calculation for the sine-Gordon case, it is possible to show that stable equilibrium positions satisfy the following conditions: For the center of mass coordinate we have $x_{cm} = \frac{(2n+1)\pi}{k}$ and for l, the expression is given by the parameter z:

$$l(z) = \frac{l_0}{\sqrt{1 - \frac{4\pi\varepsilon l_0 z}{M_0}\left[\frac{2\tanh(\frac{\pi z}{2}) - \frac{\pi}{2}z}{\sinh(\frac{\pi z}{2})\tanh(\frac{\pi z}{2})}\right]}}, \tag{6}$$

$$k(z) = \frac{z}{l_0}\sqrt{1 - \frac{4\pi\varepsilon l_0 z}{M_0}\left[\frac{2\tanh\left(\frac{\pi z}{2}\right) - \frac{\pi}{2}z}{\sinh\left(\frac{\pi z}{2}\right)\tanh\left(\frac{\pi z}{2}\right)}\right]} \tag{7}$$

where $z \in [0, \infty)$ [5].

Here, the most interesting aspect of the procedure is that it is possible also calculate the "curvature" of the effective potential $V(x_{cm}, l)$ along the direction x_{cm} evaluated in the stable equilibrium points:

$$|c| = \frac{2\pi\varepsilon l^2 k^3}{\sinh\left(\frac{\pi l k}{2}\right)}. \tag{8}$$

It is not difficult to show theoretically that $|c|$ depends on k and has a maximum for a value of k such that the distance between the perturbation extrema $\varepsilon\cos(kx)$ is comparable with l_0. In other words, for values of k over a certain finite interval ($1.5 \lesssim k \lesssim 4$), the kink moves in a curved potential with extraordinary large barriers. In fact, the kink motion is very difficult for values of k where $|c|$ is maximum ($k_m = 1.6$). For large and small values of the the wavelength of the perturbation outside of this interval, the kink moves in a effective potential almost flat.

Now, let us discuss the length scale competition phenomenon using the stability analysis approach. As we mentioned before, the stability analysis can be solved exactly for a certain function $F(x)$ with several zeros [1]. However, we are going to consider the case when $F(x)$ is also a periodic function. Let us analyze the stability conditions when the potential is given by $U(\phi) = (\phi^2-1)^2/8$ and $\gamma = 0$ in (1).

We define

$$F_{II}(x) = \frac{1}{2}A\tanh(Bx)\left[\varepsilon_1 + \varepsilon_2\cosh^{-2}(Bx)\right]. \tag{9}$$

where $\varepsilon_1 = A^2 - 1$ and $\varepsilon_2 = 4B^2 - A^2$. Depending on the values of the parameters, $F_{II}(x)$ can have either one or three zeros. When $A^2 > 1$ and $4B^2 < 1$, $F_{II}(x)$ has three zeros that correspond to one unstable equilibrium point at $x = 0$ and two stable equilibria at $x = \pm d$ ($A^2 > 1$, $4B^2 < 1$) with

$$d = \frac{1}{B}\text{arccosh}\left[\sqrt{\frac{A^2-4B^2}{A^2-1}}\right].$$

In this case, we are able of construct a periodic function $F(x)$ using $F_{II}(x)$ of the form

$$F(x) = F_{II}(x + 2\,n\,d), \quad \text{for} \quad (2n-1)d < x \le (2n+1)d,$$

with $n = 0, \pm1, \pm2, ...,$

Based on the discussion in the previous section (see Sec.2.1), we can see that if the distance between the zeros (given by d) is much smaller than the kink's width ($l_0 = 2$), the kink practically does not feel the potential barriers and it moves in almost flat potencial. On the other hand, if d is much larger than l_0, then the kink moves in a periodic potential [3, 25].

However, for certain middle values of d, comparable with the kink's width, the kink remains trapped inside some potential wells created by $F(x)$. In this case, the kink behaves as a finite size body into a "rectangular hole" with high walls and the same size of the body. Thus, the mobility of the kink is practically zero. It is clear that these results are in agreement with those obtained by means of the collective coordinate approach.

2.3. Non-existence of the translational mode

Let us consider again the ϕ^4 potential $U(\phi) = (\phi^2-1)^2/8$ in (1) but now we will discuss the phenomena of non-existence of the translational mode for some values of the parameters of the external force $F(x)$.

Let us take a function $F(x)$ that possess, at least, one zero x_* between a minimum x_*^m and a maximum x_*^M with the properties that $\left.\frac{\partial F(x)}{\partial x}\right|_{x_*} > 0$ and $F(x_*^m)$ and $F(x_*^M)$,

$$F(x) = \frac{1}{2}(4B^2-1)\sinh(Bx)/\cosh^3(Bx). \tag{10}$$

The exact solution for a kink which center of mass is placed in an equilibrium position $x = 0$ when $F(x)$ is given by (10) is $\phi_k = \tanh(Bx)$.

Solving the stability problem we get that the discrete eigenvalues of the spectral problem (2) are given by the expansion:

$$\Gamma_n = -\frac{1}{2} + B^2(\Lambda + 2\Lambda n - n^2), \tag{11}$$

where $\Lambda(\Lambda+1) = \frac{3}{2B^2}$. The integer part of Λ ($[\Lambda]$) yields the number of the internal kink modes.

When $B^2 = 1/4$, the perturbation $F(x)$ disappears and we have $\Lambda = 2$ which corresponds to a two famous internal modes of ϕ^4 equation: the translational mode and the internal shape mode [1]. But when $B^2 > 3/4$ we can show that $\Lambda < 1$ and therefore, the translational mode does not exist anymore [5]. This is due to the parameter B controls the slope of the function $F(x)$ in the zero $x = 0$ and the values of $F(x_*^m)$ and $F(x_*^M)$. When B grows, the slope of $F(x)$ in the zero and the values of $|F(x_*^m)|$ and $|F(x_*^M)|$ are increased. In other words, for larger values of B, the kink will be "tighter" inside the potential well created by $F(x)$. In fact, there exists a value when the kink has not possibility of motion. The critical value for $|F(x_*^m)|$ and $F(x_*^M)$ is $|F_c(x_*^M)| = |F_c(x_*^m)| = \frac{2}{3\sqrt{3}}$.

This phenomena can be appeared even when the function $F(x)$ possess more than one zero. Consider the function (9) as the external function $F(x) = F_{II}(x)$ but when $A^2 < 1$ and $4B^2 > 1$. In this later case, $F(x)$ possess three zeros. The zero $x = 0$ has the property that $\frac{\partial F}{\partial x}\big|_{x=0} > 0$ which represents a stable equilibrium position for the kink.

The stability problem produces the following condition for the discrete eigenvalues, $\Lambda(\Lambda+1) = 3A^2/2B^2$, while the distance between the zeros is given by

$$d = \frac{1}{B}\operatorname{arccosh}\left[\sqrt{\frac{4B^2 - A^2}{1 - A^2}}\right]. \tag{12}$$

It is clear that the condition for the non-existence of the translational mode is,

$$4B^2 > 3A^2. \tag{13}$$

Thus, we can conclude that even if $F(x)$ possess several zeros, we can find some values for the parameters such that the translational mode does not exist and therefore the kink can be trapped firmly inside one of the wells created by $F(x)$.

2.4. Kink break-up

Another phenomena related with the extended character of the kink-soliton and its internal dynamics is the kink break-up [28]. This phenomenon is produced by an instability in the internal shape mode of the kink and it has been reported in connection with the possibility of kink escape of potential wells by an additional kink–antikink pair creation and soliton explosions [28, 29]. In order to illustrate the breaking-up process, we will return to the same equation of the former section,

$$\phi_{tt} + \gamma\phi_t - \phi_{xx} - \frac{1}{2}\phi + \frac{1}{2}\phi^3 = F(x),$$

where

$$F(x) = \frac{1}{2}(4B^2 - 1)\sinh(Bx)/\cosh^3(Bx). \tag{14}$$

When $4B^2 < 1$, the zero of the function $F(x)$ ($x = 0$) has the following property $\frac{\partial F}{\partial x}\big|_{x=0} < 0$ (see Fig.1). Thus, the point $x = 0$ is an unstable equilibrium position. Now, if the translational mode is unstable but the shape modes are stables the kink center of mass is pulled out of its equilibrium position and it will move away of this position without any change on its shape structure. However, if

$$B^2 < \frac{11 - \sqrt{117}}{8}, \tag{15}$$

the first internal shape mode can be unstable ($\Gamma_1 < 0$) leading to a "kink break-up". In fact, the kink breaks into one kink and two antikinks [2, 24, 28]. In general, when $F(x)$ has a zero x_* such

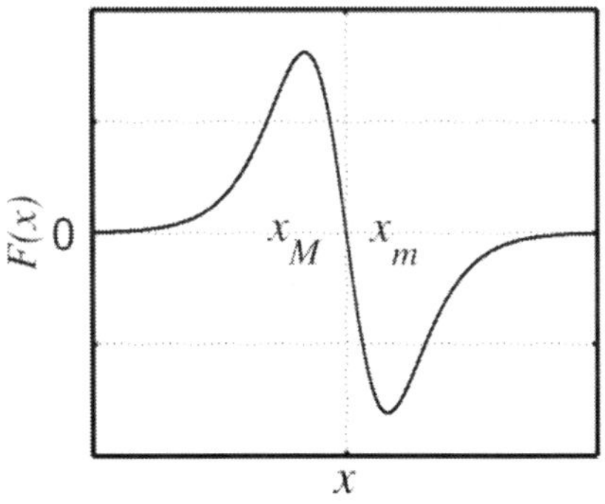

Figure 1. Function $F(x)$ given by (14) for $4B^2 < 1$.

that $\frac{\partial F}{\partial x}\big|_{x=x_*} < 0$ and this point is between the values x_M and x_m that are a maximum and a minimum of $F(x)$ respectively, then exists a critical value F_c such that if $|F(x_M) - F(x_m)| > F_c$, the kink will break up.

The latter consideration is important for a kink moving in a medium where $F(x)$ has several zeros, maxima, and minima. If the properties described above are satisfy, the kink could develop an instability near a zero of $F(x)$ and create an additional kink-antikink pair.

3. Kink propagation in inhomogeneous media

The former section gives us a vision of how the internal modes instabilities, the extended character of the kink-soliton, and the not appearance of some internal modes can affect the transport properties of solitons in inhomogeneous media. All the above considerations allow us to analyze the propagation of kink-solitons when are perturbed by different external forces.

3.1. Kink propagation in periodic media

In this section we investigate the equation

$$\phi_{tt} + \gamma\phi_t - \phi_{xx} - \frac{1}{2}\phi + \frac{1}{2}\phi^3 = A\sin(kx), \tag{16}$$

as an example of the phenomena formerly discussed. We will show that the kink motion depends on the parameter k that defines the distance between the zeros of $F(x)$.

The Cauchy problem is the following

$$\begin{aligned} \phi(x,0) &= \tanh\left[\frac{x}{2\sqrt{1-v_0^2}}\right], \\ \phi_t(x,0) &= -\frac{v_0}{2\sqrt{1-v_0^2}}\cosh^{-2}\left[\frac{x}{2\sqrt{1-v_0^2}}\right]. \end{aligned}$$

These initial conditions describe a kink with an initial velocity of center of mass v_0. This problem can not be treated directly with the stability analysis approach but the qualitative behavior can be described based on the exactly solvable cases instead. For values of $k >> 1$, the kink moves as if $F(x)$ does not exist, i.e. the kink moves in an effective potential completely flat. An example of this, is shown in Fig.2. This is due to the fact that the kink width is larger than the potential wells. Therefore, the potential wells does not affect the kink.

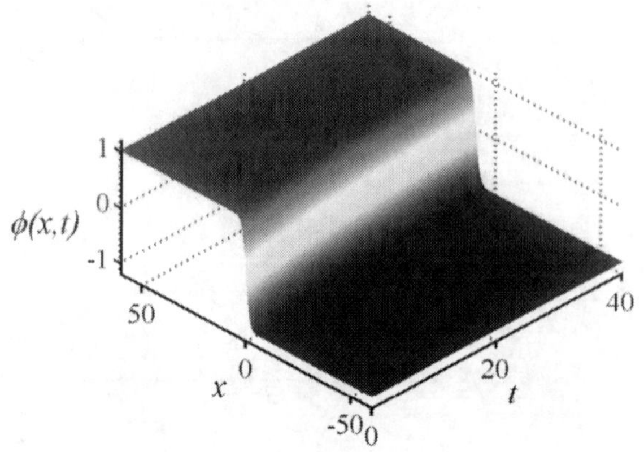

Figure 2. A kink moves in an effective potential created by $F(x) = A\sin(kx)$ for a value of $k = 10$ in Eq.(16).

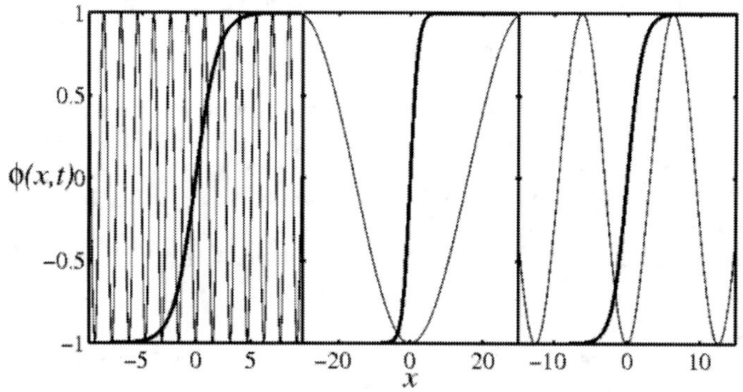

Figure 3. Comparative scheme of the effective potential generated by an external force $F(x) = A\sin(kx)$(pale line) and the kink soliton width (bold line). Left: If $k >> 1$ the kink soliton does not feel the potential wells due to its width is larger that the separation between potential wells. Middle: If $k << 1$ the kink soliton behaves as a point particle in an effective potential. Right: The kink width is comparable to the distance between equilibrium points for middle values of k.

If $k << 1$, the kink moves in a periodic effective potential. However, we must remark that the kink has mobility. It can move freely inside potential wells and with enough velocity to jump between them. This phenomenon can be seen in Fig.3 (Middle) from the point of the effective potential. Figure 4 shows the numerical simulation of (16) with $k = 0.01$.

Nevertheless, for certain middle values of k, the kink remains trapped inside a potential well with a similar extension to the kink width (Fig.3(right)). Figure 5 shows a numerical simulation of this effect. In this case, the kink remains "stuck" inside the potential well. The kink has no possibility of motion.

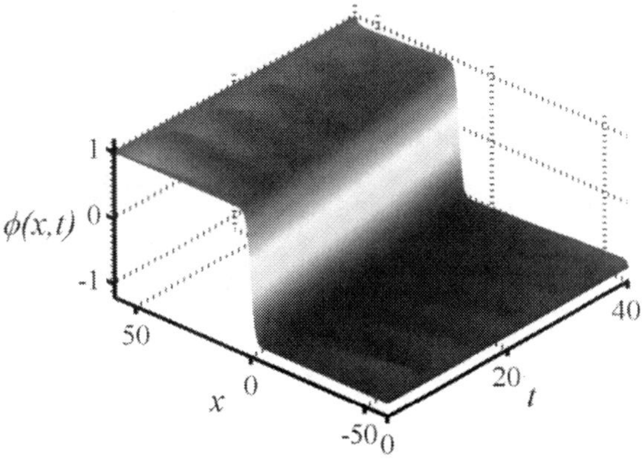

Figure 4. Kink motion in an effective potential created by $F(x)$ for a value of $k = 0.01$ in Eq.(16).

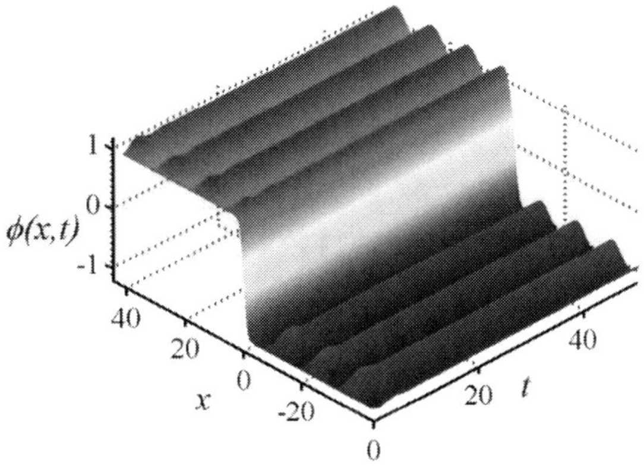

Figure 5. A kink remains trapped in a potential well for $k = 0.5$ (Eq.(16)).

3.2. Ratchet-like motion

In this section we will investigate the following equation:

$$\phi_{tt} + \gamma\phi_t - \phi_{xx} - \frac{1}{2}\phi + \frac{1}{2}\phi^3 = F_1(x) + F_2(t), \tag{17}$$

where $F_1(x) = A\sin(kx) - 2\mu A\sin(2kx + \theta)$, and $F_2(t) = f_0\sin(\omega t + \delta)$.

Note that, in the case of a point particle, the force $F_1(x)$ generates an asymmetrical effective potential that would allow ratchet-like motion (for example with $\mu = -1/4, \theta = \pi/2$) [3, 14].

Let us see first what happen when $f_0 = 0$. $F_1(x)$ is slightly different of $F(x) = A\sin(kx)$, but still is a periodic function with zeros and extrema. As on the simple case $F(x) = A\sin(kx)$, when $k >> 1$, the kink moves as if the effective potential was flat (See Fig. 6 and Fig.3)(Left)). In Fig.7 we observe that the kink remains trapped in a potential well for $k = 0.5$ with $F(x) = A\sin(kx)$, as we have seen in section 3.1.

If A is larger, the zeros of $F_1(x)$ that correspond to unstable equilibrium points for the kink, can be also points where the internal modes are unstable. This leads to a kink destruction

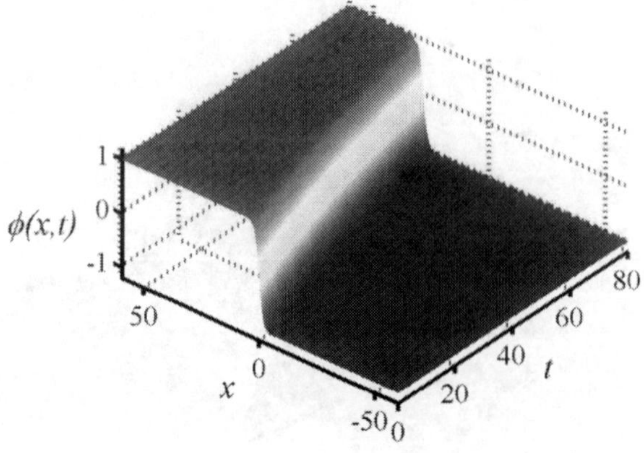

Figure 6. For $k = 10$ in $F_1(x)$ (Eq. (17)), the kink moves freely as if the effective potential was flat.

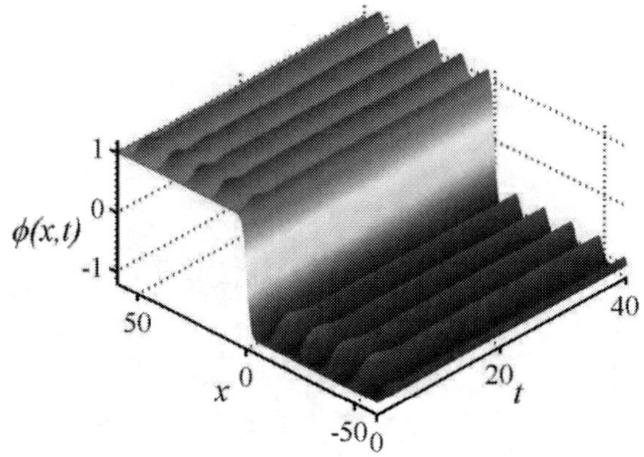

Figure 7. The kink remains trapped in the potential for $k = 0.5$ (Eq.(17)).

when the kink moves in this medium. But still when $F_1(x)$ is not sufficient to make the kink unstable, the internal mode instabilities can be activated by introducing a time dependent force $F_2(t) = f_0 \sin(\omega t + \delta)$ (See Fig.8).

3.3. Kink propagation driven by a wave field

In this section we will be interested in what happen when the inhomogeneous periodic perturbation as $F(x) = A\sin(kx)$ "moves" in the space with a certain velocity. This is equivalent to say that the kink is being perturbed by an external wave. We consider the perturbed ϕ^4 equation with dissipation and the external force is given by $A\sin[k(x - wt)$,

$$\phi_{tt} + \gamma\phi_t - \phi_{xx} + \frac{\partial U}{\partial u} = A\sin[k(x - wt)]. \quad (18)$$

The potential wells and barriers created by the perturbation $F(x) = A\sin(kx)$ now are

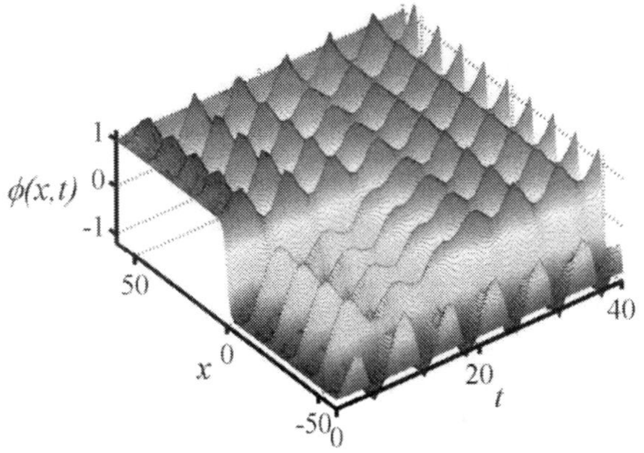

Figure 8. Kink destruction by internal modes instabilities when $k = 0.5$, $\omega = 1$, and $f_0 = 0.4$ in Eq.(17).

moving. We are interesting in the following Cauchy problem:

$$\begin{aligned} \phi(x, t = 0) &= \tanh\left[\frac{x}{2}\right], \\ \phi_t(x, t = 0) &= 0. \end{aligned}$$

In other words, the kink is initially static. We want to study if $F(x,t) = A\sin[k(x - wt)]$ produces kink motion or not. In fact, the kink can be carried by a external wave [10, 11]. To obtain the "optimal conditions" for this transport we must take into account all the phenomena discussed above.

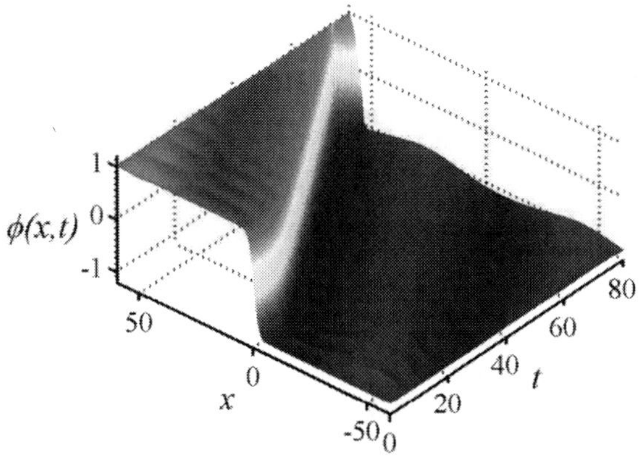

Figure 9. The kink is transported by the wave field when $k = 0.09$ in Eq. (18).

The plane wave is a periodic function and it can create equilibrium positions where the kink can be trapped during the transportation. The distance between the zeros of the function must be larger than the characteristic kink width. Thus, the potential wells are sufficiently wide as to assure that the kink feels them as a equilibrium states. However the periodic wave can create unstable equilibrium points too. If the wave amplitude is too large, these points are not only unstable for the center of mass of the kink, but the kink shape modes can be unstable too. Therefore, the kink cannot be transported in a stable way.

We have seen also that the kink can be trapped by a well due to the length-scale competition phenomenon. We have also seen that when a kink is trapped in a zero of $F(x)$ between a minimum and a maximum (if the difference between these extrema is large enough), the translational mode does not exist. Hence, we must increase the wave amplitude A, if we want to assure that the kink remains stood inside the potential well during the transportation by an external wave. In short, in order to obtain an instability on the internal modes, we must have an amplitude A large but not so much. Numerically we have encountered that the following parameters allow an efficient transport of the kink:$A = 0.2$, $k = 0.09$ and $w = 1$. The real velocity of the kink is going to depend of the damping coefficient γ.

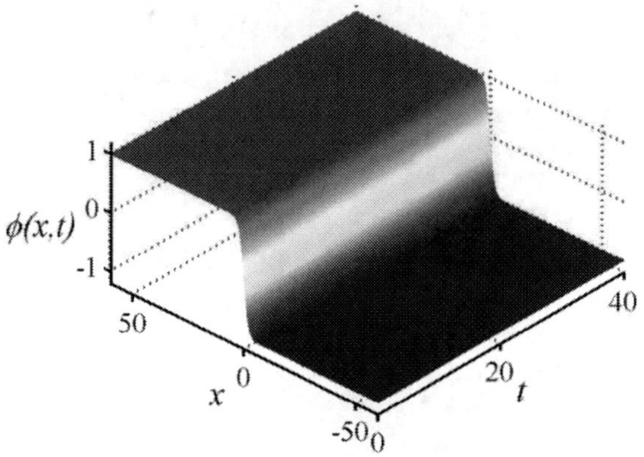

Figure 10. For $k = 3$ in Eq. (18) the kink remains at rest.

We must mention that a perturbation like $F(x,t) = A\sin[k(x - wt)]$ produces a spatiotemporal symmetry breaking [10, 11] in a similar way as a ratchet system. In fact, there are some values for external wave parameters that cause the kink remain static without any kind of motion. For example, for large values of k, the kink does not feel the potential wells and barriers created by the perturbation. Hence, the kink remains at the initial position where it has been placed. In Fig.9 we observe that for a k small enough to make the kink remains trapped inside a well, a moving external wave transports a kink with it. However in Fig. 10 we observe that for a $k >> 1$, the external wave cannot transport the kink.

4. Discussion and conclusions

The propagation of a kink in a inhomogeneous medium can be affected by different phenomena that are connected with the fact that a kink is a deformable extended object with a rich internal dynamics. In general, the soliton could be considered as a deformable object that could be trapped by holes of finite size or could move freely even when an effective potential with wells and barriers exists.

Even when the effective potential is such that a point particle could move in a ratchet-like process, the kink would be broken-up in a system with several kink and antikink, if the internal modes are unstable.

But the most extraordinary phenomenon is the kink propagation forced by the presence of a wave field. The results presented here allow to optimize this process. The wave parameters can be selected in such way that the kink stay stabilized in a potential well created by the wave.

In this case, the kink can move at an extraordinary speed. This velocity can be in the order of the maximum velocity of the kink in the Klein-Gordon equation without external perturbations and no dissipation.

Acknowledgements
We thank Jhoan Toro-Mendoza and J.J. Suárez for their help on the manuscript preparation.

References

[1] González J A and Holyst J A 1992 *Phys. Rev.* **B45** 10338–10343
[2] González J A, Bellorín A and Guerrero L E 2002 *Phys. Rev.* **E 65** 065601
[3] Braun O M and Kivshar Y S 2004 *The Frenkel-Kontorova model. Concepts, Methods and Applications.* (Springer-Verlag)
[4] Sánchez A and Bishop A R 1998 *SIAM Rev.* **40** 579
[5] González J A, Cuenda S and Sánchez A 2007 *Phys. Rev.* **E75** 03661–1–03661–7
[6] Cuenda S and Sánchez A 2005 *CHAOS* **15** 023502
[7] Cross M C and Hohenberg P C 1993 *Rev. Mod. Phys.* **65** 851–1112
[8] Nicolis G and Prigogine I 1977 *Self-Organization in Nonequilibrium Systems* (J. Wiley and sons)
[9] Haudin F, Elías R G, Rojas R G, Bortolozzo U, Clerc M G and Residori S 2009 *Phys. Rev. Lett.* **103** 128003
[10] Zheng Z, Cross M C and Hu G 2002 *Phys. Rev. Lett.* **89** 154102–1–154102–4
[11] González J A, Marcano A, Mello A and Trujillo L 2006 *Chaos, Solitons and Fractals* **28** 804–821
[12] Gammaitoni L, Peter, Hanggi, Jung P and Marchesoni F 1998 *Rev. Mod. Phys.* **70** 223–287
[13] Borromeo M and Marchesoni F 2005 *CHAOS* **15** 026110–1–026110–11
[14] Reimann P 2002 *Phys. Rep.* **361** 57–265
[15] Reimann P 2001 *Phys. Rev. Lett.* **86** 4992–4995
[16] Morales-Molina L, Quintero N R, Sánchez A and Mertens F G 2006 *Chaos* **16** 013117–1–013117–13
[17] Salerno M and Quintero N R 2002 *Phys. Rev. E* **65** 025602–1–025602–4
[18] Julicher J, Ajdari A and Prost J 1997 *Rev. Mod. Phys.* **69** 1269–1282
[19] Chacón R and Quintero N R 2007 *Biosystems* **88** 308315
[20] Astumian R D 1997 *Science* **276** 917–922
[21] Kivshar Y S and Malomed B A 1989 *Rev. Mod. Phys.* **61** 763–915
[22] Ustinov A, Coqui C, Kemp A, Zolotaryuk Y and Salerno M 2004 *Phys. Rev. Lett.* **93** 087001–1–087001–4
[23] Falo F, Martínez P J, Mazo J J and Cilla S 1999 *Europhys. Lett* **45** 700–706
[24] González J A and de A Mello B 1996 *Phys. Scr.* **54** 14
[25] González J A, Bellorín A and Guerrero L 1999 *Phys. Rev.* **E60** R37–R40
[26] Elphick C, Meron E and Spiegel E A 1990 *SIAM J. Appl. Math.* **50** 490–503
[27] Malomed B, Frantzeskakis D, Nistazakis H, Yannacopoulos A and Kevrekidis P 2002 *Phys. Lett.* **A295** 267–272
[28] González J A, Bellorin A and Guerrero L 2007 *Chaos Solitons and Fractals* **33** 143–155
[29] González J A, García-Nustes M A, Sánchez A and McClintock P V E 2008 *New J. Phys.* **10** 113015–1–113015–19

Received xxxx 20xx; revised xxxx 20xx.

Equivalent synchronization in driven and in autonomous chaotic systems

M. G. Cosenza[1], **O. Alvarez-Llamoza**[2] **and G. Paredes**[3]

[1] Centro de Física Fundamental, Universidad de Los Andes, Mérida, Venezuela
[2] Departamento de Física, FACYT, Universidad de Carabobo, Valencia, Venezuela
[3] Laboratorio de Física Aplicada y Computacional, Universidad Nacional Experimental del Táchira, San Cristóbal, Venezuela

E-mail: `mcosenza@ula.ve`

Abstract. It is shown that the synchronization behavior of a system of chaotic maps subject to either an external forcing or a coupling function of their internal variables can be inferred from the behavior of a single element in the system, which can be seen as a single drive-response map. From the conditions for stable synchronization in this single driven-map model with minimal ingredients, we find minimal conditions for the emergence of complete and generalized chaos synchronization in both driven and autonomous associated systems. Our results show that the presence of a common drive or a coupling function for all times is not indispensable for reaching synchronization in a system of chaotic oscillators, nor is the simultaneous sharing of a field, either external or endogenous, by all the elements. In the case of an autonomous system, the coupling function does not need to depend on all the internal variables for achieving synchronization and its functional form is not crucial for generalized synchronization. What becomes essential for reaching synchronization in an extended system is the sharing of some minimal information by its elements, on the average, over long times, independently of the nature (external or internal) of its source.

1. Introduction

Chaos synchronization is a fundamental phenomenon in dynamical systems (Pecora & Carroll 1990, Pikovsky *et al.* 2002). Its study has provided insights into many natural processes and motivation for practical applications such as secure communications and control of dynamical systems (Boccaletti *et al.* 2002, Uchida *et al.* 2005, Argyris *et al.* 2005). Chaos synchronization is commonly found in unidirectionally coupled systems, where a distinction can be made between a drive or forcing subsystem and another driven or response subsystem that possesses chaotic dynamics (Pikovsky *et al.* 2002). Complete synchronization occurs when the state variables of the drive and the response subsystems converge to a single orbit in phase space. More generally, generalized synchronization of chaos arises when a functional relation different from the identity is established between the drive and the response subsystems (Rulkov et al 1995, Abarbanel *et al.* 1996, Kapitaniak *et al.* 1996, Hunt *et al.* 1997, Parlitz & Kokarev 1999, Zhou & Roy 2007).

The auxiliary system approach (Abarbanel *et al.* 1996) shows that when a response and a replica subsystems are driven by the same signal, then the orbits in the phase spaces of the response and replica subsystems become identical and they can evolve on identical attractors, if their initial conditions lie on the same basin of attraction of the driven-response system. By extension, an ensemble of identical chaotic oscillators can also be synchronized by a common drive. In this case, generalized synchronization implies that the elements in the ensemble become synchronized into a single trajectory that evolves differently from that of the drive. The specific functional form of the drive is not determinantal; the basic mechanism that leads to synchronization is the sharing of the same input by the oscillators for all times.

On the other hand, there has been interest in the investigation of chaotic synchronization and other forms of collective behaviors emerging in autonomous systems consisting of networks of mutually interacting nonlinear units without the presence of external influences (Newman *et al.* 2006, Manrubia *et al.* 2004). Synchronization in globally coupled oscillators is relevant in many chemical and biological systems and have been experimentally investigated (Wang *et al.* 2000, S. De Monte *et al.* (2007), Taylor *et al.* 2009). In autonomous systems where the units are coupled through their mean field, the units and the mean field can synchronize to a common chaotic trajectory (Kaneko 1989). This process can be regarded as the analogous of complete synchronization in driven systems of elements subject to a common drive. Recently, the concept of generalized synchronization of chaos occurring in driven systems has been applied to the context of autonomous systems (Alvarez-Llamoza & Cosenza 2008). This means that, under some circumstances, the chaotic state variables in an autonomous dynamical system can be synchronized to each other but not to a shared coupling function containing information from those variables.

The occurrence of both forms of chaos synchronization in driven and in autonomous systems suggests that the nature of the common influence acting on the elements in a system is irrelevant; it could consist of an external forcing, or an autonomous interaction function (Alvarez-Llamoza & Cosenza 2008). At the local level, each element in the system is subject to a field, either external or endogenous, that eventually induces some form of synchronization between that field and the element. Thus, the local dynamics can be seen as a single element subject to a drive evolving as the common field in each case.

In this paper we present a simple and unified scheme for the study of chaos synchronization in classes of driven and in autonomous dynamical systems. Our method is based on the reported analogy between a single driven map and a system of globally coupled maps (Parravano & Cosenza 1998). This analogy provided an explanation of dynamical clustering in globally coupled systems (Parravano & Cosenza 1998, Manrubia *et al.* 2004) and of stability of steady states in systems with delayed interactions (Masoller & Marti 2005). We show that the synchronization behavior of a system of chaotic maps subject to either an external forcing or a coupling function of their internal variables can be inferred from the behavior of a single driven map in the system. When this single drive-response system reaches either complete or generalized synchronization, the local analogy implies that an ensemble of identical maps subject to a common (external or internal) field that behaves as this drive, should also synchronize in a similar fashion. We show that the conditions for stable synchronization in a single driven map also describe the regions where synchronized states occur in both driven and autonomous extended systems.

By studying a single driven-map model with minimal ingredients, we find minimal conditions for the emergence of chaos synchronization in both driven and autonomous associated systems. Our results show that the presence of a common drive or a coupling function for all times is not indispensable for reaching synchronization in an extended system of chaotic oscillators, nor is the simultaneous sharing of the drive or the coupling function by all the elements in the system. In the case of an autonomous system, the coupling function does not need to depend on all the internal variables for achieving synchronization and, in particular, its functional form is not crucial for generalized synchronization. What seems essential for reaching synchronization in a system is the sharing of some minimal information by the elements, on the average, over long times, independently of its source.

In §2, the single driven map model is presented and the stability conditions for complete and generalized chaotic synchronization are analyzed. The synchronization behavior of a system of homogeneously driven maps is explored in §3 and that of a system of maps subject to an asynchronous drive is shown in §4. An autonomous system of homogeneously coupled maps is considered in §5. The behavior of an autonomous system of heterogeneously coupled maps is shown in §6. Conclusions are presented in §7.

2. Single driven map

Our work is motivated by the practical aspect of searching for minimal requirements for the emergence of chaos synchronization in dynamical systems. As a minimal model, we consider a single map driven with a probability p,

$$x_{t+1} = \begin{cases} w(x_t, y_t), & \text{with probability } p \\ f(x_t), & \text{with probability } (1-p), \end{cases} \qquad (1)$$
$$y_{t+1} = g(y_t),$$

where $f(x_t)$ and $g(y_t)$ describe the dynamics of the driven and the drive variables, x_t and y_t, respectively, at discrete time t and the coupling relation between them is chosen to be of the diffusive form

$$w(x_t, y_t) = (1-\epsilon) f(x_t) + \epsilon g(y_t)\,, \qquad (2)$$

where ϵ is the coupling strength. Since we are focused on chaotic synchronization, for the driven dynamics we shall choose $f(x_t) = b + \ln|x_t|$, where b is a real parameter and $x_t \in (-\infty, \infty)$. This logarithmic map exhibits robust chaos, with no periodic windows and no separated chaotic bands, on the interval $b \in [-1, 1]$ (Kawabe & Kondo 1991). Robustness is an important property in practical applications requiring reliable operation under chaos in the sense that the chaotic behavior cannot be modified by small perturbations of the system parameters.

The linear stability condition for synchronization is determined by the Lyapunov exponents of the two-dimensional system Eq. (1). These are defined as $\Lambda_x = \lim_{T\to\infty} \ln L_x$ and $\Lambda_y = \lim_{T\to\infty} \ln L_y$, where L_x and L_y are the magnitude of the eigenvalues of $[\prod_{t=0}^{T-1} \mathbf{J}(x_t, y_t)]^{1/T}$, and $\mathbf{J}(x_t, y_t)$ is the Jacobian matrix for the system Eq.(1), calculated along an orbit. A given orbit $\{x_t, y_t\}$ from $t = 0$ to $t = T-1$ can be separated in two subsets, according to the source of the x_t variable, either coupled or uncoupled, that we respectively denote as $A = \{\{x_t, y_t\} : x_t = w(x_{t-1}, y_{t-1})\}$ possessing pT elements, and $B = \{\{x_t, y_t\} : x_t = f(x_{t-1})\}$ having $(1-p)T$ elements. We get

$$\left(\prod_{t=0}^{T-1} \mathbf{J}\right)^{1/T} = \begin{pmatrix} \prod\limits_{t: x_t\in A} w_x \prod\limits_{t: x_t\in B} f'(x_t) & K \\ 0 & \prod\limits_{t=0}^{T-1} g'(y_t) \end{pmatrix}^{1/T}, \qquad (3)$$

where $w_x = \frac{\partial w}{\partial x} = (1-\epsilon) f'(x)$, and K is a polynomial whose terms contain products of w_x, ϵ, and $g'(y_t)$ to be evaluated along time. Then $L_x = [\prod_{x_t\in A} w_x \prod_{x_t\in B} f'(x_t)]^{1/T}$ and $L_y = [\prod_{t=0}^{T-1} g'(y_t)]^{1/T}$. Thus we get

$$\Lambda_x = p \ln|1-\epsilon| + \lim_{T\to\infty} \frac{1}{T}\left[\ln \prod_{x_t\in A} |f'(x_t)| + \ln \prod_{x_t\in B} |f'(x_t)|\right] \qquad (4)$$

$$\Lambda_y = \lim_{T\to\infty} \frac{1}{T} \sum_{t=0}^{T-1} \ln|g'(y_t)| = \lambda_g\,, \qquad (5)$$

where λ_g is the Lyapunov exponent of the map $g(y_t)$. We shall assume chaotic drives $g(y_t)$; thus $\Lambda_y > 0$. Synchronization takes place when the Lyapunov exponent corresponding to the driven map is negative (Rulkov *et al.* 1995); i.e., $\Lambda_x < 0$. For a given set of parameter values, there is a definite value of the probability p at which the exponent Λ_x may change its sign, from positive to negative, signaling the onset of synchronization and the appearance of a contracting direction in the dynamics of the single driven map system Eq. (1). Thus, the drive just needs to act a fraction p of the time in order to achieve synchronization in the system Eq. (1).

When $g = f$, the condition $\Lambda_x < 0$ implies complete synchronization, where $x_t = y_t$. In this case we get

$$\begin{aligned} \Lambda_y &= \lambda_f , \\ \Lambda_x &= p \ln |1-\epsilon| + \lambda_f , \end{aligned} \tag{6}$$

where λ_f is the Lyapunov exponent of the map f.

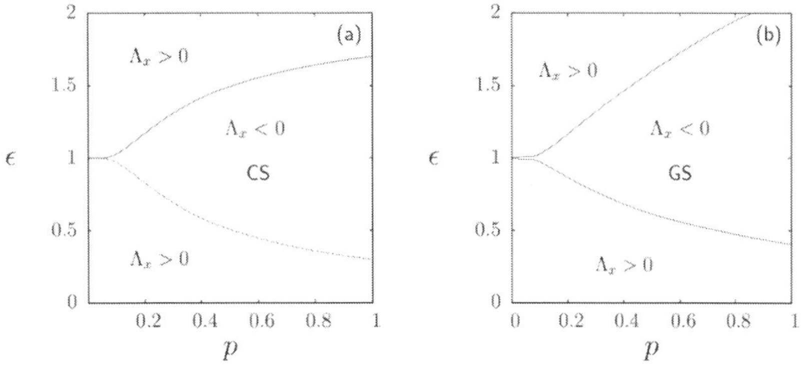

Figure 1. Boundaries $\Lambda_x = 0$ on the plane (p, ϵ) for synchronization in the single driven map Eq. (1) with $f = -0.7 + \ln |x_t|$. (a) Complete synchronization (CS) with $g = f = -0.7 + \ln |y_t|$. (b) Generalized synchronization (GS) with $g(y_t) = 0.5 + \ln |y_t| \neq f$.

Figure 1(a) shows the stability boundary, given by $\Lambda_x = 0$, for the completely synchronized states of the system Eq. (1) on the space of parameters (p, ϵ) for a drive $g = f$. On the other hand, if $g \neq f$, the condition $\Lambda_x < 0$ corresponds to generalized synchronization, characterized by $x_t \neq y_t$. Figure 1(b) shows the stability boundary, given by $\Lambda_x = 0$, for the generalized synchronized states of the system Eq. (1) on the space of parameters (p, ϵ) for a drive $g \neq f$.

3. Homogeneously driven systems

The auxiliary system approach (Abarbanel *et al.* 1996) implies that a driven map can synchronize on identical orbits with another, identically driven map. By extension, let us consider a system of N intermittently and homogeneously driven maps, defined as

$$\forall i,\ x_{t+1}^i = \begin{cases} (1-\epsilon) f(x_t^i) + \epsilon g(y_t), & \text{with probability } p, \\ f(x_t^i), & \text{with probability } (1-p) \end{cases} \tag{7}$$

where x_t^i $(i = 1, 2, \ldots, N)$ gives the state of the ith map at discrete time t, f describes the local dynamics, and ϵ is the strength of the coupling to the drive $g(y_t)$.

Since each map in the system Eq. (7) is subject to the same external influence (or lack of it) at any time, the properties of this system can be characterized from the behavior of the individual local dynamics. Thus, the occurrence of stable synchronization in the single driven map Eq. (1) should lead to synchronization in the extended system of maps Eq. (7), even when the drive acts intermittently in both cases. A completely synchronized state in the system Eq. (7) is given by $x_t^i = x_t = y_t$, $\forall i$, and it can occur when $g = f$. On the other hand, if $g \neq f$, generalized synchronization, characterized by the condition $x_t^i = x_t \neq y_t$, $\forall i$, may also arise in this system for $p \leq 1$.

Synchronization in an extended system can be characterized by the asymptotic time-average $\langle\sigma\rangle$ (after discarding a number of transients) of the instantaneous standard deviations σ_t of the distribution of state variables x_t^i, defined as

$$\sigma_t = \left[\frac{1}{N}\sum_{i=1}^{N}\left(x_t^i - S_t\right)^2\right]^{1/2}, \tag{8}$$

where S_t is the instantaneous mean field of the system,

$$S_t = \frac{1}{N}\sum_{i=1}^{N} x_t^i. \tag{9}$$

Stable synchronization corresponds to $\langle\sigma\rangle = 0$. Here we use the numerical criterion $\langle\sigma\rangle < 10^{-7}$.

Figure 2 shows σ_t as a function of time for the intermittently driven system Eq. (7) subject to chaotic drives $g = f$ and $g \neq f$, corresponding to complete and generalized synchronization, respectively.

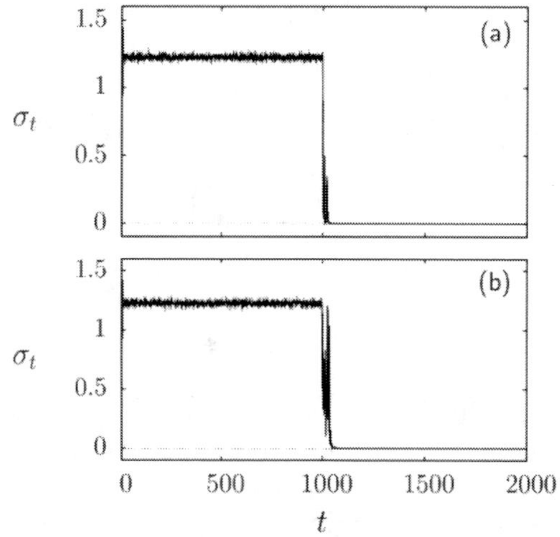

Figure 2. σ_t vs. t for the systems of driven maps Eq. (7) with $f(x_t^i) = -0.7 + \ln|x_t^i|$, for different forms of the drive $g(y)$. (a) $g = -0.7 + \ln|y_t| = f$; complete synchronization. (b) $g(y_t) = 0.5 + \ln|x_t| \neq f$; generalized synchronization. In both cases, $N = 10^4$, $\epsilon = 0.6$, $p = 0.6$, initial conditions were randomly and uniformly distributed such that $x_i \in [-10, 3]$ and the drive $g(y_t)$ is applied starting at $t = 1000$.

For a given value of the coupling strength ϵ, there is a threshold value of the probability p required to reach either type of synchronization. In each case, the condition $\Lambda_x < 0$ for synchronization in the single driven map Eq. (1) implies stable complete or generalized synchronization in the intermittently driven system Eq. (7). For the same values of parameters as in the single driven map, the region for the complete synchronized state on the space of parameters (p, ϵ) of the system Eq. (7) with a drive $g = f$ is the same as in Figure 1(a). Similarly, Figure 1(b) describes the region for generalized synchronization corresponding to the system Eq. (7) subject to a drive $g \neq f$.

4. Heterogeneously driven systems

The equivalence between a single driven map and a system of driven similar maps also suggests that, under some circumstances, the collective behavior of an extended system of interacting elements can be inferred by considering the dynamics of a single element at the local level. As an application of this idea, we consider a system of heterogeneously driven maps defined as

$$x^i_{t+1} = \begin{cases} (1-\epsilon)f(x^i_t) + \epsilon g(y_t), & \text{with probability } p, \\ f(x^i_t), & \text{with probability } (1-p). \end{cases} \tag{10}$$

The parameter p is the probability of interaction of a map with the drive g at a time t. The driven elements are randomly chosen with a probability p, so that not all the maps in the system receive the same external influence at all times. Thus, the average fraction of driven elements in the system Eq. (10) at any given time is p. In comparison, the forcing of the elements in the homogeneously driven system Eq. (7) is simultaneous and uniform; each map receives the same influence from the drive g at any t with probability p.

When the heterogeneously driven system Eq. (10) gets synchronized, we have $x^i_t = x_t$. However, the synchronized solution exists only if $g = f$. Therefore, only complete synchronization $x^i_t = x_t = y_t$ can occur in this system. Figure 3 shows σ_t as a function of time for the system Eq. (10) subject to a drive $g = f$.

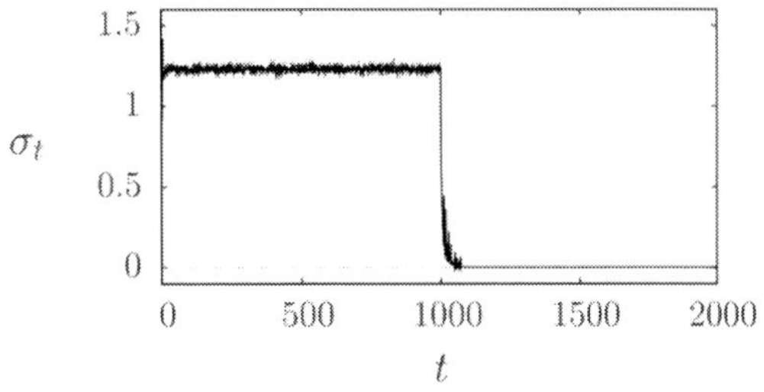

Figure 3. σ_t vs. t for the systems of driven maps Eq. (10) with $f(x^i_t) = -0.7 + \ln|x^i_t|$, for $g = -0.7 + \ln|y_t|$; complete synchronization. Parameters are $N = 10^4$, $\epsilon = 0.6$, $p = 0.6$; initial conditions were randomly and uniformly distributed such that $x_i \in [-10, 3]$ and the drive $g(y_t)$ is applied starting at $t = 1000$.

At the local level, each map in the heterogeneously driven system Eq. (10) is subject, over long times, to an external forcing g with probability p. Thus, as in the homogeneously driven system Eq. (7), the synchronization behavior of system Eq. (10) can also be studied from the behavior of the single map driven with probability p, Eq. (1). In particular, if the system of maps Eq. (10) driven with $g = f$ reaches a complete synchronized state for some values of parameters, then for this same set of parameters the single driven map Eq. (1) subject to the same drive should eventually exhibit a synchronized state similar to that of system Eq. (10). Thus, the condition $\Lambda_x < 0$ for complete synchronization in the single driven map Eq. (1), that implies stable complete synchronization in the homogeneously driven system Eq. (7), also describes the stability of the complete synchronized state $x^i_t = x_t = y_t$, $\forall i$, in the heterogeneously driven system Eq. (10). The stability boundary $\Lambda_x = 0$ in Fig. 1(a) for the driven map with $g = f$ coincides with both, the boundary that separates the region where complete synchronization occurs on the space of parameters (p, ϵ) for the homogeneously driven system Eq. (7) and the boundary for complete synchronization in the heterogeneously driven system Eq. (10).

Journal of Physics: Conference Series **246** (2010) 012009 doi:10.1088/1742-6596/246/1/012009

5. Autonomous systems with homogeneous coupling

The dynamics of the single driven map Eq. (1) can also be compared with the dynamics of an autonomous system of maps that share a global interaction with a probability p,

$$\forall i,\ x_{t+1}^i = \begin{cases} (1-\epsilon)f(x_t^i) + \epsilon H(x_t^j : j \in Q_t), & \text{with probability } p, \\ f(x_t^i), & \text{with probability } (1-p) \end{cases} \tag{11}$$

where ϵ represents the magnitude of the coupling to the global interaction function H; and Q_t is a subset having $q \leq N$ elements of the system that may be chosen at random at each time t. Each map receives the same information from the feedback coupling function H at any t with probability p. When this autonomous system gets synchronized at some values of parameters, we have $x_t^i = x_t$. Thus for this same set of parameters, the single driven map subject to a forcing that satisfies $g(y_t) = H(x_t^j = x_t : j \in Q_t)$ for long times should exhibit a synchronized state similar to that of the autonomous system Eq. (11). Thus, besides complete synchronization where $x_t^i = x_t = H$, other synchronized states, characterized by $x_t^i = x_t \neq H$ should also occur in the autonomous system, Eq. (11), for appropriate values of parameters. These states can be described as generalized synchronization in autonomous dynamical systems (Alvarez-Llamoza & Cosenza 2008).

As an example of complete synchronization in the system Eq. (11), consider a partial mean field coupling function defined as

$$H(x_t^j : j \in Q_t) = \frac{1}{q}\sum_{j=1}^{q} f(x_t^j), \tag{12}$$

where $q < N$ maps are randomly chosen at each time t. In this case the autonomous system Eq. (11) can be expressed in vector form as

$$\mathbf{x}_{t+1} = \begin{cases} \left((1-\epsilon)\mathbf{I} + \frac{\epsilon}{q}\mathbf{G}_t\right)\mathbf{f}(\mathbf{x}_t), \text{ with probability } p, \\ \mathbf{I}\ \mathbf{f}(\mathbf{x}_t), \quad \text{with probability } (1-p), \end{cases} \tag{13}$$

where the N-dimensional vectors $\mathbf{x}_t$ and $\mathbf{f}(\mathbf{x}_t)$ have components $[\mathbf{x}_t]_i = x_t^i$ and $[\mathbf{f}(\mathbf{x}_t)]_i = f(x_t^i)$, respectively, $\mathbf{I}$ is the $N \times N$ identity matrix, and $\mathbf{G}_t$ is an $N \times N$ matrix that at each time t possesses q randomly chosen columns that have all their components equal to 1 while the remaining $N - q$ columns have all their components equal to 0. The case $q = N$ and $p = 1$ corresponds to the usual mean field global coupling (Kaneko 1989). The linear stability analysis (Waller & Kapral 1884) of the complete synchronized state $f(x_t^i) = f(x_t) = H$ yields

$$\left|\left[(1-\epsilon) + \frac{\epsilon}{q}\alpha_k\right]^p e^{\lambda_f}\right| < 1, \tag{14}$$

where $\alpha_k = \delta_{0k}q$ $(k = 0, 1 \ldots, N-1)$ are the set of eigenvalues of the matrix $\mathbf{G}_t$ for any t, with the zero eigenvalue having $(N-1)$-fold degeneracy. The eigenvector corresponding to $k = 0$ is homogeneous. Thus only perturbations of $\mathbf{x}_t$ along the other eigenvectors may destroy the coherence. Thus, condition Eq. (14) with $k \neq 0$ becomes

$$p \ln|1-\epsilon| + \lambda_f < 0, \tag{15}$$

which is the same condition for stability of complete synchronized states in the driven map, Eq. (6), when $g = f$. Thus the boundary that separates the region where complete synchronization occurs on the space of parameters (p, ϵ) for the autonomous system Eq. (11), with H given by Eq. (12) for any value of q, coincides with the stability boundary $\Lambda_x = 0$ in Fig. 2 for the driven system with $g = f$.

Figure 4(a) shows the bifurcation diagram of the mean field S_t as well as $\langle\sigma\rangle$ as functions of the probability p for the autonomous system with the coupling function H given by Eq (12), with a fixed value of ϵ. The mean field S_t is chaotic for all values of p. There is a critical value of p above which $\langle\sigma\rangle = 0$ and $f(x_t^i) = f(x_t) = H$, indicating that the system is completely synchronized in a chaotic state for that range of the probability of interactions.

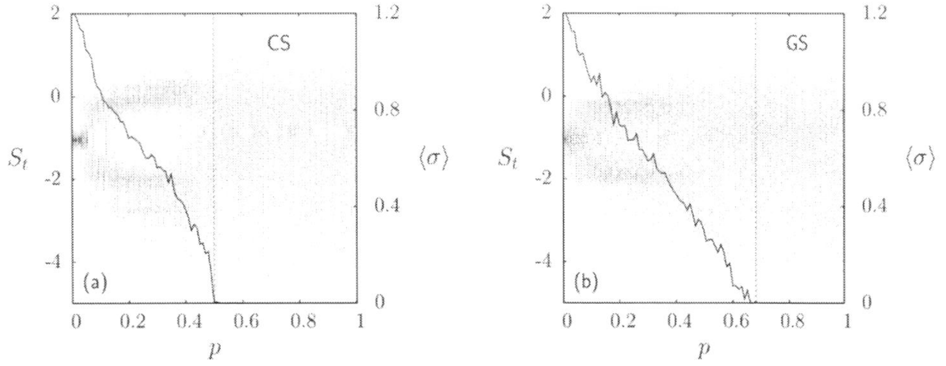

Figure 4. Left vertical axis: bifurcation diagram of S_t as a function of p for the homogeneously coupled autonomous system Eq. (11). Right vertical axis: $\langle\sigma\rangle$ vs p, continuous line. For each value of p, S_t and $\langle\sigma\rangle$ were calculated at each time step during a run starting from random initial conditions on the local maps, uniformly distributed on the interval $[-10, 3]$, after discarding the transients. (a) H_t given by Eq. (12) with $q = 0.5N$; the region where the complete synchronization state occurs is labeled CS. (b) H_t given by Eq. (16) with $k = 1.2$ and $q = 0.5N$; the region where generalized synchronization occurs is indicated by GS. In both cases, $f(x) = -0.7 + \log|x|$, $\epsilon = 0.5$, and $N = 10^4$.

For other functional forms of the coupling function H it is possible to find generalized synchronized states in the autonomous system Eq. (11). For example, consider the coupling function

$$H(x_t^j : j \in Q_t) = k + \frac{1}{q}\sum_{j=1}^{q} f(x_t^j), \tag{16}$$

where k is a constant and $q < N$ elements are chosen at random at each time t.

Figure 4(b) shows both the bifurcation diagram of S_t and the quantity $\langle\sigma\rangle$ as functions of the probability p for the autonomous system Eq. (11) with the coupling function H given by Eq (16), with a fixed value of ϵ. Again, there is a value of p above which $\langle\sigma\rangle$ vanishes, signaling the onset of a synchronized chaotic state. In this region $f(x_t^i) = f(x_t) \neq H = k + f(x_t)$, independently of the value of q, corresponding to generalized synchronization in the autonomous system Eq. (11).

We have verified that the regions of the parameter space (p, ϵ) where complete or generalized synchronized states emerge in the autonomous system with intermittent global interactions, Eq. (11), are the same regions that correspond to those states for the single driven map in Figure 1.

6. Autonomous systems with heterogeneous coupling

In analogy to a system of heterogeneously driven maps, Eq. (10), we can consider the following autonomous coupled map system,

$$x_{t+1}^i = \begin{cases} (1-\epsilon)f(x_t^i) + \epsilon H(x_t^j : j \in Q_t), & \text{with probability } p, \\ f(x_t^i), & \text{with probability } (1-p). \end{cases} \quad (17)$$

The parameter p is the probability of interaction of an element with the coupling function at time t. Thus the average fraction of fully connected elements at any given time is p. For $p = 1$, the autonomous systems Eq. (17) and Eq. (11) are identical.

Not all the maps in the system Eq. (17) experience the same global interaction at a given time. However, at the local level, each map in this autonomous system is subject, on the average, to the same coupling function H with probability p over long times. Similarly, each map in the autonomous homogeneously coupled system Eq. (11) experiences the coupling H with a probability p in time. This is also analogous to the condition of the local dynamics in the heterogeneously driven system of maps, Eq. (10), if g exhibits the same temporal evolution as H. Then, the synchronization behavior of the system Eq. (17) should also be similar to the behavior of the single map driven, Eq. (1) subject to a drive that satisfies $g(y_t) = H$ for asymptotic times.

When the autonomous system Eq. (17) reaches a synchronized state, we have $f(x_t^i) = f(x_t)$. However, this synchronized solution exists only if $H = f(x_t)$. Therefore, as in the case of the heterogeneously driven system, only complete synchronization $(x_t^i) = f(x_t) = H$ arises in the system Eq. (17).

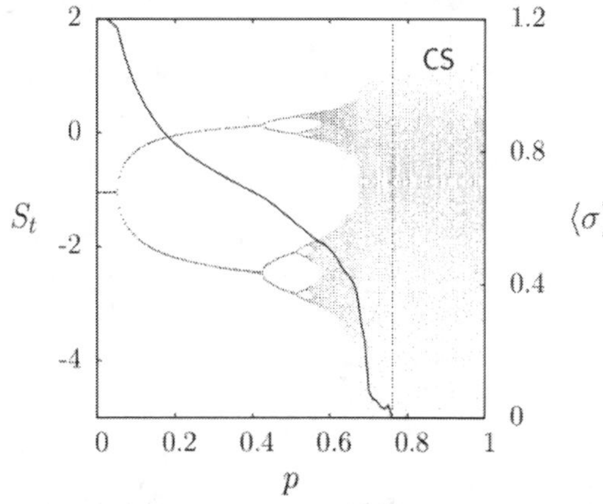

Figure 5. Left vertical axis: bifurcation diagram of S_t as a function of p for the partially coupled autonomous system Eq. (17) with H_t defined in Eq. (12) and $q = 0.5N$. The region where the complete synchronization state occur is labeled CS. Right vertical axis: $\langle\sigma\rangle$ vs p, continuous line. For each value of p, S_t and $\langle\sigma\rangle$ were calculated at each time step during a run starting from random initial conditions on the local maps, uniformly distributed on the interval $[-10, 3]$, after discarding the transients. Here, $f(x) = -0.7 + \log|x|$, $\epsilon = 0.4$, and $N = 10^5$.

As an example of complete synchronization in the autonomous system Eq. (17), we choose the form Eq. (12) as the coupling function $H(x_t^j : j \in Q_t)$ Figure 5 shows S_t and the quantity $\langle\sigma\rangle$ as functions of the probability p, for a fixed value of ϵ. For the given values of parameters, there is a threshold value of p at which $\langle\sigma\rangle = 0$, indicating that the elements become synchronized; in this case, complete synchronization, $f(x_t^i) = f(x_t) = H$, independently of the value of q. Thus, the reinjection of a coupling function containing partial information of the system to a fraction of randomly selected elements suffices to achieve complete synchronization in the system.

Beyond the common regions for complete or generalized synchronization, the driven and the autonomous systems may exhibit different collective behaviors in the parameter space. This is particularly manifested in the autonomous system with inhomogeneous coupling. Besides complete synchronization, the instantaneous mean field S_t in Figure 5 (left vertical axis) shows more complex collective behaviors arising in the system Eq. (17). For the parameter value $b = -0.7$, the local maps are chaotic. However, for some range of the parameters p and ϵ, S_t reveals the existence of global periodic attractors. In Figure 5, collective periodic states of S_t at a given value of the parameter p appear as sets of small vertical segments which correspond to intrinsic fluctuations of the periodic orbits of the mean field. Global attractors of period 1, 2, 4, and 8 are observed in the mean field as the parameter p is varied, and they become better defined as the system size N is increased. The amplitudes of the global periodic motions of S_t do not decrease with an increase of N. As a consequence, the variance of S_t itself does not decay as N^1 with increasing N but rather it saturates at some constant value related to the amplitude of the collective period. This is a phenomenon of nontrivial collective behavior, where macroscopic quantities in a spatiotemporal dynamical system exhibit spontaneous ordered evolution coexisting with local chaos (Chaté & Manneville 1998). This phenomenon arises in the heterogeneously coupled autonomous system, rather than in the homogeneously coupled one. In the first case, the coupling function presents spatial fluctuations than can be interpreted as a distributed noise. In fact, it has been shown that local noise can enhance the emergence of ordered collective behavior in coupled chaotic map systems (Shibata *et al.* 1999). Note that the emergence of collective periodic behavior in our case shown in Figure 5 cannot be attributed to the presence of periodic windows in the local dynamics since the logarithmic map possesses robust chaos for $b \in [1, 1]$ (Alvarez-Llamoza et al. 2007).

Although complete or generalized synchronization of chaos in both, driven and autonomous systems, can be characterized from the knowledge of the dynamical response of a single driven map, the transient behavior to reach such a state may be different in each case. Figure (6) shows the average time T_s required to attain complete or generalized synchronization as a function of p in the driven and the autonomous systems considered in this paper. The time T_s for generalized synchronization is larger than that for complete synchronization in an extended system subject to a field, either external or endogenous. In general, a homogeneous, synchronous field acting on a system is more efficient for achieving complete synchronization. The curves T_s vs. p for complete synchronization in either a driven or an autonomous system subject to a homogeneous field are practically indistinguishable. The same curve is also obtained for a single driven map possesing similar parameter values, as one may expect from the local dynamics analogy. Varying the size of the system N does not significantly affect the curves in Figure (6).

Note that for achieving synchronization in either the homogeneous or heterogeneous autonomous system, the coupling function H does not need to depend on all the internal variables; nor must it be active for all times or act simultaneously on all the elements. What becomes essential for synchronization is that all the elements in an autonomous system share some minimal information, on the average, over time. This is the same process that leads to synchronization in the associated driven systems Eq. (7) and Eq. (10), respectively, and it is also the mechanism for synchronization in the single driven map, Eq. (1). Thus, the nature of the common influence, either external or internal, affecting the local units becomes irrelevant for the emergence of synchronization in driven or autonomous systems.

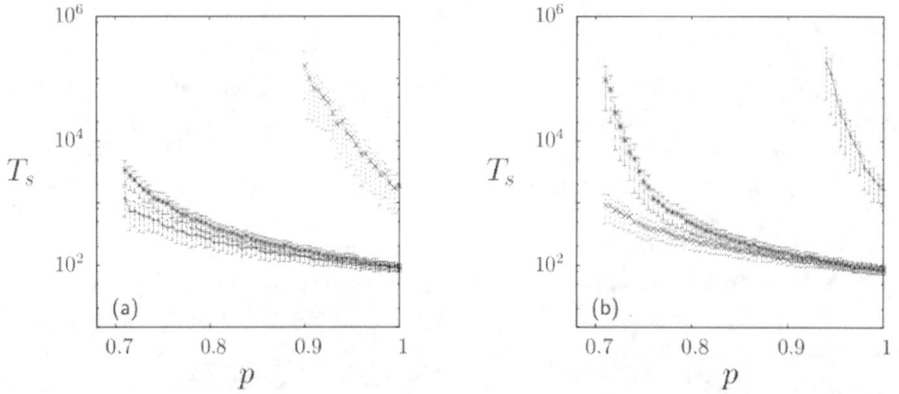

Figure 6. Average time T_s to reach synchronization as a function of the probability p. The error bars on each curve correspond to standard deviations resulting from 100 realizations of random initial conditions for each value of p. (a) Driven systems with fixed $\epsilon = 0.4$ and $N = 10^4$. Top curve: homogeneously driven system, Eq. (7), with $f = -0.7 + \log|x|$, $g = 0.5 + \log|y|$ (generalized synchronization). Middle curve: heterogeneously driven system, Eq. (10), with $f = g = -0.7 + \log|x|$ (complete synchronization). Bottom curve: homogeneously driven system, Eq. (7), with $f = g = -0.7 + \log|x|$ (complete synchronization). (b) Autonomous systems with fixed $\epsilon = 0.4$ and $N = 10^4$. Top curve: autonomous homogeneously coupled system, Eq. (11), with H given by Eq (16), $q = 0.5N$ and $k = 1.2$ (generalized synchronization). Middle curve: autonomous heterogeneously coupled system, Eq. (17), with H given by Eq. (12), $q = 0.5N$ (complete synchronization). Bottom curve: autonomous homogeneously coupled system, Eq. (11), with H given by Eq. (12), $q = 0.5N$ (complete synchronization).

7. Conclusions

We have shown that the synchronization behavior of system of chaotic maps subject to either an external forcing or a coupling function of their internal variables can be inferred from the behavior of a single element in the system. The local dynamics can be seen as a single driven map. When this single drive-response system reaches either complete or generalized synchronization, the local analogy implies that an ensemble of identical maps subject to a common field, either external or internal, that behaves as this drive, should also synchronize in a similar manner.

By considering the conditions for stable synchronization in a single driven-map model with minimal ingredients, we have found some minimal, equivalent conditions for the emergence of complete and generalized chaos synchronization in both driven and autonomous associated systems.

For the homogeneously driven or coupled autonomous systems, the sharing of the influence by the elements takes place only a fraction of the time. In the case of the heterogeneously driven or coupled autonomous system, the influence is shared only by a fraction of elements at any time. Our results show that the presence of a common drive or a coupling function for all times is not indispensable for reaching synchronization in an extended system of chaotic oscillators, nor is the simultaneous sharing of the drive or the coupling function by all the elements in the system. In the case of an autonomous system, the coupling function does not need to depend on all the internal variables for achieving synchronization and, in particular, its functional form is not crucial for generalized synchronization. What matters for reaching synchronization in a system is the sharing of some minimal information by the elements, on the average, over long times, independently of its source. This result suggests that these systems possess an ergodicity property.

Future extensions of this work include the investigation of some quantity, such as the transfer entropy (Schreiber, 2000) for measuring this minimal amount of information required for either type of synchronization, and the possibility of prediction of other forms of collective behaviors or patterns observed in dynamical networks, from the knowledge of the occurrence of these phenomena in an associated single driven-response system.

Acknowledgments

This work was supported by Consejo de Desarrollo Científico, Humanístico y Tecnológico, Universidad de Los Andes, Venezuela, through grant C-1579-10-05-B.

References

Abarbanel, H. D. I., Rulkov, N F & Sushchik, M. M. 1996 Generalized synchronization of chaos: The auxiliary system approach. *Phys. Rev. E* **53**, 4528-4535.

Alvarez-Llamoza, O. & Cosenza, M. G. 2008 Generalized synchronization of chaos in autonomous systems. *Phys. Rev. E* **78**, 046216-1-046216-6.

Alvarez-Llamoza, O., Tucci, K., Cosenza, M. G., & Pineda, M. 2007 Random global coupling induces synchronization and nontrivial collective behavior in networks of chaotic maps. *Eur. Phys. J. Special Topics* **143**, 245-247.

Argyris, A., Syvridis, D., Larger, L., Annovazzi-Lodi, V., Colet, P., Fischer, I., Garcia-Ojalvo, J., Mirasso, C. R., Pesquera, L. & Shore, K. A. 2005 Chaos-based communications at high bit rates using commercial fibre-optic links. *Nature (London)* **438**, 343-346.

Boccaletti, S., Kurths, J., Osipov, G., Valladares, D. L., & Zhou, C. S. 2002 The synchronization of chaotic systems. *Phys. Rep.* **366**, 1-101.

Chaté, H. & Manneville, P., 1992 Emergence of effective low-dimensional dynamics in the macroscopic behaviour of coupled map lattices. *Europhys. Lett.* **17**, 291-296.

De Monte, S., d'Ovidio, F., Danø, S., Sørensen, P. G. 2007 Dynamical quorum sensing: population density encoded in cellular dynamics. *Proc. Natl. Acad. Sci.* **104**, 18377-18381.

Hunt, B. R., Ott, E. & Yorke, J. A. 1997 Differentiable generalized synchronization of chaos. *Phys. Rev. E* **55**, 4029-4034.

Kaneko, K. 1989 Chaotic but regular posi-nega switch among coded attractors by cluster-size variation *Phys. Rev. Lett.* **63**, 219-223.

Kapitaniak, T., Wojewoda, J.& Brindley, J. 1996 Synchronization and desynchronization in quasi-hyperbolic chaotic systems. *Phys. Lett. A* **210**, 283-289.

Kawabe, T. & Kondo Y. 1991 Fractal transformation of the one-dimensional chaos produced by logarithmic map. *Prog. Theor. Phys.* **85**, 759-759.

Manrubia, S. C., Mikhailov, A. S. & Zanette, D. H. 2004 *Emergence of dynamical order: synchronization phenomena in complex systems.* Singapore: World Scientific.

Masoller, C. & Marti, A. 2005 Random delays and the synchronization of chaotic maps. *Phys. Rev. Lett.* **94**, 134102-1-134102-4.

Newman, M. E. J., Barabasi, A. L. & Watts, D. J. 2006 *The Structure and dynamics of networks.* Princeton, NJ: Princeton University Press.

Parlitz, U. & Kokarev, L. 1999. In *Handbook of Chaos Control* (ed. H. Schuster), p. 271. Weinheim: Wiley-VCH.

Parravano, A. & Cosenza, M. G. 1998 Driven maps and the emergence of ordered collective behavior in globally coupled maps. *Phys. Rev. E* **58**, 1665-1671.

Pecora, L. M. & Carroll, T. L. 1990 Synchronization in chaotic systems. *Phys. Rev. Lett.* **64**, 821-824.

Pikovsky, A., Rosenblum, M. & Kurths, J. 2002 *Synchronization: a universal concept in nonlinear sciences.* Cambridge, UK: Cambridge University Press.

Rulkov, N. F., Sushchik, M. M., Tsimring, L. S. & Abarbanel, H. D. I. 1995 Generalized synchronization of chaos in directionally coupled systems. *Phys. Rev. E* **51**, 980-994.

Schreiber, T. 2000 Measuring information transfer. *Phys. Rev. Lett.* **85**, 461-464.

Shibata, T., Chawanya, T., & Kaneko, K. 1999 Noiseless collective motion out of noisy chaos. *Phys. Rev. Lett.* **82**, 4424-4427.

Taylor, A.F., Tinsley, M. R., Wang, F., Huang, Z. & Showalter, K. 2009 Dynamical quorum sensing and synchronization in large populations of chemical oscillators. *Science* **323**, 614-617.

Uchida, A., Rogister, F., Garcia-Ojalvo, J. & Roy, R. 2005 Synchronization and communication with chaotic laser systems. *Prog. Opt.* **48**, 203-341.

Waller, I. & Kapral, R. 1984 Spatial and temporal structure in systems of coupled nonlinear oscillators. *Phys. Rev. A* **30**, 2047-2055.

Wang, W., Kiss, I. Z., & Hudson, J. L. 2000 Experiments on arrays of globally coupled chaotic electrochemical oscillators: synchronization and clustering. *Chaos* **10**, 248-256.

Zhou, B. B. & Roy, R. 2007 Isochronal synchrony and bidirectional communication with delay-coupled nonlinear oscillators. *Phys. Rev. E* **75**, 026205-1- 026205-5.

Chaotic synchronization on directed networks

D.J. DeCastro[1] **and K. Tucci**[2]

[1]Centro de Física Fundamental, Universidad de Los Andes, Mérida, Mérida 5251, Venezuela
[2]SUMA-CeSiMo, Universidad de Los Andes, Mérida, Mérida 5251, Venezuela

E-mail: kay@ula.ve

Abstract. The phenomenon of synchronization occurring in coupled chaotic maps on a directed random network is studied. The network is characterized by the average degree of its nodes and the fraction of directed links. It is found that the required coupling strength so that the chaotic synchronization emerges is smaller when the fraction of directed links is increased. In addition, the system undergoes a transition from an asynchronous phase to a synchronous one at some critical values of its parameters. The critical boundary separating the synchronous from the asynchronous regime is calculated on the parameter space of the system, given by the coupling strength and the fraction of directed links of the network. The phase transition between the two regimes is of second order for all values of the fraction of directed links, and the critical exponent depends of it.

The phenomenon of chaotic synchronization in dynamical networks is a well studied[1, 2, 3, 4]. There are many papers where dynamical units are defined on a undirected[5, 6], directed[7, 8] and complex[9, 10] networks. This phenomenon was also been studied by using continuous[11, 12] and discrete[13, 14] chaotic local dynamics. It is known that collective behavior of a system depends on properties of the substrate over which the dynamic takes place[15, 16]. In this paper, we explore how the directed links of the network affect the phenomenon of chaos synchronization. We consider a system of chaotic coupled maps defined on Erdös-Rényi random networks[17] of size N with a mean degree of links per node $\bar{k}$, and a fraction q of directed links. We show that there is a phase transition between chaotic spatiotemporal and synchronized phases, at a critical value of the parameter measuring the coupling strength between the units. This critical value of the coupling decreases when the fraction of directed links increases, and additionally the critical exponent of the transition also decreases.

Networks can be represented by their adjacency matrix $\mathbf{M}$, whose elements m_{ij} are given by

$$m_{ij} = \begin{cases} 1 & \text{if there is a link from node } i \text{ to node } j \\ 0 & \text{elsewhere} \end{cases} \quad , \quad i,j = 1,2,\ldots,N \,. \tag{1}$$

The fraction of directed links q is defined as

$$q = 1 - \frac{\sum_{ij} m_{ij} m_{ji}}{\sum_{ij} m_{ij}} \quad , \quad i,j = 1,2,\ldots,N \,. \tag{2}$$

Let us consider a diffusively coupled map system on a connected random network with $N = 10^4$ nodes, described by

$$x^i_{t+1} = (1-\epsilon) f(x^i_t) + \frac{\epsilon}{k_i} \sum_{j \in \nu_i} f(x^j_t) \quad , \quad i = 1,2,\ldots,N \,; \tag{3}$$

where, x_t^i is the state of the ith element at discrete time t, ϵ is the coupling strength, ν_i is the set of neighbors of the ith element, k_i is the cardinality of ν_i, and $f(x_t)$ is a chaotic map that expresses the local dynamics. The mean degree of the network is given by

$$\bar{k} = \frac{1}{N}\sum_{i=1}^{N} k_i \, , \tag{4}$$

We set $\bar{k} = 9$. The local dynamic is given by the logarithmic map[18, 19],

$$f(x) = b + \ln(x) \, , \tag{5}$$

where b is the map parameter. We fix the parameter $b = -0.7$, that is, in the region where the map presents a robust chaotic regimen; The initial conditions are randomly chosen in the interval $[-10; 10]$.

The synchronized state is characterized by the asymptotic time-average $\langle\sigma\rangle$ of the instantaneous standard deviations σ_t of the distribution of map variables x_t^i , defined as

$$\sigma_t = \left[\frac{1}{N}\sum_{i=1}^{N}(x_t^i - \langle x_t\rangle)^2\right]^{1/2} \, , \tag{6}$$

where $\langle x_t\rangle$ is the instantaneous mean of the values x_t^i. We discard the first $t = 10^4$ iterations and perform the time-average using the next 1000 iterates.

Figure 1 shows the asymptotic time-average of the instantaneous standard deviations $\langle\sigma\rangle$ as a function of the coupling strength ϵ and the fraction of directed links q.

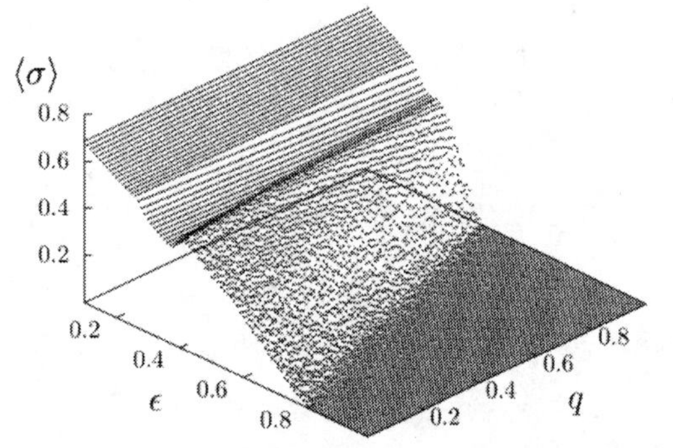

Figure 1. Asymptotic time-average of the instantaneous standard deviations $\langle\sigma\rangle$ as a function of the strength coupling ϵ and the fraction of directed links q; with $N = 10^4$ and $\bar{k} = 9$. All points are averaged over 6 realizations.

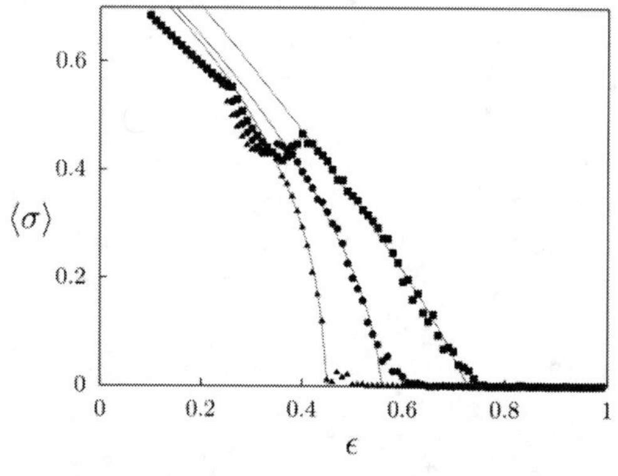

Figure 2. Asymptotic time-average of the instantaneous standard deviations $\langle\sigma\rangle$ as a function of the strength coupling ϵ. Squares $q = 0.1$, circles $q = 0.5$ and triangles $q = 0.9$. The continuous lines represent the fit of the equation (7) close to ϵ_c.

Note that, when the coupling strength ϵ increases, there is a transition in the system from a chaotic spatiotemporal ($\langle\sigma\rangle > 0$) to a synchronized ($\langle\sigma\rangle = 0$) phase. In figure 2 it is shown the asymptotic time-average of the instantaneous standard deviations $\langle\sigma\rangle$ as a function of the coupling strength ϵ with $q = 0.1$, $q = 0.5$ and $q = 0.9$.

The variation of the order parameter $\langle\sigma\rangle$ near the critical value of coupling strength ϵ_c can be characterized by a critical exponent β as

$$\langle\sigma\rangle = (\epsilon_c - \epsilon)^{\beta} \,. \tag{7}$$

We have calculated numerically the critical values of the coupling strength ϵ_c for the onset of synchronization in directed random network. The continuous lines in Figure 2 are the fitting of equation (7) for the given values of q.

Figure 3 shows the resulting critical boundary ϵ_c for the transition from chaotic spatiotemporal to synchronized regime, as well as the phase diagram of the system in the parameter space (ϵ, q).

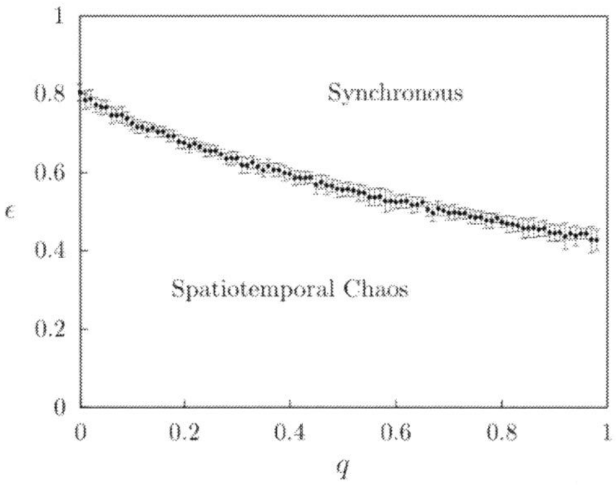

Figure 3. Phase diagram in the parameter space (ϵ, q). The numerically calculated critical boundary ϵ_c is shown with its error.

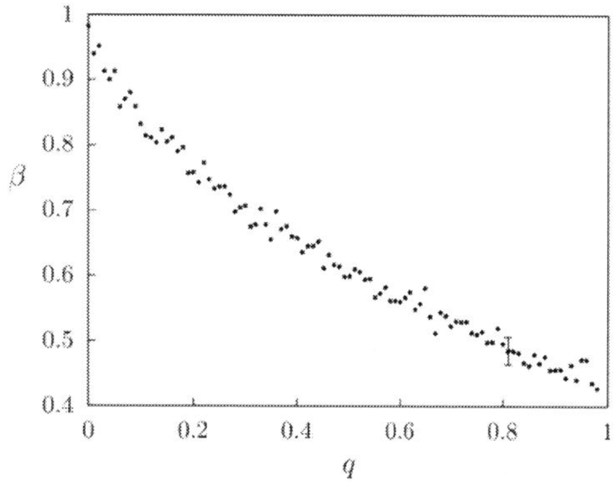

Figure 4. Numerically calculated critical exponent β as a function of q with a typical error bar.

Note that the critical value of the coupling strength ϵ_c defines a boundary in the phase diagram (ϵ, q) between the two regimes. Also, it can appreciate that the strength coupling necessary to synchronize the system is smaller when the fraction of directed links is increased.

Finally, Figure 4 shows the relation between the critical exponent β and the fraction of directed links q. It can see that, as q increases, the exponent β becomes smaller and the corresponding phase transition from spatiotemporal chaos to synchronized behavior gets more abrupt.

In summary, we have studied how the directed links affect the synchronization phenomenon on a Erdös-Rényi random networks. Although we use one class of networks and a simple dynamic for the nodes, we expect that the essential properties of the transition between synchronized-chaotic spatiotemporal phases is captured by this model.

By varying the fraction of directed links in the random networks, the behavior of the transition to synchronization can be studied in the regime between undirected lattices and completely directed random networks. The critical boundary separating synchronous and spatiotemporal chaos regimes was calculated on the parameter space (ϵ, q) of the system. We have found that the character of this transition is always of second order, that is, the critical exponent that characterize the transition is in the interval $\beta \in (0, 1)$ for all values of fraction of directed links q. The form how a phenomena appear depends on properties of the substrate over which the dynamic is carried out.

In this case, the spatiotemporal chaos-synchronous phase transition on a Coupled Chaotic Maps Lattice depends of the fraction of directed links of the network. In order to the synchronization phenomenon arises in the system, a smaller coupling strength, ϵ, is required when the fraction of directed links of the network, q, increases; and the phase transition becomes sharper.

Acknowledgments

The author thank M.G. Cosenza and O. Alvarez Llamoza for their valuable contributions. This work was supported by grant C-1696-10-05-B from the Consejo de Desarrollo Científico, Humanístico y Tecnológico of Universidad de Los Andes.

References

[1] Pecora L. M. and Carroll T. L. 1990 *Phys. Rev. Lett.*; **64** 821
[2] Pikovsky A, Rosenblum, M and Kurths J *Synchronization: a universal concept in nonlinear sciences* 2002 (Cambridge: Cambridge University Press)
[3] Boccaletti S, Kurths J, Osipov G , Valladares D L and Zhou C S 2002 *Phys. Rep.*; **366** 1
[4] Strogatz S. *Sync: the emerging science of spontaneous order* 2003 (New York: Hyperion)
[5] Alvarez-Llamoza O, Tucci K, Cosenza M G and Pineda M *Euro. Phys. J. ST*; **143** 245
[6] Motter A E, Zhou C and Kurths J 2006 *Phys. Rev. Lett*; **96** 034101
[7] Alvarez-Llamoza O and Cosenza M G 2009 *Int. J. Bif. Chaos*; **12**
[8] Bauer F, Atay F M and Jost J 2010 *Europhysics Letters*; **89** 20002
[9] Li X 2005 *Int. J. Comp. Cognition*; **3** 16
[10] Arenas A, Díaz-Guilera A, Kurths J, Moreno Y, Zhou C 2008 *Phys. Rep.*; **469** 93
[11] Rosenblum M G, Pikovsky A S, Kurths J 1996 *Phys. Rev. Lett.*; **76** 1804
[12] Agiza H N and Yassen M T 2001 *Phys. Lett. A*; **278** 191
[13] Jost J and Joy M P 2002 *Phys. Rev. E*; **65** 016201
[14] Sinha S 2002 *Phys. Rev. E*; **66** 016209
[15] Belykh I, Hasler M, Lauret M and Nijmeijer H 2005 *Int J. Bif. Chaos*; **15** 3423
[16] Atay F.M, Biyikoglu T, Jost J 2006 *Physica D, 224*; **224** 35
[17] Erdös P and Rényi A 1963 *Publ. Math. Inst. Hung. Acad. Sci.* **8** 455
[18] Kawabe T and Kondo Y 1991 *Prog. Theor. Phys.*; **85** 759
[19] Kawabe T and Kondo Y 1991 *Prog. Theor. Phys.*; **86** 581

Mass condensation on networks

B Waclaw[1], J Sopik[2], W Janke[3] and H Meyer-Ortmanns[2]

[1]SUPA, James Clerk Maxwell Building, University of Edinburgh, Edinburgh EH9 3JZ, UK
[2]School of Engineering and Science, Jacobs University Bremen, 28725 Bremen, Germany
[3]Institut für Theoretische Physik, Universität Leipzig, 04009 Leipzig, Germany

E-mail: `h.ortmanns@jacobs-university.de`

Abstract. We construct classical stochastic mass transport processes for stationary states which are chosen to factorize over pairs of sites of an undirected, connected, but otherwise arbitrary graph. For the special topology of a ring we derive static properties such as the critical point of the transition between the liquid and the condensed phase, the shape of the condensate and its scaling with the system size. It turns out that the shape is not universal, but determined by the interplay of local and ultralocal interactions. In two dimensions the effect of anisotropic interactions of hopping rates can be treated analytically, since the partition function allows a dimensional reduction to an effective one-dimensional zero-range process. Here we predict the onset, shape and scaling of the condensate on a square lattice. We indicate further extensions in the outlook.[1]

1. Introduction

A variety of stochastic processes out-of-equilibrium may be summarized under the name of stochastic mass transport models, mass transport in the sense of microscopic or macroscopic dynamics that involves spatial stochastic transport of some generic conserved quantity called 'mass'. The meaning of mass in this context ranges from cars in macroscopic traffic, to ants, to molecular motors in intracellular traffic. The transport is classically described in terms of probabilities for hopping events from one site to another. Such processes are of much interest from a theoretical as well as from a practical point of view. Since they are in general out-of-equilibrium and specified in terms of dynamical rules without any energy concept, it is difficult to predict and classify the possible stationary states and to identify transitions between phases like a liquid phase or a phase with a condensate that are associated with the different stationary states. In particular one observes the phenomenon of spontaneous symmetry breaking already in one-dimensional systems, in which the symmetry breaking manifests itself in the spontaneous formation of a condensate: a finite fraction of constituent particles condenses onto a finite extension in space (sometimes even on a single site) in the thermodynamic limit, in which the number of particles along with the volume is sent to infinity.

The practical relevance of these stochastic processes comes from the variety of condensation transitions which are ubiquitous in nature. To these belong the formation of large aggregates in aggregation and fragmentation processes [1], coalescence of shaken balls in granular flow [2], the formation of hubs in networks [3], or, last but not least, the formation of jams in traffic on various scales [4].

[1] Talk presented by H Meyer-Ortmanns

One well-known stochastic process is the zero-range process (ZRP) in which hopping rates are specified for particles via the probabilities to hop from one site to a neighboring site. The rates depend only on the departure site (and are therefore of zero range). The stationary states and the phase structure can be predicted analytically. The ZRP has been generalized in various directions, treating mass, space and time as continuous variables or allowing for internal degrees of freedom of what is called the mass, or adding annihilation and creation of particles in addition to their diffusion (for a review see for example [5]).

From a systematic point of view a natural generalization is towards the inclusion of next-neighbor interactions in the sense that the hopping rates not only depend on the departure site, but also on the neighboring sites, more precisely on the difference in occupation numbers $m \geq 1$ of such pairs. Evans *et al.* [6] formulated such a model that leads to pair-factorized steady states on a ring topology. From the exact knowledge of the stationary states they predicted the formation of a condensate above a certain critical mass density with the characteristic feature that the condensate was extended over a range that scales with $\sqrt{N}$ if N denotes the system size.

In this paper we review results of our work [7, 8, 9] which focusses on stochastic transport that leads to pair-factorized steady states on arbitrary graphs. More precisely, we look for dynamical rules that lead to stationary probability distributions of occupation numbers, which fully factorize over the links of an arbitrary graph. The rules are specified in terms of weight functions that finally determine the hopping rates and include the ZRP as special case. Once we know the stationary states in the whole parameter space, the next step is to derive the static and (if possible) the dynamic features which characterize the different phases. Here we start with the simplest topology, that of a ring, and analyze there the transition from a 'liquid', characterized by a uniform particle distribution, to a so-called 'condensed' phase, that is a phase with a condensate. We then study in detail how the shape of the condensate can be tuned by an appropriate variation of the local and ultralocal contributions to the hopping rate. We also predict the scaling of the condensate's extension with the system size and observe properties that remind us to finite-size scaling of first- and second-order phase transitions. For two dimensions we report on numerical results for isotropic interactions and analytical results for anisotropic ones. The analytical treatment becomes possible due to the dimensional reduction of the partition function to an effective one-dimensional ZRP. We conclude with a summary and outlook to possible extensions.

2. Pair-factorized steady states on arbitrary graphs

Usually in transport phenomena one would try to describe a given physical process by a set of hopping rates and predict the stationary states as solution of the corresponding master equation. If at all, this is often only approximately analytically possible. Here we proceed in the reverse way. Roughly speaking, we start from a probability distribution which is assumed to factorize over the edges of a given graph and ask which hopping rates lead to such a distribution as their stationary state. More precisely, let us consider connected, undirected, but otherwise arbitrary graphs with sites i, with k_i the degree of site i, i.e. the number of edges emerging from node i, and edges $\langle i,j \rangle$. Let M be the total number of particles of unit mass which is assumed to be conserved during the particle's diffusion, and m_i denote the occupation number of site i, $m_i \geq 0$. Let us split a hopping event into a product of a rate $u_i(m_i, m_{j_1}, m_{j_2}, ..., m_{j_{k_i}})$ which determines the probability that one particle leaves site i and depends on nodes $j_1, j_2, ...j_{k_i}$ adjacent to i, and a function $W(j \to i) \equiv W_{ji}$ that is proportional to the probability to choose site i as new target when coming from site j. For example, on a one-dimensional ring, one chooses for the particle's hopping a target to the right site with probability r, and to the left site with probability $1-r$. If we denote by $w(\vec{m})$ the probability for a distribution of occupation numbers $\{\vec{m}\} = \{m_1, ..., m_N\}$, $m_i \in \{0, 1, 2, ..., M\}$, $i \in \{1, ..., N\}$, we now ask which hopping rates $u_i(m_i, m_{j_1}, m_{j_2}, ..., m_{j_{k_i}})$

lead to a probability distribution $w(\vec{m})$ that factorizes over the pairs (edges) of a graph according to

$$w(\vec{m}) = \prod_{\langle i,j\rangle} g_{ij}(m_i, m_j) \prod_{i=1}^{N} k_i^{m_i} \delta\left(\sum_{i=1}^{N} m_i - M\right). \tag{1}$$

Here $\langle i,j\rangle$ denotes pairs of nearest neighbors, $g_{ij}(m_i, m_j)$ is a symmetric but otherwise arbitrary weight function that may depend on the link $\langle i,j\rangle$; the δ-function ensures conservation of the overall mass M (it will be dropped below). The factors $k_i^{m_i}$ are split off for convenience. Our answer is a possible class of hopping rates given as

$$u_i(m_i, m_{j_1}, m_{j_2}, ..., m_{j_{k_i}}) = \prod_{j\in\{j_1,j_2,...,j_{k_i}\}} \frac{g_{ij}(m_i - 1, m_j)}{g_{ij}(m_i, m_j)}, \tag{2}$$

where $\{j_1, j_2, ..., j_{k_i}\}$ denotes the set of neighbors of node i, provided the transition matrix W_{ji} satisfies the condition

$$k_i = \sum_j W_{ji} k_j \ , \tag{3}$$

so that the probability current is site-wise conserved. (This choice is certainly not the only possible one, since, for example, the symmetry condition on $g(m,n)$ was released in [6].) For the proof one considers the balance equation for a conserved probability current and shows that the pair-factorized steady state (1) solves this equation provided u is given by (2) and W satisfies (3). For details we refer to [9]. Although this solution is not the most general one, it contains important special cases like the ZRP, and regular topologies like all-to-all coupling or the square lattice which we shall consider below for anisotropic hopping. It should be noticed that the weight factors and therefore the hopping rates may depend on the edge $\langle i,j\rangle$, a fact which we shall exploit in section 4, where we choose $g_{ij}(m,n) = g_y(m,n)$ for all edges in the y-direction and $g_{ij}(m,n) = g_x(m,n) = \sqrt{f(m)f(n)}$ for links $\langle i,j\rangle$ in x-direction with $f(m)$ in general an arbitrary function depending on one site (so that we would separately expect fully factorized states in x-direction and pair-factorized states in y-direction), in section 4 chosen to be $f(m) = 1$ for simplicity.

Moreover it should be noticed that local hopping rates $u_i(m; m_{j_1}, m_{j_2}, ..., m_{j_{k_i}})$ can go along with non-local target functions $W(j \to i)$, this means that hopping events with non-local hopping are also covered by the proof. This fact we have actually exploited in our numerical simulations of the condensed phase. When we are only interested in the stationary states, we may use an algorithm to converge to this state as fast as possible. Here it turned out that a direct simulation of the hopping events usually takes longer by an order of magnitude than the direct simulation of the probability distributions of occupation numbers via a Metropolis algorithm. In the direct simulation of the hopping rates we select randomly a site i with occupation number m_i and nearest neighbors $n(i)$ and calculate the rate u_i for the given choice, next we leave the site with a probability proportional to this u_i and choose the target site with a probability proportional to the target function W_{ji}. In the indirect method we update the occupation number probabilities $w(\vec{m})$ via the Metropolis algorithm instead, that is, we start from a configuration of occupation numbers $\vec{\alpha} = \{m_1, ..., m_i, ..., m_k, ..., m_N\}$, choose randomly a site j and let a particle jump from i to j to end up at the configuration $\vec{\beta} = \{m_1, ..., m_i - 1, ..., m_j + 1, ..., m_N\}$. The new configuration is then accepted according to the Metropolis rule with probability prob$(\vec{\alpha} \to \vec{\beta})$ $= \min\{1, w(\vec{\beta})/w(\vec{\alpha})\}$. If the update is accepted, it corresponds to the non-local hopping from i to j and a choice of $W_{ji} = 1/(N-1)$ for a random selection of j among all sites (different from i) of the graph. The Metropolis algorithm ensures detailed balance, in this updating current conservation is therefore realized in a special way as equilibrium process, while a nonvanishing

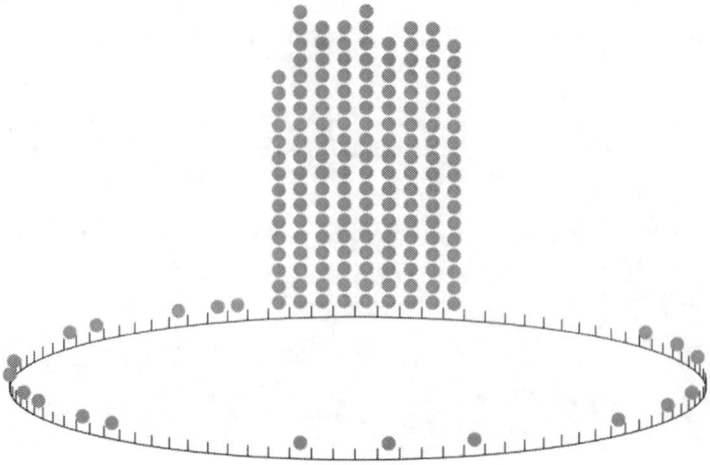

Figure 1. Schematic: Condensate of rectangular shape on a ring

conserved current would be the generic case, as it can be implemented by directly simulating the hopping events. The advantage is that the simulation of weights via the Metropolis algorithm amounts to non-local hopping, since i and j need not be nearest neighbors; it favors the melting of metastable 'precursors' of local condensates and therefore facilitates their merging to one big final condensate, but the stationary distribution of particles in the end is the same for a vanishing current or a current of nonzero constant velocity.

Exact knowledge of the stationary states of our dynamical rules is a useful first step similarly to the knowledge of the partition function in equilibrium dynamics. It remains to characterize the features of these states, or more precisely, the phases which they characterize (liquid or condensed) and inherent correlations between occupation numbers at different sites. While the liquid phase here turns out to be accessible in the grand-canonical formulation, the condensed phase is not. In addition, a reformulation in terms of a transfer matrix is available for one dimension, and there the model can be mapped to an SOS model [6], but to our knowledge the transfer-matrix formalism is not available in higher dimensions. Therefore, in the next section, we restrict our analytical study of the phase structure and the static properties of the condensate to a ring topology.

3. Condensation on a ring

Imagine we have a ring with N sites, and M balls available to distribute over the sites of the ring. We choose dynamical rules for their hopping between the sites which depend on the number of balls on the departure site as well as on the neighboring sites. Let us choose homogeneous rules over the ring so that no site is distinguished. What we were able to predict [7, 9] is not only the onset of condensation of balls over a limited extension of the ring above some critical density, but also the shape of the condensate, whether rectangular, bell-shaped or delta-like (cf. Figures 1, 2, and 3, respectively), and the scaling with the size N of the ring were predicted and numerically confirmed. The actual calculations are rather involved and cannot be presented here in detail, but we shall summarize the results and point on the main assumptions entering the analytical calculations.

Journal of Physics: Conference Series **246** (2010) 012011 doi:10.1088/1742-6596/246/1/012011

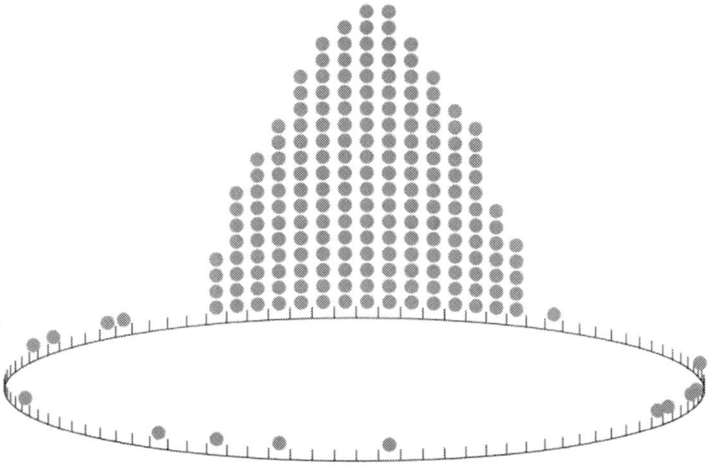

Figure 2. Schematic: Condensate of bell shape on a ring

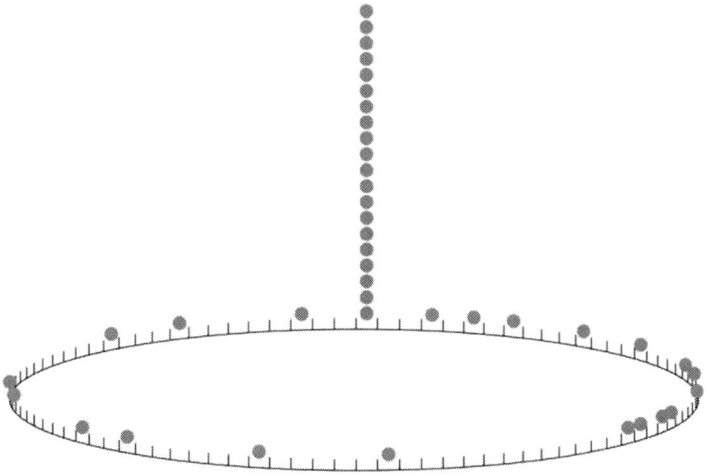

Figure 3. Schematic: Localized condensate on a ring as it also occurs for a ZRP

As special case of (2), when the hopping rate has the form

$$u(m_i|m_{i-1}, m_{i+1}) = \frac{g(m_i - 1, m_{i-1})}{g(m_i, m_{i-1})} \frac{g(m_i - 1, m_{i+1})}{g(m_i, m_{i+1})}, \tag{4}$$

with $g(m, n)$ being a symmetric non-negative but otherwise arbitrary function, the steady state factorizes over pairs of sites according to

$$P(m_1, \ldots, m_N) = \prod_{i=1}^{N} g(m_i, m_{i+1}) \delta \left(\sum_{i=1}^{N} m_i - M \right), \tag{5}$$

where the ring geometry implies $m_{N+1} \equiv m_1$. The difference as compared to the model of Ref. [6] is that there $r = 1$ and no assumption on the symmetry of $g(m, n)$ was made. The parameter

r changes only the net current of particles. When $r = 1/2$, the current is zero and the system is in equilibrium. Since the steady state (5) does not depend on r, its static properties can be calculated by formally treating the system as if it was in equilibrium with the probability of a microstate given by Eq. (5).

The criterion for condensation is non-analyticity of the grand-canonical partition function

$$Z_N(z) = \sum_{\{m_i\}} z^{\sum_i m_i} \prod_i g(m_i, m_{i+1}) = \mathrm{Tr}\, T^N, \tag{6}$$

where $T_{mn} \equiv z^{(m+n)/2} g(m,n)$ and z is the fugacity. Assuming z_c is the finite radius of convergence of $Z_N(z)$, the critical value of the density $\rho = M/N$ then follows from

$$\rho_c = \lim_{N\to\infty, z\to z_c^-} \frac{z}{N}\frac{\partial \ln Z_N(z)}{\partial z} = \frac{\sum_m m\phi_m^2}{\sum_m \phi_m^2}, \tag{7}$$

where ϕ is an eigenvector of T_{mn} associated with the largest eigenvalue $\lambda_{\max}$, for $z = z_c$. At the critical point, $Z_N \cong \lambda_{\max}^N$ for large N.

To investigate how the introduction of site-site interactions influences the properties of condensation, we stick to the following choice:

$$g(m,n) = K(|m-n|)\sqrt{p(m)p(n)}. \tag{8}$$

When $K(x) = 1$, $g(m,n)$ factorizes and we recover the ZRP. Here we assume that both $K(x)$ and $p(m)$ are some positive, decaying functions of x and m, respectively. The choice (8) is motivated by studies on the SOS model in the context of surface roughening [10, 11] and corresponds to the energy $E = -\ln K(|m-n|) - (1/2)\left[\ln p(m) + \ln p(n)\right]$ of an interface within a 1+1-dimensional surface (where the surface refers to the envelope of occupation numbers). In Ref. [6], the following choice was proposed:

$$K(x) = e^{-Jx}, \quad p(m) = e^{U\delta_{m0}}, \tag{9}$$

with parameters J and U generating an effective surface stiffness and a pinning potential, respectively. For the weights (9), we obtain the critical density as a function of J and U as follows:

$$\rho_c = \frac{e^{J_0} - 1}{(e^{J_0} - e^{-2(J-J_0)})(e^{2(J-J_0)} - 1)}, \tag{10}$$

with $J_0 = U - \ln(e^U - 1)$. If $J < J_0$, the critical density is infinite.

The generic case — In order to calculate the width W and the envelope of the condensate, one cannot use the partition function (6), because it is not defined in the condensed phase. Instead, for the weight functions (8), we use an approach which allows to calculate both quantities in the case when $K(x)$ decays exponentially or faster in x, and $p(m) \to 1$ for $m \to \infty$, that is for local interactions with exponential decay and an ultralocal part that approaches a constant above some occupation number $m_{\max}$. We make the assumption that the system factorizes into two terms, one coming from the condensate and one from the background, filled up to the critical particle density. The condensate then extends over W sites occupied by $M' = M - N\rho_c$ particles on the average, average occupation numbers $\langle m_i \rangle$ grow with N, and the uniform background is characterized by $\langle m_i \rangle = \rho_c$. Since fluctuations in the background are finite and have no long-range correlations, the mass M' cannot fluctuate more than $\sim \sqrt{N}$ so that we treat it as constant. We can therefore assume that the probability of having the condensate extended over W sites is given by

$$\begin{aligned} P(W) &\approx Z_b(W) Z_c(W) \\ &\propto \exp\left(-W \ln \lambda_{\max} + \ln Z_c(W)\right), \end{aligned} \tag{11}$$

where $Z_b(W) = \lambda_{\max}^{N-W}$ is the partition function for the background at the critical point $\rho = \rho_c$, $\lambda_{\max}$ the maximal eigenvalue of the transfer matrix, and $Z_c(W)$ is the partition function of the condensate extended over W sites and having exactly M' particles. In the calculation of $Z_c(W)$ we have furthermore extended the summation over negative occupation numbers, although this is physically not meaningful, but it allows doing the calculation, and the induced error is negligible in the end if we compare with results of the numerical simulations. The factor Z_c can be explicitly calculated when it is independent of the ultralocal part of the weight, i.e., independent of $p(m)$. This happens in particular for the choice of weight functions as it was used in [6]. Otherwise we use what we called the fixed-envelope approximation. Here we approximate Z_c via the probability $P(W)$ of having a condensate extended over W sites with unknown but fixed envelope $h(t)$, for which we use a form that is inspired by numerical simulations of the condensation. The condensate's extension follows in these cases from the maximum of the probability $P(W)$.

As it turns out, within the same fixed-envelope approximation scheme, when local and ultralocal weights are "short-ranged"[2], we can tune the extension to scale with the system size as N^α with $0 \le \alpha \le 1/2$. When both parts of the weights are long-ranged, the condensate gets localized to a single site as in the ZRP, cf. Figure 3. For a long-range local part and a short-range ultralocal part, the condensate takes a rectangular form (Figure 1), its height scales proportionally to the system size, while its extension remains constant: features that remind us to finite-size scaling of a first-order phase transition. Therefore hopping rates, leading to pair-factorized steady states, do not necessarily lead to extended condensates, and if they do so, the shape of the condensate is non-universal. For any details of the calculation we refer to [9].

4. Anisotropic hopping

A generalization of the methods used in the previous section to higher dimensions is not straightforward. Let us combine a ZRP in one dimension with a process that in isolation would lead to a pair-factorized steady state in the second dimension, but both processes coupled via the constraint of overall mass conservation. This implies that $g(m,n)$ lives on edges only in one, say vertical, direction. On horizontal links, the weight is assumed for simplicity to be constant and equal to 1, but in general one may choose any weight $f(m_{i,j})$ depending only on a single site. An extrapolation of our results in one dimension suggests a localized condensate in one direction and an extended condensate in the second direction, although the competition between the different interactions could in principle lead to a spreading of a condensate in both directions, or a uniform distribution in one direction and a condensate in the other direction. The former scenario is actually numerically observed for a certain range of mass density values $\rho_1 < \rho < \rho_2$, cf. Figure 4. Below ρ_1 we find a 'liquid' uniformly spread over the lattice, and above ρ_2 we find a wall of uniform height in one direction, but localized in the other direction, cf. Figure 5. These qualitative features can be understood and analytically reproduced by means of dimensional reduction of the partition function in two dimensions to an effective ZRP in one dimension. Let us assume that we have a two-dimensional lattice with $N = L \times L$ sites, periodic boundary conditions and the following steady state:

$$w(\vec{m}) = \prod_{i=1}^{L}\prod_{j=1}^{L} g(m_{i,j}, m_{i,j+1})\, \delta\left(\sum_{i,j} m_{i,j} - M\right). \tag{12}$$

[2] We call short-range interactions local weight factors $K(x) = K(|m-n|)$ and ultralocal factors $p(m)$ which decay faster than any power in their argument, and long-range interactions those which decay as a power of their argument (x or m, respectively). So the relevant characteristics for both local and ultralocal weight factors is the range of interactions in occupation-number space.

The hopping rate assumes now the form

$$u(m_{i,j}|\ldots) = \frac{g(m_{i,j}-1, m_{i,j-1})}{g(m_{i,j}, m_{i,j-1})} \frac{g(m_{i,j}-1, m_{i,j+1})}{g(m_{i,j}, m_{i,j+1})}, \tag{13}$$

the particles can jump either in horizontal or vertical direction. The current is defined by any W obeying equations (3). Although we will not address dynamical issues here, for definiteness let us assume that the particles can jump only to the right with probability $W(i,j \to i,j+1) = p$ or to the top with probability $W(i,j \to i+1,j) = 1-p$, while $W(i,j \to i,j-1) = W(i,j \to i-1,j) = 0$. Moreover, let the weight $g(m,n)$ be chosen according to (9). We will argue that ρ_1 is the critical density of the corresponding pair-factorized state in one dimension and also calculate ρ_2.

First we observe the following. In the steady state we can treat the probability $w(\vec{m})$ as the weight of a microstate of a system being in equilibrium, therefore we can formally write the partition function of the system in the canonical formulation as

$$\begin{aligned} Z_{c2d}(N=L^2, M) &= \sum_{\{m_{i,j}\}} \prod_{i=1}^{L} \prod_{j=1}^{L} g(m_{i,j}, m_{i,j+1}) \delta\left(\sum_{i,j} m_{i,j} - M\right) \\ &= \sum_{M_1=0}^{M} \cdots \sum_{M_L=0}^{M} \prod_{i=1}^{L} Z_{c1d}(L, M_i) \delta\left(M - \sum_i M_i\right), \end{aligned} \tag{14}$$

where

$$Z_{c1d}(L, \tilde{M}) = \sum_{m_1=0}^{\tilde{M}} \cdots \sum_{m_L=0}^{\tilde{M}} \prod_{i=1}^{L} g(m_i, m_{i+1}) \delta\left(\tilde{M} - \sum_i m_i\right) \tag{15}$$

is the partition function of a one-dimensional pair-factorized steady state. Since (14) has the same functional form as the partition function for the zero-range process [12], the partition function Z_{c1d} may be seen as the weight $p(m) \equiv Z_{c1d}(L,m)$ that is now associated with the total mass $m = M_i$ along the ith vertical line.

In [7] we have shown that for $m > \rho_c L$, where ρ_c is the critical density for condensation in the one-dimensional system, $Z_{c1d}(L,m)$ behaves as $\sim \exp(-c\sqrt{m})$ with some $c > 0$. This means that for large m, the hopping rate of the corresponding ZRP, $u(m) = p(m-1)/p(m) = Z_{c1d}(L, m-1)/Z_{c1d}(L,m)$, behaves as $u(m) \cong 1 + c/(2\sqrt{m})$. For such a hopping rate it is known [12] that the ZRP exhibits a condensation transition. The condensate occupies a single site and the fluid-phase distribution is a stretched exponential distribution. Since m denotes now the mass along a vertical line, the mapping from the ZRP back to the anisotropic model allows us to predict the spontaneous symmetry breaking to the state with a condensate so that one of the masses M_i will grow to pick up all the difference $\Delta M = M - \rho_1 L^2 = L^2(\rho - \rho_1)$ between the total mass M and the mass $\rho_1 L^2$ in the critical background. Thus, not each line carries its own condensate, but the condensate is localized onto a single line if there is a condensate at all, while all other lines have a mass density close to the critical value. To predict the critical density ρ_1 we observe that (14) in the grand-canonical formulation,

$$\begin{aligned} Z_{g2d}(L,z) &= \sum_M Z_{c2d}(L,M) z^M = \sum_{\{M_i\}} \prod_{i=1}^{L} Z_{c1d}(L, M_i) z^{M_i} \\ &= \left(\sum_M Z_{c1d}(L,M) z^M\right)^L, \end{aligned} \tag{16}$$

becomes just a power of a grand-canonical partition function in one dimension. Since the

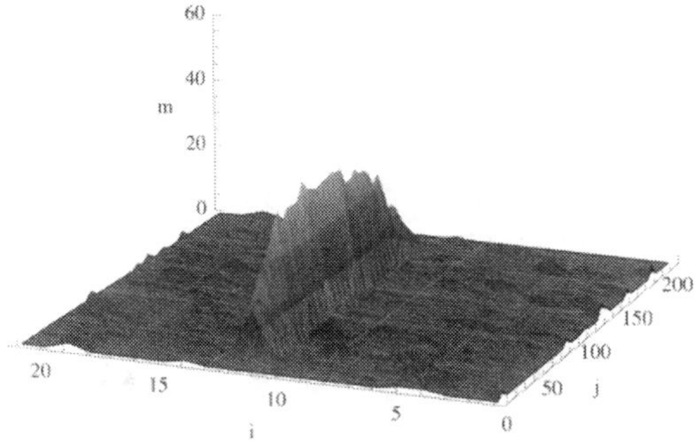

Figure 4. Condensate localized in x-direction and extended in y-direction for $\rho_1 < \rho < \rho_2$.

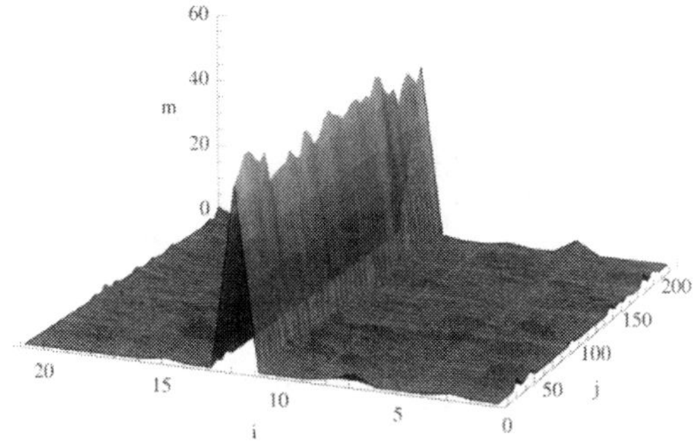

Figure 5. Condensate localized in x-direction, but uniformly spread out like a liquid in y-direction for $\rho > \rho_2$.

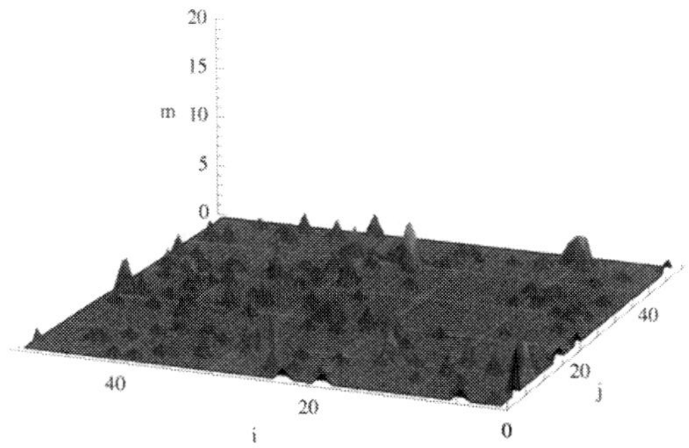

Figure 6. Liquid phase for isotropic hopping in two dimensions.

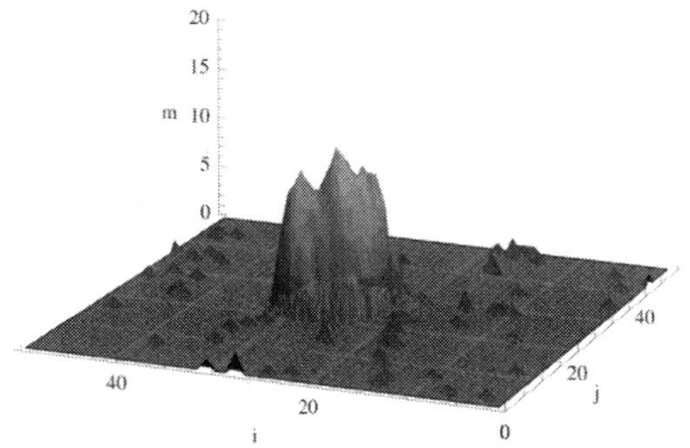

Figure 7. Condensate with quadratic support for isotropic hopping in two dimensions.

condensation transition is determined by the non-analytic behavior of Z_{g2d}, the critical density ρ_1 must be the same as ρ_c of the one-dimensional pair-factorized steady state. For $J = U = 1$ it reads $\rho_1 \approx 0.24$, as follows from [7].

It remains to determine how the particles are distributed along the lines which carry the condensate. Using results from [7] for the pair-factorized state in one dimension we know that the width W grows $\sim L\sqrt{\rho - \rho_1}$. This is the shape we see in Figure 4 for $\rho_1 < \rho < \rho_2$. When the density exceeds the values ρ_2 for which the density exceeds the width W and becomes equal to the linear size L, there are no empty sites left so that $g(m,n)$ effectively behaves as $\exp(-J|m-n|)$, a case for which we know from [7] that the system is always in a liquid state. This explains the uniform shape in y-direction in Figure 5, while it stays localized in x-direction. Both Figures 4 and 5 display only part of the lattice (20×200 sites) on which the actual simulations have been performed, these were lattices with 200×200 sites.

For interactions with isotropic weights $g_i(m,n)$, $i \in \{x,y\}$ in x- and y-direction we numerically observe the natural two-dimensional generalization of an extended condensate on a ring topology, but an analytic treatment was not possible so far, cf. Figure 6 for the liquid phase and Figure 7 for the condensed phase. It remains a challenge to derive the static features of the condensate on generic network topologies, in particular in higher dimensions.

5. Conclusions and outlook

The answer to our question which hopping rates lead to pair-factorized steady states on arbitrary graphs was given in terms of a class of weight functions. These weight functions may depend on the links of the graphs and need not be local, so that anisotropic hopping and hopping to remote target nodes are allowed. The topologies we have studied so far for deriving the phase structure were regular ones: the ring topology and a square lattice. The next step is to consider our dynamical rules on general network topologies and study the effect of explicit symmetry breaking due to the network topology, for example. For the ring and square topology it was possible to predict beyond the very onset of the phase transition several properties of the condensate: its shape as a function of the interaction between occupation numbers on different sites, its extension and scaling with the system size. For anisotropic interactions in two dimensions we derived and numerically confirmed two phase transitions: the first one from a liquid phase to a phase with a condensate localized in one direction and extended but still present in the second direction, the second transition from the former condensed phase to a phase with a condensate in one direction which is, however, uniformly spread out like a liquid in the second direction. From an experimental point of view our results are of interest for experiments in surface science [13, 14], in which condensates are observed of a certain shape and for which it may be tempting to trace back the underlying rules for particle hopping which are compatible with the observed shape and the scaling of the condensate's extension.

From a theoretical point of view it is challenging to generalize our pairwise factorizing steady states to states which fully factorize over triangles or plaquettes and to see which hopping rates lead to such steady states. Exactly known states like those we have considered in this article may always serve as starting point for approximations where weak correlations are involved, but may be neglected to zeroth order. Another direction of interest are dynamical features. We called the very formation of condensates on a finite ring spontaneous symmetry breaking. On the other hand, strictly speaking a symmetry cannot spontaneously break in a finite volume, neither here. Therefore any condensate will melt after some time and build up again at another site. To estimate this time as a function of the system size is a challenge for future calculations.

Acknowledgments

One of us (H. M.-O.) would like to thank the organizers of this conference for a very interesting meeting in a pleasant atmosphere. The collaboration of H.M.-O. and W.J. was supported by the German Research Foundation (DFG) through a joint research grant under Grant Nos. ME 1332/17-1 and JA 483/27-1.

References

[1] Majumdar S N, Krishnamurthy S and Barma M 1998 *Phys. Rev. Lett.* **81** 3691; 2000 *J. Stat. Phys.* **99** 1
[2] Liu C-H, Nagel S R, Schecter D A, Coppersmith S N, Majumdar S, Narayan O and Witten T A 1995 *Science* **269** 513; Coppersmith S N, Liu C-H, Majumdar S, Narayan O and Witten T A 1996 *Phys. Rev.* E **53** 4673
[3] Krapivsky P L, Redner S and Leyvraz F 2000 *Phys. Rev. Lett.* **85** 4629
[4] Chowdhury D, Santen L and Schadschneider A 2006 *Phys. Rep.* **329** 199
[5] Evans M R 2000 *Braz. J. Phys.* **30** 42
[6] Evans M R, Hanney T and Majumdar S N 2006 *Phys. Rev. Lett.* **97** 010602
[7] Waclaw B, Sopik J, Janke W and Meyer-Ortmanns H 2009 *J. Phys. A: Math. Theor.* **42** 315003
[8] Waclaw B, Sopik J, Janke W and Meyer-Ortmanns H 2009 *Phys. Rev. Lett.* **103** 080602
[9] Waclaw B, Sopik J, Janke W and Meyer-Ortmanns H 2009 *J. Stat. Mech.* **P10021**
[10] Chui S T and Weeks J D 1981 *Phys. Rev. B* **23** 2438; Burkhardt T W 1981 *J. Phys. A: Math. Gen.* **14** L63
[11] van Leeuwen J M J and Hilhorst H J 1981 *Physica A* **107** 319
[12] Evans M R and Hanney T 2005 *J. Phys.* A **38** R195
[13] Bréchignac C, Cahuzac P, Carlier F, Colliex C, Leroux J, Masson A, Yoon B and Landman U 2002 *Phys.*

Rev. Lett. **88** 196103; Yoon B, Akulin V M, Cahuzac P, Carlier F, de Frutos M, Masson A, Mory C, Colliex C and Bréchignac C 1999 *Surf. Science* **443** 76
[14] Nötzel R 1996 *Semicond. Sci. Technol.* **11** 1365

XI Latin American Workshop on Nonlinear Phenomena IOP Publishing
Journal of Physics: Conference Series **246** (2010) 012012 doi:10.1088/1742-6596/246/1/012012

A local adaptive algorithm for emerging scale-free hierarchical networks

Ignacio J. Gomez Portillo[1], **and Pablo M. Gleiser**[1,2]

[1] Centro Atómico Bariloche, Instituto Balseiro, Bariloche, 8400 Río Negro, Argentina
[2] CONICET

E-mail: gleiser@cab.cnea.gov.ar

Abstract. In this work we study a growing network model with chaotic dynamical units that evolves using a local adaptive rewiring algorithm. Using numerical simulations we show that the model allows for the emergence of hierarchical networks. First, we show that the networks that emerge with the algorithm present a wide degree distribution that can be fitted by a power law function, and thus are scale-free networks. Using the LaNet-vi visualization tool we present a graphical representation that reveals a central core formed only by hubs, and also show the presence of a preferential attachment mechanism. In order to present a quantitative analysis of the hierarchical structure we analyze the clustering coefficient. In particular, we show that as the network grows the clustering becomes independent of system size, and also presents a power law decay as a function of the degree. Finally, we compare our results with a similar version of the model that has continuous non-linear phase oscillators as dynamical units. The results show that local interactions play a fundamental role in the emergence of hierarchical networks.

1. Introduction

There is an increasing interest in the scientific community to analyze systems as complex networks, that is, at an abstract level where the description is simply a set of vertices and edges. This approach has received considerable attention, since the same generic laws that allow for the emergence of complex network structures seem to govern a vast diversity of system, including social, biological and technological networks [1, 2, 3, 4, 5]. Many of these systems can be described by small world networks that are characterized by the presence of highly clustered modules and short path lengths between nodes [6]. Also, when the degree distribution presents a power law behavior the network can be described by scale-free network models [7]. Ravasz *et al.* [8, 9] noted that for a large number of real networks the scale free and modular characteristics are not exclusive. These systems are characterized by overlapping clusters of nodes that cluster into larger ones through a hierarchical organization, and thus are referred to as hierarchical networks. Examples include technological networks such as the Internet at the autonomous system level, social networks, semantic networks, biological networks such as metabolic and protein interaction networks [8, 9] and also anatomical and functional networks in the brain [10, 11, 12].

In the present work we analyze which are the basic ingredients necessary for the emergence of hierarchical networks. In order to tackle this issue we focus our interest in a novel approach where the evolution of the network structure is intimately coupled to the dynamical state of the nodes.

In particular, Gong van den Berg and van Leeuwen [13, 14, 15] proposed recently an adaptive rewiring algorithm, and showed that starting from an initial random system, small world networks emerge by adaptively rewiring chaotic units according to their dynamic coherence [14, 15]. In these works the choice of chaotic dynamics for the microscopic units was inspired in the neural sciences. In fact, the algorithm was proposed as a candidate model for morphogenesis [14] and also shown to present characteristics that can be used to dynamically encode sensory information in neural systems [15]. A similar form of adaptive evolution was considered in a network of coupled non-linear phase oscillators by Gleiser and Zanette [16]. They also observed the emergence of small world networks. In this case, the system presents an evolution that strongly resembles the synchronization phenomenon of flashing fireflies, with a rich diversity of synchronized oscillator groups that are directly associated with network clusters [16]. The fact that an initial random system is able to spontaneously evolve to a small world network structure, either when the microscopic units present chaotic or oscillatory dynamics, shows that the algorithm is robust, thus presenting a plausible mechanism for the emergence of large scale complex network structures.

Gong and van Leeuwen also presented a generalization of the algorithm that allows for network growth. and observed the emergence of scale-free networks in a system with chaotic units [13]. Recently, we presented a similar version of this growing network model using non-linear phase oscillators as dynamical units. We also observed the emergence of scale-free networks, and proposed the system as an adaptive model for the complex network structures observed in human brain functional networks [10, 11, 12, 17, 18]. Again the algorithm proved to be robust, allowing for the emergence of scale-free networks either when the microscopic units present chaotic or oscillatory dynamics.

In this work we analyze a growing network model similar to the one presented by Gong and van Leeuwen [13]. By restricting the amount of information available to the new incoming nodes we analyze under which condition hierarchical structures emerge (or not). By comparing with a model with phase oscillators as dynamical nodes we show that the relevant ingredients that allow for the emergence of hierarchical structures do not seem to depend on the details of the dynamics of the nodes. The results show that the restriction of the information available to the new nodes, from a global to a local point of view, plays a relevant role in the emergence of hierarchical structures.

2. The Model

We model a growing adaptive network, where the microscopic units are logistic maps, $f(x) = 1 - ax^2$, where $a = 1.796$ so that the activity is chaotic. The units are coupled through the following equation:

$$X^i_{t+1} = (1-\epsilon)f(X^i_t) + \frac{\epsilon}{M_i}\sum_{j=1}^{N} W_{ij} f(X^j_t) \tag{1}$$

where X^i_t is the activity of unit i at time step t, and ϵ is the coupling constant. The weights W_{ij} define the adjacency matrix of the interaction network: $W_{ij} = 1$ if unit i interacts with unit j, and 0 otherwise. The number of neighbours of i is $M_i = \sum_j W_{ij}$. Interactions are symmetric, so that $W_{ij} = W_{ji}$ and the network is a non-directed graph.

The algorithm for the evolution of the system is as follows. Begin with a small random network with N_0 nodes linked by L_0 connections per node.

(i) Add a new node i_n with k_{in} connections to k_{in} different nodes randomly chosen in the current network.

(ii) Set the initial activity of each node with random numbers $z \in [-1, 1]$.

(iii) Calculate the state of the system according to Eq. (1) over a time interval of length T.

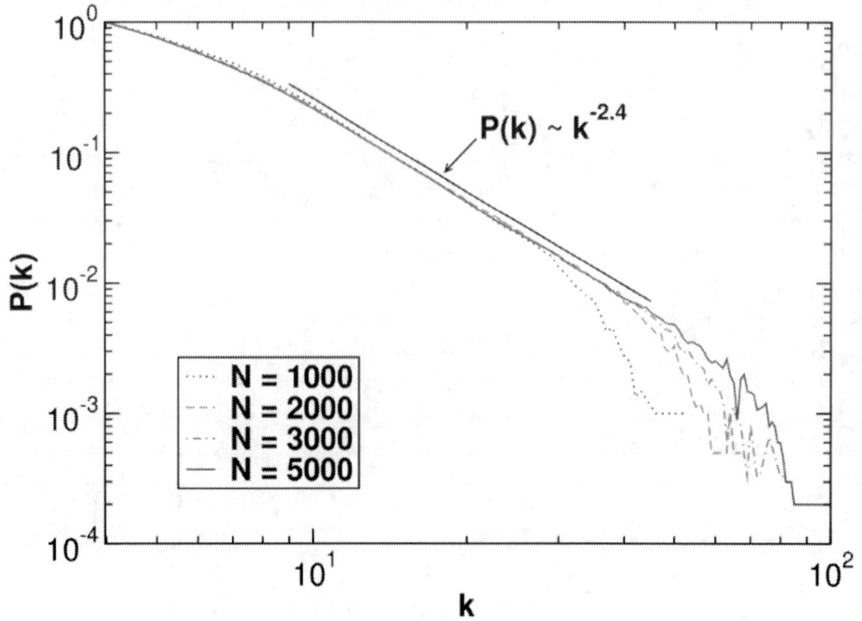

Figure 1. Degree distribution $P(k)$ vs. degree k for four system sizes $N = 1000$, 2000, 3000 and 5000 when $k_{in} = 4$. The straight line is a guide to the eye and shows the power law behavior $P(k) \sim k^{-2.4}$.

(iv) At $T+1$ calculate the value of $d_{ij} = |X_n^i - X_n^j|$ for the first and second neighbours of i. Detect the oscillator j_1 for which d_{ij_1} is minimum amongst all the second neighbours of i. Also detect, amongst the neighbours of i, the oscillator j_2 for which d_{ij_2} is maximal.

(v) If j_1 is one of the neighbours of oscillator i, then make no changes to the connections. Otherwise, replace the link between i and j_2 by the link between i an j_1.

(vi) Go to step 2) and repeat the algorithm K_0 times.

(vii) Go to step 1)

3. Numerical simulation results

We begin with random networks with $N_0 = 20$ nodes and $k_0 = 6$ connections per node. The coupling constant is fixed to $\epsilon = 0.5$. All the curves correspond to averages over ten different realization of the algorithm.

3.1. Degree distribution

In Fig. 1 we present the behavior of the degree distribution $P(k)$ vs. degree k for four different system sizes, $N = 1000$, 2000, 3000 and 5000, when each new node has $k_{in} = 4$ links. The degree distribution presents a slow decay followed by a crossover to a faster decay. Note that as N increases the crossover moves to larger values of k. This finite size behavior suggests that for sufficiently large system sizes only the slow decay will be observed. In that case the degree distribution can be approximated by a power law function. The straight line presented as a guide to the eye in Fig. 1 shows the best approximation given by $P(k) \sim k^{-2.4}$.

We note that this finite size behavior has been observed in real systems, such as brain functional networks. In fact, Zalesky *et al.* [19] presented recently an extended analysis of brain networks at different ranges of scales and grey matter parcellations, showing that comparision of brain networks across different studies should be made with reference to the spatial scale of parcellation. For example, Achard *et al.* [18] used functional magnetic resonance imaging (fMRI) time series to create undirected graphs with $N = 90$ nodes, and found that the best fit for the degree distribution of these networks was given by an exponentially truncated power law $P(k) \sim k^{-1}e^{(-k/k_c)}$. On the other hand, Eguíluz *et al.* [17] build large functional brain networks with up to $N = 31.503$ nodes, and observed a power law degree distribution $P(k) \sim k^{-\alpha}$, with $\alpha \approx 2.0$, and with no cutoff. The results obtained with the model show how these different functional shapes of the degree distribution could be a consequence of finite size effects.

In their original study, where global rewiring rules were used, Gong and van Leeuwen considered a network where each new node has $k_{in} = 14$ links. They showed that in this case the degree distribution of the self-organized network can be fitted by a power law [13]. In order to analyze the effect of increasing the number of connections of the new nodes, in our new model with local rewiring dynamics, we present in Fig. 2 the behavior of $P(k)$ vs. k for networks with system size $N = 3600$ and three different values of incoming links, $k_{in} = 4$, 8, and 14. Note that all the curves present the same qualitative behavior described in Fig. 1, where a slow decay is followed by a crossover to a faster decay. Clearly, as k_{in} grows the curves are displaced to larger k values. The data collapse presented in the inset of Fig. 2 shows that this behavior can be characterized simply by scaling the degree k with k_{in} as k/k_{in}.

It is also interesting to compare the results when the dynamics of the units is given by non-linear phase oscillators with local rewiring rules as in Ref. [12]. In this case, the degree distribution seems to converge to a power law decay $P(k) \sim k^{-\alpha}$, with $\alpha = 2.1$ for increasing system sizes. It is worth stressing that the change in the dynamical units leads to important differences in the spatiotemporal dynamics [12, 13, 14, 15, 16]. Nevertheless, in both cases a similar qualitative behavior of the degree distribution, with a slow decay followed by a crossover to a faster decay is observed. Also the crossover moves to higher degree values as the system size grows. This result shows that the emergence of wide degree distributions is robust for the local rewiring algorithm when the dynamic of the nodes is changed from continuous non-linear phase oscillator to discrete chaotic maps.

3.2. Preferential attachment

Gong and van Leeuwen [13] observed that in the growing network model with chaotic units a preferential attachment mechanism was present. They noted that in contrast to models where the connectivity distribution of the entire network is available to each node the preferential attachment mechanism in this case is an emergent property based on information about network dynamics. It is an interesting question if this emergent preferential attachment mechanism will still be present when the information available to the nodes in the system is changed from the global view, where the dynamical state of all the nodes in the system is available to each node, as in Ref. [13], to a restricted local environment as in the present model.

In this section we present a simple qualitative analysis that shows the presence of a preferential attachment mechanism. A detailed quantitative study and further results will be presented elsewhere. We use the LaNet-vi network visualization tool [20, 21], that uses a recursive algorithm to obtain all the k-cores of the network, which are the connected maximal induced subgraphs which have minimum degree greater than or equal to k. Through the pruning of connections the nodes are labeled according to their shell index k, indicating that they belong to the k-core but not to the k+1 core. This tool is suited for the analysis of large sparse networks with a non trivial hierarchical structure, since the vertices are shown in different two dimensional annular shells which correspond to their actual shell index and provide a centrality measure.

In Fig. 3 we show the graphical representation of a system with $N = 5000$ nodes and $k_{in} = 4$ and an initial random network with $N_0 = 20$ nodes with degree $k_0 = 6$. The column on the left shows the labels for the degree of the nodes. Larger values of degree are indicated with

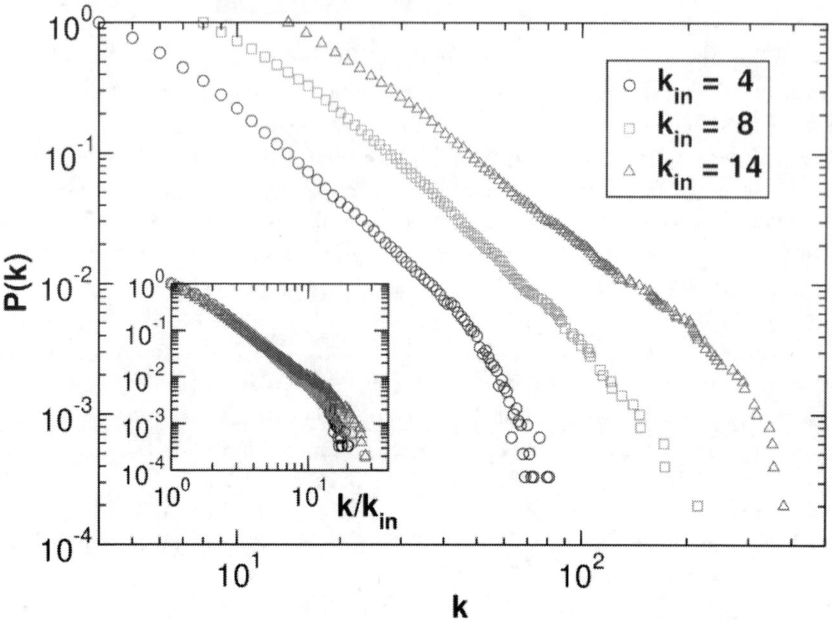

Figure 2. Degree distribution $P(k)$ vs. degree k for system size $N = 3600$, when each new node has $k_{in} = 4$, 8 and 14. The inset presents the data collapse obtained by scaling k/k_{in}.

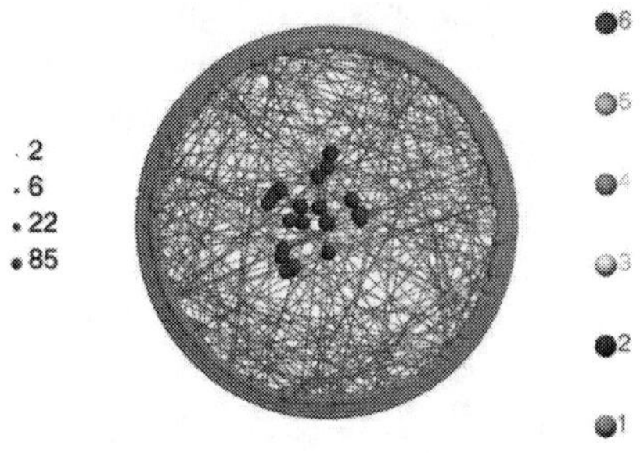

Figure 3. Graphical representation of a system with $N = 5000$ nodes using the LaNet visualization tool. A clear degree-coreness correlation is revealed by a central core formed only by hubs.

increasing node size. The column on the right indicates the coreness value with a color code. Note that only two layers with different coreness values are visible. The outer layer with coreness 4 is a consequence of the growing rules of the model. Since the incoming nodes have a fixed incoming degree $k_{in} = 4$, the coreness sets with values lower than this will be empty. Also, the pruning algorithm will remove first the last node that has entered the system (that has degree $k_{in} = 4$), then it will remove the node added at the preceding step (that also has $k_{in} = 4$), and so on up to the initial vertices. The central nodes with higher coreness correspond to the nodes of the random initial network with initial degree $k_0 = 6$. The figure clearly shows that there is a degree-coreness correlation. In fact, the central core of the network is formed only by hubs, that is by the nodes that have the highest degree. Our particular choice of the random initial network

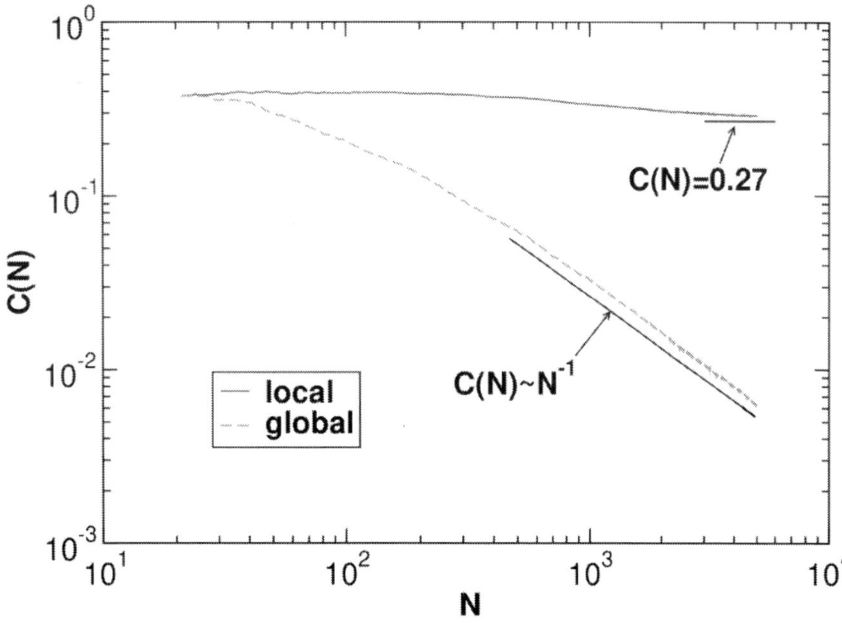

Figure 4. Clustering coefficient $C(N)$ vs. system size N for local and global rewiring dynamics when $k_{in} = 4$. The straight lines are guides to the eye and show the slow convergence to a constant value for local dynamics, $C \approx 0.27$ and the power law decay for global dynamics, $C \sim N^{-1}$.

permitted us to show that even when the information available to the new nodes is restricted to a local subset a preferential attachment mechanism is still present as in the work of Gong and van Leeuwen[13]. We stress that this is an emergent property based only on local information about the dynamics of the nodes in the network and not on their connectivity distribution.

3.3. Clustering coefficient

A quantitative characterization of the hierarchical structure can be obtained through the analysis of the clustering coefficient C, that measures the average number of neighbours of a given node which are in turn mutual neighbours [6]. In these networks C is independent of system size N, and also, the presence of a hierarchy of nodes with different degrees of clustering can be quantified by the scaling law $C(k) \sim k^{-\beta}$ [8, 9].

Figure 4 shows the behavior of C as a function of system size N both for local and global rewiring dynamics. The behaviors are clearly different. On the one hand, for the global rewiring dynamics the clustering coefficient presents a power law decay $C(N) \sim N^{-1}$. This clearly shows that the networks that emerge with this rewiring rule are not hierarchical. It is worth stressing that a power law decay of C is also present in other non-hierarchical networks such as the random network model, where $C(k) \sim N^{-1}$ or the Barabási-Albert scale free model where $C(k) \sim N^{-0.75}$ [7]. On the other hand, for the local rewiring dynamics C seems to converge to a constant value $C \approx 0.27$. We stress that this is one of the behaviors that characterizes real hierarchical networks, where C is independent of system size N [8, 9]. Note that even when the convergence to a constant value is very slow, the difference in the value of the clustering coefficients for $N = 5000$ is larger than an order of magnitude, indicating the dramatic change that takes place in the network structure when considering the local rewiring rule.

Using deterministic scale-free networks, Dorogovtsev, Goltsev and Mendes[22] noted that the

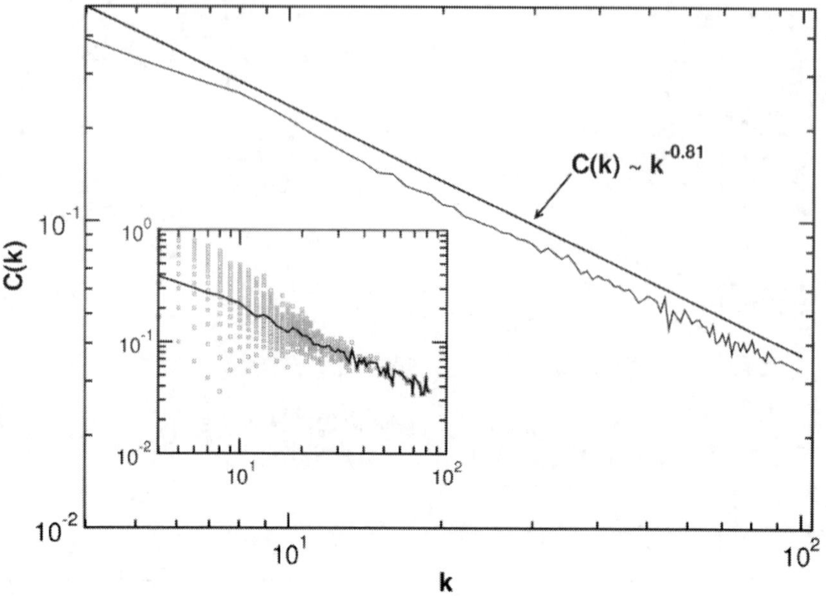

Figure 5. Clustering coefficient $C(k)$ vs. degree k. The straight line is a guide to the eye, and shows the power law behavior $C(k) \sim k^{-0.81}$. The inset shows the clustering for each node (circles) and the average curve $C(k)$ as a continuous line for a single realization.

coexistence of a hierarchy of nodes with different degrees of clustering can be quantified by the scaling law $C(k) \sim k^{-\beta}$ with $\beta = 1$, in contrast to the k-independent clustering predicted by the random network model and the Barabási-Albert scale free model. Ravasz and Barabási [7] also noted that β is not a universal exponent, and may depend on the parameters of the model. In Fig. 5 we present the behavior of C as a function of degree k for a system with $N = 5000$ nodes averaged over ten different realizations of the local rewiring algorithm. We find that the system presents another characteristic of hierarchical networks. In fact, the clustering presents a slow decay, that can be approximated by the power law function $C(k) \sim k^{-0.81}$.

In the inset we show results for a single realization. The clustering C_i for each node i is indicated with a small square, and the average curve $C(k)$ as a continuous line. Note that the dispersion for small k is large, nevertheless a clear decay of C is observed for increasing values of k. Note also that there are no large deviations between the behavior of a single realization and the averaged curve. We stress that similar qualitative results are obtained when the dynamics of the units is given by non-linear phase oscillators [12]. In fact, for the global rewiring rules C vs. N presents a decaying behavior, while it converges to a constant value for the local rewiring dynamics. Also C presents a decaying behavior that can be approximated by a power law for local rewiring dynamics. Again we find evidence that the networks that emerge are robust when the dynamic of the nodes is changed from continuous non-linear phase oscillator dynamics to discrete chaotic maps.

4. Discussion

In this work we analyzed a growing network model with chaotic dynamical units. We showed that a modification of the global rewiring algorithm presented by Gong and van Leeuwen [13] to a local rewiring algorithm can have dramatic consequences in the emerging network structure. In particular, for the global rewiring algorithm the clustering coefficient presents a power law decay as in random networks. The result shows that the network that emerge with this algorithm are not hierarchical. On the other hand, for the local rewiring algorithm the clustering coefficient converges to a constant value as the system size grows, as observed in real hierarchical networks. Also, the coexistence of a hierarchy of nodes with different degrees of clustering is revealed by a power law decay of the clustering as a function of the degree.

It is worth stressing that the qualitative behavior of other network characteristics, such as the degree distribution seem to be robust under the change from global to local dynamics. In both cases wide degree distributions which can be approximated by power laws emerge. Also, we observed that an emerging preferential attachment mechanism is still present.

We have recently analyzed a similar algorithm where both global and local rewiring dynamics are considered in a system formed by continuous non-linear phase oscillators [12]. A comparison with the results obtained in this work shows that even when the spatiotemporal dynamics of these systems are clearly different, the dynamics of the nodes do not seem to play fundamental role in the emergence of hierarchical networks. It is the restriction of the information available to the new nodes from global to local that allows for the emergence of networks with a hierarchical structure.

Ravasz and Barabási [9] have noted that real networks that do not present a hierarchical structure such as the Internet at the router level and the Power Grid network share a geographic organization as a common feature. As a consequence there is a cost associated to the links, which does not appear in other networks with a hierarchical structure. We believe this could be related to the emergent behavior observed for the different rewiring dynamics. In fact the global rewiring dynamics can be interpreted as an optimal rewiring algorithm, since the new nodes can always make the best rewiring available based on information of the whole system. If costs for rewiring were present then the algorithm could allow for the best cost to be chosen, again by using the information available on the whole system.

Finally we note that the local and global rewiring rules of the model correspond to extreme cases. In real systems most probably a combination of both mechanisms will be present. However, we stress that the presence (or absence) of a hierarchical structure will reveal which of the mechanisms dominates the emergence of the network structure.

Acknowledgments P.M.G. acknowledges financial support from project SECTYP-UNCUYO 06/C298.

5. References

[1] Strogatz S H 2001 *Nature* **410** 268
[2] Dorogovtsev S N and Mendes J F F 2002 *Adv. Phys.* **51** 1079
[3] Albert R and Barabási A L 2002 *Rev. of Mod. Phys.* **74** 47
[4] Newman M E J, Barabási A L and Watts D J 2006 *The Structure and Dynamics of Networks* (Princeton University Press) p 624
[5] Boccaletti S, Latora V, Moreno Y, Chavez M and Hwanga D U 2006 *Phys. Rep.* **424** 175
[6] Watts D J and Strogatz S H 1998 *Nature* **393** 440
[7] Barabási A-L and Albert R 1999 *Science* **286** 509
[8] Ravasz E, Somera A L, Mongru D A, Oltvai Z N and Barabási A L 2002 *Science* **297** 1551
[9] Ravasz E and Barabási A L 2003 *Phys. Rev. E* **67** 026112
[10] Zhou C, Zemanová L, Zamora G, Hilgetag C C and Kurths J 2006 *Phys. Rev. Lett.* **97** 238103
[11] Ferrarini L, Veer I M, Baerends E, van Tol M J, Renken R J, *et al.* 2008 *Human Brain Mapping* **30** 2220
[12] Gomez Portillo I J and Gleiser P M 2009 *PLoS ONE* **4(9)**:e6863
[13] Gong P and van Leeuwen C 2003 *Physica A* **321** 679
[14] van den Berg D and van Leeuwen C 2004 *Europhys. Lett.* **65(4)** 459
[15] Gong P and van Leeuwen C 2004 *Europhys. Lett.* **67** 328
[16] Gleiser P M and Zanette D H 2006 *Eur. Phys. J. B* **53** 233
[17] Eguíluz V M, Chialvo D R, Cecchi G A, Baliki M and Apkarian A V 2005 *Phys. Rev. Lett.* **94** 018102
[18] Achard S, Salvador R, Whitcher B, Suckling J and Bullmore E 2006 *The Journal of Neuroscience* **26(1)** 63

[19] Zalesky A, Fornito A, Harding I H, Cocchi L, Yücel M, Pantelis C and Bullmore E T 2010 *NeuroImage* **50** 970
[20] LArge NETwork VIsualization tool
`http://xavier.informatics.indiana.edu/lanet-vi/`
[21] Alvarez-Hamelin J I, DallAsta L, Barrat A and Vespignani A 2005 *Preprint* arXiv.org:cs/0504107
[22] Dorogovtsev S N , Goltsev A V and Mendes J F F 2002 *Phys. Rev. E* **65** 066122

On the Topology of Optical Transport Networks

[1] **J.P. Cárdenas,** [1]**A. Santiago,** [1] **J.C. Losada,** [1]**R.M. Benito and** [2]**M.L. Mouronte**

[1]Grupo de Sistemas Complejos, Departamento de Física, Universidad Politécnica de Madrid, 28040 Madrid (Spain).
[2]Telefónica I+D, c/Emilio Vargas 6, 28043 Madrid (Spain).

E-mail: rosamaria.benito@upm.es

Abstract. Synchronous Digital Hierarchy (SDH) is the standard technology for information transmission in broadband optical networks. Unlike systems with unplanned growth, such as those of natural origin or the Internet network, the SDH systems are strictly planned as rings, meshes, stars or tree-branches structures designed to connect different equipments. In spite of that, we have found that the SDH network operated by Telefónica in Spain shares remarkable topological properties with other real complex networks as a product of its evolution since 1992. In fact, we have found power-law scaling in the degree distribution ($N \cdot P(k) = k^{-\gamma}$) and small-world networks properties. The complexity found in SDH systems was reproduced by two models of complex networks, one of them considers real planning directives that take into account geographical and technological variables and the other one is based in the compatibility among SDH equipments.

1. Introduction

During the last decade many complex systems have been studied from a network-theoretical perspective [1]. The results of this research shown the presence of particular, common and non intuitive statistical properties in these systems when they are abstracted to a network [2, 3]. The Internet, social organizations and metabolic networks are examples of complex systems that display scaling in the distribution of connectivities, transitivity in the organization of nodes and the *small world* phenomena [4].

Taking into account the universality of the named *complex networks* we have developed an exhasutive analysis of an artificial technological system: the Spanish Synchronous Digital Hierarchy (SDH) telecommunication network. The SDH and its North American equivalent SONET are two closely related multiplexing protocols for transferring multiple digital bit streams over the same optical fiber [5]. For the last two decades these protocols have been the standard technology for information transmission in broadband optical networks. The SDH system, developed to replace the Plesiochronous Digital Hierarchy (PDH) system at the end of the 80s, transports different traffic types, such as voice, video, multimedia and data packets (as those generated by IP) over the same fiber wire without synchronization problems. This property, along with others such as its bandwidth scalability, infrastructure scalability (both nationally and internationally), proven operation administration management and provisioning functionality, and robust protection and restoration, are the ingredients for its success [6]. Many studies of SDH networks have been carried out since their inception, most of them focused on the capability of SDH equipments and their adaptability to new scenarios [7]. However few studies as the one developed by us have analyzed SDH systems from a network-theoretical perspective. Our results [8, 9] , developed at national and provincial levels, have shown that the SDH system operated by Telefónica in Spain displays the non-trivial topology previously described for complex networks.

The understanding of the origin of complexity observed in this and other complex systems has emerged as a useful paradigm for describing the structure of their organizations. Under this perspective, in this paper we present part of the empirical analysis of the Spanish SDH and show two approachs to the understand of the mechanisms behind the complex topology observed.

The organization of the paper is as follows. In the next section we present the results obtained from the empirical analysis of the SDH network. The application of two models of complex networks as simulators of SDH systems is shown in the third section. Finally, in the fourth section we sum up the study highlights and present the concluding remarks.

2. Empirical analysis of the Spanish SDH network

In order to administer the use of the fiber infrastructure in an efficient way, SDH networks are end-to-end circuits strictly designed as ring, mesh, bus and other motifs connecting equipments with different link capacities (Mbps/Gbps). Equipments are its basic elements and correspond to three different types [5]: terminal multiplexer, add-drop multiplexer and digital cross-connect. From a hierarchical point of view the SDH network of Telefónica España is divided into four layers with different topologies. The lower layer corresponds to the Access Network and the three upper layers correspond to the Carrier Network. A more detailed description of the network is described in Sec. 3.

Under the perspective of network theory, the SDH network of Spain can be seen as a graph $G_{SDH} = (V, E)$ composed by a set V of nodes and a set E of connecting edges. We define a node of the network as any SDH equipment, as well as any equipment belonging to other technologies connected by optical systems to a SDH equipment. Likewise, we define a link as any Synchronous Transport Module (STM-N) optical system that interconnects two SDH equipments or SDH equipments with equipments belonging to other technologies. It is worth remarking that in our study of the Spanish network all the nodes belonging to any of the layers, either the Carrier Network or the Access Network are taken into account.

The connectivity degree k of a node in a network is defined as the number of nodes directly connected to it. In the case that we are considering in this paper, the degree k of an equipment of the SDH network is defined as the number of STM-N systems directly connected to it. The degree distribution of the network, $N \cdot P(k)$, gives the distribution of the number of equipments with k connections.

In the left part of Fig. 1 we present the $N \cdot P(k)$ distributions for the entire Spanish SDH network and for the provincial network of Sevilla. It can be observed that in the log-log plots both distributions follow a power law. This means that the distributions of the connectivity of the SDH networks follow the function $N \cdot P(k) \sim k^{-\gamma}$ where γ is the scaling exponent. This behaviour reveals a high level of inhomogeneity in the number of STM-N systems connected to the equipments of the SDH network. Thus, it can be noticed that most of the equipments have few connections but there are rare nodes with many connections.

Nevertheless, what should be remarked is that the number of equipments decays asymptotically as the node degree k to the power of a constant γ, denoting the scale free character of the networks. We perform the same analysis for other provincial networks and the results are similar (for more details see references [8, 9]).

In the search of other complex properties in the SDH system, we have carried out an analysis of the distribution of nodes taking into account their number of ports (*i.e.*, different connection capacities). The right part of Fig. 1 shows the distribution of equipments as a function of their number of ports (np) for the two largest SDH provincial networks of Spain: Madrid and

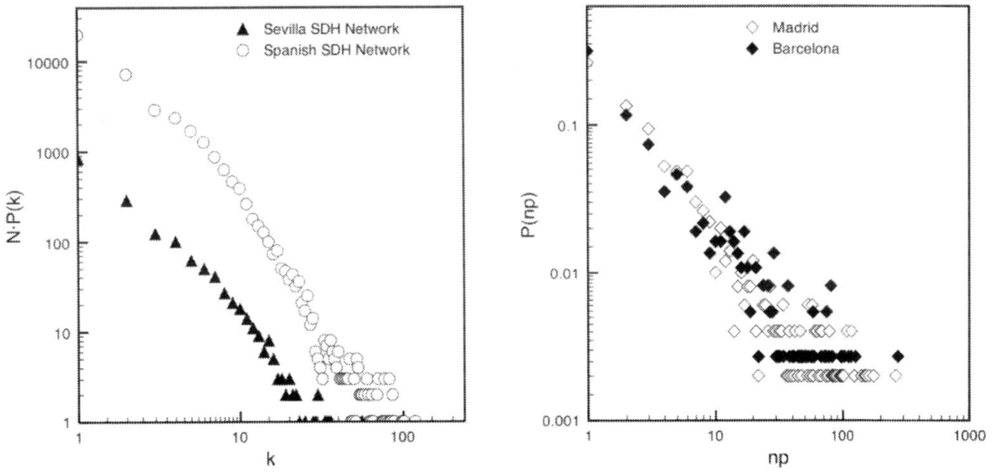

Figure 1. *Left*: Distributions of the number of nodes with k connections, $N{\cdot}P(k)$, in two SDH networks: the entire network of Spain and the provincial network of Sevilla. *Right*: Probability distribution of equipments with different number of ports, $P(np)$, for Madrid and Barcelona.

Barcelona. As in the case of connectivity, nodes with few ports are more frequent than those with many ports and again we have found that the distribution fits a power law, which could be another sign of self-organization in the process of network growth. It should be emphasized that the fitting to a power law is not as accurate as in the case of the degree distributions.

To further characterize the topology of the SDH network, we have also analyzed other statistical properties. We compute the average path length ($\langle l \rangle$) between pairs of nodes in order to investigate the presence of the *small world* property associated to a low distance between nodes. In order to characterize the non-trivial relationships between nodes, we also analyze the clustering of the network that indicates its degree of transitivity. The *small world* networks properties, associated with a high clustering and a low value of the average path length between pairs of nodes, have been observed in Spanish SDH networks. In fact, for the entire Spanish SDH network both values correspond to $\langle C \rangle = 0.04$ and $\langle l \rangle = 11.35$ respectively. It is worth noting that the clustering coefficient is significantly larger than those obtained for random networks of similar size and average connectivity, $\langle C_{rand} \rangle = 0.00005$. The mean value of the path length between nodes is also in good agreement with those of other complex technological networks [3, 2] and close to the value obtained in random networks [10], $\langle l_{rand} \rangle >= 9.75$, where the *small world* property has been reported. The same network analysis was carried out at provincial level, obtaining analogous results (for more details see references [8, 9]).

3. Modeling the topology of SDH systems

With the aim to understand the origin of the complex properties of the SDH system we have developed two network models described in this section. One of the models corresponds to an *ad hoc* algorithm that considers real planning directives using by Telefónica in Spain that take into account geographical and technological variables. The other model, Compatibility Attachment Model (CAM), is a general purpose network model that considers the compatibility among SDH equipments. For a complete description of *ad hoc* and CAM models see references [9] and [11] respectively.

3.1. *Ad hoc model*

The transmission networks of SDH technology are built as hierarchical networks of equipments interconnected through bidirectional links (synchronous links of different speeds). In SDH networks the equipments display different roles depending on their characteristics: transmission capacity, scalability, etc. However, these roles are not rigid, and equipments of the same class or model can play different roles depending on how the network has been built. While obviously every SDH network is built depending on the particular needs of the telecommunications operator to which it belongs, the specific characteristics of the SDH technology mean that the design of the networks follows in all cases similar schemes. Thus these general guidelines can be translated into a growth algorithm of the network model.

In our proposed *ad hoc* model we adopt the four hierarchical levels described in Sec. 2, and based on this hierarchy we distinguish five SDH equipment classes: A (national level), B (regional level), C (interconnection level), D (access level) and E (client equipment connected to SDH equipment by SDH connections).

The proportion of equipments in each class is variable in real SDH networks but tends to remain in a certain range. In the *ad hoc* model we assign to each class a share based on empirical data of real SDH networks collected in 2007.

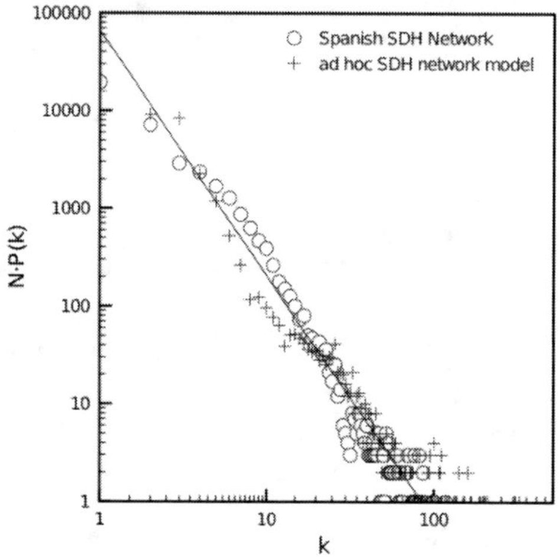

Figure 2. Comparison between the distributions of the number of nodes with k connections, $N{\cdot}P(k)$, obtained empirically for the entire Spanish SDH network and those generated by the *ad hoc* model. The straight line corresponds to the power law fit, $N \cdot P(k) \sim k^{-\gamma}$, for the Spanish SDH network (γ=2.51, R^2=0.95).

We consider links as all the SDH synchronous optical connections with speeds of 155 Mbps or higher. Other properties of realistic SDH connections such as actual traffic volume or traffic capacity have been abstracted away in the model for the sake of simplicity, hence all the connections between equipments are considered equal.

In real SDH networks each equipment is characterized by a number of installed ports that serve to establish SDH connections (see the empirical analysis detailed in Sec. 2), as well as by a maximum number of ports that can be installed. The number of installed and potential ports of an equipment determines the task followed when additional connectivity must be added to the network. If all the installed ports are being used and the maximum capacity has not been reached, then an additional port can be installed in the equipment, otherwise a new equipment

Table 1. Comparison between statistical parameters of the Spanish SDH network and its respective simulation using the *ad hoc* model.

Network	N	$\langle C \rangle$	$\langle l \rangle$
Spanish SDH network	38739	0.04	11.3
ad hoc model	38739	0.08	10.4

must be added to the considered region. In the *ad hoc* model we assume that all the models belonging to a given class are equivalent in these respects, thus we characterize each equipment class by a number of initially installed ports and a number of maximally supported ports.

Concerning geographical location, it is worth noting that each equipment in real SDH networks posses an ID that signals its address. In the *ad hoc* model we consider that the region where the network is deployed is divided into territorial units denominated *provinces*, and each province is divided into lower-level territorial units called *sectors*. Although traffic demand is an important factor behind the growth of real SDH networks we will consider for the sake of simplicity that the entire Spanish territory has a uniform demand and therefore the probability that a SDH equipment is added to any region will be considered uniformly distributed.

Fig. 2 provides a comparison between the real Spanish SDH network and the networks generated by the *ad hoc* model. In both cases the network size is $N = 38739$ nodes and the ensemble size in the numerical simulation is $S = 2000$ networks. Remarkably, the model reproduces the *scale-free* distribution of connectivity. It is worth noting that the distribution generated by the *ad hoc* model differs from the empirically obtained distribution in the range of low connectivity degrees, approximately in the interval $k = [3, 15]$. Despite of the differences the model can reproduce the global topology observed in the real SDH network and predicts pretty well the number of equipments located in low hierarchies (*i.e.* equipment classes E, D and C). Equipments with low degree, $k = 1$ or $k = 2$, classes E and D, that represent around 60% of the equipments present in the SDH network, are in good agreement with the real network. However, equipments with connectivity $k = 3$, classes D and C, are overestimated by the model.

In addition to the reproduction of the scaling in the degree distribution, the networks generated by the *ad hoc* model exhibit the *small world* effect like the real one. Table 1 shows the values of $\langle C \rangle$ and $\langle l \rangle$ obtained with the model in comparison with those obtained empirically from the Spanish SDH network.

3.2. Compatibility attachment model

CAM [11] simulates the growth of a network in a dynamical process that considers the incorporation of a new node at each iteration. The new nodes, in this case SDH equipments, will be linked to the older ones when compatibility exists among them. In the CAM SDH model, two equipments are considered compatible when *the difference between their characters is less than a certain value, given by the compatibility threshold.* The character of an equipment represents the set of attributes that determine its nature and intrinsic properties (*i.e.*, class, model, etc.). This character is time invariant and corresponds to a random number generated by a probability density function (PDF). In the CAM the compatibility always is tested between two equipments (the added one and one of the equipments already present in the network). It should be noted that the new added equipment can be connected with more than one of the old equipments.

We tested the CAM in a regional context. Fig. 3 shows the distribution of number of equipments as a function of their connectivity in the SDH network of Sevilla and in an ensemble of $S = 1000$ networks generated by the CAM.

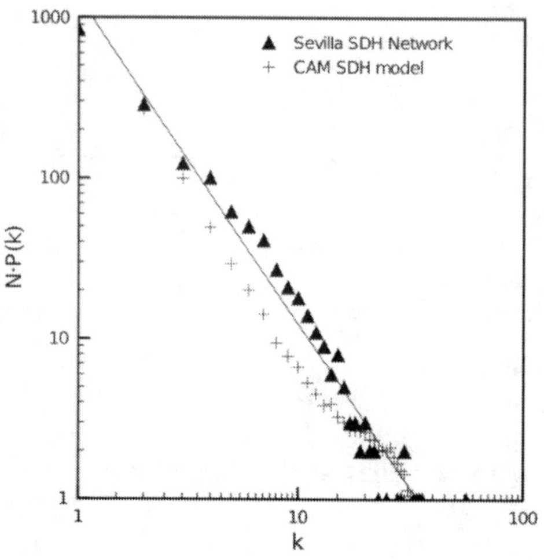

Figure 3. Comparison between the distributions of the number of nodes with k connections, $N \cdot P(k)$, obtained empirically in Sevilla and those generated by the CAM. The straight line corresponds to the power law fit, $N \cdot P(k) \sim k^{-\gamma}$, for the SDH network of Sevilla province (γ=1.98, R^2=0.93).

Table 2. Comparison between statistical parameters of the SDH network of Sevilla and its respective simulation using the CAM.

Network	N	$\langle C \rangle$	$\langle l \rangle$
Sevilla SDH network	1652	0.07	6.2
CAM	1600	0.07	4.3

As can be seen the CAM reproduces the scaling in the distribution of connectivities, as well as other topological properties such as the average clustering, $\langle C \rangle$, and the average path length, $\langle l \rangle$, as can be observed in Table 2.

We have to remark that the statistical properties of the networks generated by both models are in good agreement with the real ones. Despite the similarities, the *ad hoc* model and the CAM SDH differ in the computational complexity (*i.e.*, computation time) of their algorithms. The computational complexity, associated to the generation of networks with different sizes, is much higher in the *ad hoc* model, specially in large simulations [11]. These results indicate that the computational simplicity of the CAM contrasts with the complexity of the *ad hoc* model which has many parameters involved. Though this simplicity enhances algorithm efficiency, it is not suitable when precise regulation of a parameter is required for modelling the growth process of SDH networks.

4. Conclusions

An extensive study of the SDH network of Telefónica España was carried out. We have used the available data in 2008 of Spanish SDH systems with the aim of analysing classical traits for complex networks: scaling in the connectivity distribution, properties of *small world* network and other topological features. The empirical results have clearly shown that the Spanish SDH network presents characteristic traits of complex systems at different scales, both national and provincial level. Irrespective of the network size, the studied networks evidence *small world* network properties and scaling in certain topological metrics such as the connectivity distribution and distribution of equipments as a function of their number of ports. These properties suggest the emergence of complexity in the system evolution.

It should be emphasized that, despite being largely dependent on a strict planning, SDH networks are also dependent on factors such as user demand, costs associated to the connections of a new node and complex intrinsic constrictions. It is therefore such mixture of planning and unpredictable events that would give rise to the observed emergent complex topology of the SDH system. The fact that the *ad hoc* model reproduces quite well the empirically observed properties of the real SDH network evidences the possibility to generate an algorithm that captures these events associated to the network evolution and predicts the future topology of the system. All these events associated to the evolution can be "compressed" into the compatibility concept behind the CAM algorithm. In fact, our results suggest that compatibility between two SDH equipments is sufficient to generate the non-intuitive topology observed. However, the CAM can not be used for design proposals because of its simplicity. Additionally, the CAM generates assortative networks, different to the uncorrelated ones observed empirically and in the *ad hoc* model.

Acknowledgment
This work was supported by the Spanish Ministry of Science, under Projects No. MTM2009-14621 and 'Ingenio Mathematica (i-MATH)' No. CSD2006-00032, and Telefónica I+D Spain.

References
[1] Barabási A -L 2009 Scale-Free Networks: A Decade and Beyond *Science* **325** 412-413
[2] Dorogovtsev S N and Mendes J F F 2003 *Evolution of Networks From Biological Nets to the Internet and WWW* (Oxford: Oxford University Press)
[3] Newman M E J, Barabási A -L and and Watts D 2006 *The Structure and Dynamics of Networks* (Princeton: Princeton University Press)
[4] Watts D J and Strogatz S H 1998 Collective dynamics of Small-World networks *Nature* **393** 440-442
[5] Perros H G 2005 *Connection-oriented networks : SONET/SDH, ATM, MPLS and optical networks* (New York: Wiley)
[6] Choy L 2002 Virtual concatenation tutorial: enhancing SONET/SDH networks for data transport *Journal of Optical Networking* **1** 18-29
[7] Cavendish D, Yun S H, Matsuda O, Murakami K and Nishihara M 2002 New Transport Services for Next-Generation SONET/SDH Systems *IEEE Communications Magazine* **40** 80-87
[8] Cárdenas J P, Benito R M, Mouronte M L and Feliu V 2009 The Effect of the Complex Topology on the Robustness of Spanish SDH Network *Fifth International Conference on Networking and Services* 86-90
[9] Santiago A, Cárdenas J P, Mouronte M L, Feliu V and Benito R M 2008 Modeling the topology of SDH networks *International Journal of Modern Physics C* **19** 1809-1820
[10] Erdős P and Rényi A 1959 On random graphs *Publ. Math. Debrecen* **6** 290-297
[11] Cárdenas J P, Mouronte M L, Benito R M and Losada J C 2009 Compatibility as Underlying Mechanism Behind the Evolution of Networks *Physica A* **389** 1789-1798

Journal of Physics: Conference Series **246** (2010) 012014 doi:10.1088/1742-6596/246/1/012014

Characterizing electrostatic turbulence in tokamak plasmas with high MHD activity

Z. O. Guimarães-Filho[1,2], G. Z. dos Santos Lima[1], I. L. Caldas[1], R. L. Viana[3], I. C. Nascimento[1], and Yu. K. Kuznetsov[1]

[1] Instituto de Física, Universidade de São Paulo, Caixa Postal 66316, 05315-970, São Paulo, SP, Brazil
[2] Equipe DSC/PIIM, Université de Provence UMR6633, F-13397 Marseille cedex 20, France
[3] Departamento de Física, Universidade Federal do Paraná, Caixa Postal 19044, 81531-990, Curitiba, PR, Brazil.

E-mail: `viana@fisica.ufpr.br`

Abstract. One of the challenges in obtaining long lasting magnetic confinement of fusion plasmas in tokamaks is to control electrostatic turbulence near the vessel wall. A necessary step towards achieving this goal is to characterize the turbulence level and so as to quantify its effect on the transport of energy and particles of the plasma. In this paper we present experimental results on the characterization of electrostatic turbulence in Tokamak Chauffage Alfvén Brésilien (TCABR), operating in the Institute of Physics of University of São Paulo, Brazil. In particular, we investigate the effect of certain magnetic field fluctuations, due to magnetohydrodynamical (MHD) instabilities activity, on the spectral properties of electrostatic turbulence at plasma edge. In some TCABR discharges we observe that this MHD activity may increase spontaneously, following changes in the edge safety factor, or after changes in the radial electric field achieved by electrode biasing. During the high MHD activity, the magnetic oscillations and the plasma edge electrostatic turbulence present several common linear spectral features with a noticeable dominant peak in the same frequency. In this article, dynamical analyses were applied to find other alterations on turbulence characteristics due to the MHD activity and turbulence enhancement. A recurrence quantification analysis shows that the turbulence determinism radial profile is substantially changed, becoming more radially uniform, during the high MHD activity. Moreover, the bicoherence spectra of these two kinds of fluctuations are similar and present high bicoherence levels associated with the MHD frequency. In contrast with the bicoherence spectral changes, that are radially localized at the plasma edge, the turbulence recurrence is broadly altered at the plasma edge and the scrape-off layer.

1. Introduction

Electrostatic turbulence is the main cause of the anomalous particle and energy transport at the plasma edge in tokamaks [1]. Usually the standard linear spectral analysis provides information about the waves at the plasma edge and the particle transport driven by these waves [2, 3]. In many tokamak experiments the correlation between the magnetic oscillations and the electrostatic turbulence is not significant [4]. However, this correlation turns to be relevant in new high-confinement regimes achieved in tokamaks [5, 6, 7, 8]. A number of experiments in tokamaks indicate that, under certain conditions, electrostatic turbulence can be modulated by Mirnov oscillations [9, 10, 11, 12, 13]. During L-H transitions in the Joint Experiment Tokamak

(JET) abrupt changes in the electric and magnetic fluctuation spectra have been observed [14]. Sudden changes in the spectra also appeared on the drift flow driven turbulence in the Large Plasma Device (LAPD) [15].

In the Tokamak Chauffage Alfvén Brésilien (TCABR) the magnetic and electrostatic frequency spectra present a peculiar partial superposition, in contrast with most tokamaks, for which such spectra are rather separated [13]. This fact might enhance the coupling, normally small, between these two kinds of fluctuations and opens the possibility of investigating this coupling. Moreover, in some TCABR regimes, the MHD activity increases at different instants of time during the discharge, and reaches high amplitudes with a narrow wave-number spectrum and a well-defined peak on the Mirnov frequency of $\sim$ 13 kHz [13, 16, 17, 18]. This increase of MHD activity occurs for two different kinds of discharge regimes [17]. In the first one, the MHD activity enhancement is observed only after the activation of a biased electrode, while in the second regime, the MHD activity increases spontaneously following changes in the edge safety factor [17]. For both regimes, the electrostatic turbulence is also much affected during the high MHD activity [16, 17, 18]. In previous works, some dynamical tools were applied to study the turbulence alterations. Namely, for the first regime a synchronization of the turbulence on the MHD frequency was observed [18], while the bicoherence enhancement was reported for the second regime [19].

Since magnetic oscillations (MHD activity) cause variations of plasma parameters, including density, temperature, and potential, an influence of these oscillations on the electrostatic fluctuations is naturally expected, depending on the strength of the MHD behavior. Usually, the fluctuating magnetic and electric fields show quite different frequency spectra without synchronization evidences between these two kinds of oscillations, as clearly shown, for example, in Figure 5 of Reference [20]. However, it is possible to distinguish the floating potential fluctuations with a broadband frequencies from those with the MHD frequencies [20, 21, 22]. In this sense, the synchronization in TCABR studied in this article could be considered as a peculiar relation between the high MHD activity and the electrostatic turbulence when the spectra of fluctuating magnetic and electric fields are partially superimposed.

The purpose of this article is to systematically describe dynamical alterations on the electrostatic turbulence observed with high MHD activity in the mentioned second TCABR regime. To do that, we apply some mathematical procedures used to identify dynamical characteristics on the turbulence time series. We use the recurrence quantification analysis and the bicoherence spectral analysis to evaluate the alterations in the radial profiles of the analyzed turbulence data. The turbulence data considered here are the floating potential fluctuations measured by Langmuir probes that will be also referred to as electrostatic turbulence. Initially we present experimental evidence of electrostatic turbulence synchronization at the Mirnov frequency, during high MHD activity in the second regime of the TCABR discharges (characterized by a spontaneous increase of MHD activity).

After that, we introduce an approach based on recurrences to study the turbulence dynamical alterations with high MHD activity, the recurrence quantification analysis (RQA) [23, 24, 25, 26]. This method is based on the so-called *recurrence plot* (RP), which is a plot showing, for a given time, the times at which a state changing its position in phase space, visits roughly the same area in the phase space. Recurrence plots have been introduced as a tool to calculate the maximum Lyapunov exponent related to a time series [27]. Later it has been shown that RPs can also be used to study non-stationarity of a time series as well as to indicate its degree of aperiodicity [28]. The quantitative characterization of recurrence plots, or recurrence quantification analysis (RQA), became an extremely useful tool in nonlinear time series analysis [29], and has been used in many applications ranging from meteorology [30], finance [31], and geophysics [32] to cardiology [33]. In particular, RPs have become important to characterize data from space [34] and fusion plasmas [24, 25, 26].

Turbulence can have recurrent behavior, as periodicities, and irregular variations. The RQA quantifies the recurrent structures in a given time series, indicating states which are again arbitrarily close after some time. This is a fundamental property of nonlinear systems [23]. We use RQA to evaluate the radial dependence of the recurrence alterations on the turbulence. For the turbulence with high MHD activity, we found changes in the recurrence structures of floating potential fluctuations in all the analyzed region. Thus, significant recurrence structures are found both in the plasma edge and in the scrape-off layer during the high MHD activity.

Furthermore, in order to complement the description of the turbulence dynamical alterations during the high MHD activity we apply the bicoherence spectral analysis to quantify the intensity of the electrostatic turbulence synchronization as well as its radial dependence. We observe that the turbulence bispectrum is much modified during the high MHD activity and, consequently, presents the same regularity of the magnetic oscillations with high bicoherence between the dominant Mirnov frequency and the remaining broadband spectrum frequencies. This high bicoherence is most pronounced near the plasma edge position around the rational magnetic surface with the safety factor $q = 3$. The result about bispectral analysis of the turbulence alterations during high MHD activity was partially published in [19], but it was included here in order to point out the differences in the turbulence radial profile alterations that can be detected when using the recurrence-based and the bispectral methods.

The rest of the paper is organized as follows: in the second Section we describe the TCABR experimental setting and the main alterations on turbulence fluctuations during high MHD activity. Section III treats the recurrence plots and the corresponding recurrence quantification analysis, as well as its application to the turbulence measured in the TCABR during ambient and high MHD activity. In Section IV we present the turbulence bicoherence analysis. The last Section contains our conclusions.

2. Experimental setup

The experiments have been done in the hydrogen circular plasma of the Brazilian tokamak TCABR [16] (major radius $R = 61$ cm and minor radius $a = 18$ cm). The plasma current reaches a maximum value of 100 kA, with duration 100 ms, the hydrogen filling pressure is 3×10^{-4} Pa, and toroidal magnetic field $B_0 = 1.1$ T. At the plasma edge the electron plasma density is $n_e \approx 3 \times 10^{18}$ m^{-3}, and the electron temperature is $T_e \approx 10$ eV. The floating electrostatic potential has been measured in the region comprising the plasma edge and the scrape-off layer ($0.9 < r/a < 1.2$) by a set of three movable electrostatic probes. Two of them were placed in two poloidal positions separated by 0.4 cm, yielding the floating potentials, φ_1 and φ_2, at the corresponding positions. The third probe, used to measure the ion saturation current I_S, was placed at a toroidal position 0.4 cm apart from the poloidal probes. The whole set was mounted at the equatorial region of the tokamak. Magnetic fluctuations were detected using a Mirnov coil located at $r = 19.5$ cm (i.e., 1.5 cm outside the plasma radius). Radial profiles of electrostatic fluctuations were obtained on shot-to-shot basis. Probe and Mirnov coil data were digitized at 1.00 and 0.25 megasample per second, respectively, and filtered to avoid anti-aliasing filter.

Figure 1 shows the time evolution of a typical example of the plasma discharge in TCABR that have been considered in this work. The plasma current [Fig. 1(a)] grows rapidly in the first milliseconds until ~ 50 kA, then increases at a slower rate until reaching a short plateau where the current stays at a 100 kA level, decaying slowly during the second half of the discharge. The central chord-integrated electron density, indicated by Fig. 1(b), exhibits a similar evolution, with a first plateau level of $n_e \sim 1.0 \times 10^{19}$ m^{-3} which roughly coincides with the slower current ramp, and is followed by a second plateau of $\sim 1.2 \times 10^{19}$ m^{-3} between 42 ms and 60 ms. During the rest of the discharge the density decreases gently until disruption. It is interesting to observe that during this second (and higher) plateau of the plasma density the magnetic activity is high.

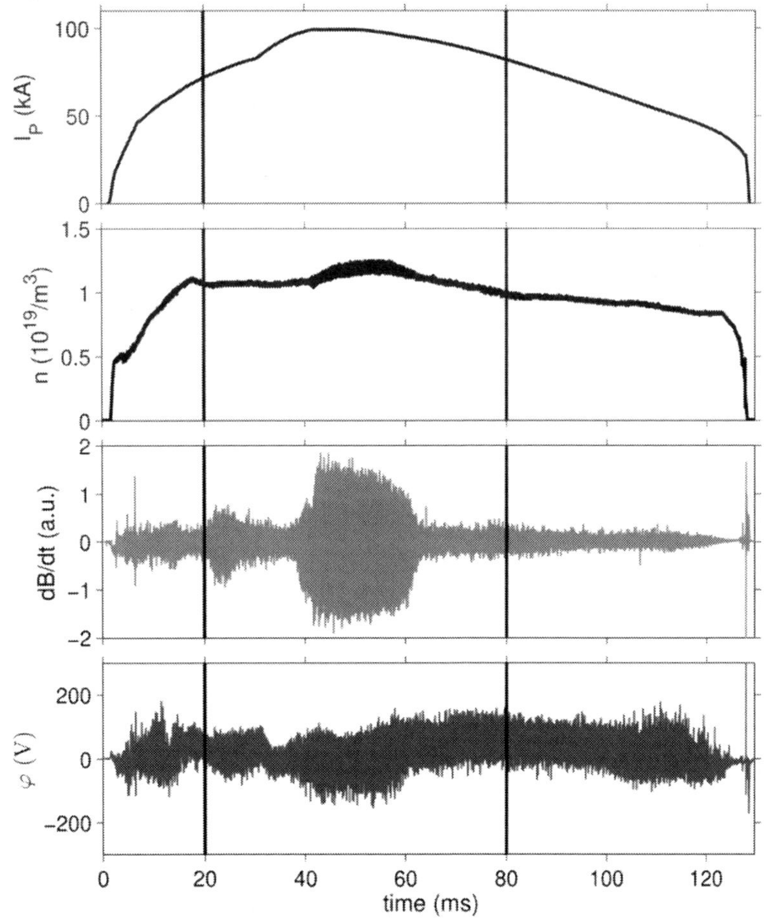

Figure 1. (color online) Time evolution of the (a) plasma current, (b) central chord-integrated plasma density, (c) MHD activity, and (d) floating potential (at $r_S/a = 0.92$) for a discharge (# 18369) of TCABR tokamak with bias polarization.

The MHD activity measured by the coils, or the fluctuations of the poloidal field time rate, $\dot{B}$, are depicted in Fig. 1(c). It is interesting to observe that, during this discharge, the MHD activity grows in two stages. A little increase around 22 ms occurs when the edge safety factor at plasma edge reaches the integer value $q(a) = 4$ (and have dominant poloidal mode $m = 4$). The more strong growth of MHD activity observed around 40 ms has, in fact, two saturated stages: the first one starts to grow at *circa* 38 ms when $q(a) \approx 3$ (and the dominant poloidal mode is $m = 3$ [17] and evolves to a more intense second stage at ~ 42 ms. This second stage, more strong and reproducible, will be analyzed in this work. After these events the magnetic activity is reduced to a noisy level. Fig. 1(d) shows the time evolution of the floating electrostatic potential φ_1 along the discharge. During the burst of magnetic activity started at 42 ms, important dynamical characteristics of the electrostatic fluctuation are strongly affected, as it will be shown later along this article. Moreover we find that, after the magnetic burst ceases, the electrostatic fluctuations resume to their former behavior with respect to these characteristics, but with different mean and standard deviation values.

The Fourier spectra of both magnetic and electrostatic fluctuations are depicted in Fig. 2

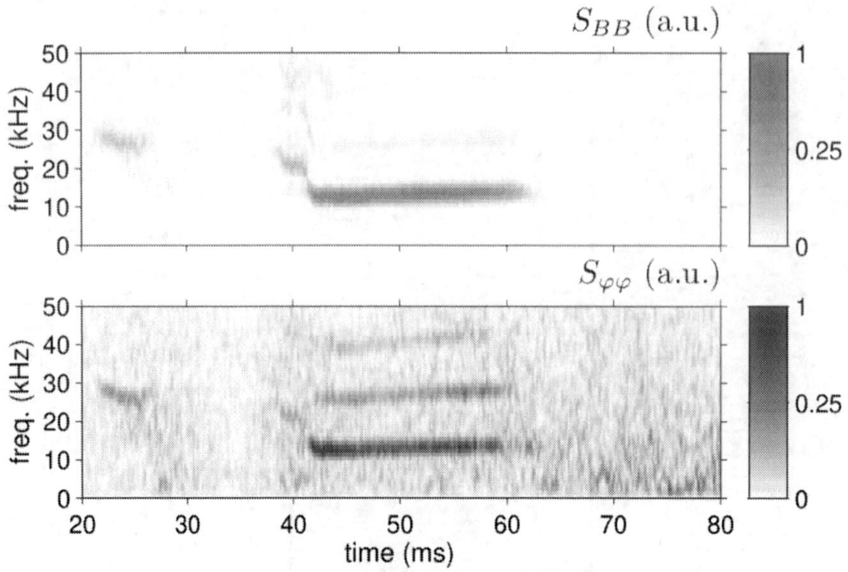

Figure 2. (color online) Fourier power spectral density (arbitrary units) for Mirnov oscillations (a) and electrostatic potential (b) at $r_S/a = 0.94$ during the discharge of Fig. 1.

for magnetic data and electrostatic measurements performed at a fixed radius $r = 17$ cm. In the spectrograms, the oscillation frequencies are plotted against the discharge duration, the corresponding power spectral densities (in arbitrary units) being plotted in a color scale. Figure 2(a) exhibits the spectrogram for Mirnov oscillations, where the periods of magnetic activity are associated with peaks of the power spectral density S_{BB}. In particular, the intense burst started at 42 ms has a sharp line with a dominant frequency of ~ 13 kHz which extends up to 60 ms, with at least two overtones corresponding to higher harmonics of the dominant frequency. The modulation in the electrostatic potential fluctuation can be seen in [Fig. 2(b)], where its spectrum presents peaks (or lines for running time) at the same frequencies of the magnetic fluctuations. It is interesting to note that the MHD activity at 22 ms, related to $q(a) \approx 4$, also modulates the floating potential fluctuations.

In addition, we verified a phase-locking between signals measured at different poloidal positions by computing the phase angle θ_{12} related to the cross-spectrum of the potential fluctuations, $\tilde{\varphi}_1(t)$ and $\tilde{\varphi}_2(t)$, measured at different poloidal positions of the plasma column (separated by 0.4 cm). The spectrogram of this phase angle is exhibited in Fig. 3(a), showing an interval for which the angle is nearly constant for the same base frequency of MHD activity f_{MHD}. Moreover, the spectrogram of phase angle between one of these signals and Mirnov oscillations [Fig. 3(b)] is another indication that this synchronization is driven by the MHD activity. Is important to mention, that similar behavior was also observed for the ion saturation current fluctuations (which are proportional to the local plasma density fluctuations, provided the temperature fluctuations are not taken into account).

The electrostatic turbulence synchronization at the MHD frequency can be characterized in a more quantitative way by considering the so-called order parameter, which is a well-known diagnostic of synchronization in problems of nonlinear dynamics [35, 36]. Let θ_{12} be the phase angle obtained from the cross-spectrum of the signals of the two Langmuir probes measuring

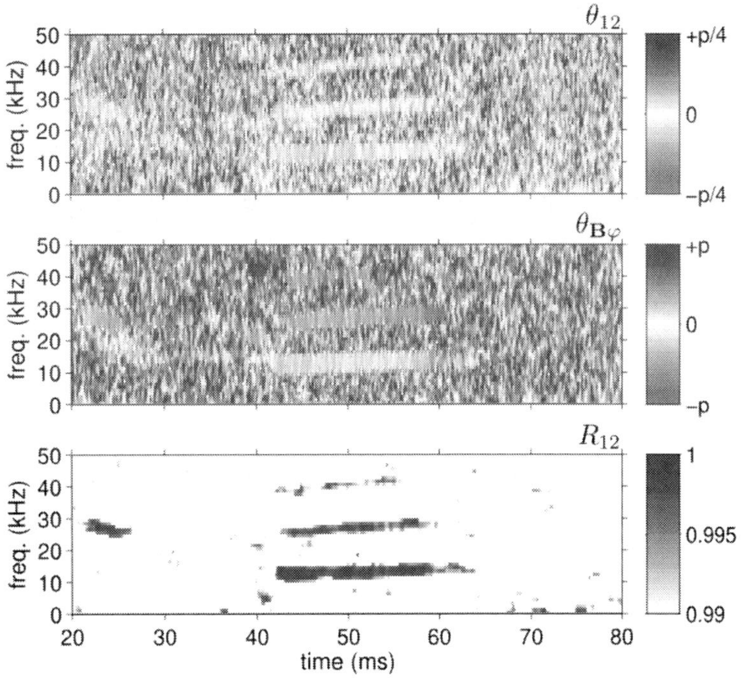

Figure 3. (color online) Spectrogram showing the time evolution of the spectral properties of (a) phase angle between floating potential data sampled at different poloidal locations and $r/a = 0.92$. (b) phase angle between magnetic oscillations and floating potential fluctuations measured at $r/a = 0.92$. (c) Time evolution of the order parameter magnitude computed for the phase angles between floating potential data sampled at different poloidal locations and $r/a = 0.92$.

floating potential, $Vf_1(t)$ and $Vf_2(t)$. The corresponding order parameter is defined as

$$R_{12}(t, f) = \left| \left\langle e^{i\theta_{12}} \right\rangle_t \right|, \tag{1}$$

which is the magnitude of a centroid phase vector, and the time average

$$\left\langle e^{i\theta_{12}} \right\rangle_t = \frac{1}{\Delta t} \int_{-\Delta t/2}^{\Delta t/2} e^{i\theta_{12}(\tau+t,f)} d\tau, \tag{2}$$

is computed over an interval of duration $\Delta t = 1.5$ ms, which is a convenient interval to obtain the order parameter with a good precision (that requires Δt to be some integer multiples of the window size used to compute the Fourier spectrogram), but the interval is small enough to allow analyzing the parameter evolution.

A completely synchronized state is such that the centroid vector is the coherent sum of all phase vectors, thus giving $R_{12} = 1$. Non-synchronized states, in general, are characterized by lower values of the order parameter. A spectrogram of the Mirnov oscillations [Fig. 2(a)] can be compared with that obtained for the order parameter magnitude $R_{12}(f)$ [Fig. 3(c)] of the floating potentials measured at two different poloidal positions. The order parameter allows us to identify the intervals and frequencies for which this synchronization occurs. A clearly noticeable synchronization band appears in the phase spectrogram, reinforcing our previous claims. Note in Fig. 3 that the early growth of MHD activity at 22 ms also leads to a temporary phase synchronization similar to the one discussed before.

3. Characterization of turbulence using dynamical recurrences

3.1. Recurrence plots

Let us suppose that an experiment yields an univariate time series obtained from measurements of a single physical variable $x_i = x(t = i\Delta)$ at a given point of space, where Δ is the inverse sampling frequency, and $i = 1, 2, \ldots N$ (N is the total number of points). An example is the floating potential measured by a probe inserted somewhere within the plasma tokamak edge. We make an embedding using the vectors [37].

$$\mathbf{x}_i = \{x_i, x_{i+\tau}, x_{i+2\tau}, \cdots, x_{i+(d-1)\tau}\}, \tag{3}$$

where d is the embedding dimension and τ is the delay. There are standard methods devised to obtain reliable estimates from both d and τ [38]. For example, a good value of the delay is provided by the first zero of the auto-correlation function of the time series. The embedding dimension, on the other hand, can be estimated from the method of false neighbors.

It is convenient to regard the vectors $\mathbf{x}_i$ as belonging to a d-dimensional reconstructed phase space. Embedding theorems warrant that this reconstructed space reflects qualitative dynamical properties of the actual phase space of the system, provided d is large enough. However, even when d is a small number, we may retain essential and valuable information about the system's dynamics. The time evolution of the vector $\mathbf{x}_i$ can thus be represented by a trajectory in this reconstructed phase space.

We define a recurrence as the event where a trajectory visits roughly the same d-dimensional volume in the reconstructed phase space. A recurrence plot (RP) depicts, at a given instant, the times at which $\mathbf{x}_i \approx \mathbf{x}_j$, where $i = 1, 2, \cdots N$ (N is the total number of points) is represented in the horizontal axis, and j, with the same range, in the vertical axis. More formally, we construct a recurrence matrix by comparing embedding vectors with each others at different times, drawing pixels whenever the distance between vectors is within some threshold ϵ [27]

$$\mathbf{R}_{i,j} = \Theta(\epsilon - ||\mathbf{x}_i - \mathbf{x}_j||), \qquad i, j = 1, 2, \cdots N, \tag{4}$$

where $\Theta(.)$ is the Heaviside unit step function, and $||\cdots||$ stands for some norm, e.g. the Euclidean norm. The RP is thus obtained by assigning a black (white) dot to the points for which $\mathbf{R}_{i,j} = 1$ (0). By construction, the recurrence matrix is symmetric ($\mathbf{R}_{i,j} = \mathbf{R}_{j,i}$), and a point is always recurrent to itself, i.e. $\mathbf{R}_{i,i} = 1$, forming the main diagonal line. Diagonal lines, in general, are noteworthy features of RP, for they are consequences of the deterministic character of the dynamics generating the time series being investigated.

The higher the threshold ϵ, the more points are recurrent to other ones and the density of pixels increases. Hence, in order to warrant clearly distinguishable pixels in a RP, the threshold is usually chosen as a fraction, say 10%, of the time series variance. Moreover, other norms can be used (the maximum norm, for example) with minor modifications. Non-thresholded RPs can also be drawn, for which the step function in (4) is replaced by some smooth function with a color scale. We shall be concerned here, however, only to thresholded RPs.

Recurrences provide a systematic way to analyze the dynamical behavior of systems, even when the equations which govern their behavior are not known in advance. This applies well, for example, to situations involving plasma turbulence due to the extremely large dimensionality of the phase space. Moreover, as a complex system, the tokamak plasma has features of both

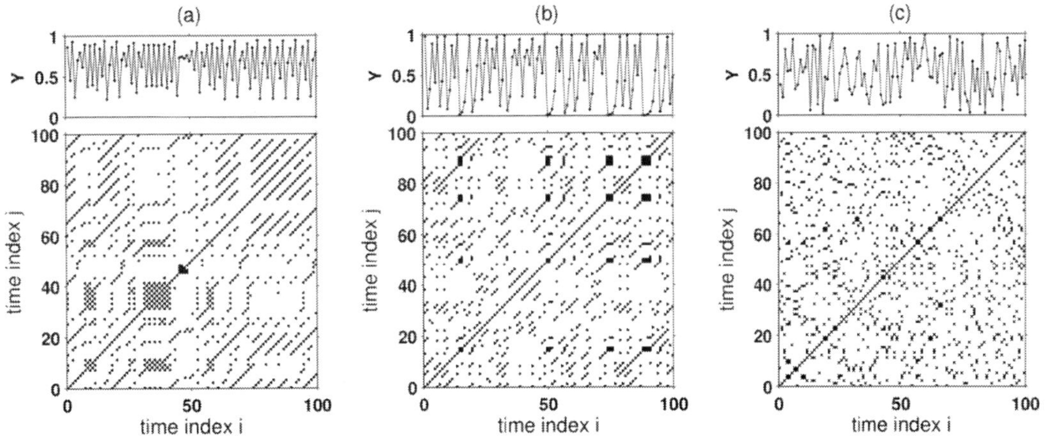

Figure 4. Time series (top) and corresponding recurrence plots (bottom) obtained from the logistic map $x \mapsto ax(1-x)$ with: (a) $a = 3.75$; (b) $a = 4.0$. In (c) we plot white noise obtained from a uniform distribution of random numbers in the interval $[0, 1]$.

deterministic and stochastic behavior, the dynamics alternating among various periodic, quasi-periodic, and chaotic states, with a pervading noise level of variable strength.

As a representative example of the usefulness of RPs for identifying dynamical regimes let us consider the logistic map $x_{n+1} = ax_n(1 - x_n)$, where $x \in [0, 1]$ and $a \in (0, 4]$, which is a well-known prototype of a low-dimensional dynamical system known to exhibit a wealth of dynamical phenomena, like periodic orbits, bifurcations, chaos, and crisis. When $a = 3.75$, for example, the time series for x_n displays an irregular (actually chaotic) dynamics [Fig. 4(a), top]. Its RP reflects the deterministic nature of this evolution, presenting many diagonal lines as well as isolated points which remind us of the irregularity of the time evolution [Fig. 4(a), bottom].

If a is raised to 4.0 in the logistic map, we know in advance that the time evolution is strongly chaotic (with Lyapunov exponent $\ln 2$, indeed), and the relative number of diagonal lines decreases in comparison with the previous case [Fig. 4(b)]. It is completely different, though, from white noise [Fig. 4(c)], which comes from a stochastic (non-deterministic) process (in this case, we considered pseudo-random numbers generated with a uniform probability distribution in the unit interval). In this case, diagonal lines are practically absent and chiefly isolated points are found (any recurrent point here would appear by chance). Hence RPs suggest a criterion for distinguishing chaos from noise in low-dimensional dynamical systems.

Dynamical regimes can be identified in RPs thanks to the existence of characteristic patterns, or structures, made of certain distributions of pixels. There are both large and small scale structures, which form a typology of dynamical behavior. Homogeneous RPs are typical of stationary series, for which the relaxation times are shorter than the series duration, as in a random time series, which is composed of scattered points uniformly distributed over the RP [Fig. 4(c)]. Periodic behavior is characterized by regularly distributed patterns in a checkerboard and diagonally oriented way. Abrupt changes or rare events manifest themselves as white bands, whereas drifted patterns are a fingerprint of non-stationary regimes, where system parameters may have suffered variations along the time series. Chaotic behavior has both structures and isolated points in variable proportions [Figs. 4(a) and (b)].

These features make RPs a useful tool for data analysis even when the time series is short, noisy, and non-stationary, which are commonly observed situations in fusion plasma experiments, specially turbulence data. The usefulness of RPs is further justified if we reckon that most of dynamical diagnostics, like Lyapunov exponents, correlation dimensions, entropies, etc. require that data be long enough, stationary and corrupted with moderately weak noise. Recurrence plots are among the few dynamical diagnostics suitable for plasma turbulent fluctuations.

3.2. Recurrence quantification analysis

Recurrence quantification analysis (RQA) is a set of quantitative procedures to identify and characterize dynamical behavior using RPs, based in three basic kinds of structures: (i) isolated points; (ii) diagonal lines; (iii) horizontal and vertical lines. The first kind are single, or isolated points, which occur if the dynamical states are rare, do not persist for any time, or fluctuate heavily. Diagonal lines are sets of pixels parallel to the main diagonal line of the RP. They happen when finite segments of a trajectory run close to other segments for later times, which is a hallmark of recurrent behavior. Hence it is not surprising that the more diagonal lines are present, the more predictable is the system. The predictability of the further states given the actual state suggests a strong link with the presence of deterministic rules generating the time series. It is important to note that it is possible to generate predictable time series even using stochastic rules, then the link between the diagonal lines in the RPs and determinism is not unique. Finally, horizontal lines (or vertical lines which, due to the symmetry of RP, amounts to the same) occur when a system state does not change or change at a very slow pace, as in laminar states of intermittent behavior.

In order to quantify such structures, RQA provides a variety of numerical diagnostics. Two of the most important ones, namely the *determinism* and the *laminarity*, are described in the following, chiefly because they are used in this work. Several other measures can be found in the literature, and many of them are available in software packages [23].

The density of black recurrence points is a RP (for which $\mathbf{R}_{i,j} = 1$) is the *recurrence rate* (RR), or

$$RR = \frac{1}{N(N-1)} \sum_{i,j=1;i\neq j}^{N} \mathbf{R}_{i,j}, \tag{5}$$

where N^2 is the total number of pixels (black or white) in a RP, for a chosen data length. We remark that the main diagonal points are excluded from the double sum, since each point is recurrent with itself.

Diagonal lines of length ℓ can be formally defined as

$$\mathbf{R}_{i+k,j+k} = 1, \quad (k = 1, 2, \cdots, \ell), \quad \mathbf{R}_{i,j} = \mathbf{R}_{i+\ell+1,j+\ell+1} = 0, \tag{6}$$

and are parallel to the main diagonal line $R_{i,i} = 1$, $i = 1, 2, \cdots, N$. In a RP diagonal lines appear with various lengths, such that we compute $P(\ell) = \{\ell_i; i = 1, 2, \cdots N_\ell\}$, which is the frequency distribution of the lengths ℓ_i of diagonal lines, and N_ℓ is the absolute number of diagonal lines, with the exception of the main diagonal line which always exist by construction.

Since the presence of diagonal lines is usually linked with deterministic behavior, the percentage of points in a RP belonging to diagonal lines are called *determinism*. Accordingly we define the *determinism* (DET) as

$$DET = \frac{\sum_{\ell=\ell_{min}}^{\ell_{max}} \ell P(\ell)}{\sum_{\ell=1}^{\ell_{max}} \ell P(\ell)}, \tag{7}$$

where ℓ_{min} is the minimum length allowed for a diagonal line, whereas the *maximum diagonal length* is $\ell_{max} = max(\{\ell_i, i = 1, 2, \cdots N_\ell\})$.

Since the lengths of the diagonal lines measure locally the time it takes for two recurrent trajectories to depart from each other more than a given threshold ϵ, the diagonal line lengths

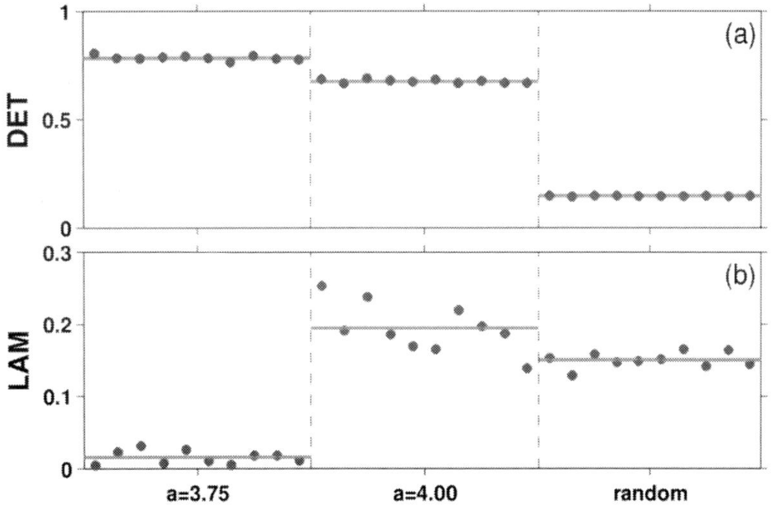

Figure 5. Determinism (a) and laminarity (b) of the time series shown in Figure 4. There were considered ten consecutive and non-overlapping windows of 1,000 points each.

are related to the Lyapunov time, or the inverse of the largest positive Lyapunov exponent of the system. Hence periodic behavior, which has a large degree of predictability, is characterized by long diagonal lines. By way of contrast, a white noise, which has no predictability at all, would exhibit no more diagonal lines than those expected by pure chance. Chaotic systems, having both short and long diagonal lines as well as scattered isolated points, represent an intermediate kind of behavior, as is well-known in nonlinear dynamics.

Vertical lines of length v are defined by:

$$\mathbf{R}_{i,j+k} = 1, \quad (k = 1, 2, \cdots, v), \quad \mathbf{R}_{i,j} = \mathbf{R}_{i,j+v+1} = 0, \tag{8}$$

whose probability distribution function is denoted $P(v) = \{v_i; i = 1, 2, \cdots N_v\}$, where N_v is the absolute number of vertical lines. The *laminarity* (LAM) is the percentage of RP points forming vertical lines

$$LAM = \frac{\sum_{v=v_{min}}^{v_{max}} vP(v)}{\sum_{v=1}^{N_v} vP(v)}, \tag{9}$$

where v_{min} is the minimum lengths of a vertical line, whereas the maximum vertical length is $v_{max} = max(\{v_i, i = 1, 2, \cdots N_v\})$. Despite the vertical lines being related with laminar states of the system, the *laminarity* also can be used to characterize intermittency, since intermittent series are formed by a sequence of laminar states interspersed with bursts of irregular activity.

As an illustration of some of these quantifiers we plot in Figure 5 the values of DET and LAM corresponding to the time series depicted in Figure 4. We computed these quantifiers for ten consecutive and non-overlapping windows with 1,000 points each. The absence of large fluctuations of DET and LAM along the time series is a compelling evidence for the stationarity. Indeed, these series are supposed to be stationary. In the first two cases the series come from a dynamical system (logistic map) with constant parameters, whereas in the third case the white noise has a time-independent mean and variance, which is a symptom of stationarity in stochastic data.

Moreover, the values of DET decrease as the time series goes from weakly chaotic (logistic map with $a = 3.75$) to completely random (white noise) [Fig. 5(a)], confirming that DET

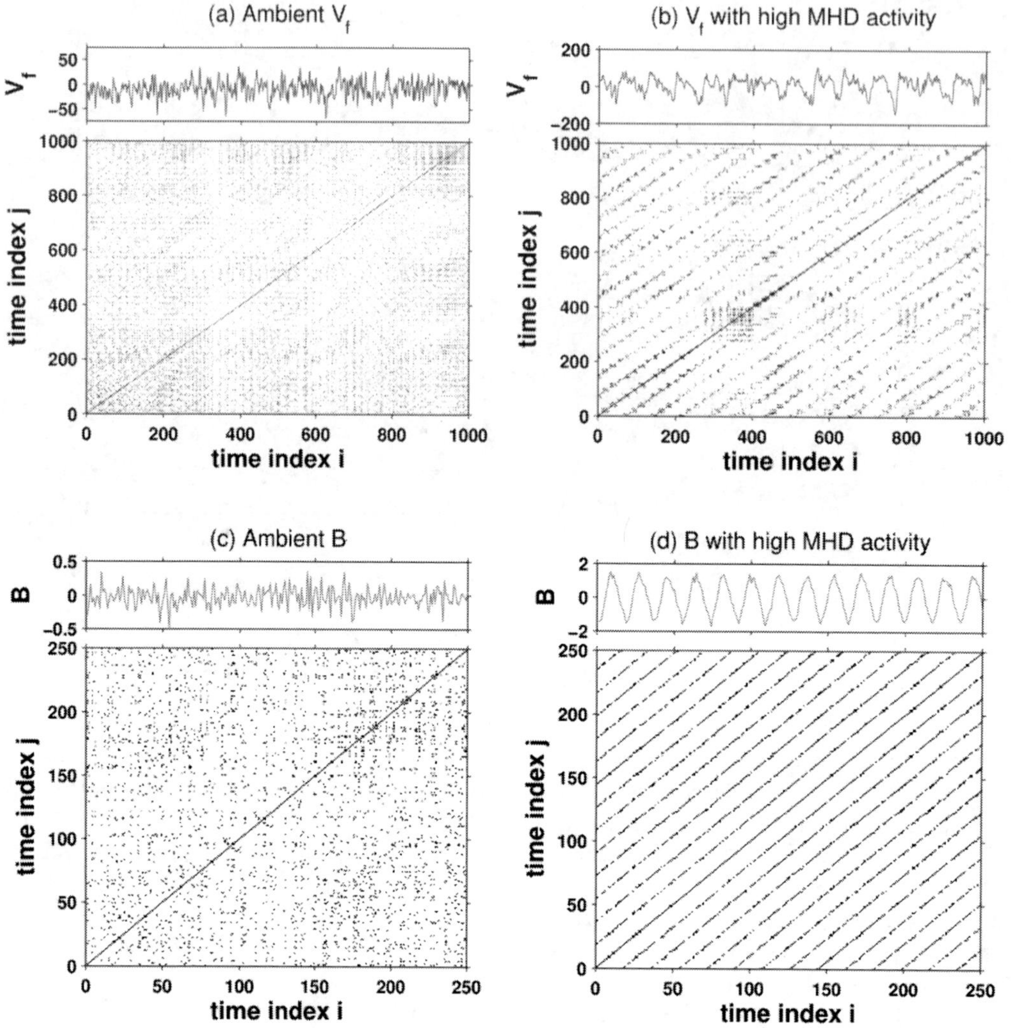

Figure 6. (color online) Recurrence Plots and correspondent time series of floating potential signals measured at $r/a = 0.94$ (for the same discharge showed in Fig. 1) during ambient (a) and high MHD activity (b) time intervals. The correspondents RPs for Mirnov oscillations are displaced in pictures (c) and (d).

measures our ability to detect the underlying determinism of the system generating the time series. A less evident measure is the laminarity LAM, as it does not display the same clearcut dependence on the deterministic character of the system from which the time series stems [Fig. 5(b)]. However, LAM can be useful if we are interested in the laminar behavior of a time series, as in the case when intermittent alternation exist between two states. In this work we mainly focus on DET.

3.3. Recurrence analysis of turbulence data

As we have seen, if the RP shows a cloud of points with a homogeneous yet irregular distribution, then the time series has a pronounced stochastic nature. On the other hand, the formation of

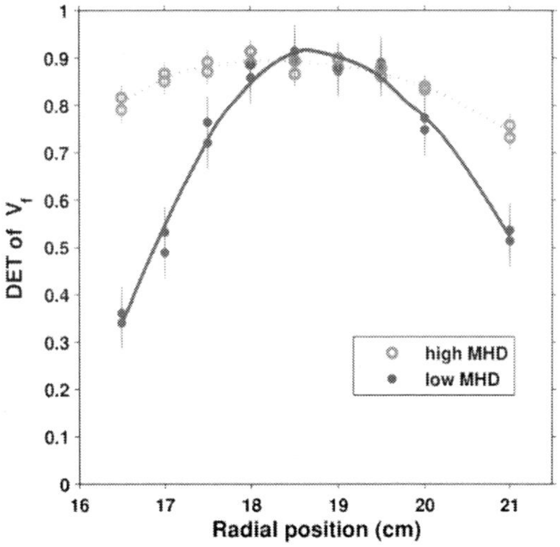

Figure 7. (color online) Radial profile of determinism of floating potential signals during high MHD activity (open red points). The blue points indicates the corresponding values during ambient MHD activity.

patterns in RPs may indicate stationary chaotic behavior. One example of changes in RPs can be seen in Fig. 6 for floating potential and magnetic signals measured at $r/a = 0.94$ (for the same discharge showed in Fig. 1) during ambient [Fig. 6(a),(c)] and high MHD activity [Fig. 6(b),(d)] time intervals. To compute these RPs, we choose the embedding dimension $m = 4$, and the time delay $\tau = 10$ μ s. We also choose the threshold ϵ for each RP to guarantee a fixed global density of black (recurrence) points. Is important to notice that the qualitative description does not depend much on these chosen parameters [24]. The deterministic content increases when we pass from $m = 1$ to 4, but saturates afterwards (see, by example, Figure 5 in [24]. We concluded that $m = 4$ is enough for our purposes. This does not imply that $m = 4$ is an optimal embedding dimension, in the sense we require, for example, for obtaining the correlation dimension through standard procedures [39], but recurrence plots with $m = 4$ already furnish results as good as with other higher-dimension (and time-consuming) embeddings.

While during the ambient MHD activity the RP is dominated by isolated points, during the high MHD activity structures can be observed aligned along the main diagonal in the RP. The structures present in the recurrence plots can be used to characterize the dynamical properties of the underlying time series [23]. One quantitative diagnostic of these structures is the determinism (DET), i.e., the fraction of recurrence points belonging to diagonal lines, which are structures parallel to the main diagonal line. Along a diagonal line, two pieces of a trajectory undergo for a certain time (the length of the diagonal) a similar evolution and visit the same region of phase space at different times. Hence the existence of many diagonal lines is a signature of determinism. The quantity DET is related with the predictability of the dynamical system, because a random process would have a recurrence plot with almost only single dots and very few diagonal lines, whereas a deterministic process has a recurrence plot with very few single dots but many long diagonal lines.

Thus, we perform this analysis and use the DET to estimate the change in the radial profile of the floating potential determinism. Figure 7 shows the DET radial profile for time intervals with ambient and high MHD activity. The obtained profile for ambient MHD activity has a maximum near the plasma radius ($r/a \sim 1.05$) indicating a large deterministic turbulence component in this region. The effect of high MHD activity on the determinism of floating potential fluctuations is to uniformize the DET (in high values) for all analyzed region. In TCABR the scrape-off layer is large, presenting a high density gradient and indications of large scale structures [10, 18]. The increase of the determinism in this region is in favor of deterministic descriptions of plasma edge turbulence based on non-linear coupling of drift-waves, for which density gradients turn out to be an important source of energy. Moreover, the long scale structures are also indicated by the persistence of these signals, characterized by their Hurst exponent, that present a radial profile similar to the radial profile of determinism as shown in Fig. 7 of Reference [40].

4. Bicoherence analysis of turbulence data

The results presented in the previous section suggest the presence of nonlinear effects, as fluctuation couplings, in the analyzed fluctuations associated to high MHD activity. The standard procedure in time series analysis to detect and quantify the amount of mode coupling in a given turbulent signal is the bispectral analysis. Mode coupling results from nonlinear interactions which may be thought of as quadratic, in a first approximation. Thus, two modes with frequencies f_1 and f_2 can interact to generate a third mode with frequency f_3, which is phase-coupled to the other two modes. In this quadratic three-mode nonlinear process the mode frequencies must obey the resonance condition $f_1 + f_2 = f_3$. However, for regular oscillations, the phase differences between three modes can remain constant giving a high bicoherence value without any nonlinear coupling if their frequencies obey the resonant condition.

Let $\phi(f)$ be the Fourier transform of the time signal at the mode frequency f. The bispectrum is defined as [41, 42]

$$\mathcal{B}(f_1, f_2) = \langle \phi(f_1)\phi(f_2)\phi^*(f_3) \rangle , \tag{10}$$

which can be further normalized by introducing the associated bicoherence

$$b^2(f_1, f_2) = \frac{|\mathcal{B}(f_1, f_2)|^2}{\left\langle |\phi(f_1)\phi(f_2)|^2 \right\rangle \left\langle |\phi(f_3)|^2 \right\rangle}, \tag{11}$$

such that $0 \leq b^2 \leq 1$ and, accordingly, can be used as a measure of the amount of bicoherence between the modes in the triplet with frequencies f_1, f_2, and f_3. If the three modes are phase-coupled, there follows that b^2 is near the unity; whereas it is near zero if the modes are barely or no coupled at all.

Since we are quantifying the intensity of mode coupling in both electrostatic and magnetic time series, we shall distinguish them by using the following notations for the bicoherence: $b^2_{\varphi\varphi\varphi}$ and b^2_{BBB}, respectively. We present the bicoherence spectrum (in a grayscale for which darker shades of gray indicate larger values of b^2) as a function of the two frequencies f_1 and f_2 for magnetic and electrostatic signals in Figs. 8(a) and (b), respectively. We have chosen the time window (between 45 and 60 ms) corresponding to the occurrence of the major burst of spontaneous MHD activity [cf. Fig. 1(c)] in order to evidence the coupling between magnetic and electrostatic modes. The most pronounced peak of bicoherence in magnetic oscillations is centered at $f_1 = f_2 \approx 13$ kHz, indicating a strong coupling between these frequencies and a third mode with $f_3 \approx 26$ kHz. Other coupling can be detected in this way along a rough line of $f_2 \approx 13$ kHz. The second peak, centered at $f_1 \approx 26$ kHz, resonates with a third mode $f_3 \approx 39$ kHz and so on.

A similar picture is displayed in the bicoherence spectrum of electrostatic oscillations [Fig. 8(b)]. There is a line of high bicoherence along $f_1 \approx 13$ kHz, with peaks roughly at the same values as in the magnetic bicoherence. These bispectral similarities indicate a coupling between the electrostatic and magnetic fluctuations. The mentioned similarities are related

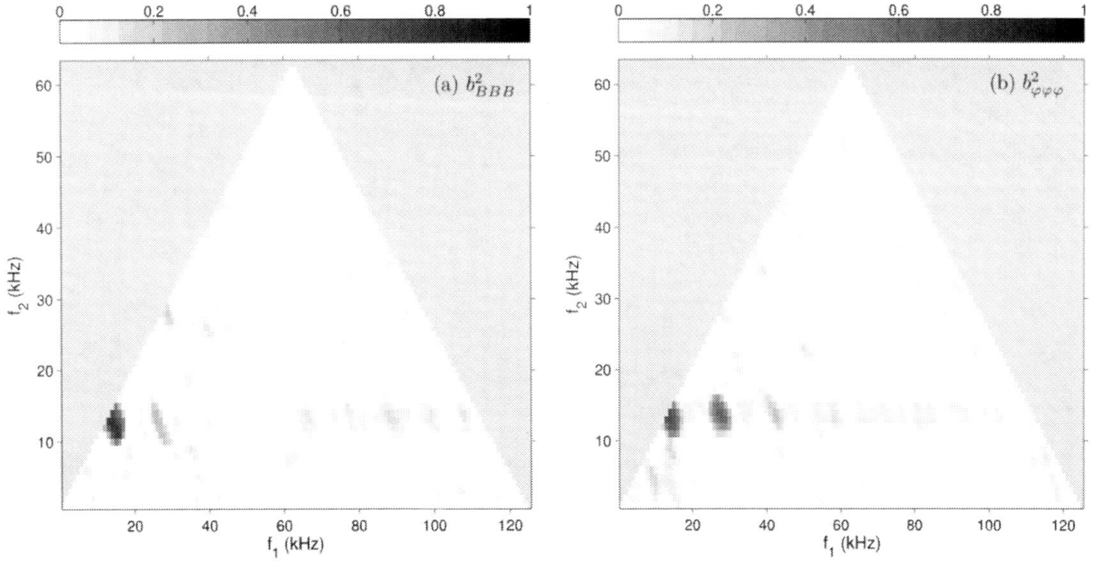

Figure 8. (color online) Bicoherence spectrum of (a) magnetic (b^2_{BBB}) and (b) electrostatic ($b^2_{\varphi\varphi\varphi}$) oscillations during high MHD activity (between 45 and 60 ms).

to prominent peaks in $f_1 = f_2 = 13$ kHz and their harmonics that appear in both magnetic and floating potential bicoherence spectra. A less intense but significative level of bicoherence between ambient turbulence and driven frequency (13 kHz) is seen in the coupling involving the continuous part of bicoherence spectrum with frequencies above the peaks region. In order to investigate the radial dependence of the coupling between electrostatic and magnetic fluctuations we plot in Fig. 9 the radial profile of the maximum values of the bicoherence for floating potential (points) during the high magnetic activity.

According to Fig. 9 the strength of bicoherence for electrostatic potential fluctuation increases as we approach a maximum radius of 17 cm, one centimeter before the plasma column radius, and decreases afterwards to a minimum value at the scrape off layer. The behavior is essentially the same for the ion saturation current measurements [19]. It is remarkable that this radial dependence only occurs during the strong MHD activity, since the corresponding values of the maximum bicoherence without magnetic activity are very low [indicated by a dashed line in Fig. 9]. This may be regarded as a consequence of coupling of electrostatic and magnetostatic oscillations. The radial position of this maximum agrees with the position with the higher value of the synchronization intensity, and is similar to the 3/1 resonant magnetic surface position. However, the radial position of the bicoherence maximum does not coincide with the position of the maximum determinism during high MHD activity (at $r \approx 18$ cm). This difference suggests that the observed recurrence structures are not determined by the mode coupling only.

The foregoing data analysis could be well simulated by a chaotic dynamical system based on three coherent coupled modes with an incoherent noise and a nonlinear periodic driven, which plays the role of a source on the MHD frequency [19]. A better physical interpretation of this forcing would require solving basic equations describing coupled magnetic and electrostatic oscillations under the conditions of the analyzed experiments [43].

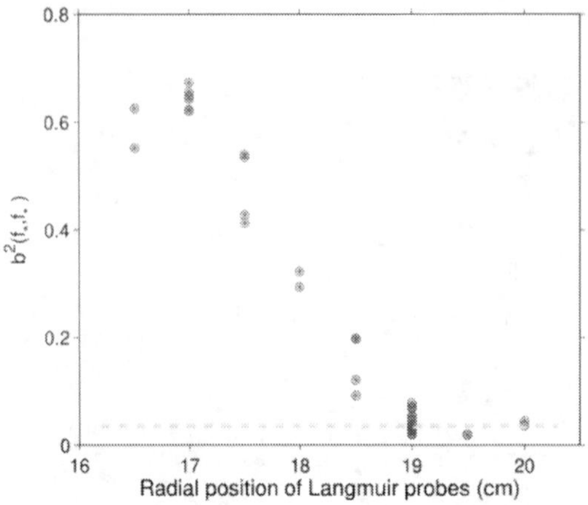

Figure 9. (color online) Radial profile of the maximum bicoherence spectra of floating potential signals (points). The dashed line indicate the corresponding values for periods of ambient MHD activity (reproduced from Fig. 4(b) of Ref. [19]).

5. Conclusions

Previous works have shown that, during TCABR tokamak discharges with high MHD activity, the electrostatic plasma edge turbulence and the turbulence driven particle transport are strongly influenced by the MHD activity at its dominant frequency [10, 16, 17]. This high magnetic activity can be obtained in two different regimes of TCABR discharges. In one regime, the enhancement of MHD activity was observed only after the activation of a biasing electrode while, in the other regime, this enhancement occurs spontaneously [17]. In this article, we considered the second TCABR regime, and reported dynamical characteristics of electrostatic plasma edge turbulence that are clearly changed by the MHD activity enhancement. Namely, in these discharges, (i) high MHD activity and the electrostatic turbulence synchronize at the MHD frequency, (ii) the recurrence structures change broadening the radial profile of plasma with high determinism, and (iii) the bicoherence involving the dominant MHD frequency increases inside plasma nearest to edge.

The recurrence quantification analysis shows that the changes in the recurrence structures can be observed in whole plasma edge region, clearly indicating that these structures are determined by the high MHD activity. On the other hand, by applying the bicoherence analysis to the TCABR Langmuir probe and Mirnov coil data, we identified similar bicoherence spectra with noticeable high values related to the MHD frequency. For the electrostatic turbulence it was possible to obtain the radial profile of the coupling strength, with a local maximum near the estimated 3/1 resonant magnetic flux surface radial position.

In comparing the results of analyses reported in this article we claim that, while the recurrence quantification analysis shows the alterations on the recurrence structures of turbulence, the bicoherence evidences the nonlinear coupling in both magnetic and turbulence fluctuations during high MHD activity. Furthermore, RQA indicates that the structure alterations occur throughout the whole edge region, thus requiring a model to describe the onset and propagation of perturbing structures that consider both the plasma edge and the scrape-off layer regions. The bicoherence alterations, for other hand, are radially localized and gives us indication about the position of the coupling between the magnetic and turbulent fluctuations.

The amplitude dependent synchronization on specific frequencies considered in this article is commonly observed in other experimental nonlinear dynamical systems, as in chaotic electronic circuits perturbed by AC tensions [44, 36, 45]. Recently this kind of synchronization has also been observed in some plasma discharges in linear devices perturbed by time-periodic electric potentials [46, 47, 48]. Thus, the above mentioned observations in TCABR can be interpreted as an indication of an onset of nonlinear coupling between the magnetic and the electric fluctuations during periods of high MHD activity.

We conjecture that a nonlinear coupling modifies the turbulence until it assumes the observed similarities. Thus, we may be observing the turbulence final state and not the turbulence transition driven by high MHD activity. Other interpretations could also be considered to explain the observed spectrum and bispectrum similarities. For instance, an interaction between a saturated nonlinear tearing mode and the turbulence, that modifies the equilibrium so as to include large magnetic islands [3, 49]. Another possibility would be to associate the onset of high MHD activity during some TCABR discharges to the generation of electromagnetic non-normal modes in drift-wave-zonal flow turbulence, as predicted in [50, 51].

Acknowledgements

This work was made possible with partial financial help from FAPESP, CNPq, CAPES, FINEP/CNEN (Brazilian Fusion Network), Fundação Araucária (Paraná), Université de Provence and PACA (Provence Alpes Côte d'Azur) Region. Z.O.G.-F. is recipient of a bourse d'accueil de la Ville de Marseille. The authors acknowledge useful discussions with J. Kurths.

References

[1] Horton W 1999 *Rev. Mod. Phys.* **71** 735
[2] Hidalgo C *et al* 2005 *Nucl. Fusion* **45** S266
[3] Pedrosa M A *et al* 2007 *Plasma Phys. Control. Fusion* **49** B303
[4] Conway G D 2008 *Plasma Phys. Control. Fusion* **50** 124026
[5] Finken K H, Evans T E, Reiter D, Spatschek K H and Suttrop W 2008 *Nucl. Fusion* **48** 024001
[6] Tokar M Z, Evans T E, Singh R and Unterberg B 2008 *Phys. Plasmas* **15** 072515
[7] Schaffer M J, Menard J E, Aldan M P, Bialek J M, Evans T E and Moyer R A 2008 *Nucl. Fusion* **48** 024004
[8] Schmitz O *et al* 2008 *Nucl. Fusion* **48** 024009
[9] Lin H 1991 *Turbulence and transport studies in the edge plasma of the TEXT tokamak* Ph.D. Thesis, Report FRCR/401, University of Texas, Fusion Rechearch Center, Austin, TX, USA
[10] Heller M V A P, Castro R M, Brasílio Z A, Caldas I L and Silva R P 1995 *Nucl. Fusion* **35** 59
[11] Devynck P *et al* 2005 *Plasma Phys. Control. Fusion* **47** 269
[12] Heller M V A P, Caldas I L, Ferreira A A, Saettone E A O and Vannucci A 2007 *J. Plas. Phys.* **73** 295
[13] Heller M V A P, Caldas I L, Ferreira A A, Saettone E A O, Vannucci A, Nascimento I C and Severo J H F 2005 *Czech. J. Phys.* **55** 1572
[14] Sharapov S E *et al* 2008 *Europhysics Conference Abstracts* (35th EPS Conference on Plasma Physics, Hersonissos) eds P Lalousis and S Moustaizis (European Physical Society) vol 32D pp P-4.071
[15] Perez J C, Horton W, Bengtson R D and Carter T 2006 *Phys. Plasmas* **13** 055701
[16] Nascimento I C *et al* 2005 *Nucl. Fusion* **45** 796
[17] Nascimento I C *et al* 2007 *Nucl. Fusion* **47** 1570
[18] Guimarães-Filho Z O, Caldas I L, Viana R L, Heller M V A P, Nascimento I C, Kuznetsov Y K and Bengtson R D 2008 *Phys. Plasmas* **15** 062501
[19] Lima G Z S, Guimarães-Filho Z O, Batista A M, Caldas I L, Lopes S R, Viana R L, Nascimento I C and Kuznetsov Y K 2009 *Phys. Plasmas* **19** 042508
[20] Yan L W *et al* 2005 *Nucl. Fusion* **47** 1673
[21] Nagashima Y *et al* 2005 *Phys. Rev. Lett.* **95** 095002
[22] Pedrosa M A *et al* 2008 *Phys. Rev. Lett.* **100** 215003
[23] Marwan N, Romano M C, Thiel M and Kurths J 2007 *Phys Rep.* **438** 237
[24] Guimarães-Filho Z O, Caldas I L, Viana R L, Kurths J, Nascimento I C and Kuznetsov Y K 2008 *Phys. Lett.* A **372** 1088
[25] Guimarães-Filho Z O, Caldas I L, Viana R L, Nascimento I C, Kuznetsov Y K and Kurths J 2010 *Phys. Plasmas* **17** 012303

[26] Zivkovic T and Rypdal K 2008 *Phys. Rev.* E **77** 037401
[27] Eckmann J P, Kamphorst S O and Ruelle D 1987 *Europhys. Letters* **4** 963
[28] Casdagli M 1997 *Physica* D **108** 12
[29] Webber C L and Zbilut J P 1994 *J. Appl. Physiol.* **76** 965
[30] Marwan N, Trauth M H, Vuille M and Kurths J 2003 *Climate Dynamics* **21** 317
[31] Holyst J A , Zebrowska M and Urbanowicz K 2001 *Eur. J. Phys.* B **20** 531
[32] Kurths J, Schwarz U, Sonett C P and Parlitz U 1994 *Nonlinear Processes in Geophysics* **1** 71
[33] Marwan N, Wessel N, Meyerfeldt U, Schirdewan A and Kurths J 2002 *Phys. Rev.* E **66** 026702
[34] March T K, Chapman S C and Dendy R O 2005 *Physica* D **200** 171
[35] Pinto S E S and Viana R L 2000 *Phys. Rev.* E **61** 5154
[36] Baptista M S, Pereira T, Sartorelli J C, Caldas I L and Kurths J 2005 *Physica* D **212** 216
[37] Takens F 1980, in *Dynamical Systems and Turbulence* eds D A Rand and L S Young, Springer Lecture Notes in Mathematics, vol 898 (New York: Springer)
[38] Kantz H and Schreiber T 1997 *Nonlinear Time Series Analysis* (Cambridge: Cambridge University Press)
[39] Grassberger P and Procaccia I 1983 *Physica* D **9** 189
[40] Rodrigues Neto C, Guimares-Filho Z O, Caldas I L, Nascimento I C and Kuznetsov Y K 2008 *Phys. Plasmas* **15** 082311
[41] Ritz C P *et al* 1988 *Rev. Sci. Instrum.* **59** 1739
[42] Milligen B P, Sánchez E, Estrada T, Hidalgo C , Brañas B, Carreras B and García L 1995 *Phys. Plasmas* **2** 3017
[43] Beyer P, Garbet X and Ghendrih P 1998 *Phys. Plasmas* **5** 4271
[44] Baptista M S, Silva T P, Sartorelli J C and Caldas I L 2003 *Phys. Rev.* E **67** 056212
[45] Baptista M S, Garcia S P , Dana S K and Kurths J 2008 *Eur. Phys. J.* **165** 119
[46] Ticos C M, Rosa Jr E, Pardo W B , Walkenstein J A, Monti M 2000 *Phys. Rev. Lett.* **85** 2929
[47] Davis M S, Nutter N G , Rosa Jr E 2007 *Int. J. Bifurcat. Chaos* **17** 3513
[48] Schröder C, Klinger T, Block D, Piel A , Bonhomme G , Naulin V 2001 *Phys. Rev. Lett.* **86** 5711
[49] Militello F, Waelbroeck F L, Fitzpatrick R and Horton W 2008 *Phys. Plasmas* **15** 050701
[50] Camargo S J, Tippett M and Caldas I L 1998 *Phys. Rev.* E **58** 3693
[51] Horton W, Correa C, Chagelishvili G D, Avsarkisov V S, Lominadze J G, Perez J C, Kim K H and Carter T A 2009 *Phys. Plasmas* **16** 092102

Journal of Physics: Conference Series **246** (2010) 012015 doi:10.1088/1742-6596/246/1/012015

Effect of turbulent fluctuations on the behaviour of fountains in stratified environments

D Freire[1], C Cabeza[1], S Pauletti[1], G Sarasúa[1], I Bove[1], G Usera[2], Arturo C Martí[1]

[1] Instituto de Física, Universidad de la República, Igua 4225, 11400 Montevideo, Uruguay

[2] Instituto de Mecánica de los Fluidos, Facultad de Ingeniería, Universidad de la República, CC 30, 11000 Montevideo, Uruguay

E-mail: marti@fisica.edu.uy

Abstract. The interaction between a turbulent fountain and its stratified environment was studied. A heavy fluid, cold water, was injected vertically upwards into a linearly stratified medium. The round heavy-fluid jet reaches a maximum height before it begins to fall due to the effect of gravity. Because of the effects of friction and mixing, the vertical momentum and density of the jet fluid decrease as it submerges to an intermediate height of zero buoyancy. At this point, the jet fluid spreads as a horizontal front, intruding into the stratified environment. The degree of fluctuation in the proximity of the injection point was studied under both unrestricted- and restricted-flow configurations at the injection, using two differently sized stainless-steel woven-wire screens at the injection port as flow-restricting means. Using visualization and velocimetry techniques, both maximum and spreading heights were found to decrease with increasing turbulence at the point of injection.

1. Introduction

Frosts significantly impact agriculture and food production worldwide. Broadly, frosts may be classified as either advection or radiation frosts. Advection frosts result from the passage of a cold, dry polar air mass over a delimited area. Although these frosts have devastating effects, they are extremely infrequent in temperate regions. In contrast, radiation freezing occurs in temperate regions —on clear, still winter or spring nights— tens of times yearly. A high frequency of radiation frosts results in considerable economic losses due to crop failure, the late spring frosts being especially harmful. The damage resulting from freezing depends on both exposure time and temperature. Temperatures below a sharp limit result in severe crop damage, with nearly all the vegetable tissues of temperate crops being virtually damaged within a couple of hours at -2 °C [1].

Under radiation frost conditions, the Earth's surface cools by radiation through the atmosphere, which is semitransparent to long-wavelength electromagnetic radiation. Once the surface temperature has decreased below that of the surrounding air, the latter cools down by radiation and, to a lesser extent, by turbulent convection and molecular conduction. The lower layers of atmospheric air — within a few tens of meters of the ground— are subject to cooling at a rate decreasing with altitude. Under these conditions, air temperature increases with altitude through the lower atmospheric layers. Moreover, the stratified structure of lower atmospheric air is stabilized as the lowest layers are the coldest and therefore the densest ones. In this situation, referred to as temperature inversion, the atmosphere can be ideally represented as a vertical succession of strata —*i.e.* horizontal layers of atmospheric air— of increasing density with increasing proximity to the Earth's surface.

Several strategies, such as the installation of heaters, sprinklers, and even the use of helicopters, may be adopted to reduce damages resulting from radiation frosts; yet, all these solutions have serious drawbacks. Another commercially available alternative, the Selective Inverted Sink (SIS) [2] operates by selective withdrawal of the densest (coldest) air layer. This device ejects cold air upward through a source nozzle, supplying the required mechanical power. The cold-air jet not only expels (air from the earth's surface) but also entrains and suctions fluid from the horizontal layers it runs through. It is estimated that the volumetric flow-rate of the jet doubles every three orifice diameters away from the nozzle [3].

Although the usefulness of the SIS may come as a surprise, its suitability for the mitigation of radiation frosts has been advocated [1,2]. With a view to contributing to the improvement of SIS design, a laboratory model was here studied. A relationship was established between the maximum height reached by an upward jet of cold air in a temperature-stratified medium and the turbulence structure of the jet stream.

Turbulent fountains. A fountain is usually defined as a jet in which the buoyancy force acts —as a result of the temperature difference between the jet fluid and the environmental fluid— in the opposite direction to that of the jet velocity. In view of the multiple applications of light-fluid jets — *i.e.* where the density of the jet fluid, at ground level, is lower than that of the environmental fluid—, they have been more widely studied than heavy-fluid jets. However, several important applications of the latter include selective withdrawal, desalination plants, and the replenishment of magma chambers [4-9]. The here-studied SIS model also constitutes an example of heavy-fluid jets.

The upward ejection of dense fluid through a nozzle placed at the base of a tank produces the desired fountain [5-7]. The fluid inside the container has a stable stratification and a density lower than or equal to that of the jet fluid. The jet entrains the environmental fluid into the initial upflow, and the fountain radius increases as the jet fluid density decreases with height. The momentum of the rising fluid decreases as a result of the opposing buoyancy force until it comes to a halt at an initial height. The downflow forming at the upper end of the fountain reduces the fountain height to a final value — on account of turbulent interaction with the upflow— and mixes with the environmental fluid.

In a homogeneous environment, the falling fluid always remains denser than the environment, and the flow spreads across the base of the tank. However, in a stratified medium the final density of the downflow depends strongly on the strength of the environment stratification. Any entrainment of environmental fluid from intermediate strata reduces the density of the downflow to equal that of the environment at some point at an intermediate height. At this point, the downward flow still has some momentum, so a small overshoot is formed before it finally intrudes into the environment. The

Journal of Physics: Conference Series **246** (2010) 012015 doi:10.1088/1742-6596/246/1/012015

thickness of the resulting outflow is comparable to the spreading height near the fountain axis but quickly decreases with increasing radial distance.

Whenever a jet fluid has negative buoyancy at the base of a tank containing a stably stratified fluid, the behaviour of the fountain flow lies between that of homogeneous and stratified media. In this work, experimental results describe the dependence of the jet evolution on the turbulence level at the point of injection. A description of the jet flow obtained by PIV is also provided.

2. Experimental setup and visualization techniques

The experimental setup shown in Fig. 1, comprises a prismatic container whose lateral walls consist of 1.5 cm-thick acrylic panes of 40 cm in width and 100 cm in height. The bottom and top walls consist of thermostated aluminium plates of 1.0 cm in thickness.

In order to ensure a constant inlet flow, feed water was discharged from an elevated tank, keeping the feed water level constant by means of an overflow system (Fig. 1, top left). Feed water, running by gravity through a pipe from the feed tank to the inlet port, was injected through a nozzle located at the container bottom. The water flow-rate at the inlet was measured (McMillan 101-8 flow meter) and controlled by means of a valve. The container was initially full of water. A vent was fitted in the top wall to allow the outflow of excess water from the container.

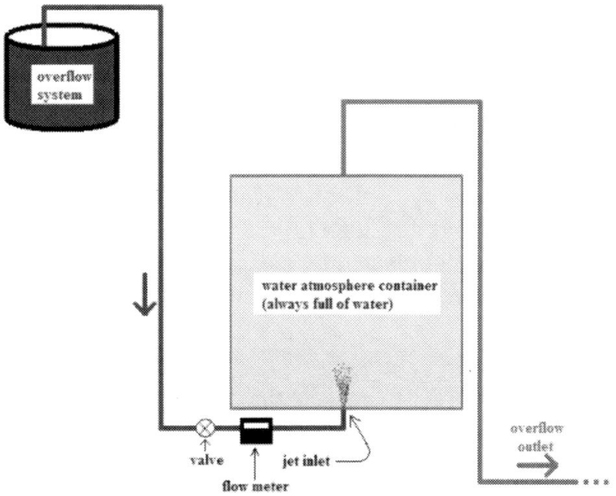

Figure 1: Experimental setup

The fluid within the container consisted of linearly stratified water at 15 ºC at the base and 40 ºC at the top. In order to obtain such stratification, different intermediate water layers were carefully heated by means of a sliding resistance. Various thermocouples were used to monitor the temperature of different water layers.

The density profile of the resulting stratified medium was allowed sufficient time (*c.a.* 2 h) to stabilize. The temperatures of the top and bottom walls were kept constant (by means of electrical resistance and external cooling water circuit). The lateral walls of the container were insulated with 10.0-cm deep panes of expanded polystyrene. Work conditions at the start of the run and the respective uncertainty values are shown in Fig. 2. The medium stratification fitted a linear model.

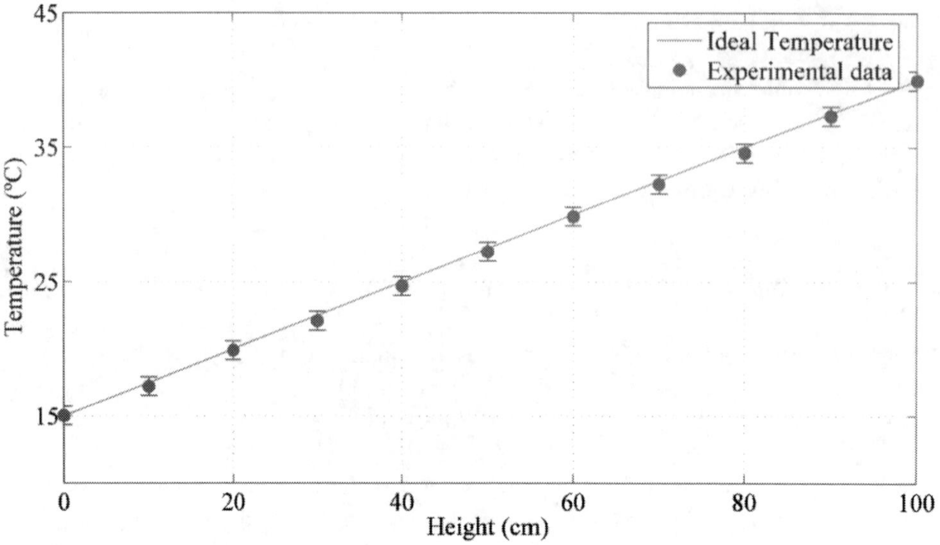

Figure 2: Characterization of the stratified environment

Jet water at 15 °C was injected through an inlet port of 8.0 mm diameter located in the centre of the base wall. Different degrees of turbulence were generated by means of a set of stainless-steel woven-wire mesh screens placed at the inlet port. Three situations have been studied: a) unrestricted flow at injection point; b) restricted flow at injection point by means of 40-mesh screen with wire diameter of 0.22 mm; and c) 80-mesh screen with wire diameter of 0.18 mm. The flow rate was kept at 0.30 $L.min^{-1}$ during all runs.

Digital Particle Image Velocimetry (DPIV) was used to determine the velocity field and characterize the turbulence intensity, by application over a small area immediately above the point of injection. Both the environmental and the jetted fluid were appropriately seeded with 50 μm diameter polyamide particles. The fluid was illuminated by means of a 500 mW green laser sheet of 2.0 mm in thickness. Images were captured by a PULNIX (TMC-6740GE) camera at 500 fps and processed in MatLab using a standard algorithm.

A dying technique was used in order to analyze the overall motion of the fluid, and to visualize its intrusion into the stratified medium, as well as measure the height at which the fluid starts to spread laterally (point of zero buoyancy) after reaching its maximum altitude. The images were captured under uniform illumination by means of a PIXELINK (PL-A741) camera.

3. Turbulence generation

Once the jet passes through the different mesh screens, the flow develops appreciable disorder. In order to study these fluctuations, several images were taken within a small region in close proximity to the point of injection. Images of the jet flow were taken along the length of its axis, starting a couple of centimetres from the injection point. A plot of mean vertical velocity profiles at different heights is shown in Figure 3 for the three studied configurations. Mean velocity was determined over 4-second intervals.

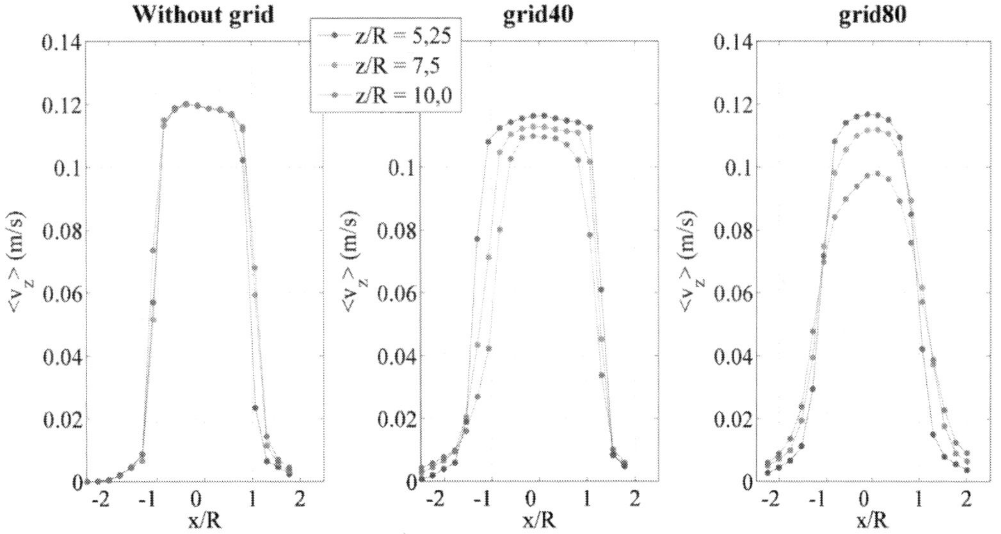

Figure 3: Mean vertical velocity profiles at different heights as a function of non-dimensional radial position under three different configurations

Standard deviations of the velocity field (σ) were calculated in order to analyze the velocity fluctuation (μ). In Figure 4, a plot of the vertical component of the velocity fluctuation (μ_z) is shown at point z/R = 5.7 and x/R = 0.12 for the three configurations. A quantitative difference was observed in the fluctuation pattern of the 40-mesh and 80-mesh screen configurations. However, no significant differences were found between the unrestricted flow and 40-mesh configurations.

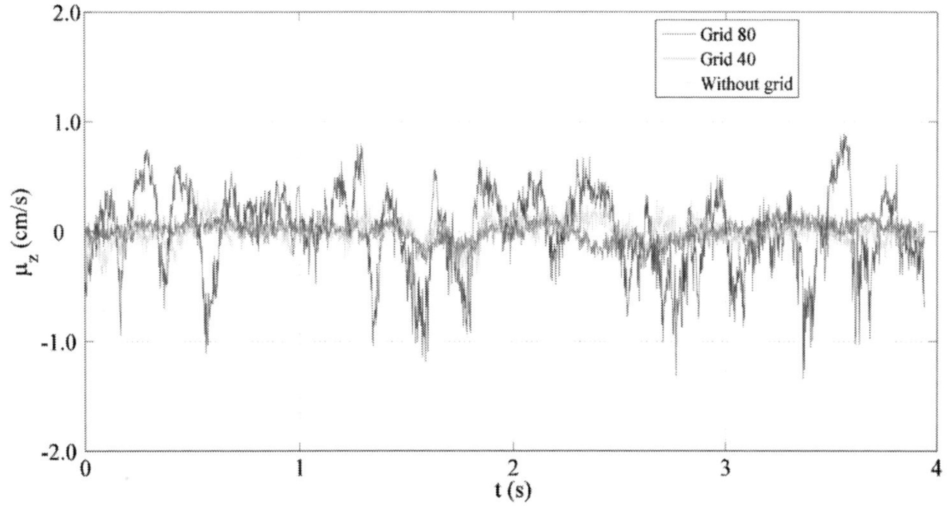

Figure 4: Fluctuation of the vertical component of the velocity (μ_z) at point z/R = 5.7 and x/R = 0.12 as function of time under three flow configurations at the jet source

A colour map of the standard deviation of the vertical component of the velocity σ_z over the capture region is shown in Fig. 5 for the three configurations studied. Asymmetries in the boundaries between the jet and the environment may be ascribed to imperfections in the mesh screen or to external fluctuations in the pipes upstream of the inlet. Such asymmetries, appearing as a result of the transition from laminar to turbulent flow, led to different fluctuation patterns in repeated runs.

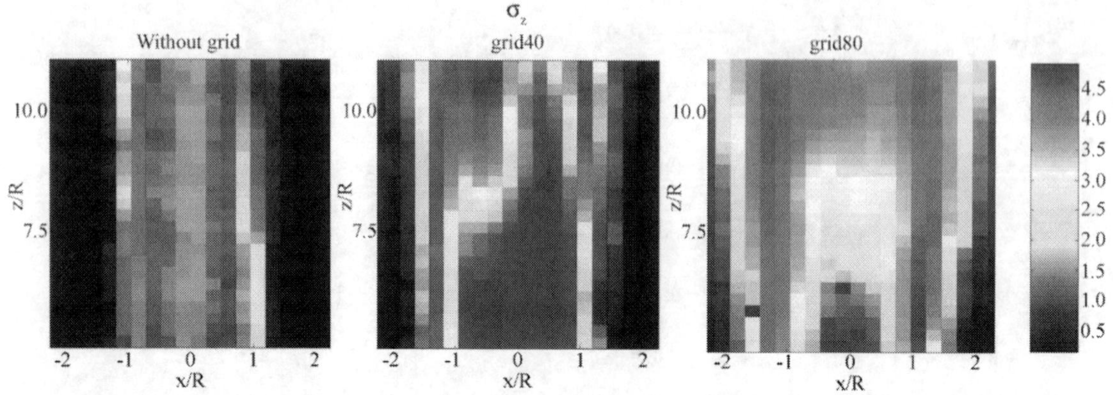

Figure 5: Standard deviation of the vertical velocity component σ_z under the different flow configurations at the injection point

Figure 6 shows the standard deviation σ_z of the vertical velocity component σ_z at different heights at the horizontal position x/R = 1.2. The standard deviation increased with height under the 40-mesh and 80-mesh configurations, and was consistently the greatest for the 80-mesh configuration. Using the 40-mesh configuration, fluctuations were within a range close to the range of fluctuations obtained under unrestricted flow at the injection point. Using unrestricted flow, the standard deviation of the velocity was, in good approximation, nearly constant.

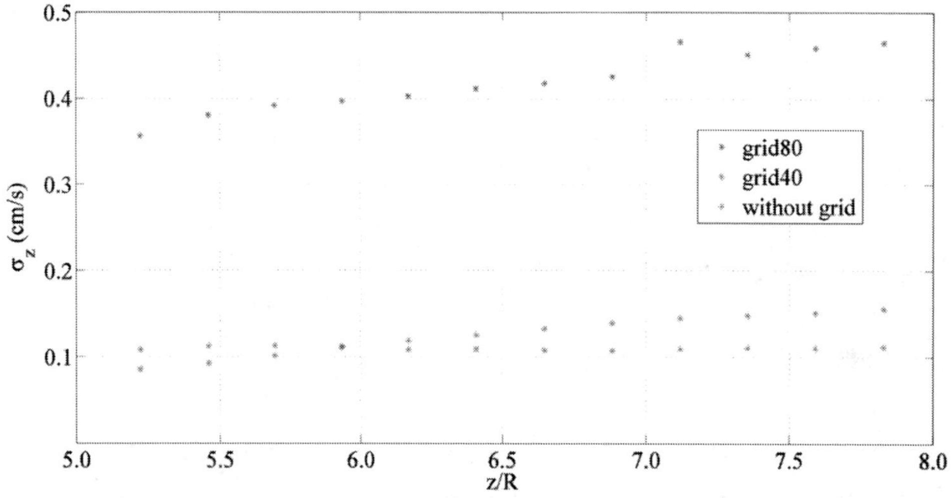

Figure 6: Standard deviation of v_z as function of the non-dimensional height z/R at x/R = 0.12

4. Characterization of the jet in the stratified medium

The evolution of a turbulent jet —dyed with KMnO4— injected into a stratified medium was studied for the unrestricted-flow and the 80-mesh restricted-flow configuration.

Images of the entire container volume were taken by means of a CCD camera at 4 fps under bright, uniform illumination. Four different stages of the jet development are shown in Figure 7. Initially, the jet intruded into the stratified environment (Figure 7a) and became less dense as it warmed. It continued to rise until it reached its maximum height (Figure 7b), whence it submerged to an intermediate height at which its density (temperature) equalled that of the environment (Figure 7c), before finally spreading laterally (Figure 7d).

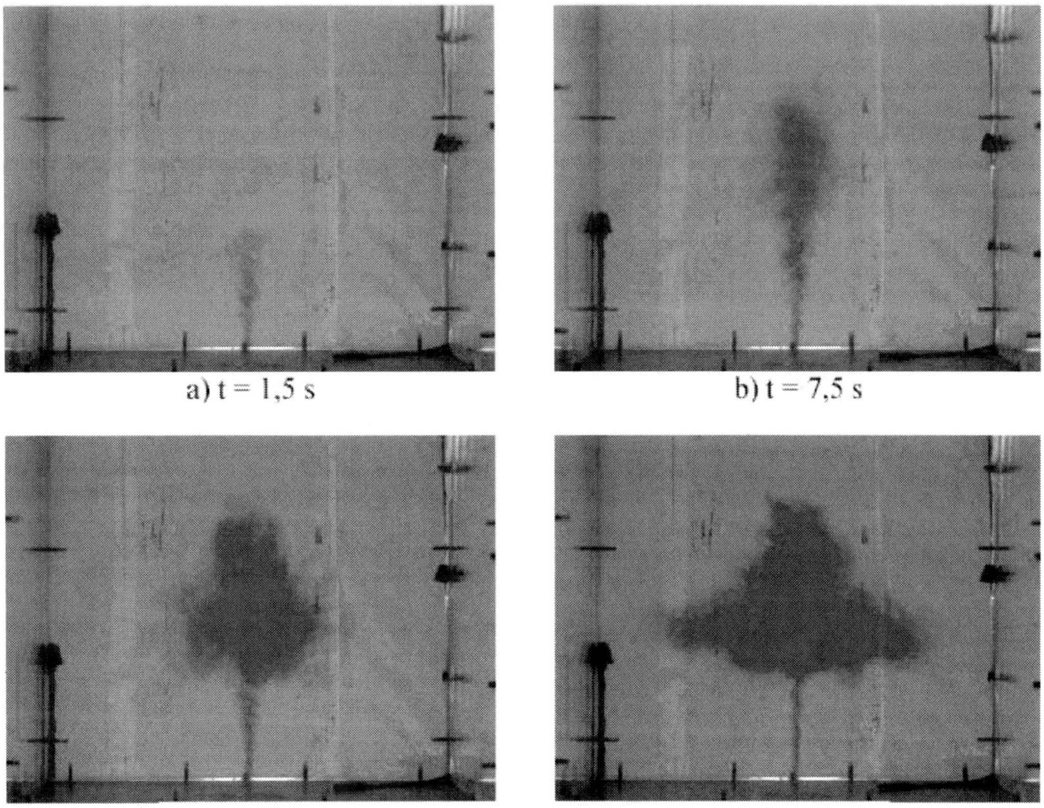

a) t = 1,5 s b) t = 7,5 s

c) t = 20 s d) t = 30 s

Figure 7: Jet evolution a) upon injection; b) at near-maximum height; c) in downward flow; d) spreading laterally at an intermediate zero-buoyancy height

Figure 8 shows the maximum height reached by the jet as a function of time for the unrestricted flow and 80-mesh configurations. Despite the similar behaviour of the two configurations, lower maximum height values were reached in the latter case. The intermediate height at which the downward flow intrudes into the environment was also studied. The evolution of the spreading height as function of time is shown in Figure 9.

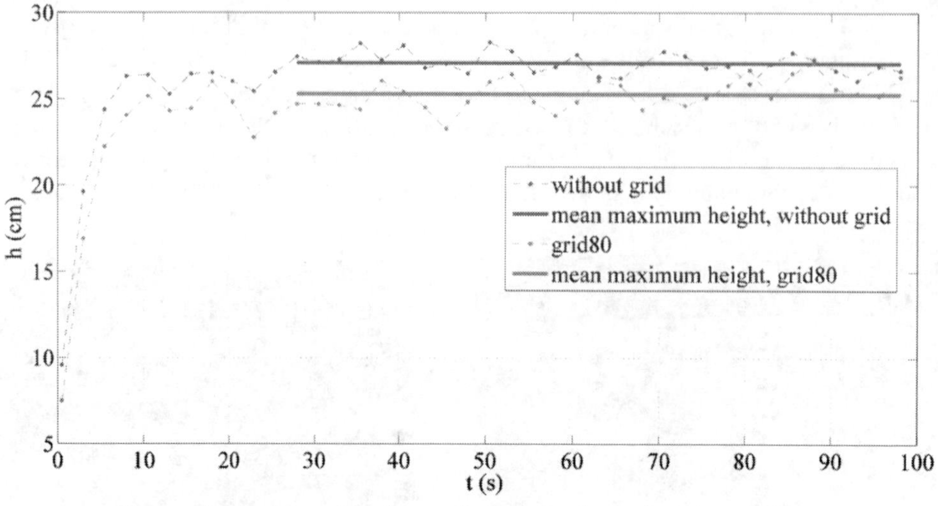

Figure 8: Maximum height reached by the jet as function of time

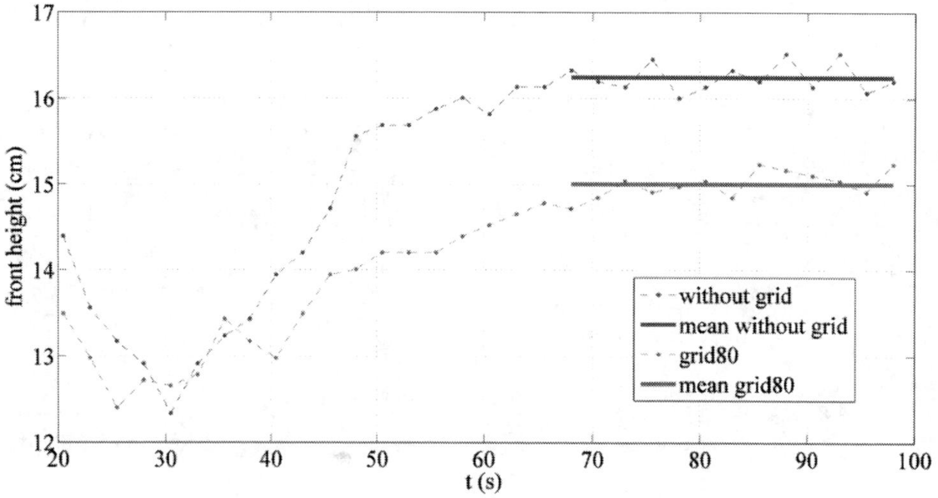

Figure 9: Intermediate height at which the spreading front develops

The power input required by the above-described SIS relates to the kinetic energy flux and, thus, to the third power of the reference velocity at injection point. Figures 8 and 9 suggest that a given spreading height may be reached using a lower momentum at injection point provided the turbulence intensity at the device outlet is reduced accordingly, resulting in a significant reduction in power input. A similar effect should be expected from a reduction of the swirl at the point of injection, an issue not addressed in this work.

Finally, the radial position of the spreading front as a function of time is illustrated in Figure 10. Since the flow rate was kept constant during all runs, the restricted and unrestricted flow configurations led to similar results.

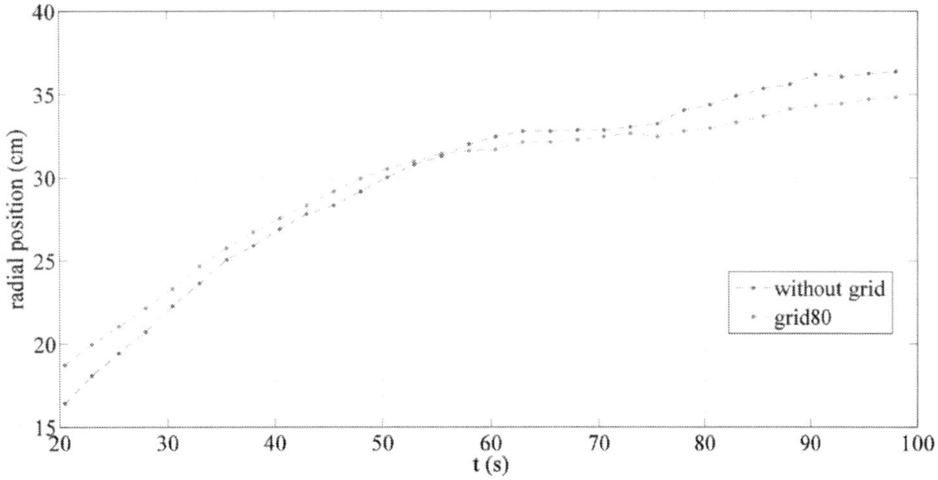

Figure 10: Spreading front radius as a function of time

5. Conclusion

Results here reported show that both the maximum height reached by a heavy-fluid jet as well as its spreading height decrease with increasing degree of turbulence at the point of injection. These results suggest that the generation of turbulence at the injection point may affect the efficiency of systems like the SIS. The inclusion of stationary or rotating blades or other mechanical device may lead to reduced jet turbulence, resulting in lower power requirements of the jetting machine without otherwise affecting the performance of an SIS.

References

[1] Augsburger H K M 2000, Frost control in temperate climates through dissipation of cold air, *Aspects of Applied Biology* **61** 1

[2] Guarga R, Mastrángelo P, Scaglione G and Supino E 2000, Evaluation of the SIS : a New Frost Protection Method Applied in a Citrus Orchard, Proceedings of the 9th. Congress of the Int. Soc. of Citriculture (Orlando, Florida). See also http://www.frostprotection.com.

[3] Albertson M L, Dai Y B, Jensen R A and Rouse H 1950, Diffusion of a submerged jet, *Trans. A.S.C.E.* **115** 639-664

[4] Baines W D, Turner J S and Campbell I H 1990, Turbulent fountains in an open chamber, *J. Fluid Mech* **212** 557-592

[5] Bloomfield L J and Kerr R C 1998 Turbulent fountains in a stratified fluid, *J. Fluid Mech.* **358** 335-356

[6] Bloomfield L J and Kerr R C 1999 Turbulent fountains in a confined stratified environment *J. Fluid Mech.* **389** 27-54

[7] Baines P G 2002 Two dimensional plumes in stratified environments *J. Fluid Mech.* **471** 315-317

[8] Baines P G 2005 Mixing regimes for the flow of dense fluid down slopes into stratified environments *J. Fluid Mech.* **538** 245-267

[9] Ansong J K, Kyba P J and Sutherland B R 2008 Fountains Impinging upon a Density Interface *J. Fluid Mech.* **595** 115-139

Role of arches in the generation of shear bands in a dense 3D granular system under shear

L Sigaud[1], A L Bordignon[2], H Lopes[2], T Lewiner[2], G Tavares[2] and W A M Morgado[3]

[1] Instituto de Física, Universidade Federal do Rio de Janeiro, P.O. 68528, 21941-972 Rio de Janeiro, RJ, Brazil
[2] Departamento de Matemática, Pontifícia Universidade Católica do Rio de Janeiro, C.P. 38071, 22452 970 Rio de Janeiro, Brazil
[3] Departamento de Física, Pontifícia Universidade Católica do Rio de Janeiro and National Institute of Science and Technology for Complex Systems, C.P. 38071, 22452-970 Rio de Janeiro, Brazil

E-mail: `lucas@if.ufrj.br`

Abstract. A model for propagation of arches on cubic lattices, to simulate the internal mobility of grains in a dense granular system under shear is proposed. In this model, the role of the arches in granular transportation presents a non-linear dependence on the local values of the stress components that can be modeled geometrically. In particular, we study a modified Couette flow and were able to reproduce qualitatively the experimental results found in the literature.

1. Introduction

Granular Systems (GS) have been the object of many studies recently, due to the many applications they present [1, 2] and to the large quantity of unusual equilibrium and nonequilibrium physical phenomena they exhibit. One of these interesting phenomena is a new model on the role played by the elastic stresses in such systems [3, 4, 5].

The internal stresses of GS are, in comparison with most continuous systems, quite unusual: the stresses are propagated via the contact forces between the grains, which are distributed unevenly in these systems. Therefore, one can see arches of forces: lines of high stress contacts which are responsible for the transmission of the internal forces. Therefore, the grains in any GS can be classified as being either part of these arches or part of the mass of relatively loose grains, the so-called soft phase [2].

The importance of arches in GS can be clearly seen in many of their unusual phenomena, like the Janssen effect [6]. In a recent work, based on the strong influence of arches in dense granular media [7], a phenomenological model for a bi-dimensional granular system, sufficiently dense so that no fluidized phase is present and thus only the stress effects of the solid phase are relevant, was presented [8]. When a shear stress was applied, the transport of granular material and the formation of shear bands were observed as expected. In the present work, we expand this phenomenological model to a simplified 3D model, in which we reproduce qualitatively some experimental results obtained by Fenistein *et al* [9, 10] and find them compatible with theoretical predictions of Török *et al* [11, 12].

2. The 2D Model

The 2D model is described in detail in a previous work [8]. There, the system is represented by a rectangular lattice, constituted of cells and their edges. Each cell corresponds to a volume of grains and is assigned a local parameter $\psi(r) = \psi_{i,j}$ (with i, j integers) associated with the local distribution of grains. The grains are not represented individually but only through the cell's granular mass, via this parameter $\psi_{i,j}$.

It is assumed there (as well as in this work) that the grains inside a given cell are part of the soft phase, i.e., their interaction is weak. On the other hand, the arches are represented by the edges of these cells, interacting with the granular material inside them and with the adjacent cells. The forces, which represent the arches, evolve with time by phenomenological rules and their magnitude is a stochastic function of time, and are given an orientation associated with the direction of local mass motion. Two possible mechanisms arise for momentum transport: via collisions and frictional contact forces (in the soft phase), and via direct normal contact (mostly along the arches).

When external shear stress was applied for long times, the appearance of a stationary state that closely resembles experimental data was obtained, including shear bands formation.

3. The 3D Model

Based on the success of our 2D model in describing the behavior of a dense granular system under shear, in particular the appearance of shear bands, we applied the same concepts in formulating a lattice-based 3D model. Now our square cells are transformed into cubic cells, trying to maintain the same properties for the formation and breaking up of the arches. We used the experimental setup of Fenistein *et al.* [9, 10] as the modeled experimental setup in these simulations, namely a modified cylindrical Couette cell with radius R and filling height H_s. Instead of letting an inner cylinder rotate with respect to an outer one, their cell has no inner wall, but exerts the shear deformation via its base, which is split into a rotating inner disk of radius R_s and an outer annulus rotating in the opposite direction. Both the cubic cell of our model and the experimental setup are shown in figure 1. This setup has attracted considerable interest because it provides access into the fundamental problem of shear band formation.

Fenistein *et al.* observed the behavior of the topmost granular layer in their experiment, with interesting results: they noticed that a shear band was formed, but the rotating disk at the topmost granular layer was smaller than the one at the base. More interestingly, the radius of the shear band at the top did not vary linearly with the height of the granular column, and it ceased to exist altogether at the topmost layer after a certain height. Török *et al.* [11, 12] proposed theoretical models to describe these results, based on minimal dissipation principles. They accounted for the experimental results, and their model proposed an explanation for the behavior of the granular system after that certain height: the shear band collapsed, forming a dome – hence the impossibility of observing the shear band looking only at the topmost layer.

Therefore we used this cubic cell lattice, with the arches propagating along the edges of the cells, to model this experimental setup, in order to verify if we could obtain, qualitatively, both the behavior observed experimentally and the theoretical predictions (in the case of the dome-like shear band formation). Following the phenomenological equations obtained for our 2D model [8], we used, for the stress propagation, equations for the forces applied by the arches on the loose grains inside the cells (soft phase) and how they change over time, both for the strengthening and softening of the arches and, in consequence, the forces they apply. 1 shows the force equation for the edge uf (see figure 1), along the x axis, of the cell i, j, k, which is the address of the cell in the lattice. It should be noted that this orienting of the arches comes from the apparent conflict of trying to describe a strictly local quantity (the stress) by means of a variable associated with a finite range in length. In fact, the granular mobility will depend on the gradients of stress, hence the necessity of orienting the edge-arches.

Journal of Physics: Conference Series **246** (2010) 012016 doi:10.1088/1742-6596/246/1/012016

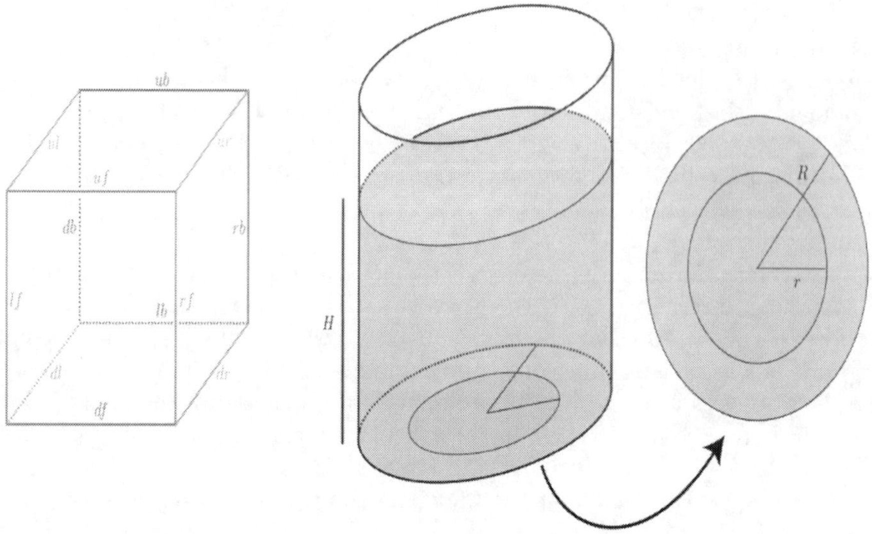

Figure 1. On the left, representation of a cubic cell of our 3D model, with the "address" of each cell edge. On the right, schematic representation of our experimental model: a cylinder filled granular material up to a height $H = H_s$ and with a rotating internal disk of radius $r = R_s$ at the base.

$$
\begin{aligned}
&F^{x(uf)}_{i,j,k}(t+\tau) = \\
&= \frac{1}{\alpha+\beta+\gamma+\lambda}\left[\alpha\left(\frac{F^{x(uf)}_{i,j-1,k}(t)+F^{x(uf)}_{i,j+1,k}(t)}{2}\right)+\right. \\
&\left.+\beta\left(\frac{F^{x(uf)}_{i,j-2,k}(t)+F^{x(uf)}_{i,j+2,k}(t)}{2}\right)+\gamma\left(\frac{F^{x(df)}_{i,j,k}(t)+F^{x(ub)}_{i,j,k}(t)}{2}\right)+\lambda F^{x(db)}_{i,j,k}(t)\right]+ \\
&+c_2\left\{\left[\frac{(\psi_{i,j+1,k}-\psi_{i,j,k})}{|\psi_{i,j+1,k}-\psi_{i,j,k}|+\epsilon}\right]e^{\left(|\psi_{i,j+1,k}-\psi_{i,j,k}|\right)}+\right. \\
&\left.+\left[\frac{(\psi_{i,j-1,k}-\psi_{i,j,k})}{|\psi_{i,j-1,k}-\psi_{i,j,k}|+\epsilon}\right]e^{\left(|\psi_{i,j-1,k}-\psi_{i,j,k}|\right)}\right\}+ \\
&+\eta^{x(uf)}_{i,j,k}(t),
\end{aligned}
\tag{1}
$$

where each term can be analyzed individually. The first term represents the contributions of the first and second edges immediately before and after the edge we are looking at (proportion constants α and β) and the other edges of this cell in the same direction x (proportion constants γ and λ). The second term represents the contributions of the soft phase media, since $\psi_{i,j,k}$ is the cell density (this means that if the cell is almost full, it would be much more difficult to continue filling it, and vice-versa; this term takes this into account). And the last term ($\eta^{x(uf)}_{i,j,k}(t)$) is the stochastic term to account for random diffusion. The force equations for the other edges are symmetric to the one above, and can be found accordingly. The transport equation reads:

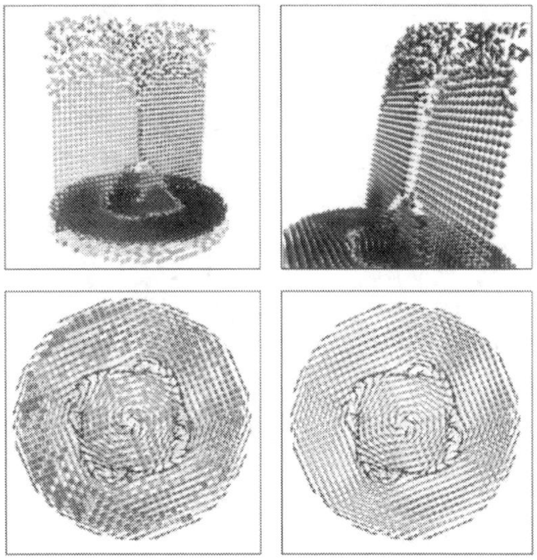

Figure 2. Images of the computer simulations perpetrated with our 3D model. The arrows represent the granular flux at a cross section of cells of the 3D lattice. The granular transport and the formation of the dome-like shear band (expected for filling heights above $H_s > R_s$) can be seen on the upper images. The bottom images show the flux at a horizontal plane of cells. Vortexes can be seen along the shear band profile.

$$\Delta\psi_{ijk} = K \ln(k_f |F_{R_{i,j,k}}| + 1) + \eta^{\psi}_{i,j,k}(t), \tag{2}$$

where K sets the time scale, k_f is the force scale, the η's are randomly generated numbers (through a normal distribution), ψ is the density function of the grain cells and F_R is the resultant of the arch forces applied on the edges of the (i, j, k) cell and calculated for all directions separately.

4. 3D Model Results

We performed the simulations using the equations above for different configurations of radiuses R_s for the rotating base disk and filling heights H_s for the granular media inside the cylinder. In order to make the shear band formation more evident, instead of keeping the cylinder steady while rotating the inner disk at the base, we make it rotate in the other direction. It can be seen, when looking at figure 2, that not only the shear band is – once again – formed, but also that its dome-like shape agrees with the theoretical predictions of the literature. Even more interestingly, when the filling height H_s is reduced to $H_s < R_s$, then the experimental results are reproduced, as one can seen in figure 3.

Thus it can be said that this arch-based model with the cubic lattice and the forces being transmitted to the loose granular media inside is quite good at reproducing qualitatively the physical phenomena at a real dense granular system. The most important thing to point out is that this provides a physical explanation for the transport properties observed in such systems – the arches (and the stresses they introduce in the systems).

Nonetheless, the model, as it is, still presents limitations. Foremost of those, and one that can be seen in figure 2, is the formation of the four vortexes at the shear band. Although it is indeed a turbulent area and vortexes are bound to appear, they should appear randomly and

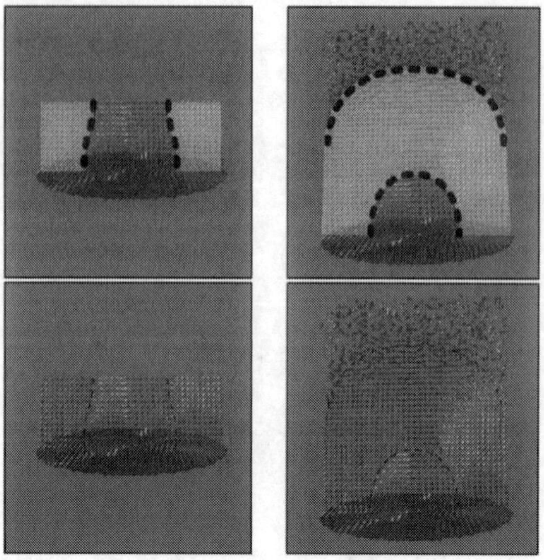

Figure 3. Comparison between lower filling heights $H_s < R_s$ and higher filling heights $H_s > R_s$. The colored shadow on the upper images are there to highlight the shear band pattern. It can clearly be seen that the shear band profile follows qualitatively the ones observed by Fenistein *et al.* [9, 10] and predicted theoretically by Török *et al.* [11, 12].

were not supposed to remain constant throughout the simulation. These, on the other hand, do not only remain constant through the entire simulation, but appear systematically at the same locations. Therefore, it is certain that it is an artificial result of our simulation. Most likely, it is a consequence of simulating a cylindrical system with a cubic (i.e. cartesian) lattice. The rotating disk at the base is not modeled with geometrical precision, and hence the appearance of vortexes at the shear band.

The other limitation that can also be seen from figure 2 is the neglect of friction between the grains and the walls of the cylinder. It was seen in our 2D model [8] that such precaution was crucial to a more accurate physical description of a real system. Without the friction in the walls, the force transmitted from the base of the cylinder to the grains at the upper layers of the filling height (after the collapse of the shear band) is almost negligible. Therefore, without the friction on the walls, the random component of the force equations become predominant and the random flux observed at the top of the simulation (see figures 2 and 3) is produced, which is clearly not physical. A second (and interesting) consequence of this limitation is that, for $R_s > R/2$, the shear band collapses, but not to the center, forming the dome, but in the direction of the walls, as can be seen in figure 4. This is consistent with the principle of minimal dissipation [12] – since there is no friction on the walls, the least energy dissipation will be achieved by collapsing to the walls. If we included the effects of friction on the cylinder walls, we would expect to see the shear band collapsing to the center, forming the dome, since the walls would be dissipating energy.

5. Conclusion

The 3D model expands on the results of our previously published 2D model, in the sense that now the qualitative behavior of experimental results and that predictions of theoretical models found in the literature were reproduced. This shows that arches are, in fact, very likely candidates for being responsible for important granular transportation properties.

Figure 4. Non-physical result obtained when $R_s > R/2$, due to the absence of friction effects on the walls of the cylinder. The shear band collapses to the wall, instead of forming the dome.

The model has some important limitations that have to be taken into account and need to be improved for obtaining quantitative results, namely the inclusion of friction effects on the walls and a modification of the lattice structure of the model. Probably, a lattice based on cylindrical coordinates would be a more friendly approach than our cubical cells lattice.

The model's limitations do not, however, diminish the significance of the arches for the grains transportation. Quite the opposite – even with such important limitations, the model is able to produce good qualitative results.

Acknowledgments
The first author was partially supported by FAPERJ and CNPq, Brazil. The remaining authors were partially supported by CNPq, Brazil.

References

[1] Bagnold R A 1941 *The Physics of Blown Sand and Desert Dunes* (London: Methuen)
[2] Duran J 1999 *Sands, Powders and Grains: An Introduction to the Physics of Granular Materials* (Springer)
[3] Jiang Y and Liu M 2003 *Phys. Rev. Lett.* **91** 144301
[4] Jiang Y and Liu M 2007 *Phys. Rev. Lett.* **99** 105501
[5] Jiang Y and Liu M 2008 cond-mat:0807.1883v1
[6] Janssen H A 1895 *Zeitschrift des Vereines Deutscher Ingenieure* **39** 1045
[7] Radjai F 2008 cond-mat.soft/0801.4722v1
[8] Bordignon A L, Sigaud L, Tavares G, Lopes H, Lewiner T and Morgado W A M 2009 *Physica A* **388** 2099
[9] Fenistein D and van Heck M 2003 *Nature (London)* **425** 256
[10] Fenistein D, van de Meent J W and van Heck M 2004 *Phys. Rev. Lett.* **92** 094301
[11] Török J, Unger T, Kertész J and Wolf D E 2007 *Phys. Rev. E* **75** 011305
[12] Unger T, Török J, Kertész J and Wolf D E 2004 *Phys. Rev. Lett.* **92** 214301

Mixing foams and grains in Hele-Shaw cells

A P B Tufaile, A Tufaile, and T A S Haddad

Escola de Artes, Ciências e Humanidades da Universidade de São Paulo, R. Arlindo Bettio, 1000, 03828-000, São Paulo, Brazil

E-mail: tufaile@usp.br

Abstract. We have observed some features of the coexistence of foams and granular materials in Hele-Shaw cells. The most part of the liquid and granular material stays at the bottom of the cell, with only a small quantity of the mixture resting on the froth. The fractal dimensions of the final states of the foams are close to the values obtained from the Random Apollonian Packing model. The disperse structure of the granular material affects the probability distribution of number of sides of the foam bubbles. The nearest neighbor distances between the peaks of the sand piles at the bottom of the cell are close to a lognormal distribution.

1. Introduction

When granular materials are mixed with foams, the grains tend either to float and stick to the bubbles or to sink in the liquid. Although some physical mechanisms of granular materials and foams mixtures are partially explored in the studies of granular media [1], wet granular matter [2], capillarity [3], foams [4], a better understanding of the behavior of granular materials in foam systems still remains an important prospect [5]. The understanding of the coexistence of foams and granular matter is difficult because the detailed knowledge of the components is insufficient to describe the whole system. For example, we have to take into account the transport of the three immiscible phases (air, liquid, and grains), or the interplay of the morphology of polyhedral bubbles and sand piles. For that reason, we have explored experimentally some aspects of this coexistence of foams and granular matter in vertical Hele-Shaw cells. We have compared two setups. The first one is based on the foam formation, while the second one is composed by the mixture of foam and granular matter. After that, we also compared some fractal properties of space filling models with the experimental data of both setups. Some adhesive properties of wet granular systems were explored in order to have an idea of the quantity of grains in these froths. Additionally, we have obtained the distribution of nearest-neighbor distances between the sand piles that form at the bottom of the cell.

2. Experimental Apparatus

The experiments were performed in two transparent Hele-Shaw cells, each one consisting of two plain parallel Plexiglas plates separated by a gap (20 x 20 x 0.2 cm^3). The first cell contains only air and an amount of commercial dishwashing liquid ($V = 6$ cm^3), and this cell is referred in this letter as setup I [6]. This liquid is manufactured by Bombril, and is used without dilution. The essential surfactant is Linear Alkylbenzene Sulfonate (LAS). The surface tension is $\gamma = 25$ dyne/cm, and the density of this detergent is $\rho = 0.95$ g/cm^3. In addition to the dishwashing liquid and air, setup II is a similar system containing 10 g of granular material. The granular material consists of polydispersed spherical glass beads with diameters d varying from 500 μm to 700 μm. The density of the glass beads is 2.6(1) g/cm^3. In both experiments, the cell is flipped upside-down generating the foam. Let us consider each flip of the Hele-Shaw cell as one iteration, and nf is the number of successive iterations. After each iteration in setup II, part of the granular material tends to remain in the foam layer because of the competition of the involved forces. Considering a single spherical particle moving through the foam, the capillary force $\gamma(d/2)$ is twenty-five times stronger than the inertial force $\rho g(d/2)^3$. The pictures of the foam were recorded with a digital camera. These pictures were converted into binary images and analyzed with a computer.

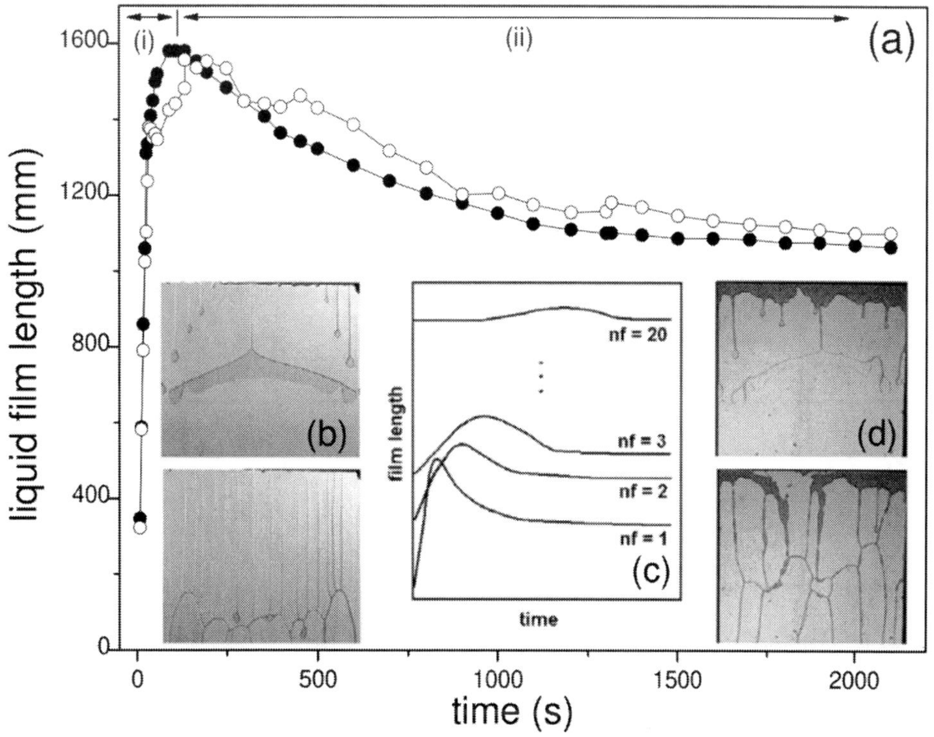

Figure 1. (a) Time evolution of the liquid film length after the first flip (nf = 1) is represent by the solid circles for setup I (just foam), and the open circles for setup II (foam and granular matter). The film network reorganizes with time following a relaxation process in (ii). (b) Two photos of the cell for setup I in the liquid stretching regime (i). (c) Time evolution of the liquid film length after each flip, with nf = 20 represents the full developed foam. (d) Two photos of the cell for setup II during the stretching regime.

3. Liquid film evolution

First, we compare the first flip in both setups, in order to have a glimpse of the involved phenomena. In figure 1(a) the solid circles represent the evolution of the liquid film length as a function of time for setup I. As soon as the cell is flipped, there is an unstable arrangement between the two phases causing the main part of the liquid to fall down through the cell wetting the walls. The expansion of the liquid filaments forming a grid of parallel lines is shown in figure 1(b). This interface expansion lasts 108 seconds in the plot of figure 1(i). After that, there is a rearrangement of the network of filaments shown in figure 1(ii). The film network reorganizes with time following a relaxation process, with smooth decay to the minimal surface. The relaxation time of setup I for the first flip is around 33 minutes. In short, the diagram for a sequence of flips is shown in figure 1(c), on which the mechanism of interface length growing is a cumulative process that depends on the number of iterations, as reported in a previous paper [6]. On the other hand, a different evolution of the foam interface in the presence of granular material for the first flip is represented by the open circles of figure 1(a). In the case of setup II, part of the liquid goes down under gravity to the bottom of the cell, as in the previous case, but a mixture of liquid and granular material remains at the top (figure 1(d)). This mixture engenders an effective cohesive interaction between the glass beads, forming slurry confined between the plates. This slurry moves down slowly through the cell. Consequently, the foam is affected by the motion of this slurry during the relaxation time, and the film length of the foam for setup II does not follow a smooth decay. The relaxation time of setup II is around 40 minutes, due to the motion of the granular material.

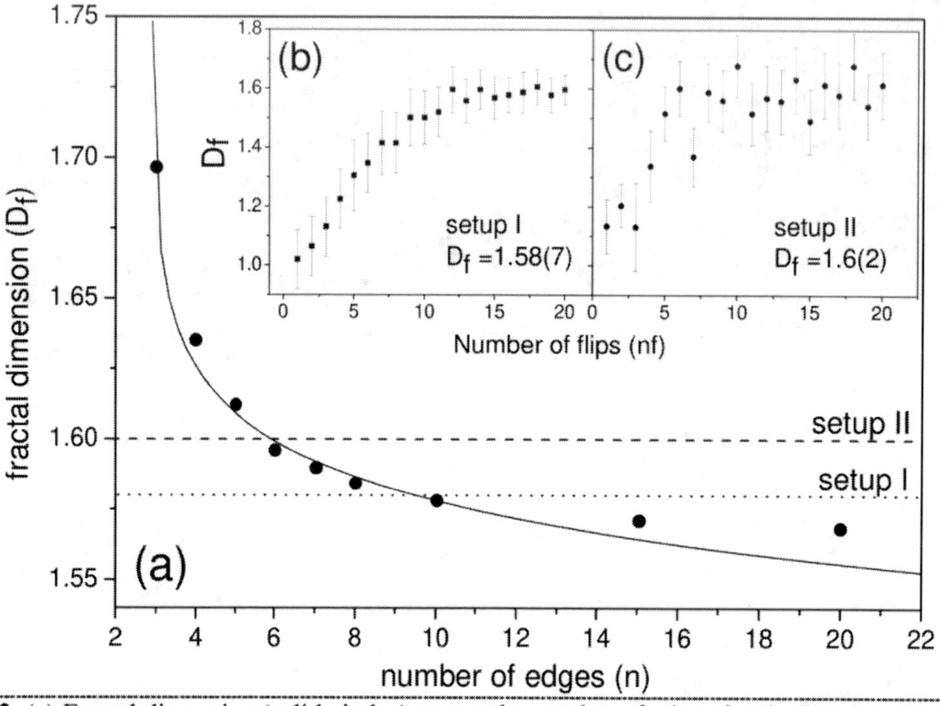

Figure 2. (a) Fractal dimension (solid circles) versus the number of edges for the Random Apollonian Packing. The continuous line is the fitted function as in equation (1). The dotted line represents the value of fractal dimension for the experimental results of setup I, and the dashed line of setup II. The insets are the fractal dimension versus number of flips for setup I (b), and for setup II (c), with the values of the fractal dimension for the foams in the stationary states in both cases.

4. Foam structure

For several iterations the developed foam is a planar network of filaments extending throughout the Hele-Shaw cell, with the main part of the liquid resting at the bottom of the cell for both setups. The completely developed foam is called a stationary state, represented by bubbles having a characteristic length related to the capillary length and by spatial homogeneity, following a kinematic model for bubble splitting process [7]. In order to capture the general features of the evolution of the foam structure, we have used some space filling models to compare fractal dimensions and differentiate the structures of foams. We can find similar approaches in Refs. [6] [8]. The Random Apollonian Packing is the case in which circles touching one another, with different sizes, are placed randomly in the plane [9]. This packing was further extended by Delaney [10] to consider a model in which cells can have a range of polygonal shapes. Considering the smoothness term $[\phi/(n\text{-}k)]$ for polygonal bubbles, we obtained a general expression relating the fractal dimension and number of bubble edges n filling the plane given by

Journal of Physics: Conference Series **246** (2010) 012017 doi:10.1088/1742-6596/246/1/012017

$$D_f = \frac{\log\left(\frac{\phi}{n-k}\right)+\beta}{\alpha} \tag{1}$$

with fitting parameters $\alpha = 17$, $\beta = 25.1$, $\phi = 360$, and $k = 2.9$. The parameter k is related to the number of side of a triangle, and is almost 3, and ϕ is the angle for the circumference in degrees. A strong dependence of scaling properties on the bubble shape was obtained, and it is shown in figure 2, along with the fractal dimension. For example, the fractal dimension $D_f = 1.634$ for squares, $D_f = 1.612$ for pentagons, and $D_f = 1.597$ for hexagons. In the case of the fractal dimension of the experimental data, we applied the box counting method to measure the fractal dimension formed by the foam for sequences of iterations. The estimates for the fractal dimension for setup I are shown in figure 2(b), and the analogous measurement of setup II is shown in figure 2(c). The dispersion of fractal dimensions of the foam in presence of the granular material is higher than in the case of pure foam. The foam structures in the stationary state in both cases have almost the same fractal dimension values, with $D_f = 1.58(7)$ for setup I, and $D_f = 1.6(2)$ for setup II. These values of fractal dimension for the foam in the stationary state are close to those obtained for the case of Random Apollonian Packing of hexagons, taking into account the error bars.

The presence of the granular material increases the stand deviation for the value of the fractal dimension of the foam, from $\sigma = 0.07$ for setup I to $\sigma = 0.2$ for setup II. A possible reason for that effect is because the particles could represent some kind of defect in pattern of the foam structure. These imperfections also have influence in the box-counting algorithm used for the computation of the fractal dimension D_f.

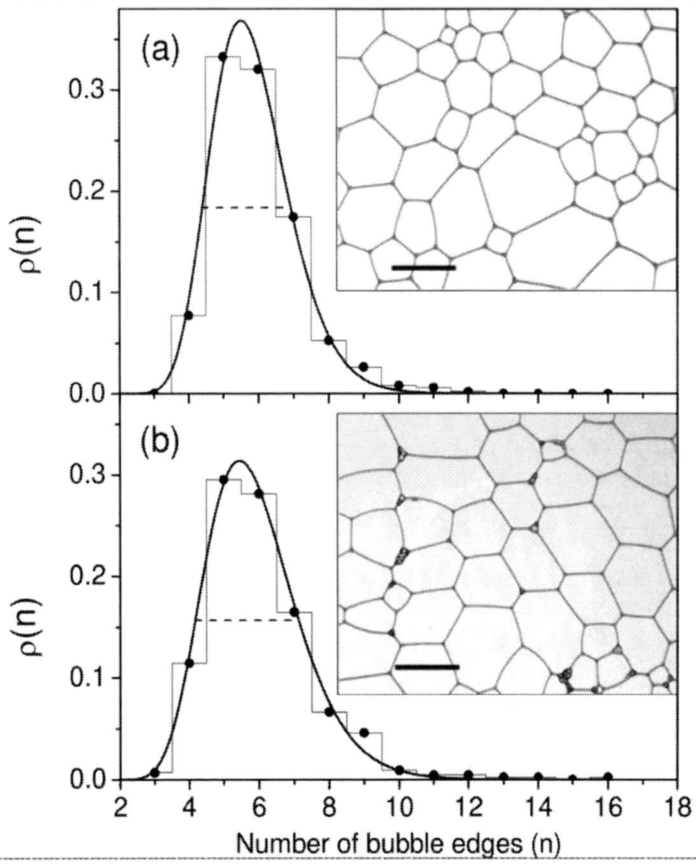

Figure 3. The probability distribution, $\rho(n)$, of the number of bubble edges is shown in (a) for setup I, and for setup II in (b). The insets are photographs of the foam shown in (a) and for the case of the foam with sand in (b), where the horizontal bars are 1 cm long. Although the appearance of the foam seams the same, the quality of a foam is represented by the width of $\rho(n)$ distribution. The thick lines represent lognormal functions fitted to the experimental data. The addition of the sand changes the quality of the foam by spreading the $\rho(n)$ distribution. Both setups have $\bar{n}$ around 5.7, hence the fitted functions agree with the fractal dimensions obtained for the Random Apollonian Packing for both setups, taking into account the error bars.

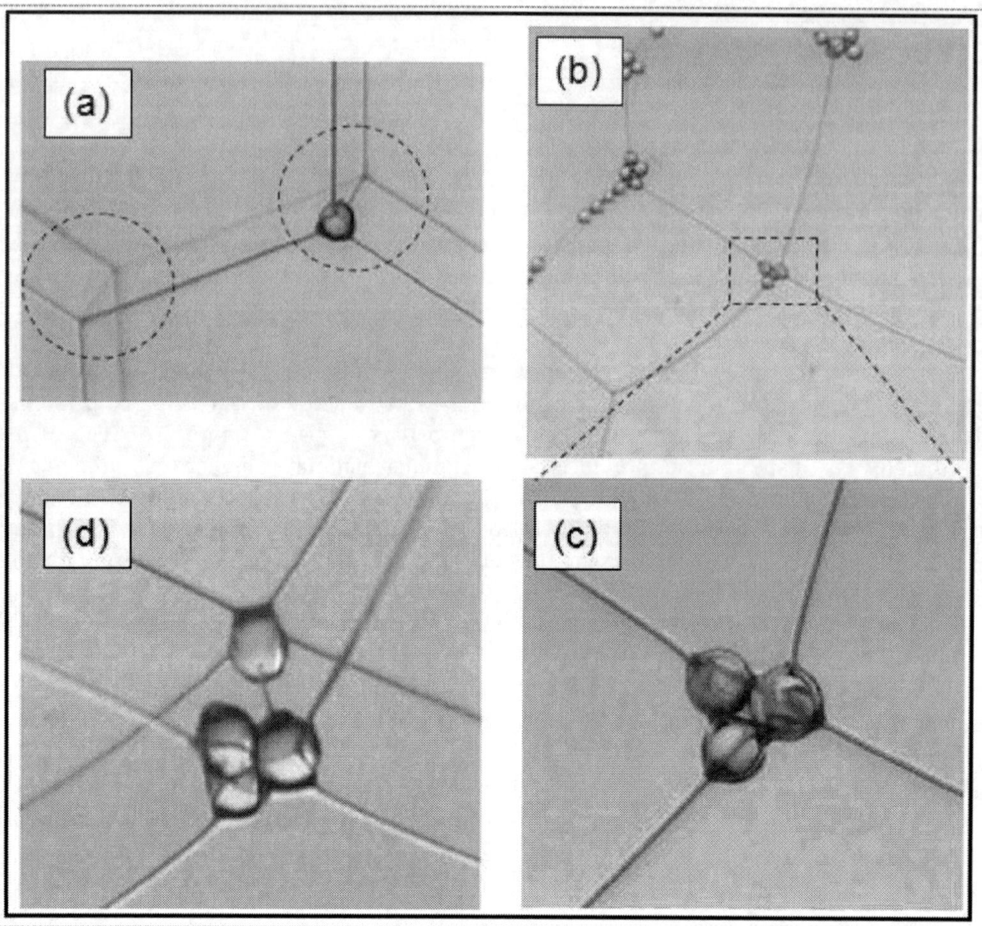

Figure 4. Close up photographs showing nodes, where the edges of the bubbles touch each other. (a) A perspective view of two nodes of degree three, following the rules of Plateau; there is a grain in one of them. All nodes observed for pure foam have degree three, as well as the majority of nodes with grains or cluster of grains. Nevertheless, as it is shown in (b), we have observed a node degree four. In (c) is shown the close up of the picture in (b). In (d) there is a perspective view of this node, showing that it is really a node connecting four bubble edges. From this view we can see the grain in the opposite plate behind the three grains seen in (c).

We also have performed a quantitative comparison between the bubble geometry in both setups. The main difference was observed in the probability distribution $\rho(n)$ of the bubble sides. For the case of setup I, the probability distribution of the number of sides of the bubbles is the discrete one shown in figure 3(a), while the data of setup II is shown in figure 3(b). By inspection, we have observed the existence of skewed size probability distributions in both cases. The thick line in each case represents the best fit of a lognormal distribution to these discrete distributions. Although the data are related to discrete distributions, the fit of a lognormal distribution in terms of a log-transformed variable enable us to compare typical parameters, such as the mean value are $\overline{n}$ and standard deviation σ. For the case of setup I we have obtained $\overline{n}$ = 5.72(2) and σ = 0.192(3). For setup II, the values are $\overline{n}$ = 5.74(3) and σ = 0.227(4). In the both cases the reduced chi-square is equal to 0.996. The disperse structure of the granular material affects the probability distribution $\rho(n)$, indicated by the fact that the standard deviation of the bubble sides distribution increases in the presence of granular materials.

The change of the detergent film configuration is an interesting example of this result. The place where detergent films meet is a Plateau border. Considering the Plateau border as a node, the degree of a node is the number of liquid films touching it. The two nodes of figure 4(a) have degree three. For the case of system I, the degree of every node is always three, following the rules of Plateau. On the other hand, the granular material in dry foam occasionally changes the degree of some nodes in the soap film, because part of the granular material spreads out through the network of liquid film, creating some grain clusters caused by the interstitial liquid bridges between grains. Figure 4(b) shows the case of a node with degree four apparently not consistent with Plateau's laws. The zooming of this node is shown in figure 4(c), and there is the perspective view of the node in which there are four films meeting at the center of figure 4(d), anchored by the glass beads. The explanation for this node with degree four is related to the spherical geometry of the grains, enabling several films touching the same grain at various angles, compatible with Plateau rules. So these nodes can be considered an emergent property due to the combination of foams and grains.

We have considered the nodes with liquid films with three grains between the plates, because nodes containing many particles do not allow defining the geometry of the Plateau borders precisely.

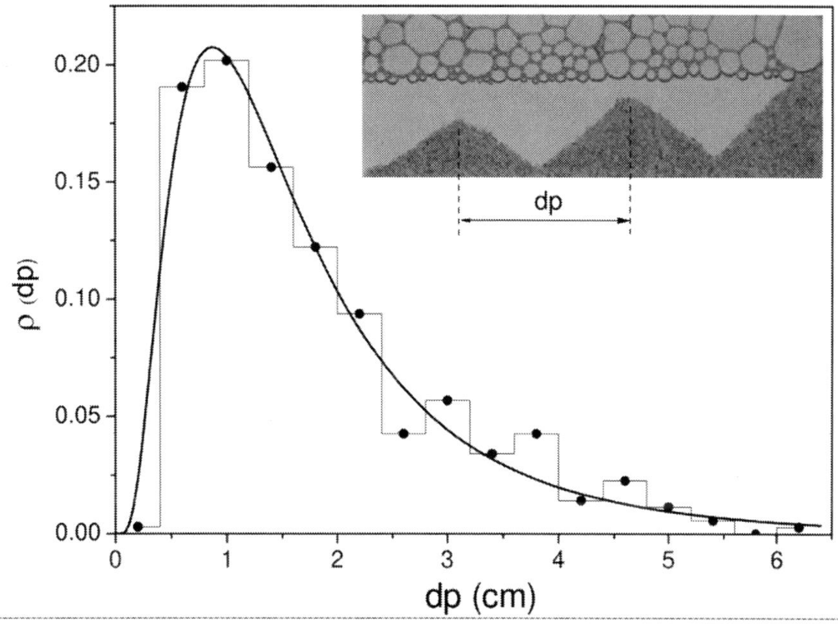

Figure 5. The probability distribution, $\rho(dp)$, of nearest neighbor distances between the peaks of the sand piles at the bottom of the Hele-Shaw cell. The lognormal function is fitted to the experimental data. The inset picture shows the sand piles totally immersed in the liquid.

In addition to the previous comments, an intriguing sand pile distribution was observed. The nearest neighbor distances *dp* between the two peaks of the sand piles at the bottom of the cell (see the inset of figure 5) exhibit a statistical distribution $\rho(dp)$, represented by the fit of the thick line of figure 5. We have obtained the plot of a lognormal distribution with $\overline{dp}$ = 1.43(6) cm, σ = 0.70(4), and reduced chi-square equal to 0.976. Due to the fact that this probability distribution was obtained from the experiment, we can conjecture that $\rho(dp)$ somehow represents the net result of the dynamics of wet granular materials passing through the foam structure.

5. Conclusion

In conclusion, we have observed some features of foams and granular materials in Hele-Shaw cells. We have found that the general evolution of the foam in the presence of the granular material is different from the case without grains, while the foam structure in the stationary state in both cases is almost the same, with their fractal dimension close to the values obtained from the Random Apollonian Packing of hexagons. The foam forms a network of ducts which transport the grain, and it is affected by the motion of this slurry during the relaxation time. Most part of the liquid and granular material stays at the bottom of the cell, with only a small quantity of the mixture resting on the froth. We have observed that at least 22% of the foam profile is composed by granular matter. Our results indicate that granular matter can alter some aspects of pattern formation in foams, such as the emergence of nodes with degree four. Plots of the number of bubble-sides followed a lognormal distribution. This distribution spans for the case of the coexistence of granular material and foams. We have measured the distance of the first neighbors of the sand pile peaks, and once again we have obtained a lognormal distribution.

Acknowledgments
This work was supported by Conselho Nacional de Desenvolvimento Científico e Tecnológico (CNPq), and Instituto Nacional de Ciência e Tecnologia de Fluidos Complexos (INCT-FCx).

References
[1] Jaeger H M, Nagel S R and Behringer R P 1996 *Phys. Today* **49(4)** 32
[2] Hornbaker D J, Albert R, Albert I, Barabási A L and Schifer P 1997 *Nature* (London) **387** 795
Herminghaus S 2005 *Adv. Phys.* **54** 221
Mitarai N and Nori F 2006 *Adv. Phys.* **55** 1
Andersen K H, Abel M, Krug J, Ellegaard C, Sondergaard L R and Udesen J 2002 *Phys. Rev. Lett.* **88** 234302
[3] Pomeau Y and Villermaux E 2006 *Phys. Today* **59(3)** 39
[4] Pugh R J 1996 *Adv. Colloid Interface Sci.* **64** 67
[5] Rubio J, Souza M L and Smith R W 2002 *Min. Eng.* **15** 139
[6] Tufaile A and Tufaile A P B 2008 *Phys. Lett.* A **372** 6381
[7] Caps H, Vandewalle N and Broze G 2006 *Phys. Rev.* E **73** 065301(R)
[8] Sauerbrei S, Hass E C and Plath P J 2006 *Discrete Dynam. Nat. Soc.* **2006** 79117
Herdtle T and Aref H 1991 *Phil. Mag. Lett.* **64** 335
[9] Manna S S and Herrmann H J 1991 *J. Phys.* A **24** L481
[10] Delaney G W, Hutzler S and Aste T 2008 *Phys. Rev. Lett.* **101** 120602
[11] de Gennes P-G, Wyart F B, Quere D and Reisinger A 2004 *Capillarity and Wetting Phenomena: Drops, Bubbles, Pearls, Waves* (New York: Springer)

The extended gaussian ensemble and metastabilities in the Blume-Capel model

Rafael B. Frigori[1,2], **Leandro G. Rizzi**[1] **and Nelson A. Alves**[1]

[1] Departamento de Física e Matemática, FFCLRP, Universidade de São Paulo, Avenida Bandeirantes 3900, CEP 14040-901 Ribeirão Preto, SP, Brazil

[2] Universidade Tecnológica Federal do Paraná, Rua XV de Novembro 2191, CEP 85902-040 Toledo, PR, Brazil.

E-mail: frigori@utfpr.edu.br, lerizzi@usp.br, alves@ffclrp.usp.br

Abstract. The Blume-Capel model with infinite-range interactions presents analytical solutions in both canonical and microcanonical ensembles and therefore, its phase diagram is known in both ensembles. This model exhibits nonequivalent solutions and the microcanonical thermodynamical features present peculiar behaviors like nonconcave entropy, negative specific heat, and a jump in the thermodynamical temperature. Examples of nonequivalent ensembles are in general related to systems with long-range interactions that undergo canonical first-order phase transitions. Recently, the extended gaussian ensemble (EGE) solution was obtained for this model. The gaussian ensemble and its extended version can be considered as a regularization of the microcanonical ensemble. They are known to play the role of an interpolating ensemble between the microcanonical and the canonical ones. Here, we explicitly show how the microcanonical energy equilibrium states related to the metastable and unstable canonical solutions for the Blume-Capel model are recovered from EGE, which presents a concave "extended" entropy as a function of energy.

1. Introduction

Thermodynamical properties of physical systems are described in Statistical Mechanics by the ensemble theory. Its development rests on the existence of both additivity and thermodynamic limit, which are essential ingredients to describe critical phenomena. The existence of a thermodynamic limit is related to the extensivity property which is required in order to obtain meaningful physical results in that limit [1]. Large enough systems with short-range interactions present the additivity property for extensive quantities like energy and entropy. Additivity means that when we consider the energy, or any other extensive quantity of a system, and divide the system into two or more macroscopic parts, the total energy is still the sum of the energies of those parts. From theoretical backgrounds, the canonical and grand-canonical ensembles approximate the microcanonical ensemble in the limit of infinitely large number of particles, where surface effects and fluctuations can be disregarded with respect to the bulk mean values [2, 3]. There is definitely more consensus around the microcanonical ensemble as the most fundamental one. Thus, the canonical and grand-canonical ensembles are taken as approximations for the microcanonical ensemble whenever the full equivalence does not hold. In cases where full consistency of statistical ensembles holds for systems that undergo phase transitions, it is found that finite-size scaling relations still place the microcanonical approach as the fundamental one [4].

There are many examples of systems whose microcanonical equilibrium properties contrast to the ones obtained in the canonical ensemble. Differences in the equilibrium properties have been described analytically for systems with long-range interactions [5, 6, 7, 8], which are consequences of the nonconcavity of the entropy as a function of energy [5, 9, 10]. The nonequivalence appears for those systems when the canonical ensemble exhibits a first-order phase transition and this has been demonstrated by means of the large deviation theory [11]. Such nonequivalence results in uncommon features like temperature discontinuity and negative specific heat in the microcanonical ensemble [5, 10, 11, 12, 13, 14].

In this paper, we will be dealing with a less popular ensemble, the extended gaussian ensemble (EGE) [15], which is constructed upon the gaussian ensemble [16, 17, 18, 19, 20]. They are known to present a smooth interpolation between their limiting behaviors, corresponding to the microcanonical and canonical ensembles. Recently, the EGE solution was obtained for the Blume-Capel (BC) model with infinite-range interactions [21]. Our numerical calculations illustrate this limiting behavior in a model with a tricritical point. This tricritical point does not coincide numerically in the two standard ensembles. Here, we explicitly show how the microcanonical energy equilibrium states related to the metastable and unstable canonical solutions for the Blume-Capel model are recovered from EGE, which presents a concave "extended" entropy. Moreover, the discontinuous microcanonical solution is also investigated and the metastable and unstable lines are also obtained from EGE.

The content of our presentation is as follows. The EGE analytical solution for the BC model is reviewed in the next section. Analytical and numerical results for thermodynamical quantities characterizing this ensemble are presented in section 3. In section 4 we summarize the results of this methodology in identifying the region of nonequivalence of ensembles only from the EGE point of view.

2. Extended gaussian ensemble and the Blume-Capel model

The EGE is defined by the condition $\gamma \neq 0$ and probability density

$$P_{\gamma,\alpha}(E) = \frac{\rho(E)\, e^{-\alpha E - \gamma (E-U)^2}}{Z_\gamma(U,\alpha)}, \tag{1}$$

where $Z_\gamma(U,\alpha)$ stands for the normalization constant, which is the corresponding partition function in EGE [15, 22, 23]. As usual, $\rho(E)$ is the density of states. Here, γ and α are parameters, and U stands for the mean energy, which is α and γ dependent, $U = U(\alpha,\gamma)$. The extended gaussian ensemble is a particular case in a class of ensembles defined by functions $g(E)$ [22, 23], where the quadratic form $g(E) = \gamma(E-U)^2$ is a convenient choice. The mean energy U can be obtained from the extended Gibbs's measure,

$$U = \int E\, P_{\gamma,\alpha}(E) dE. \tag{2}$$

This ensemble has two important limits. The canonical results correspond to $\gamma \to 0$ and in this case the parameter α identifies the inverse thermodynamical temperature, $\alpha = 1/k_B T$. On the other hand, the limit $\gamma \to \infty$ corresponds to the microcanonical case. This can be seen as $\lim_{\gamma\to\infty} \sqrt{\pi/\gamma} Z_\gamma(U,\alpha)$ through the use of the Dirac's delta sequence in the gaussian form [24]. The finite γ case presents an intermediary thermal description between the known limiting ensembles. All thermodynamical quantities can be defined in this ensemble as well. The extended thermodynamical potential is analogously defined to the canonical ensemble, $\Phi_\gamma(U,\alpha) = -\ln Z_\gamma(U,\alpha)$. From here, the derivative at fixed value γ gives the mean energy,

$$\left(\frac{\partial \Phi_\gamma}{\partial \alpha}\right)_\gamma = U, \tag{3}$$

which parallels that of the usual canonical approach. The extended heat capacity is [15, 19] $C_\gamma = -\alpha^2(\partial U/\partial\alpha)_\gamma$.

Let us now consider the Blume-Capel model [25, 26], a particular case of the Blume-Emery-Griffiths model [27]. Here, we work with its infinite-range interaction version,

$$H(S) = \Delta\sum_{i=1}^{N} S_i^2 - \frac{J}{2N}\left(\sum_{i=1}^{N} S_i\right)^2, \tag{4}$$

where $S_i = \{0, \pm 1\}$. The couplings $J > 0$ and Δ are the exchange and crystal-field interactions, respectively. This is a well studied model, which presents a rich phase diagram in the $(\Delta/J, T/J)$ plane. It exhibits a first-order phase transition line, tricritical point, and a second-order phase transition line. The critical properties of the BC model have been studied analytically in both the microcanonical [5] and canonical ensembles [27, 28]. The BC model has become an outstanding model for what concerns the nonequivalence of ensembles. The canonical and microcanonical ensembles do not yield the same phase diagram for the first-order critical line [5]. The canonical tricritical point occurs at $(\Delta/J, T/J) \simeq (0.46209812, 1/3)$, which gives origin to the canonical first-order phase transition line for larger values of Δ/J (and lower values of T/J). The microcanonical solution identifies the tricritical point at $(\Delta/J, T/J) \simeq (0.46240788, 0.33034383)$.

The EGE analytical solution [21] is obtained from the extended partition function by means of Hubbard-Stratonovich transformations. To this end, we introduce the order parameters magnetization $M = \sum_{i=1}^{N} S_i = N_+ - N_-$, and the quadrupole moment $Q = \sum_{i=1}^{N} S_i^2 = N_+ + N_-$, where N_+ and N_- are, respectively, the number of sites with up and down spins. Also, let N_0 be the total number of zero spins, then $N = N_+ + N_- + N_0$ is the total number of spins in the system. Therefore, the extended partition function reads

$$Z_\gamma(U,\alpha) = \sum_{N_0=0}^{N}\sum_{N_+=0}^{N-N_0} \frac{N!}{N_0!N_+!N_-!} e^{-\alpha(\Delta Q - \frac{J}{2N}M^2) - \gamma(\Delta Q - U - \frac{J}{2N}M^2)^2}. \tag{5}$$

Here, it is convenient to work with the intensive quantities $q = Q/N$ and $m = M/N$. Let us also define $K = J/2\Delta$ and $\varepsilon = U/\Delta N$ the energy per site, following the authors in [5].

First of all, we recall the EGE solution for $\gamma \to \infty$, which yields the known microcanonical ensemble results. The extended partition function must be well-behaved in this limit. Thus, it is required in equation (5) that $\varepsilon = q - Km^2$, which becomes a constraint equation for the mean energy ε as $\gamma \to \infty$. The thermodynamic limit $N \to \infty$ can be evaluated and one obtains the thermodynamical potential per site

$$\varphi(\varepsilon,\alpha,m) = \varepsilon\alpha\Delta + \left[q\ln\left(\frac{\sqrt{q^2-m^2}}{2(1-q)}\right) + \frac{m}{2}\ln\left(\frac{q+m}{q-m}\right) + \ln(1-q)\right], \tag{6}$$

as a function of mean energy per site ε and, parameters α and mean magnetization m. The term inside the brackets is the correct microcanonical entropy $-s_{\text{micro}}(\varepsilon, m)$ obtained in [5] as a function of ε and m. The entropy is an even function of m and a nonconcave function of independent variables m and ε. The nonconcave behavior is the necessary condition for metastability and instability in the canonical states [29] and gives origin to the nonequivalence of ensembles (microcanonical and canonical). The consequences on the thermodynamical quantities specific heat and specific susceptibility have been analyzed in [5, 21].

The solution for finite γ describes up to what extension one recovers the microcanonical stable states. The full equivalence of these results with the stable states of the microcanonical ensemble is achieved for *finite* γ only for Δ/J between the canonical and the microcanonical

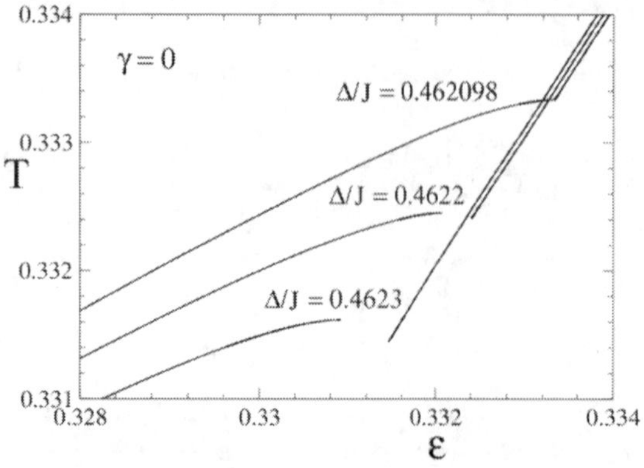

Figure 1. EGE temperatures obtained in the canonical limit for some values of Δ/J. $\Delta/J = 0.462098$ is in both canonical and microcanonical second-order phase transition regions. $\Delta/J = 0.4622$ and 0.4623 are in the canonical first-order phase transition region but in the microcanonical second-order one.

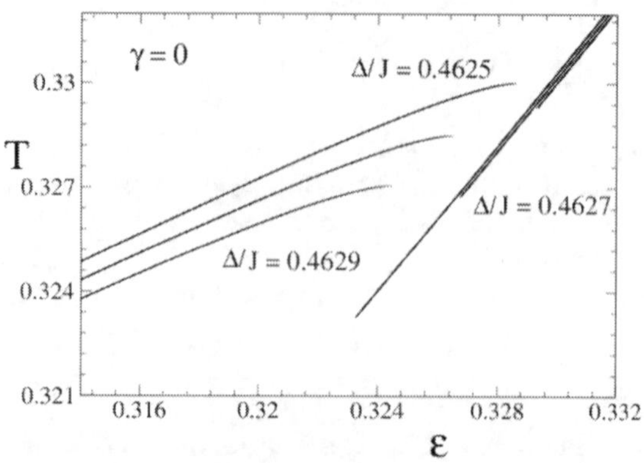

Figure 2. EGE temperatures obtained in the canonical limit for some values of Δ/J. The values of Δ/J are in both canonical and microcanonical first-order phase transition regions.

tricritical points. For couplings Δ/J larger than the microcanonical tricritical point, i.e., on the microcanonical first-order transition line, one needs $\gamma \to \infty$ for such full recovery of the microcanonical states [21].

The finite γ solution for the thermodynamical potential per site reads

$$\begin{aligned} \varphi_\gamma(\varepsilon, \alpha, m, q) &= q \ln\left(\frac{\sqrt{q^2 - m^2}}{2(1-q)}\right) + \frac{m}{2} \ln\left(\frac{q+m}{q-m}\right) + \ln(1-q) \\ &+ \gamma\Delta^2(q - Km^2 - \varepsilon)^2 + \alpha\Delta(q - Km^2). \end{aligned} \tag{7}$$

The EGE analytical solution poses a clear way of studying the formation of stable energy-dependent microcanonical states as the parameter γ increases. Furthermore, it is possible to draw the EGE solutions to observe the discontinuous microcanonical transition. This will be illustrated in the next section.

3. Results and discussion

The lack of ensemble equivalence is a consequence of a nonconcave entropy. As discussed in [29], the possible stable states, labeled by ε, are the ones defined by the microcanonical entropy $s_{\text{micro}}(\varepsilon)$. If those values of ε are not global minima of the canonical thermodynamical potential per site as we vary the temperature, then we do not find such correspondence between these states in both ensembles. Local minima and local maxima in the canonical thermodynamical potential per site are, respectively, metastable and unstable states.

To see how the nonconcave entropy for the BC model induces this nonequivalence of ensembles, we initially evaluate the caloric curve T *versus* ε, in the canonical limit. To this end, we calculate the equilibrium solutions of $\varphi_\gamma(\varepsilon, \alpha, m, q)$ for $\gamma \to 0$. The equilibrium solutions are easily obtained in this limit and are shown in figures 1 and 2. Figure 1 shows caloric curves for different values of Δ/J. Notice that the point $\Delta/J = 0.462098$, although close to the canonical tricritical point, is still in the second-order phase transition region according to both canonical

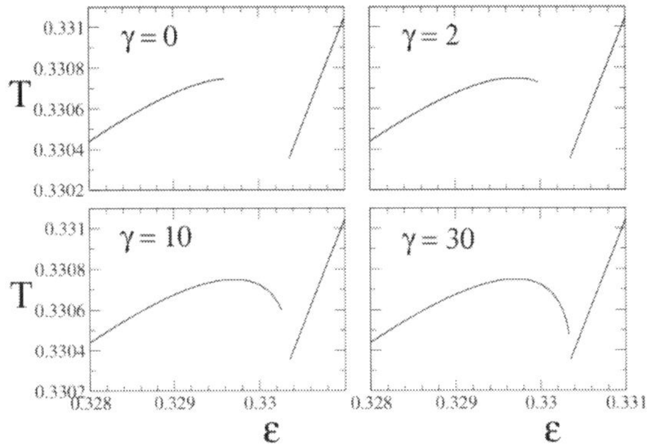

Figure 3. EGE temperatures for some values of γ with $\Delta/J = 0.462407$. A finite $\gamma \simeq 4950$ is necessary to recover all microcanonical states.

and microcanonical predictions. The points $\Delta/J = 0.4622$ and 0.4623, are in the canonical first-order phase transition region, but they are in the second-order region from the microcanonical point of view. Figure 2 shows caloric curves for Δ/J in a region where both ensembles present first-order phase transition lines. Notice that we have not drawn in those figures horizontal lines corresponding to the canonical Maxwell construction. Such constructions can be appreciated in [5] for some values of Δ/J. The missed stable microcanonical states are the ones related to the gaps in ε in both figures. We anticipate that the presentation of those stable microcanonical states is left to figure 4, while in figure 5 we included all the equilibrium solutions.

The EGE equilibrium states for finite γ can be obtained by solving the extremum equations for $\varphi_\gamma(\varepsilon, \alpha, m, q)$. We call attention to the fact that the microcanonical constraint for the specific quantities is not enforced here, the variables ε, m and q are treated as independent variables in this approach. The linear term in γ, equation (7), can be seen as a constrained equation leading to the microcanonical ensemble only for $\gamma \to \infty$.

As done in [21], the desired solutions must satisfy equation (3), $\partial\varphi_\gamma/\partial\alpha = \varepsilon\Delta$, and the extremum conditions $\partial\varphi_\gamma/\partial m = 0$, and $\partial\varphi_\gamma/\partial q = 0$, for a fixed ε. The stability criterion in the canonical ensemble for the BC model is nicely discussed in [13]. We must observe that the solutions for the three equations above are not stable for all γ. Since the analytical expression φ_γ comes from a saddle-point approximation, one needs to study the stability of those EGE solutions as a function of m and q. To this end, the determinant of the Hessian matrix,

$$d(m,q) = \det \begin{pmatrix} \frac{\partial^2\varphi_\gamma}{\partial m^2} & \frac{\partial^2\varphi_\gamma}{\partial m \partial q} \\ \frac{\partial^2\varphi_\gamma}{\partial q \partial m} & \frac{\partial^2\varphi_\gamma}{\partial q^2} \end{pmatrix},$$

is analyzed in the T versus ε plane as a function of γ. This amounts to exploring which points $\{m, q\}$ minimize φ_γ for fixed T and ε, and satisfy the condition $d(m,q) \geq 0$.

Figure 3 collects our results of stable points for different values of γ. That behavior is illustrated for $\Delta/J = 0.462407$, a coupling close to the microcanonical tricritical point and on the canonical first-order critical line, but still on the second-order critical line according to the microcanonical calculations. This figure shows that, as γ increases, one recovers the microcanonical solution. It can be shown that for sufficiently large γ, $s_\gamma(\varepsilon)$ becomes entirely concave and continuous on ε [23, 30],

$$\frac{\partial^2 s_\gamma}{\partial \varepsilon^2} < 0. \tag{8}$$

In other words, the addition of the term in γ to the thermodynamical potential changes the energy range where its nonconcavity is observed. This energy range is reduced as γ increases. Equation (8) can be used to find the necessary condition on γ to obtain the microcanonical

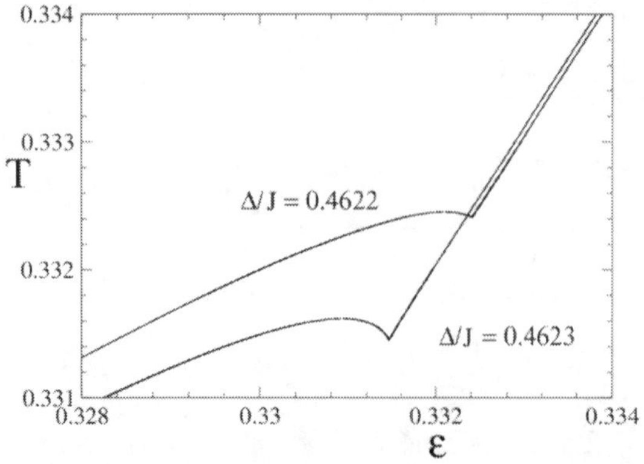

Figure 4. EGE temperatures obtained in the microcanonical limit for $\Delta/J = 0.4622$ and 0.4623. The couplings Δ/J are in the microcanonical second-order phase transition region.

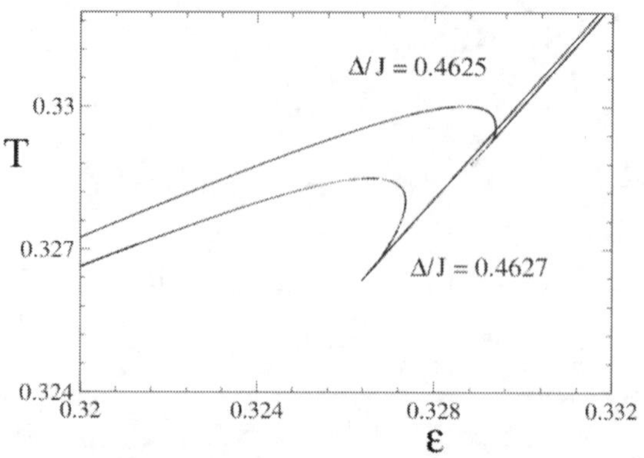

Figure 5. EGE temperatures obtained in the microcanonical limit for $\Delta/J = 0.4625$ and 0.4627. The couplings Δ/J are in the microcanonical first-order phase transition region.

solution,

$$\gamma > \frac{-1}{2\Delta T^2(\varepsilon)c(\varepsilon)}. \tag{9}$$

Here, $c(\varepsilon)$ is the specific heat, a negative quantity for energy ranges in figure 3 such that $dT/d\varepsilon < 0$. Figure 3 shows that, for the coupling $\Delta/J = 0.462407$, a finite value of γ is able to recover all stable microcanonical states. Actually, $\gamma \simeq 4950$ is the minimum value necessary for this end. To obtain the corresponding stable microcanonical states for $\Delta/J = 0.4622$ and 0.4623 in figure 1, one needs a rather small value for γ. It can be shown [21] that $\gamma \sim 30$ is sufficient for such couplings Δ/J. The numerical solutions are presented in figure 4.

For Δ/J larger than 0.46240788, the microcanonical tricritical point, the recovery of all microcanonical states is achieved only for $\gamma \to \infty$. Different of a continuous recovery of microcanonical states, as shown in figures 3 and 4, the caloric curves on this microcanonical first-order phase transition line sets a new physical situation. Here, one observes jumps in the temperature, the analogous of the latent heat in the canonical ensemble. The energy where this transition occurs can be obtained by defining a vertical line corresponding to the Maxwell construction for the microcanonical ensemble [13, 31]. Moreover, the microcanonical hysteresis effect is also present. This effect in the thermodynamics of a self-gravitating system has been discussed by Chavanis [31]. Our figure 5 illustrates this situation for $\Delta/J = 0.4625$ and 0.4627. Although for such values of Δ/J the full recovery of *all* microcanonical states is only possible in the limit $\gamma \to \infty$, we can still obtain valuable estimates about those states by choosing a very large numerical value for γ. Moreover, we were able to draw the continuation lines that define two branches, the high and low temperature branches, which produce the unstable and metastable lines of microcanonical states. The solutions depicted in figure 5 show that $T(\varepsilon)$ is a multivalued function of ε, signaling a microcanonical first-order phase transition, in addition to the standard profile of the canonical first-order phase transition when one draws horizontal lines corresponding to the Maxwell's construction.

4. Conclusions

The EGE approach leads to analytical expressions for the extended free energy and entropy in a simple way, and quantifies the nonequivalence of ensembles. The nonconcavity of the microcanonical entropy function means that the system contains several energy-dependent stable states, revealed in the microcanonical ensemble, which have their counterpart in the canonical ensemble, but they are unstable or metastable in this latter ensemble. By microcanonical entropy we do not refer to the standard Legendre transform of the canonical thermodynamical potential, which always produces a concave function of energy. The new term in γ added to the canonical free energy turns such unstable or metastable points into equilibrium points in the extended ensemble [22, 23, 29, 32] because it presents a concave "extended" entropy. Moreover, it is explicitly shown for the BC model that not only the stable states present in the microcanonical approach can be obtained from EGE but also the unstable and metastable ones. This is illustrated in figure 5, which shows the complexity of the microcanonical phase diagram and the usefulness of EGE for such study.

Acknowledgments

The authors acknowledge support from the Brazilian agencies FAPESP, CAPES and CNPq.

References

[1] Touchette H 2002 *Physica* A**305** 84
[2] Gross D H E, *Microcanonical Thermodynamics: Phase Transitions in Small Systems.* Lecture Notes in Physics, vol. 66, World Scientific, Singapore, 2001
[3] Gross D H E 1997 *Phys. Rep.* **279** 119
[4] Kastner M and Promberger M 2001 *J. Stat. Phys.* **103** 893
[5] Barré J, Mukamel D and Ruffo S 2001 *Phys. Rev. Lett.* **87** 030601
[6] Mukamel D, Ruffo S and Schreiber N 2005 *Phys. Rev. Lett.* **95** 240604
[7] Barré J, Bouchet F, Dauxois T and Ruffo S 2005 *J. Stat. Phys.* **119** 677
[8] Campa A, Ruffo S and Touchette H 2007 *Physica* A**385** 233
[9] Bouchet F and Barré J 2005 *J. Stat. Phys.* **118** 1073
[10] Casetti L and Kastner M 2007 *Physica* A**384** 318
[11] Ellis R S, Haven K and Turkington B 2000 *J. Stat. Phys.* **101** 999
[12] Ruffo S 2008 *Eur. Phys. J.* B**64** 355
[13] Barré J, Mukamel D and Ruffo S *Lecture Notes in Physics*, Vol. **602**, 45 (Springer, 2002)
[14] Ellis R S, Touchette H and Turkington B 2004 *Physica* A**335** 518
[15] Johal R S, Planes A and Vives E 2003 *Phys. Rev.* E**68** 056113
[16] Hetherington J H 1987 *J. Low Temp. Phys.* **66** 145
[17] Hetherington J H and Stump D R 1987 *Phys. Rev.* D**35** 1972
[18] Stump D R and Hetherington J H 1987 *Phys. Lett.* B**188** 359
[19] Challa M S S and Hetherington J H 1988 *Phys. Rev. Lett.* **60** 77
[20] Challa M S S and Hetherington J H 1988 *Phys. Rev.* A**38** 6324
[21] Frigori R B, Rizzi L G and Alves N A 2010 *Eur. Phys. J.* B, to appear, arXiv:0910.0500
[22] Costeniuc M, Ellis R S, Touchette H and Turkington B 2005 *J. Stat. Phys.* **119** 1283
[23] Costeniuc M, Ellis R S, Touchette H and Turkington B 2006 *Phys. Rev.* E**73** 026105
[24] Lukkarinen J 1999 *J. Phys. A: Math. Gen.* **32** 287
[25] Blume M 1966 *Phys. Rev.* **141** 517
[26] Capel H W 1966 *Physica* (Amsterdam) **32** 966; 1967 **33** 295; 1967 **37** 423
[27] Blume M, Emery V J and Griffiths R B 1971 *Phys. Rev.* A**4** 1071
[28] Mukamel D and Blume M 1974 *Phys. Rev.* A**10** 610
[29] Touchette H, Costeniuc M, Ellis R S and Turkington B 2006 *Physica* A**365** 132
[30] Costeniuc M, Ellis R S, Touchette H and Turkington B 2007 *Prob. Geom. Integr. Syst.* **55** 131
[31] Chavanis P H 2006 *Int. J. Mod. Phys.* B**20** 3113
[32] Costeniuc M, Ellis R S and Touchette H 2007 *Phys. Rev.* E**74** 010105(R)

Journal of Physics: Conference Series **246** (2010) 012019 doi:10.1088/1742-6596/246/1/012019

Long-Range k-core Percolation

Cristian F Moukarzel[1] **and T Sokolowski**[2]
[1] CINVESTAV del IPN, Depto. de Física Aplicada,
97310 Mérida, Yucatán, México.
[2] Fachrichtung Theoretische Physik, Universität des Saarlandes,
66041 Saarbrücken, Germany.
E-mail: `cristian@mda.cinvestav.mx`

Abstract. Bootstrap (or k-core) Percolation with $k = 3$ is studied numerically on two-dimensional lattices with long-range links whose lengths r_{ij} are distributed according to $P(r) \sim r^{-\alpha}$. By varying the decay exponent α the topology of these networks can be made to range from two-dimensional short-range networks to ∞-dimensional random graphs. The 3-core transition is found to be of first-order character with a divergent correlation length for $\alpha < 2.75$ and of second order for larger α. Whenever the transition is first-order an associated critical corona is found to exist. The correlation length exponent ν defined from the corona correlation length above the first order transition is estimated as $\nu \approx 1/2$ for $\alpha = 0$, and only shows a weak α dependence for $\alpha \leq 2.50$. The second-order transition at large α is found to be in the universality class of two-dimensional Percolation.

1. Introduction

k-core [1] (or Bootstrap Percolation [2]) generalizes Percolation [3] by requiring that a site have k or more neighbors in a cluster in order to be connected to it. Equivalently the k-core (or 'core' for simplicity) is the maximal subset of a graph with minimal coordination k. The core identifies a subset of 'strongly connected sites', a concept that finds many applications in Network Analysis, including Biological Evolution [4], Dynamics of Avalanches [5, 6, 7], Community Identification[8], and Modelling of Neuron Dynamics [9], among many others.

In mean-field the k-core undergoes, for all $k > 2$, a *first-order critical* transition when the average coordination γ reaches a critical value γ_c. At this transition the density P_∞ of sites in the largest core cluster (the order parameter) has a power-law singularity on top of a discontinuity, of the form

$$P_\infty = \begin{cases} 0 \quad \text{for} \quad \gamma < \gamma_c \\ P_c \quad +a(\gamma - \gamma_c)^{1/2} \quad \text{for} \quad \gamma > \gamma_c \end{cases} \tag{1}$$

In two and three dimensions however, numerical results and analytic arguments indicate that, depending on k and on the lattice coordination, k-core has either a trivial transition at $p = 1$ or a critical point at a non trivial p_c but in the universality class of Percolation [10, 11, 12, 13]. In four dimensions k-core it is believed to have a non-trivial ($p_c \neq 1$) but non-critical first-order transition [14].

Recently it has been shown [15, 16] that a critical correlation-length can be defined for k-core in terms of the *corona*, which is the subset of the core with exactly k neighbors in the core.

Analytical calculations in mean-field show that the corona correlation length diverges at γ_c^+ in the usual power-law fashion $\xi \sim (\gamma - \gamma_c)^\nu$, with a thermal exponent ν that is either 1/2 or 1/4. Schwartz, Liu and Chayes (SLC) [15] pointed out a number of similarities between k-core and the Jamming transition [17] of repulsive spheres in finite dimensions, and suggested that k-core is a simplified model for Jamming. At the Jamming transition of frictionless spheres, the average coordination z jumps discontinuously from zero to the isostatic [18] value $z_c = 2d$ at the Jamming density ϕ_c and behaves as

$$z = z_c + a(\phi - \phi_c)^\beta, \qquad \text{for} \quad \phi \geq \phi_c, \tag{2}$$

where β has been estimated numerically to be close to 1/2 in 2 and 3 dimensions [17].
While packings of stiff spheres are isostatic [19, 18] with $z = z_c$ in any dimension, compressed packings of deformable spheres develop an excess number of contacts $\delta z = z - z_c$ and become hyperstatic. Wyart [20] has argued that for weakly hyperstatic packings a length scale $l^* \sim (\delta z)^{-1}$ can be defined such that the system behaves essentially as an isostatic one below l^*. This length scale l^* would therefore diverge as $(\delta\phi)^{-1/2}$ on approach to the unjamming transition. Additionally, numerical simulations [21] near the unjamming transition find a dynamical length that diverges with a thermal exponent which is near 1/4.
SLC also remark that the k-core connectivity condition is satisfied in a jammed packing, since a sphere in d dimensions must be in contact with at least $d+1$ other spheres in order to be itself locally jammed. Based on these similarities, SLC propose that k-core is a good description of the Jamming transition. Difficulties with this interpretation include the fact that k-core is only first-order in MF but not in two and three dimensions, while the Jamming transition is first order in two and three dimensions.

The aim of the present work is to investigate how the critical behavior of k-core changes from second-order to first-order critical, as space dimensionality increases. As an alternative to studying lattices of various dimensions, which is time-consuming and only allows for small linear lengths in large dimensions, in this work k-core Percolation with $k = 3$ is studied numerically on two-dimensional networks with random long-range (LR) links. By adequately choosing the length-distribution of LR links, networks can be built that have an effective dimension between two and ∞ [22, 23, 24, 25].

2. LR networks

Our LR networks are built by taking a 2-dimensional square array of $N = L^2$ points, and adding to it a total of $\gamma N/2$ randomly located links, where $\gamma > 0$ is a parameter specifying the *average coordination.* A second parameter $\alpha \geq 0$ determines the *a priori* length-distribution of links through

$$P_{ij} \sim r_{ij}^{-\alpha}, \tag{3}$$

where r_{ij} is the Euclidean distance between i and j in the two dimensional array.
For $\alpha \to \infty$, only near neighbors are connected, and the resulting network, although being disordered, has a two-dimensional topology if γ is above its percolation threshold. For $\alpha \to 0$ all pairs ij are connected with the same probability, and a Random Graph [26] is obtained (See Fig. 1).

2.1. Numerical implementation

In order to implement this link distribution numerically, we first choose a random site i and then a site j in such a way that (3) is satisfied. If j is already connected to i then this link is rejected and a new site j is chosen, otherwise j is connected to i.
Consider the following two ways to generate link-length distribution (3) numerically: Method A consists in calculating and storing beforehand, for each relative vector $\{i, j\}$ on the 2D lattice,

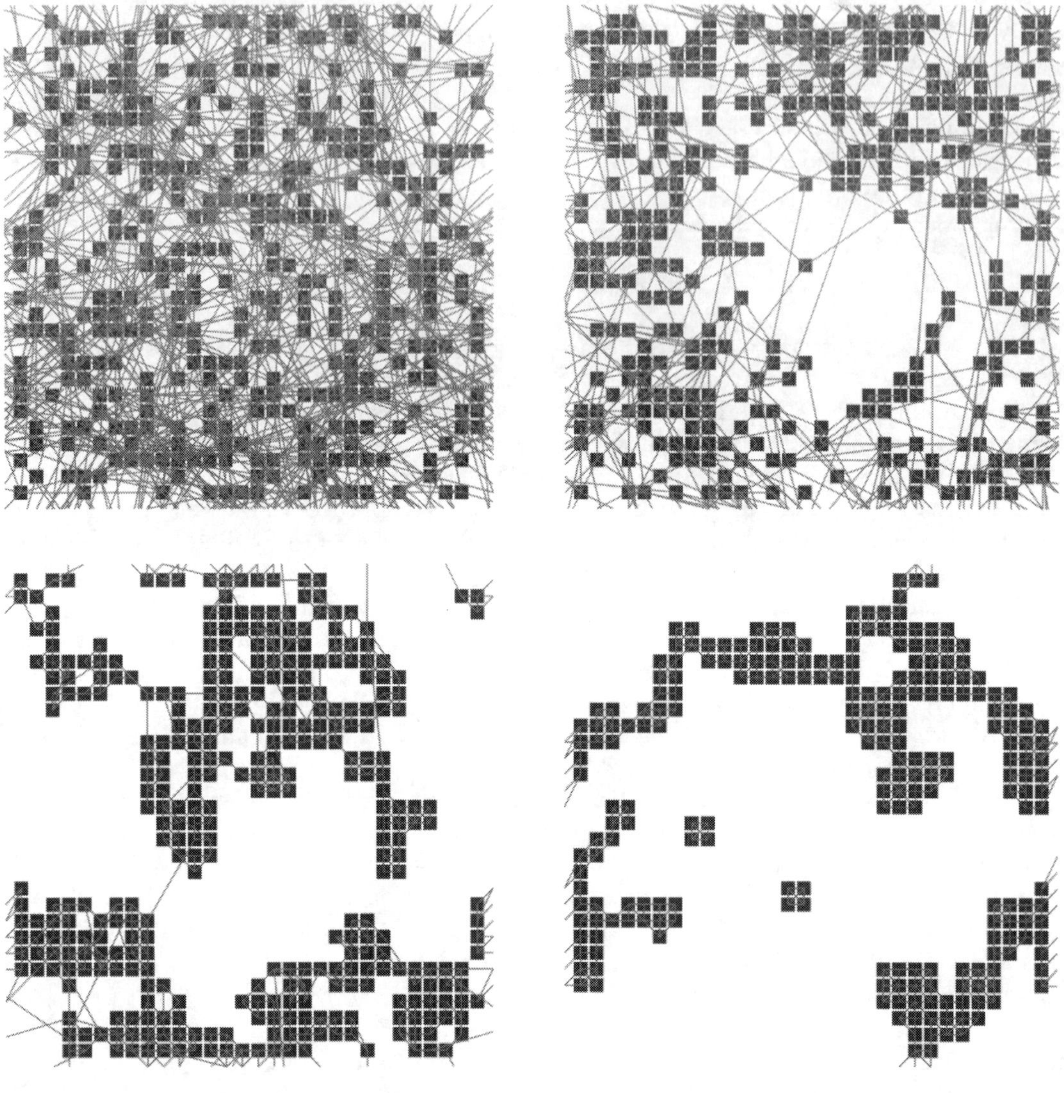

Figure 1. Sites in the 3-core (blue squares) for 30×30 networks whose LR links have a decay exponent α respectively equal to $0, 2, 4$ and 10 (from left to right and from top to bottom). The average coordination is $\gamma = 3.5$ in all cases. Shown in red are the links in the core. Helicoidal boundary conditions were used.

the normalized probability (3) for the given value of α, and then choosing one vector $\{i, j\}$ using those probabilities. The choosing step, however, takes time proportional to N for *each* new link if α is small, making Method A very inefficient in general. Method B consists in first generating a random length ℓ between 1 and $L/2$ with probability $P(\ell) \sim \ell^{-(\alpha-d+1)}$ (the extra ℓ^{d-1} factor with respect to (3) takes into account the number of lattice points inside a thin shell of radius ℓ) and then a random direction $\vec{e}$. Site j is chosen to be the one whose position r_j is closest to $\vec{r}_i + \ell * \vec{e}$. This alternative method however is not efficient for large values of α and γ, since very short lengths ℓ are generated most of the times, therefore when all near neighbors of i are already connected to i, the rejection rate is very large.

We have chosen to implement a hybrid of methods A and B, that circumvents the disadvantages mentioned above. The set of lattice vectors $\{i, j\}$ is divided into two classes: those respectively shorter and longer than a prespecified length ℓ_0. For all short vectors, the probabilities P_{ij} are calculated according to (3) and stored, and from them the total weight P_s of short vectors is obtained:

$$P_s = \sum_{ij|\ell_{ij}<\ell_0} P_{ij}, \tag{4}$$

where the sum runs over short vectors only. For each new link connection, a random number R is first generated and if $R < P_s$, method A is used to choose among the list of short vectors. If $R > P_s$ then method B is used, i.e. a random length $\ell > \ell_0$ is generated, etc. We have found that this hybrid algorithm with a limit length $\ell_0 = 1.5$ works efficiently at all values of α and γ analyzed so far.

2.2. *Link dilution*

For each fixed α, we determine and store the γ-dependence of the k-core on a given sample, i.e. a realisation of links, by starting with a large coordination and reducing it in small steps (deleting links) until the k-core disappears. Notice however that a network with coordination γ_0 obtained by *random link-dilution* of a higher-coordination newtork is *not* statistically equivalent to one built with coordination γ_0 by the method above. The reason for this is that (3) gives more weight to short links, so that close neighbors are connected first. As γ increases during the building procedure, longer links begin to appear, since most short links are already connected and double connections are not accepted. This shows that the actual link distribution depends not only on α through (3) but also on γ, since networks with large coordination will necessarily have more long links. This effect is stronger the larger α is. In view of this, and in order to avoid the "irreversibility" or history dependence that random dilution would produce, we chose to implement the dilution process by eliminating links in exactly the opposite order as they were connected. This ensures that a network with coordination γ has the same statistical properties, no matter if it was obtained by adding or diluting links.

3. Numerical results

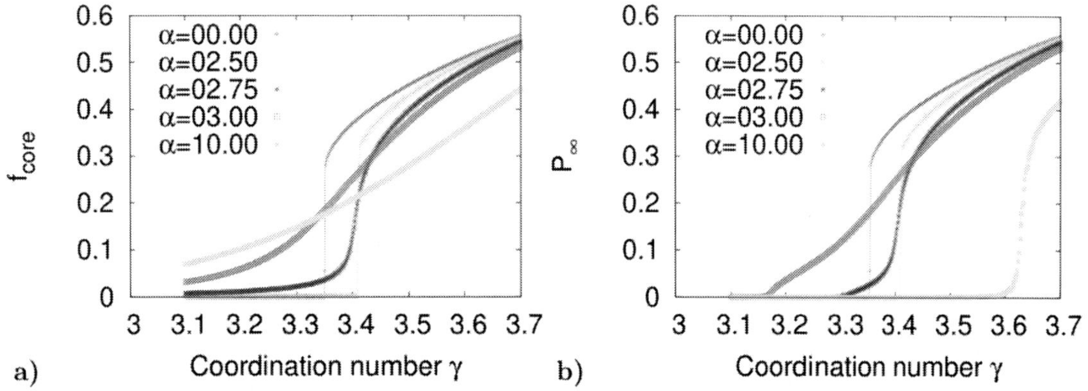

Figure 2. a) The fraction f_{core} of sites in the k-core is discontinuous for small α, but non-singular for large α. b) The order parameter P_∞ shows a singularity at γ_c for all α, being discontinouus for small α. All data are for $N = 10^7$ sites.

Early studies [27] of critical phenomena on d-dimensional LR networks suggest that in general, for a system undergoing a *second-order* phase transition, three different regimes can be expected as a function of α: a MF regime at low α in which classical behaviour is found, a large α limit in which d-dimensional behaviour is expected, and a possible intermediate regime where critical indices depend on α.

Bootstrap Percolation on LR networks could potentially display an even more complex behavior,

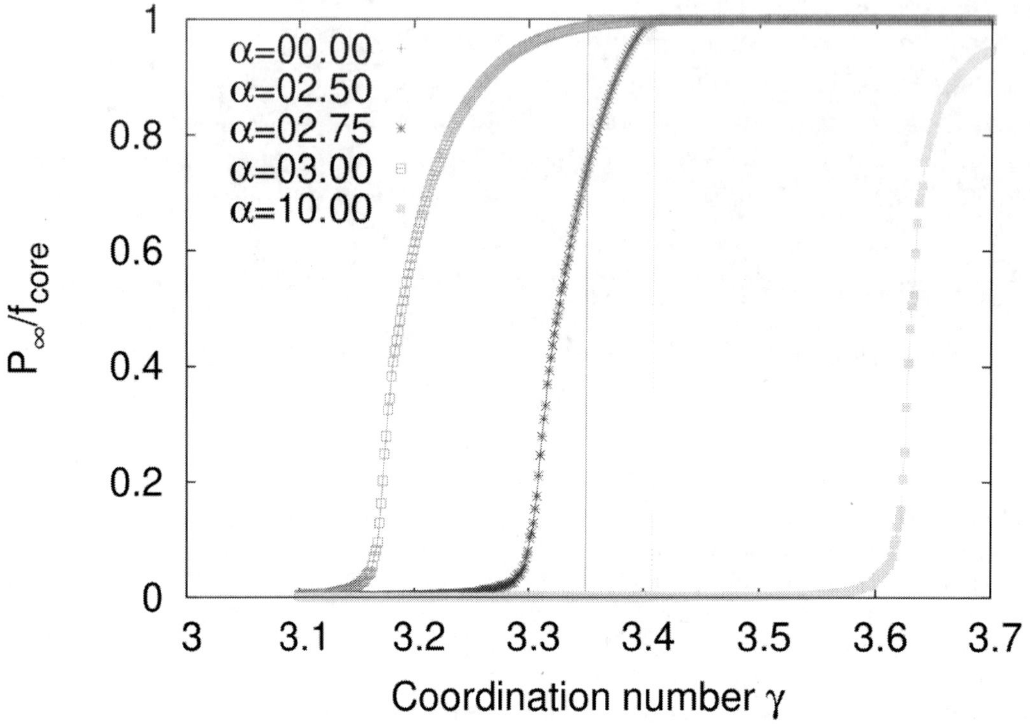

Figure 3. The ratio P_∞/f_{core} indicates the relative importance of the largest core cluster, and clearly displays the character (first-order or second-order) of the transition. All data are for $N = 10^7$ sites.

since in addition to the above mentioned different critical regimes, a crossover from first-order to second-order phase transition occurs as α is increased. In this work we concentrate on understanding the crossover from first-order to second-order transition that happens at a critical decay rate α_0.

We have analyzed k=3 Bootstrap Percolation on networks with $\alpha = 0$ to $\alpha = 10$, which enables us to identify two main regimes: for $\alpha < 2.75$ the 3-core transition is first-order critical, as on Random Graphs [16], while for $\alpha > 2.75$ there is a second-order percolation transition. In order to extract critical indices from finite-size scaling (FSS) we have done measurements on networks made of $N = 10^3, 10^4, 10^5, 10^6$ and 10^7 sites. A sample average was done in each case over $10^9/N$ independent realisations of LR links. For each value of α we measured a number of observables that are relevant for the characterization of the k-core transition occurring when the average coordination γ is varied. We report here mainly our results regarding the behavior of the k-core order parameter and corona correlation-length near $\gamma_c(\alpha)$.

3.1. Size of the core and order parameter

The fraction f_{core} of sites in the k-core is shown in Fig. 2a. For small α this quantity suffers a discontinuity as a function of the coordination γ, but behaves regularly for large α. Because of the

tree-like topology of these LR networks for small α, there are no (small) loops and therefore no finite k-core clusters. This implies that the k-core transition happens discontinuously instead of progressively nucleating from the union of many smaller clusters as in a second-order transition. The order parameter P_∞, which is the fraction of sites in the *largest* k-core cluster, is shown in Fig. 2b. These data indicate that the k-core transition is discontinuous for small α and becomes continuous for large α. Further results to be described later are consistent with the fact that the critical decay exponent separating first- and second-order transition regimes is located around $\alpha_0 \approx 2.75$. In addition, in a certain α interval it appears possible that two different transitions, one of second order and the other of first order, exist.
The ratio P_∞/f_{core} (Fig. 3) is particularly interesting. This ratio goes to ≈ 1 above the transition for all values of α, i.e. for first-order as well as for second-order transitions, indicating that the core becomes essentially a single connected cluster. The manner in which this quantity grows, however, depends on the type of transition, i.e. on α. It does so discontinuously for $\alpha < 2.75$, and continuously for $\alpha > 2.75$. This defines the critical decay rate for the crossover from first-order to second-order transition as roughly $\alpha_0 \approx 2.75$.

3.2. The Corona

The corona is defined [15] as the subset of the k-core that has exactly k neighbors in the k-core, and plays an important role for the critical properties of the discontinuous k-core transition [15, 16]. The total fraction of corona sites f_{corona} jumps discontinuosuly at the first-order k-core transition for $\alpha < 2.75$ (Fig. 4a). The corona comprises a finite fraction of the k-core for all values of the coordination γ for which a k-core exists. As seen in Fig. 4a, the

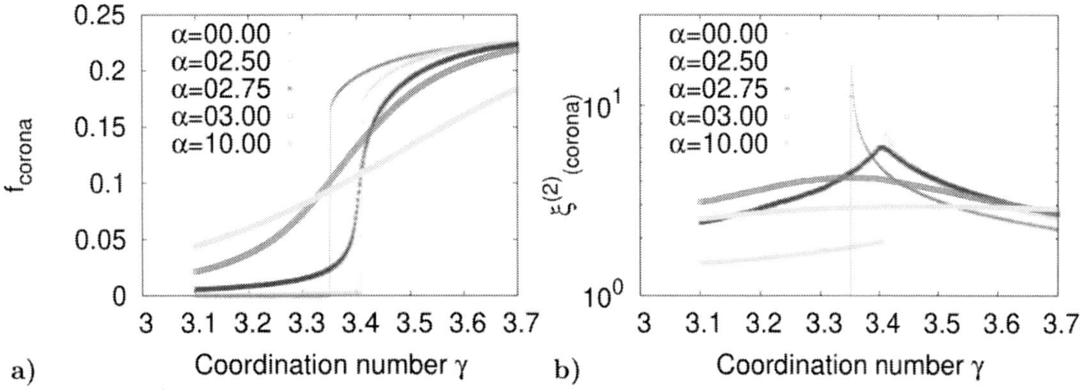

Figure 4. a) The fraction of sites with exactly three neighbors in the core, or *corona*, is responsible for the critical properties at the first-order transition [15, 16]. b) The average chemical extension of the corona, or corona correlation length, is singular at the first-order transition, but remains finite for all γ for large α. Data are for $N = 10^7$ sites.

corona ceases to behave singularly as soon as the k-core transition becomes continuous for large α.

3.3. Corona correlation length

A correlation length ξ_{corona} can be defined [15, 16] in terms of the decay rate of the number $\mathcal{N}_{crn}(\ell)$ of corona sites ℓ steps away from a randomly chosen k-core site via a path of corona sites.

$$\xi_{corona}^{-1} = \lim_{\ell\to\infty} -\frac{1}{\ell}\log \mathcal{N}_{crn}(\ell), \tag{5}$$

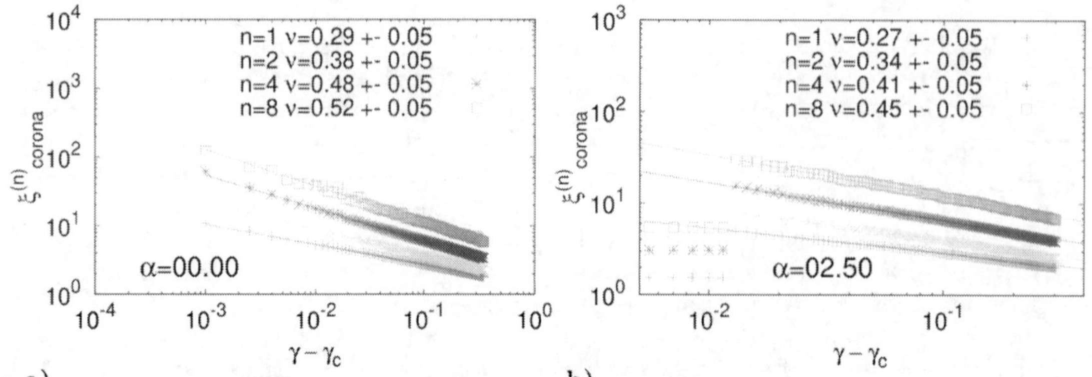

a) **b)**

Figure 5. a) Correlation lengths $\xi^{(n)}_{corona}$ as defined in (6), for $n = 1, 2, 4, 8$, from systems of size $N = 10^7$ with $\alpha = 0$. Straight lines are fits of the form $\xi^{(n)} \sim (\gamma - \gamma_c)^{-\nu_n}$. b) Same as a) but for $\alpha = 2.50$.

It has been shown that, for random graphs [16] ($\alpha = 0$ in our networks) or on Bethe lattices [15], this length diverges above the k-core transition as $\xi_{corona} \sim |\gamma - \gamma_c|^{-\nu}$. However there seems to be a discrepancy as to whether $\nu = 1/2$ [16] or $\nu = 1/4$ [15].
Alternatively, it is possible to define $\xi^{(n)}_{corona}$ from the n-th moment of $\mathcal{N}_{crn}(\ell)$ as [28, 29]:

$$\xi^{(n)}_{corona} = \left(\frac{\sum_\ell \ell^n \mathcal{N}_{crn}(\ell)}{\sum_\ell \mathcal{N}_{crn}(\ell)} \right)^{1/n}, \tag{6}$$

where often the second moment $n = 2$ is used. The γ-dependence of $\xi^{(2)}_{corona}$ is shown in Fig. 4b for several values of α. It is seen that $\xi^{(2)}_{corona}$ is discontinuous at γ_c when $\alpha < 2.75$.
If the ℓ-dependence of $\mathcal{N}_{crn}(\ell)$ is purely exponential, then (6) and (5) should give equivalent results for the estimation of the critical exponent ν, since in that case $\xi^{(n)}_{corona} \propto \hat{\xi} \quad \forall n$. However, if one more generally assumes that

$$\mathcal{N}_{crn}(\ell) = \ell^{-\theta} e^{-\ell/\hat{\xi}}, \tag{7}$$

then

a) If $\theta < 1$:

$$\xi^{(n)}_{corona} = \hat{\xi}, \quad \Rightarrow \quad \nu_n = n \quad \forall n. \tag{8}$$

b) If $1 < \theta < 2$ on the other hand:

$$\xi^{(n)}_{corona} \sim (\hat{\xi})^{1-\frac{\theta-1}{n}}, \quad \Rightarrow \quad \nu_n = \nu \left(1 - \frac{\theta - 1}{n}\right). \tag{9}$$

Fig. 5 displays $\xi^{(n)}_{corona}, n = 1, 2, 4, 8$ as obtained from the sample-averaged layer-mass distribution of corona clusters $\mathcal{N}_{crn}(\ell)$, as a function of coordination number γ near γ_c for two representative values of α in the first-order regime. For $\alpha > 2.75$, on the other hand, ξ_{corona} does not seem to grow with N at any γ_c, indicating that the corona is composed of finite clusters when the k-core transition is of second order.
For each $\xi^{(n)}_{corona}$, the critical exponent ν_n can be estimated [1] by fitting the γ-dependence of ξ above γ_c as $\xi^{(n)}_{corona} \sim (\gamma - \gamma_c)^{-\nu_n}$. There is a clear n-dependence in the estimated critical

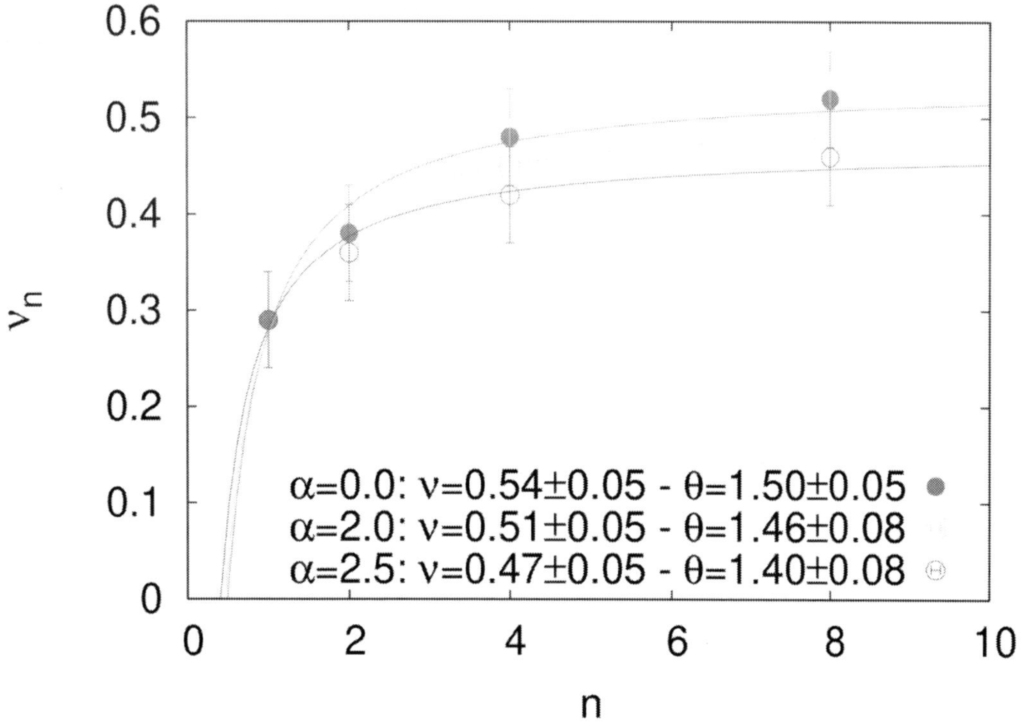

Figure 6. The critical exponent $\nu_n(\alpha)$ obtained from fits of the γ-dependence of $\xi^{(n)}$ near γ_c in Fig. 5 is fitted for their n-dependence according to (9). The estimates for ν and θ that result are in all three cases consistent with $\nu = 1/2$ and $\theta \approx 3/2$.

exponent ν_n, which is not consistent with (8). This suggests that $\theta > 1$ in (7). In order to check if this is the case, the exponents ν_n calculated from the fits in Fig. 5 (and an additional value $\alpha = 2.0$) were fitted for their n-dependence using (9), which would hold if $1 < \theta < 2$. The results are shown in Fig. 6 for $\alpha = 0, 2$, and 2.5. All three cases are roughly consistent with $\nu \approx 1/2$ and $\theta \approx 3/2$.

The fact that $\theta > 1$ and therefore that ξ^n do not scale with the same critical exponent is rather unusual, so it is convenient to provide independent evidence of this fact. By fitting $\mathcal{N}_{crn}(\ell)$ using (7) near γ_c (Fig. 7a) a power-law decay exponent $\theta = 1.46 \pm 0.05$ is obtained, which is consistent with the result derived from the analysis of the γ-dependence of the moments of $\mathcal{N}_{crn}(\ell)$. The plot at the center (Fig. 7b) shows that, after subtracting the exponential decay, the remaining ℓ-dependence is very accurately described by a power law $\ell^{-\theta}$.

However, notice that analytical results [16] for k-core in the $\alpha = 0$ case indicate that $\theta = 0$ should be observed instead. On the other hand, in MF percolation models, the connectivity function $\mathcal{N}_{crn}(\ell)$ usually decays exponentially [30], i.e. $\theta = 0$. Our numerical finding $\theta \approx 3/2$ in (7) is therefore rather puzzling and deserves further study.

3.4. The second-order Percolation transition at large α

For large values of the decay exponent α, the topology of our networks becomes two-dimensional and therefore we should recover the known results for Bootstrap Percolation in two dimensions,

[1] Admittedly, only very roughly, since γ_c has to be estimated as well.

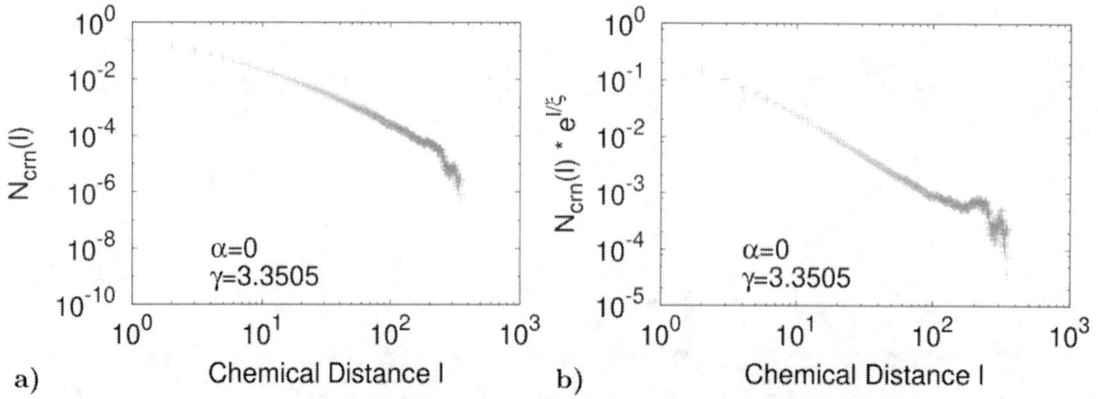

Figure 7. a) The number $\mathcal{N}_{crn}(\ell)$ (red crosses) of corona sites ℓ steps away from a given k-core site is described acceptably well by (7), as this plot shows. In this plot data for $\alpha = 0$ and $\gamma_c = 3.3505$ are shown. The green line is a fit of the form (7), which results in $\theta = 1.46 \pm 0.05$ and $\xi \approx 10^2$. b) The same data as in a), except for the fact that the exponential dependence $e^{-\ell/\xi}$ has been subtracted, showing that the remaining ℓ-dependence is to a very good approximation a power-law decay $\ell^{-\theta}$.

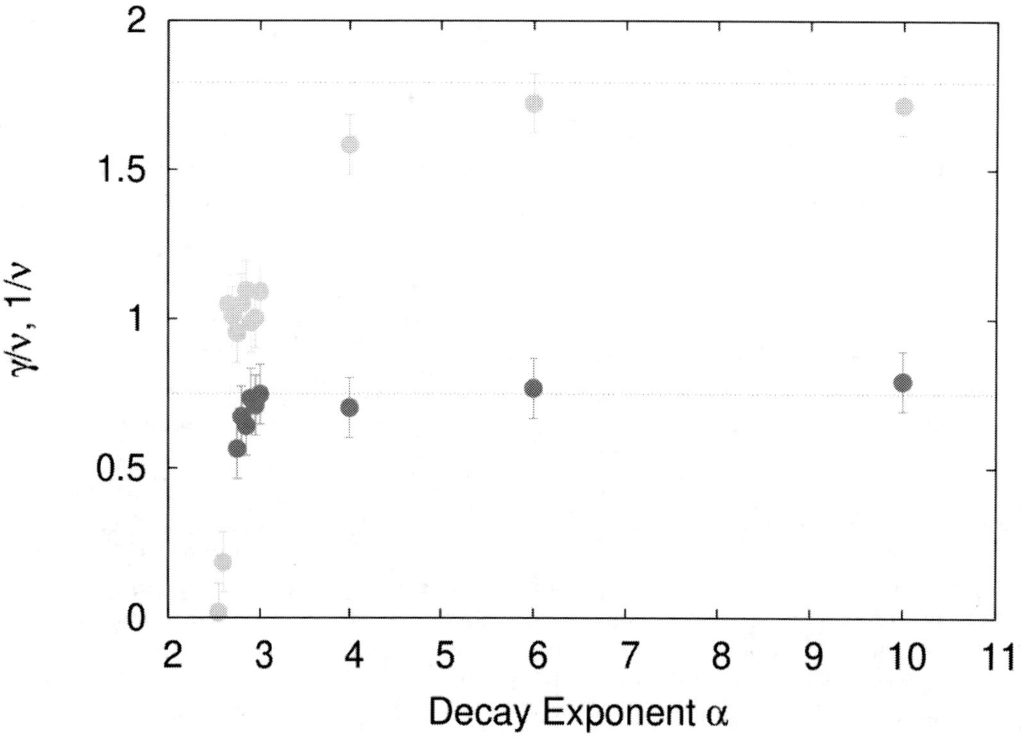

Figure 8. Critical scaling exponents, at the second-order transition, for the mass S of a finite core cluster (green dots) and for $\partial \log P_\infty / \partial \gamma$ (red dots) for various values of α. These quantities scale respectively as $L^{\tilde{\gamma}/\nu}$ and $L^{1/\nu}$ at γ_c. Dotted lines indicate the expected (2d Percolation) values of these scaling exponents when α is large.

i.e. critical indices in the universality class of usual Percolation. This is verified in this section for the critical indices ν and γ.
The correlation length exponent ν can be estimated, for a second-order transition, from FSS of $\partial \log P_\infty/\partial\gamma$, which at γ_c scales as $L^{1/\nu}$ [31]. For large α one expects to recover $\nu = 4/3$ of 2d Percolation [3]. Similarly, the average mass S of a finite connected cluster (disregarding the largest cluster) behaves at the 2d Percolation transition as $S_L(\gamma_c) \sim L^{\hat{\gamma}/\nu}$, where $\hat{\gamma} = 43/18$ is the susceptibility exponent [3].
FSS exponents for S (green circles) and $\partial \log P_\infty/\partial\gamma$ (red circles) are shown in Fig. 8 for various values of the decay exponent α. Dashed lines indicate the expected values from 2d Percolation. For $\alpha > 4.0$ both scaling indices rapidly approach their expected 2d Percolation values, which is consistent with the notion [32] that k-core is in the universality class of usual Percolation whenever it has a second-order transition.
These scaling exponents decrease strongly around $\alpha \approx 2.75$ which is consistent with other numerical evidence indicating that the second-order transition no longer exists for $\alpha < 2.75$.

3.5. Two transitions ?

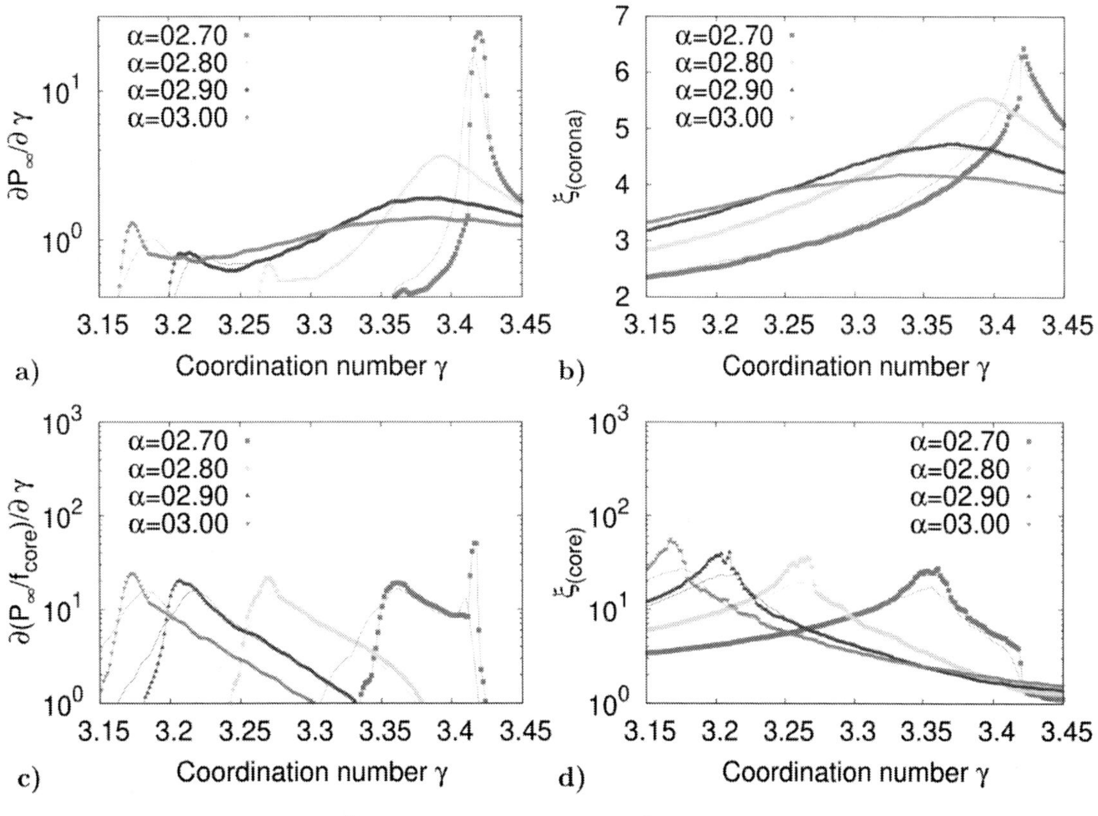

Figure 9. Data for $N = 10^7$ (symbols) and $N = 10^6$ (thin lines) help identify which of the maxima grow with N, in each of these four plots.

Data for $\partial P_\infty/\partial\gamma$ in Fig. 9a show two distinct maxima for each α in the range $2.7 \leq \alpha \leq 3.0$. In order for a peak in this quantity to be associated with a phase transition, it must grow with increasing N. The leftmost maximum in each curve on this plot indicates the incipient

second-order transition. This identification is supported by the fact that the k-core correlation length ξ_{core}, given by the average gyration radius of a finite k-core cluster [3], diverges (Fig. 9d) around the same value of the coordination γ. Therefore a second-order transition with a diverging correlation length seems to exist for all $\alpha \geq 2.7$. An additional criterion supporting this conclusion is the fact that $\partial(P_\infty/f_{core})/\partial\gamma$, which more clearly indicates the second-order transition, also shows growing peaks at the same locations, as Fig. 9c indicates.
The rightmost peaks in Fig. 9a, on the other hand, indicate the location of the potential first-order transition. It is seen that these peaks grow only slowly with N for $\alpha \leq 2.8$, perhaps not at all for $\alpha = 2.9$, and definitely do not grow with N for $\alpha = 3.0$. Looking at ξ_{corona} in Fig. 9b a similar conclusion is reached. The corresponding peaks in ξ_{corona} grow slowly with N for $\alpha = 2.8$ and possibly 2.9 but not for $\alpha = 3.0$. This evidence would locate the upper limit of the first-order transition regime somewhere between $\alpha = 2.7$ and $\alpha = 3.0$. Notice that our results are compatible with the existence of a small α-region around $\alpha = 2.70$ where two phase transitions (one of second order and the other of first order) might occur as a function of γ. Also interesting is the fact that the location of the second-order transition has a strong α-dependence, while that of the first-order transition only a weak one.

4. Discussion

We have numerically studied 3-core Percolation on two-dimensional lattices with long-range links whose length distribution is $P(l) \sim l^{-\alpha}$. For $\alpha < \alpha_0 \approx 2.75$, a first-order critical transition is found, in accordance with previously known results in mean-field[33, 1, 34]. The corona, that is the subset with exactly three neighbors in the core, is critical at this first-order transition. A correlation length defined from the average chemical extension of the corona is found to diverge with a critical exponent $\nu \approx 1/2$. The decay of connectivity with chemical distance in the corona is found to be well described by (7) near γ_c, with $\theta \approx 3/2$, in slight discrepancy with recent analytical results [16] predicting $\theta = 0$.
For $\alpha > \alpha_0$ on the other hand we observe a second-order transition. Critical indices measured for this transition are consistent with known percolation indices when α is large.
Despite the relatively large system sizes analyzed, the limit α_0 between the first- and second-order regimes could not be determined very precisely, and is estimated to be $\alpha_0 = 2.75 \pm 0.25$. There might as well be a small region around $\alpha = 2.7$ in which two transitions happen: a second order transition followed by a first-order one at a larger γ. However this could well be an artifact due to finite size effects.

Acknowledgments

Financial support from CONACYT, Mexico, under project 48783-F is acknowledged. TS kindly thanks the 'Studienstiftung des deutschen Volkes' for their financial support of his stay at CINVESTAV. CFM wishes to thank the "Computational Physics for Engineering Materials" group of ETH Zürich, for their hospitality during his stay.

[1] Pittel B, Spencer J and Wormald N 1996 *J. Comb. Theory Ser. B* **67** 111–151
[2] Leath P L, Chalupa J and Reich G R 1979 *Bull Amer Phys Soc* **24** 362–362
[3] Stauffer D and Aharony A 1994 *Introduction to Percolation Theory* 2nd ed (Bristol: Taylor and Francis)
[4] Klimek P, Thurner S and Hanel R 2009 *J. Theor. Biol.* **256** 142–146
[5] Farrow C, Duxbury P M and Moukarzel C F 2005 *Physical Review E* **72**
[6] Farrow C L, Shukla P and Duxbury P M 2007 *J. Phys. A-Math. Theor.* **40** F581–F587
[7] Iwata M and Sasa S I 2009 *J. Phys. A-Math. Theor.* **42** 075005
[8] Saito K, Yamada T and Kazama K 2008 *IEICE Trans. Fundam. Electron. Commun. Comput. Sci.* **E91A** 3304–3311
[9] Tlusty T and Eckmann J P 2009 *J. Phys. A-Math. Theor.* **42** 205004

[10] Vanenter A C D 1987 *J. Stat. Phys.* **48** 943–945
[11] Aizenman M and Lebowitz J L 1988 *J. Phys. A-Math. Gen.* **21** 3801–3813
[12] Schonmann R H 1990 *J. Stat. Phys.* **58** 1239–1244
[13] Chaves C M and Koiller B 1995 *Physica A* **218** 271–278
[14] Parisi G and Rizzo T 2008 *Phys. Rev. E* **78** 022101
[15] Schwartz J M, Liu A J and Chayes L Q 2006 *Europhys. Lett.* **73** 560–566
[16] Goltsev A V, Dorogovtsev S N and Mendes J F F 2006 *Physical Review E* **73**
[17] O'Hern C S, Silbert L E, Liu A J and Nagel S R 2003 *Phys. Rev. E* **68** 011306
[18] Moukarzel C F 1998 *Phys. Rev. Lett.* **81** 1634–1637
[19] Moukarzel C F 2001 *Granul. Matter* **3** 41–52
[20] Xu N, Wyart M, Liu A J and Nagel S R 2007 *Phys. Rev. Lett.* **98** 175502
[21] Silbert L E, Liu A J and Nagel S R 2005 *Phys. Rev. Lett.* **95** 098301
[22] Kleinberg J M 2000 *Nature* **406** 845–845
[23] Jespersen S and Blumen A 2000 *Phys. Rev. E* **62** 6270–6274
[24] Moukarzel C F and de Menezes M A 2002 *Phys. Rev. E* **65** 056709
[25] Moukarzel C F 2005 *Physica A* **356** 157–161
[26] Bollobás B 2001 *Random Graphs* 2nd ed (Cambridge: Cambridge University Press)
[27] Fisher M E, Nickel B G and Ma S 1972 *Phys. Rev. Lett.* **29** 917–&
[28] Nguyen B G 1987 *J. Stat. Phys.* **46** 517–523
[29] Hara T 1990 *Probab. Theory Relat. Field* **86** 337–385
[30] Bunde A and Havlin S (eds) 1996 *Fractals and Disordered Systems* 2nd ed (Heidelberg: Springer Verlag)
[31] Mailhot A, Plumer M L and Caille A 1994 *Phys. Rev. B* **50** 6854–6859
[32] Branco N S and Silva C J 1999 *Int. J. Mod. Phys. C* **10** 921–930
[33] Chalupa J, Leath P L and Reich G R 1979 *J Phys-C-Solid State Phys* **12** L31–L35
[34] Moukarzel C, Duxbury P M and Leath P L 1997 *Phys. Rev. E* **55** 5800–5811

The role of energy in the dynamics of the HMF model

Carolina B. Tauro, Cesar G. Maglione and Francisco A. Tamarit

Facultad de Matemática, Astronomía y Física, Universidad Nacional de Córdoba, Ciudad Universitaria, 5000 Córdoba, Argentina.
Instituto de Física Enrique Gaviola, UNC-CONICET

E-mail: tauro@famaf.unc.edu.ar

Abstract. In this work we analyze the Hamiltonian Mean Field model, which consists of a set of N interacting rotators that move around an unitary circle. This model presents a very complex dynamical behavior and, in particular, the canonical predictions do not necessarily coincide with the temporal averages obtained along numerical simulations. Recently, a topological explanation for the existence of these anomalies has been introduced. In this work we further analyze this hypothesis by considering the role of the energy of the system in the dynamics of the model.

The Hamiltonian Mean Field (HMF) model [1] can be considered a paradigmatic complex system. Despite its apparent extreme simplicity, its dynamical behavior has revealed a great amount of surprinsing outcomes, never expected before in a ferromagnetic infinite range model [2]. The system consists of N *rotators*, each one moving on its own unitary circle. Mechanically, each rotator is completely described by the angle θ_i it forms with the x axis and the corresponding generalized momentum p_i, and the system is ruled by the Hamiltonian

$$H = \frac{1}{2}\sum_{i=1}^{N} p_i^2 + \frac{1}{2N}\sum_{i,j=1}^{N}[1 - \cos(\theta_i - \theta_j)]. \tag{1}$$

The first term represents the kinetic energy of the system and the second one the interacting potential energy. The second sum runs over all pairs of particles, indicating the infinite range character of these interactions. Note that, for fixed kinetic energy, this second term is minimized when all the particle rotate syncrhonously with the same angle.

This model is nothing but a kinetic version of the infinite range XY ferromagnetic model. In that sense, one can associate to each rotator a planar local magnetization $\vec{m}_i = (\cos\theta_i, \sin\theta_i)$ and then define a global order parameter as, $\vec{M} = \frac{1}{N}\sum_{i=1}^{N}\vec{m}_i$. From a thermodynamical point of view, the free energy of the model can be very easily evaluated in the canonical ensemble [1] yielding a second order phase transition between a high temperature paramagnetic phase (characterized by $|\vec{M}| = 0$) and a low temperature ferromagnetic phase (whit $|\vec{M}| \neq 0$). The scenario is very different when considering its dynamical evolution. One can numerically integrate the Hamilton equations of motion, given by,

$$\dot{\theta}_i = \frac{\partial H}{\partial p_i} = p_i, \qquad \dot{p}_i = -\frac{\partial H}{\partial \theta_i} = -\frac{1}{N}\sum_{j=1}^{N}\sin(\theta_i - \theta_j), \quad i = 1, \ldots, N, \tag{2}$$

and then compare the canonical predictions with temporal mean values of certain physical observables. Surprisingly, one finds that, depending on the initial preparation of the system, the expected equivalence fails to hold. This phenomenology has been observed for energies just below the critical transition, which occurs at $U_c/N = 3/4$ [1, 2]. Consider for instance the case of the temperature, which in true equilibrium must coincide with twice the kinetic energy per particle $T_k = 2\frac{\langle K\rangle}{N}$. Let us assume that the system is in thermal contact with a thermal bath at certain temperature T. The expected equivalence between the canonical results and those obtained through a microcanonical numerical integration is not always valid for the HMF model. When the simulations are performed with energies just below U_c and start very far from equilibrium, the system can eventually get trapped into long standing out of equilibrium trajectories. A simple way of preparing the system far from equilibrium is by putting initially the system in a "water–bag "configuration, in which all the angles have the same value ($|\vec{M}| = 1$) while the momenta are chosen at random from an uniform distribution. Because $V/N = (1 - |\vec{M}|^2)/2$, the initial potential energy per particle is zero and the system has only kinetic energy $U = K$. The time evolution of $T_k = 2 < K > /N$ shows an initial drastic cooling down, but instead of reaching the expected equilibrium value T, it attains a quasi-stationary value $T_{qs}(N) < T$ which depends on the system size N in such a way that, if the system were infinitely large, it would get trapped into this quasi-stationary solutions forever [1, 2].

On the other hand, from a topological point of view, the potential energy per particle, V/N, foliates the configuration space of the system into the family $\{\Sigma_v\}_{v\in\mathbb{R}}$ of equipotential hypersurfaces, where $\Sigma_v = \{(\theta_1, \dots, \theta_N) \in \mathbb{R}^N | V/N(\theta_1, \dots, \theta_N) = v\}$ (with $0 \le v \le 1/2$). The configurational space is an N-dimensional torus parametrized by the N angles. In the ordered phase the system is confined to move in half a torus, characterized by critical points that correspond to saddle nodes. At the critical value $v_c = 1/2$ a drastic topological change occurs and the sytems can then visit the complete torus. This drastic change is precisely related to the second order phase transition the system suffers at $U_c/N = 3/4$. This approach, which has been vastly used in the context of critical phenomena to relate topological changes in configurational spaces with phase transitions in finite systems [3], is also connected to the anomalous dynamical behavior observed in the HMF model.

In [4, 5] it was showed that the complex nonequilibrium quasi-stationary regime observed just below the critical energy can also be understood in terms of the drastic topologic change observed at $v_c = 1/2$. In order to verify this hypothesis, the system was prepared in a water–bag initial configuration, with all the rotators aligned along the x axis ($\theta_i = 0$ for all i), the momenta drawn from an uniform distribution (actually we used regularly spaced momenta) and energy per particle $U/N = 0.69$ (just below the transition point $U_c/N = 3/4$). Under these conditions, the system has initially zero potential energy and suddenly decreases its kinetic energy (temperature) as much as possible, settling at the flat top of its potential energy function ($v_c = 1/2$). Then, it starts to wander among different saddle nodes of the critical manifold $v_c = 1/2$, all of them characterized with only two negative directions and $N - 2$ marginal ones. To do this analysis, one has to evaluate at each time step of a numerical integration the modulus of the N derivatives of V/N and identify the maximum over $i = 1, \dots, N$, through

$$\lambda = N \max_i \left| \frac{\partial(V/N)}{\partial\theta_i} \right|. \tag{3}$$

Note that $\lambda = 0$ implies $\partial(V/N)/\partial\theta_i = 0$ for all the particles. In other words, each time λ vanishes we can assert that the system passes over a critical point of V/N. These results confirmed that, in the quasi-stationary trajectories, the system is trapped in the critical submanifolf $v_c = 1/2$. Moreover, the time $\Delta\tau$ elapsed between two successive critical points for fixed energy $U/N = 0.69$ has a power-law distribution, indicating a very complex pattern of wandering on the configurational space. A different scenario appears when the system is initially prepared also in a water-bag confirguration but with $U/N = 10.0$, in the disordered phase. Now one observes a periodic pattern of visits to the critical submanifold $v_c = 1/2$.

As mentioned above, in [4] only the cases $U/N = 0.69$ and $U/N = 10$ were considered. In this paper we extend the previous work [4] in order to analyze the role of the energy of the system. In particular, we consider here systems with different energies, ranging from $U/N = 0.60$ (in the ordered phase) to $U/N = 2.0$ (in the disoredered phase) and follow the same protocol described in [4]. The initial momenta have been always chosen from an uniform random distribution and the angles were fixed along the x axis. The numerical simulations were performed by using a fourth-order symplectic algorithm [6] in system of $N = 500$ rotators (we have verified that for larger sizes the behavior is the same). Along each simulation we evaluated λ as a function of t for a single sample.

In Fig. 1 we can clearly identify three different regimes. In the top plot we present the behavior of $\lambda(t)$ in the disordered phase, with energies per particle $U/N = 0.78$, 1.0 and 2.0. We observe that the system periodically touches the critical submanifold $v_c = 1/2$ (each time $\lambda = 0$) and that the period Δt decreases as U/N increases. In the middle plot we describe the behavior observed just below the critical point. In the three cases analyzed, $U/N = 0.74$, 0.71 and 0.68, the system wanders in the configurational space touching the critical submanifold. But, unlike the previous case, now the behavior of λ has lost the periodicity. Instead we verify that the system has a complex pattern of wandering, as described in [4] for $U/N = 0.69$. In the bottom plot we present the behavior of $\lambda(t)$ also in the ordered phase but when the system is prepared with energies $U/N = 0.60$, 0.63 and 0.66. Now, it is initially attracted to the critical submanifod $v_c = 1/2$ but after a few visits, it finally scapes. Once the system scapes, it will never again touch a critical point. At very low energies (which are not presented here) the system wanders always far away from the critical submanifold $v_c = 1/2$.

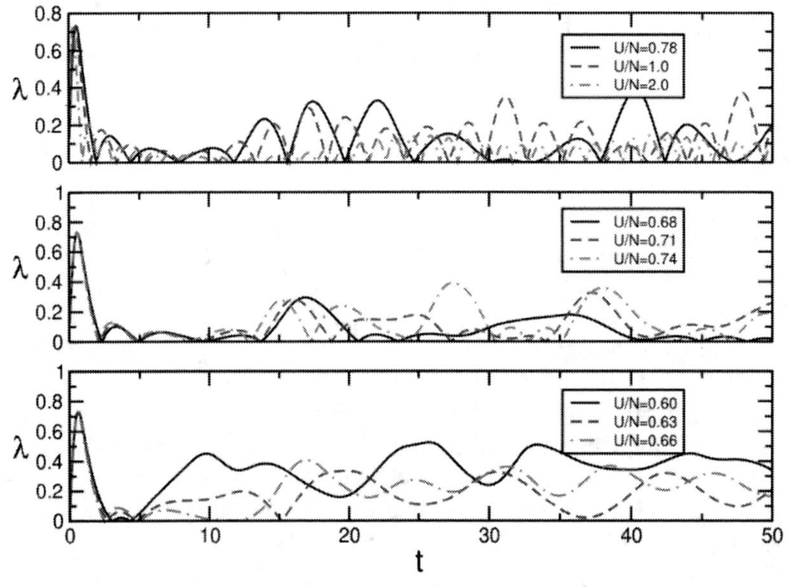

Figure 1. Time evolution of λ for three different regimes and water–bag initial conditions: disordered phase (top), ordered phase with energies just below U_c/N (middle) and ordered phase with intermediate energy values (bottom).

As described in the previous figure, in the high energy phase the system presents a regular pattern of wandering on the configurational space, touching the critical submanifold $v_c = 1/2$

in a periodic manner. In Fig. 2 we present, in a log-log plot, a study on the dependency of the period Δt on the energy of the system. We verify that, for large values of U/N, the period decays as a power law $\Delta t \propto t^{-\alpha}$, with $\alpha \approx 0.60$, indicating that the periodic behavior will always exist, independently of the energy of the system.

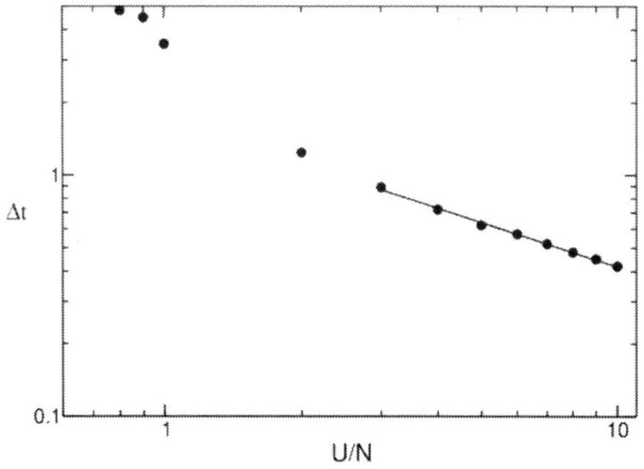

Figure 2. The period Δt between two sucessive visits to the critical sub-manifold $v_c = 1/2$ as a function of U/N in a log-log plot. The solid line corresponds to a power-law fitting of the last eight points.

Concluding, in this work we have extended the results presented in [4] by analyzing the role of the energy of the HMF in the behavior of $\lambda(t)$. Taking into account that each time the system touches the critical submanifold the parameter λ equals zero (indicating that the system passes over a saddle node of the the potential energy function per particle) we were able to identify four different regimes. First, in the high energy disordered phase, the system visits periodically the submanifold $v_c = 1/2$ and the period decreases as a power-law as U/N increases. In the ordered phase we found three different behaviors. Just below the critical point U_c/N the system visits the manifold $v_c = 1/2$ in a very irregular way. For intermediate values of U/N the system is iniatially attracted by the critical sub-manifold but finally scapes. In the low energy regime the system wanders in the configurational space without visiting the critical submanifold.

This work was partially supported by grants from CONICET, FONCyT Grant No. PICT-2005 33305 and SeCyT, Universidad Nacional de Córdoba, Argentina.

References

[1] Antoni M and Ruffo S 1995 *Phys. Rev.* E **52** 2361
[2] Latora V, Rapisarda A and Ruffo S 1998 *Phys. Rev. Lett.* **80** 692
Latora V, Rapisarda A and Ruffo S 1999 *Phys. Rev. Lett.* **83** 2104
Latora V and Rapisarda A 2001 *Chaos, Solitons Fractals* **13** 401
Latora V, Rapisarda A and Tsallis C 2001 *Phys. Rev.* E **64** 056134
Latora V, Rapisarda A and Tsallis C 2002 *PHYSICA* A **305** 129
Anteneodo C and Tsallis C 1998 *Phys. Rev. Lett.* **80** 5313
Tamarit F and Anteneodo C 2000 *ibid.* **84** 208
Zanette D H and Montemurro M A 2003 *Phys. Rev.* E **67** 031105
Yamaguchi Y Y, Barré J, Bouchet F, Dauxois T and Ruffo S 2004 *PHYSICA* A **337** 36
[3] Casetti L, Cohen E G D and Pettini M 1999 *Phys. Rev. Lett.* **82** 4160
[4] Tamarit F A, Maglione C G, Stariolo D A and Anteneodo C 2005 *Phys. Rev.* E **71** 036148
[5] Anteneodo C and Tamarit F A 2006 *PHYSICA* A **371** 135
[6] Yoshida H 1990 *Phys. Lett.* A **150** 262

Journal of Physics: Conference Series **246** (2010) 012021 doi:10.1088/1742-6596/246/1/012021

Orientational order in systems with competing interactions

Daniel A Stariolo[1] **and Daniel G Barci**[2]

[1] Departamento de Física, Universidade Federal do Rio Grande do Sul and National Institute of Science and Technology for Complex Systems
CP 15051, 91501-970 Porto Alegre, RS, Brazil

[2] Universidade do Estado do Rio de Janeiro, Rua São Francisco Xavier 524, 20550-013, Rio de Janeiro, RJ, Brazil

E-mail: daniel.stariolo@ufrgs.br

Abstract. Quasi two dimensional systems with competing interactions usually display complex patterns in the relevant order parameter. In many cases these patterns are analog to liquid-crystal phases, showing smectic, nematic or hexatic order. We show that order parameters suitable for the characterization of these phases in systems with nearly isotropic competing interactions emerge naturally from an analysis of a Landau model. We describe with some detail the nematic case, which characterizes orientational order of striped domain walls. The Landau model presents an isotropic-nematic transition of the Kosterlitz-Thouless type. Although for the perfectly isotropic model long range nematic order is absent in infinite systems, we show that in real systems of finite size nematic order of domain walls can be observed.

1. Introduction

In many natural systems complex patterns originate due to the presence of competing interactions at different scales. Examples range from classical systems like magnetic ultrathin films [1], diblock copolymers [2] and colloidal suspensions [3], to quantum systems like quantum Hall samples [4] and high T_c superconductors [5]. These pattern forming systems may present positional as well as orientational order, the origin of which depends on microscopic interactions. In the case of magnetic ultrathin films, of a few atomic monolayers thick, strong anisotropy perpendicular to the film plane forces the magnetization to be out of plane. In this case short range ferromagnetic interaction competes with antiferromagnetic dipolar interaction. As a consequence, the system shows ordered striped phases, of alternating magnetization along one direction, as is seen in Figure 1.

In Figure 2 similar patterns can be observed in a completely different system, a block copolymer. In this case, elongated polymer chains with different types of building "blocks" compete between them and produce phase segregation, giving rise to complex patterns in two and three dimensions.

Figure 3 shows a schematic representation of microemulsion phases. Microemulsions are mixtures of anfiphilic molecules which have an hydrophilic head and an hydrophobic tail. When in contact with water and oil, as depicted in the figure, they tend to phase separate giving rise to different kinds of ordered patterns with an order similar to liquid crystals [7, 8].

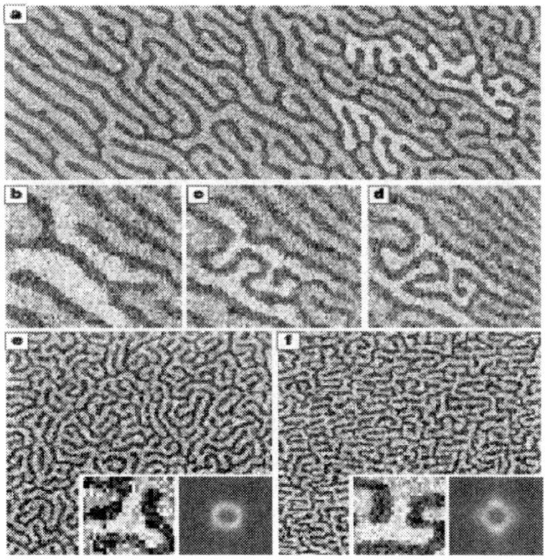

Figure 1. Fe/Cu(001) ultrathin films. The images, taken with scanning electron microscopy with polarization analysis (SEMPA), show a sequence of characteristic patterns and topological defects. From [1].

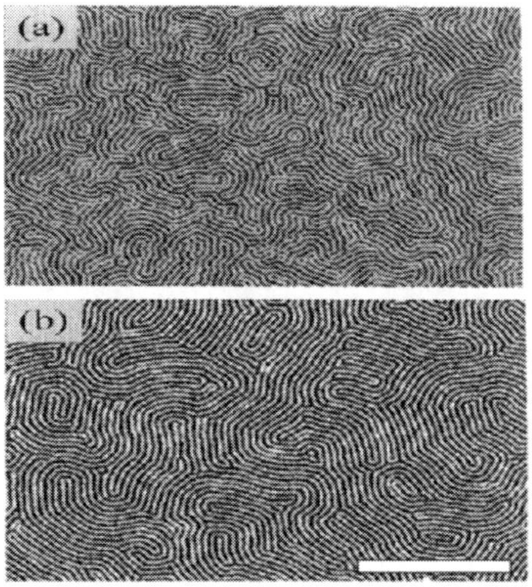

Figure 2. SBS block copolymers annealed on a bare silicon wafer and a PS brush-coated wafer respectively. From [6].

The equilibrium pattern in these systems depends in general in details of the interactions, like anisotropies, anchoring with a substrate, whether interactions are short or long range. Usually the kinematics shows several well defined time scales, and the ultimate equilibrium pattern is difficult to attain due to very long relaxation times. Despite these important characteristics, even the ideal equilibrium phases in simple situations are poorly understood. In general, there is a characteristic temperature T_C where the system begins to phase separate and a characteristic length scale sets in, usually called the modulation length, l_m. Below this characteristic temperature the equilibrium pattern adopts a labyrinth-like structure, like in

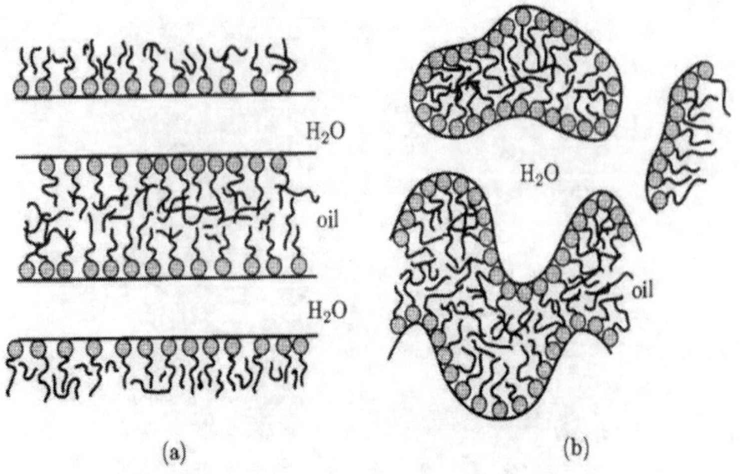

Figure 3. (a) A lamellar microemulsion phase with water, oil and surfactant layers. (b) Schematic representation of a random bicontinuous phase in which there is a random surfactant surface separating oil and water regions. From [9].

Figure 2. These patterns show neither positional nor orientational order, but a well defined stripe-like structure with a characteristic modulation length is easily recognized. Upon further lowering temperature or changing concentration of one species orientational order of the stripes begins to be observed at a temperature T_N. Domains of elongated stripes form and a new characteristic length appears in the system, l_s. Locally positional order develops together with orientational order of stripe domains walls. This regime may show long range orientational order but only short range positional order. In the presence of anisotropies or anchoring a further transition to a state with full positional order can happen at a temperature T_S.

Although this phenomenology is well known, the theoretical description of this sequence of phase transitions and the changing orders associated with them is usually done in terms of analysis of elasticity of domain walls [10, 11] instead of, e.g. microscopic approaches. A long time ago Brazovskii [12] proposed a Landau-type model for systems with nearly isotropic competing interactions. He obtained the important result that, although the Landau expansion in the model had only square and quartic powers of the order parameter, a first order phase transition induced by fluctuations was observed, contrary to what could be expected a priory. The essential feature behind this behavior is that the dominant low energy mode is not the usual $k = 0$ one, but a mode with $k \neq 0$, which in fact defines the modulation length of the order parameter. Brazovskii's analysis had a strong influence on subsequent work on these kind of systems, for example in explaining the low temperature behaviour of diblock copolymers [13] and the also well known Swift-Hohenberg model for the onset of instability in Raleigh-Bénard convection [14, 15]. Nevertheless, analysis of the Brazovskii's model were limited to the isotropic-stripe transition, which was the last one described above in the possible sequence of phase changes. The low temperature stripes or modulated phase is strongly unstable in two dimensions, as recognized long time ago by Swift and Hohenberg [14]. Nevertheless, stripes with long range orientational order are seen in experiments in many different systems belonging to the Brazovskii class. Clearly, a more comprehensive theoretical description of orientational phases in systems with isotropic or nearly isotropic competing interactions is needed.

An extension of the Brazovskii model, capable of describing phases with orientational but

Journal of Physics: Conference Series **246** (2010) 012021 doi:10.1088/1742-6596/246/1/012021

not necessarily positional order, has recently been discussed by us [16, 17]. We have identified a suitable order parameter of domain wall orientation, similar to the director of nematic liquid crystals. Interestingly, this order parameter emerges naturally from a proper analysis of the Landau expansion for the kind of systems of the Brazovskii class. The resulting Landau free energy describes the two relevant transitions from isotropic to nematic and from nematic to stripes, in agreement with experimental observations.

Here we first describe a nematic-like order parameter suitable for quantifying orientational order of domain walls in two dimensional stripe-forming systems. Then we outline the derivation of a Landau free energy where this order parameter emerges naturally from symmetry considerations. We present and discuss the results of a mean field calculation of this free energy and show that the model has a nematic-isotropic second order phase transition [16]. The consequences of fluctuations around the mean field solution are also discussed [17]. We show that, although in two dimensional systems with isotropic competing interactions there is no true long range nematic order, in typical experimental scales nematic long range order can be observed.

2. The nematic order parameter

The kind of order we want to describe is exemplified in Figure 4.

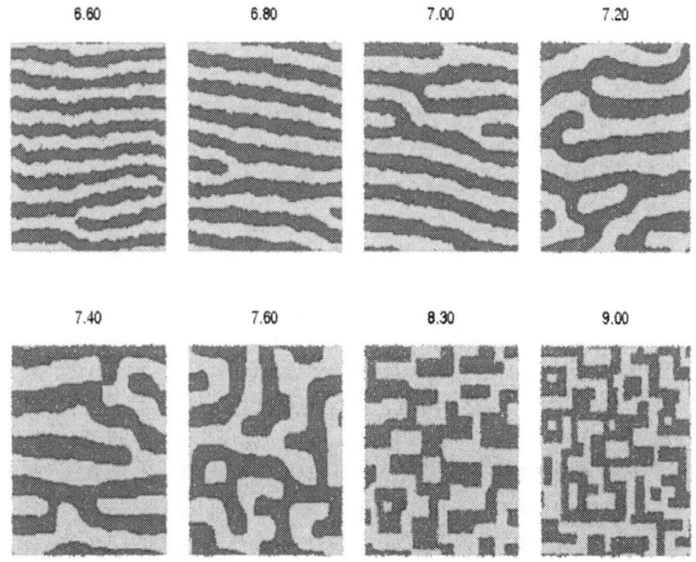

Figure 4. Computer simulation of the temperature evolution of patterns in a two dimensional Heisenberg ferromagnet with perpendicular anisotropy and competing dipolar interactions [18].

Competition between short range attraction and long range repulsion, or short range ferromagnetic and long range antiferromagnetic interactions usually gives rise to stripe-like patterns of the order parameter, which is generically represented by a density or magnetization $\phi(\vec{x})$.

Domain walls are observed at the transitions between positive and negative values of the order parameter. In order to quantify the degree of orientation of domain walls one can define the *director field*, defined as the gradient of the order parameter:

$$\mathbf{N}(\vec{x}) \equiv \nabla\phi(\vec{x}) = (\partial_x\phi, \partial_y\phi) \tag{1}$$

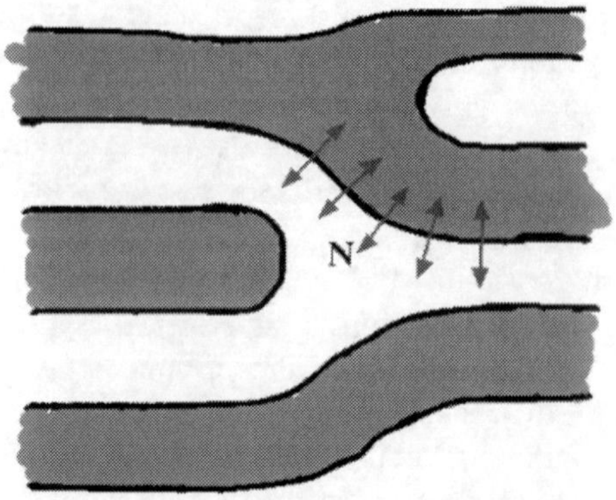

Figure 5. Schematic representation of a stripe phase with a dislocation. The arrows represent the director field, which is a local measure of orientation of domain walls.

Defined in this way the director is a vector which defines a vector field over the area of the two dimensional system. But this is not exactly the kind of order parameter we are seeking for. It is clear from Figure 5 that orientational order does not depend on up-down or left-right vector directions, both orientations along the perpendicular to the domain walls are equivalent. One way to incorporate this symmetry is to define a *tensor* order parameter, similar to the tensor order parameter of nematic liquid-crystals. In terms of the gradient of the density ϕ, the local tensor order parameter is defined as:

$$Q_{ij}(\vec{x}) \equiv \phi(\vec{x}) \left(\partial_i \partial_j - \frac{1}{2} \partial^2 \delta_{ij} \right) \phi(\vec{x}), \tag{2}$$

where $i, j = \{x, y\}$. This tensor is symmetric and traceless, and in two dimensions it has only two independent elements, which essentially represent the mean orientation of domain walls and the strength of the orientational order.

In order to get some feeling of the physical content of this order parameter it is useful to write it in reciprocal space. Introducing the Fourier transforms:

$$\tilde{Q}_{ij}(\vec{k}) = \int d^2x \, Q_{ij}(\vec{x}) \, e^{-i\vec{k}\cdot\vec{x}} \qquad\qquad Q_{ij}(\vec{x}) = \int \frac{d^2k}{(2\pi)^2} \tilde{Q}_{ij}(\vec{k}) \, e^{i\vec{k}\cdot\vec{x}}, \tag{3}$$

we get

$$\tilde{Q}_{ij}(\vec{k}') = \int d^2k \left(k_i k_j - \frac{1}{2} k^2 \delta_{ij} \right) \tilde{\phi}(\vec{k}) \tilde{\phi}(\vec{k}' - \vec{k}) \tag{4}$$

The mean global order parameter is given by:

$$\begin{aligned} \langle Q_{ij} \rangle &= \int d^2x \; Q_{ij}(\vec{x}) = \langle \tilde{Q}_{ij}(\vec{k}' = 0) \rangle \\ &= \int d^2k \left(k_i k_j - \frac{1}{2} k^2 \delta_{ij} \right) \langle \tilde{\phi}(\vec{k}) \tilde{\phi}(-\vec{k}) \rangle \end{aligned} \tag{5}$$

Choosing the x-axis as the principal axis, the only relevant element of the tensor is:

$$\langle Q_{xx} \rangle = \int d^2k \, k^2 \, \cos(2\theta) \, C(\vec{k}) \tag{6}$$

where $k_x = k\cos\theta$ and $k_y = k\sin\theta$ and $C(\vec{k})$ is the structure factor of the system. Written in this way the orientational order parameter quantifies the degree of anisotropy of the domain pattern. In a completely isotropic phase, for example a liquid phase or a mosaic of domains with no preferential direction, the corresponding isotropy in the structure factor will be reflected in a zero value of the orientational order parameter. This is illustrated in Figure 6.

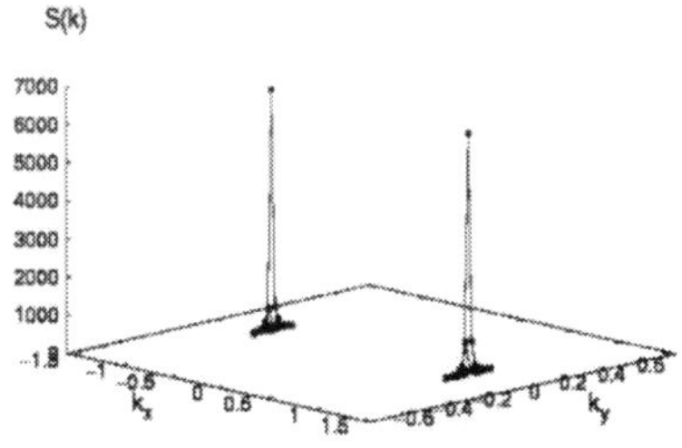

$\langle Q\rangle \sim 1$

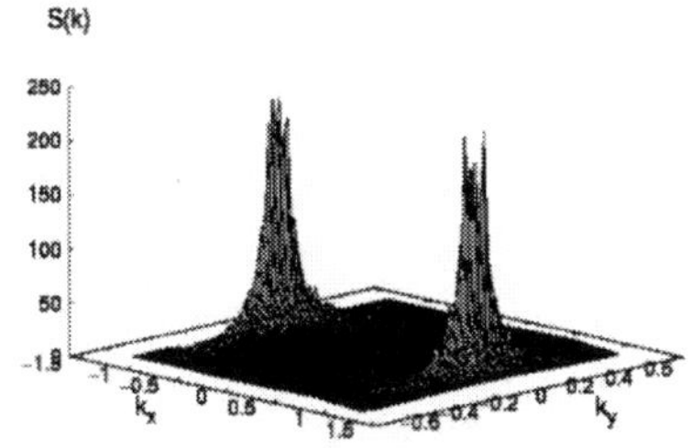

$\langle Q\rangle \sim 0.8$

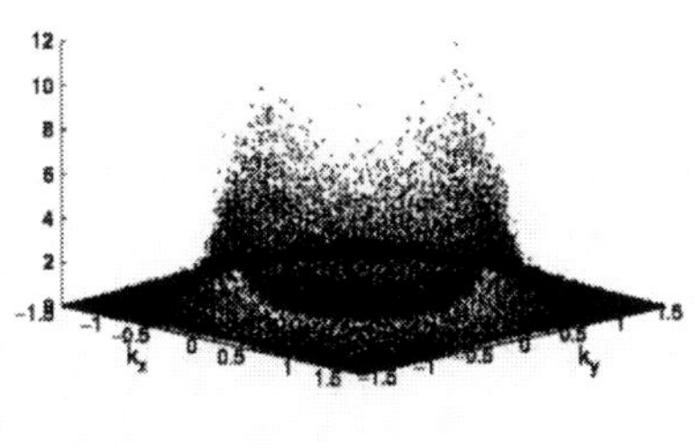

$\langle Q\rangle \sim 0.1$

Figure 6. Structure factor of a model with competing exchange and dipolar interactions in $d = 2$ for three different temperatures. The pictures from top to bottom are characteristic of the stripes, nematic and isotropic phases respectively. Taken from [19].

The three figures represent the structure factor of a model for ultrathin magnetic films with strong perpendicular anisotropy, which has short range ferromagnetic interactions competing with long range dipolar interactions [19]. The top figure shows $C(\vec{k})$ for a very low temperature, where the equilibrium state of the system presents almost perfectly straight stripes of up and down magnetization. $C(\vec{k})$ presents two very narrow peaks at a wave vector corresponding to the modulation length of the stripes. The system has both orientational and positional order. The second figure corresponds to a higher temperature. The two peaks are considerably broaden and positional order has been lost. Nevertheless orientational order is still present, as confirmed by the high value of the tensor order parameter $\langle Q \rangle \approx 0.8$. This corresponds to a nematic-like phase. The third figure shows an almost rotationally invariant structure factor. This isotropy reflects an absence of orientational order and corresponds to a disordered high temperature phase.

3. The extended Brazovskii model

The dominance of a mode with a non-zero wave vector at low energies can be studied through a phenomenological model of the Landau-Ginzburg type, first analyzed by Brazovskii [12]:

$$\mathcal{H} = \int \frac{d^2k}{(2\pi)^2}\, \phi(\vec{k}) \left(r_0 + \frac{1}{m}\,(k-k_0)^2 + \ldots \right) \phi(-\vec{k}) +$$

$$+\, u_0 \int \frac{d^2k_1}{(2\pi)^2}\frac{d^2k_2}{(2\pi)^2}\frac{d^2k_3}{(2\pi)^2}\frac{d^2k_4}{(2\pi)^2}\, \phi(\vec{k}_1)\phi(\vec{k}_2)\phi(\vec{k}_3)\phi(\vec{k}_4)\, \delta(\vec{k}_1+\vec{k}_2+\vec{k}_3+\vec{k}_4) \tag{7}$$

This model, when considered at mean field level, gives rise to a isotropic-stripes transition which is second order. Below T_c the order parameter varies in a sinusoidal way:

$$\phi(\vec{x}) = A\ \cos(\vec{k}_0 \cdot \vec{x}). \tag{8}$$

Brazovskii showed that including fluctuations in a self-consistent way, the model leads to a first order phase transition induced by fluctuations. The low temperature striped phase, dominated by a mode with k_0 wave vector, has positional long range order in the form of two symmetric delta function peaks in the structure factor, at least near the transition. At lower temperatures other modes come into play and the profile of the order parameter changes developing sharp domain walls as $T \to 0$. Although applied to some two dimensional systems, it is known that the low temperature stripe phase is strongly unstable in this case. Angular fluctuations of the dominant wave vector of modulus k_0 does not cost energy, as shown in Figure 7. Some kind of anisotropy in the interactions or pinning to a substrate are necessary in order to stabilize the phase.

Nevertheless, it was recently shown that orientational order of domain walls can persist even in the case of ideal isotropic interactions [16, 17]. Let's analyze the implications of isotropy for the interaction term, which in general will be a function $u(\vec{k}_1, \vec{k}_2, \vec{k}_3, \vec{k}_4)$, with wave vectors k_1, k_2, k_3 and k_4 as shown in Figure 7:

- Isotropy forces the wave vectors to lay on a circle of radius k_0 and hence $u(\vec{k}_1, \vec{k}_2, \vec{k}_3, \vec{k}_4) = u(\theta_1, \theta_2, \theta_3, \theta_4)$.
- Conservation of total momentum allows to eliminate one of the four $k's$.
- Fixing two of them, the other two are automatically fixed by the requirement that the four vectors lay on the circle.
- Because of rotational invariance, the interaction can only depend on the difference between the two independent angles:

$$u(\theta_1, \theta_2, \theta_1 + \pi, \theta_2 + \pi) = u(\theta_1 - \theta_2) = u(\theta)$$

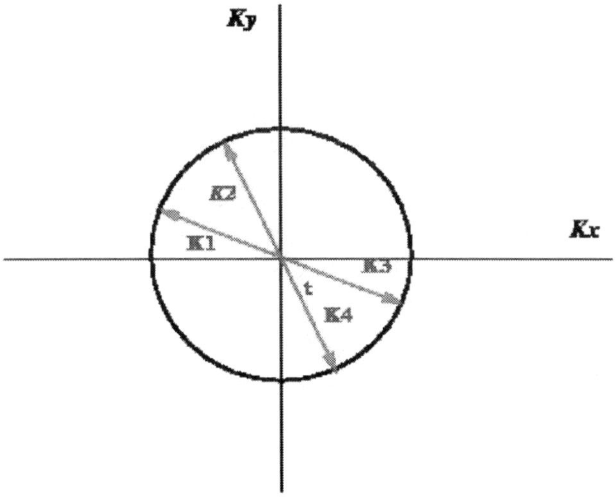

Figure 7. The wave vectors present in the quartic interaction term must lie on a circle of radius k_0 for perfectly isotropic interactions. This fact alone determine the form of the quartic term of the Landau expansion, as explained in the text.

- Finally, index permutation invariance in $u(\vec{k}_1, \vec{k}_2, \vec{k}_3, \vec{k}_4)$ implies $u(\theta) = u(\theta+\pi)$, then u can be expanded in Fourier series:

$$u(\theta) = u_0 + u_2 \cos(2\theta) + u_4 \cos(4\theta) + \dots \tag{9}$$

In the previous expression for the interaction we recognize the first, constant term, as that present in the original Brazovskii's model, eq. (7). But symmetry considerations led us to conclude that an infinite number of additional terms must be considered in the expansion, and it turns that all of them are relevant in a Renormalization Group sense [16]. From the form of eq. (9) one sees that all the cosine terms share a nematic symmetry, i.e. they are invariant with respect to rotations $\theta \to \theta + \pi$. Keeping the first two terms, we can rewrite the Brazovskii energy in the form:

$$\mathcal{H} = \mathcal{H}_{Br} + u_2 \int \frac{d^2k}{(2\pi)^2} Tr\, \tilde{Q}^2, \tag{10}$$

where $\mathcal{H}_{Br}$ is given by (7), $Tr\,\tilde{Q}^2$ is the trace of the nematic tensor squared, and $\tilde{Q}_{ij}(\vec{k})$ is given by eq. (4).

An analysis of this Hamiltonian in the self-consistent or Hartree approximation gives the following results [16, 17]:

- If $u_2 = 0$ we recover the Brazovskii's Hamiltonian, which has a low temperature smectic phase with a stripe-isotropic first order transition induced by fluctuations. In two dimensions this solution is unstable, and then $T_{Br} \to 0$.
- If $u_2 > 0$ the nematic term has a repulsive character, and the only possible solution for the nematic order parameter is $\langle Q(T)\rangle = 0$.
- When $u_2 < 0$ the nematic energy is attractive and the model shows a *nematic-isotropic* phase transition. The transition is second order in the mean field approximation, with $\langle Q(T)\rangle \propto (T_c - T)^{1/2}$ and the critical temperature is given by $T_c = 2/(mk_0^2)\,\sqrt{u_0/|u_2|}$.

A representation of the mean field phase diagram for the extended model can be seen in Figure 8. We see that the stripe phase is depressed to zero temperature. This is because of strong fluctuations in the orientation of the dominant wave vector k_0 in two dimensions. In reference [20] a criticism to the results in [16] was raised, based on the fact that the stripe solution is strongly unstable for isotropic interactions in two dimensions. In the reply to that comment [21] we clarified the controversy, which was probably funded in a misinterpretation of our main conclusion, i.e. the existence of a *nematic* phase with a finite critical temperature, in addition to the stripe (smectic-like) phase which is depressed to zero temperature as a consequence of fluctuations. As we said before, this fluctuations can be made weaker by means of anisotropies originated e.g. by an underlying lattice which breaks rotational symmetry. In this case it is expected that T_{Br} will have a finite value.

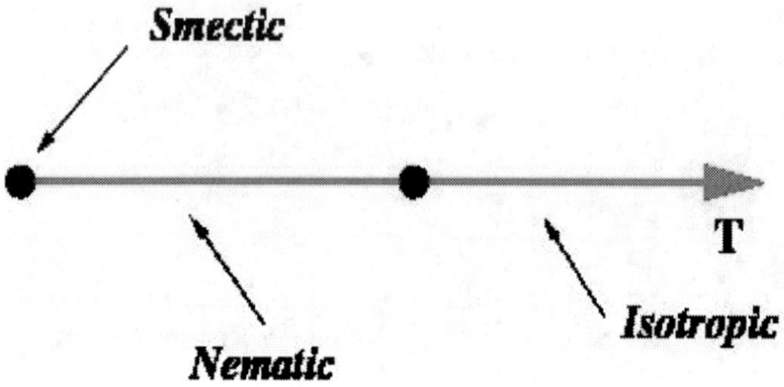

Figure 8. Mean field phase diagram of model (10).

We must also consider the effect of fluctuations on the mean field nematic solution. Considering Gaussian fluctuations around the saddle point solution we arrive to a correction term contributing to the free energy of the form [17]:

$$\delta F = \rho_s(T) \int d^2x \, |\vec{\nabla}\varphi(\vec{x})|^2, \tag{11}$$

where $\varphi(\vec{x})$ represents angular fluctuations of the director and $\rho_s(T)$ is the temperature dependent stiffness. This free energy term has the form of the two dimensional XY model, and then shares all of its physical properties. In particular it implies that the phase transition is of the Kosterlitz-Thouless type [22]. The low temperature nematic phase has not true long range order, because correlations decay algebraically. Instead, it presents *quasi-long-range order*. In particular, the nematic order parameter is depressed below its mean field value in a way given by:

$$\langle Q \rangle = \langle Q \rangle_{MF} \ \exp(-W) \tag{12}$$

with

$$W = \frac{T}{4\pi\rho_s} \ln\left(\frac{L}{a}\right), \tag{13}$$

where L is the linear dimension of the system and a a microscopic interparticle distance. It is then clear that in the thermodynamic limit $L/a \to \infty$ we have $\langle Q(T) \rangle \to 0$ logarithmically, and the there is no long range order at any finite temperature. Nevertheless, in some real systems, the actual situation if far from the ideal thermodynamic limit. For example, in ultrathin films of *Fe/Cu(001)* studied in [1] typical scales are $L \approx 10^{-3}m$ and $a \approx 10^{-10}m$. At room temperature, where most of the experiments are done, $W \approx 7/32$ and then $\langle Q \rangle \approx 0.8 \ \langle Q \rangle_{MF}$. Our conclusion is that, even in cases with perfectly isotropic interactions it is possible to observe a low temperature nematic phase, with long range order up to the scales characteristic of many experimental samples. The experimental characterization of these orientational phases in diverse systems as magnetic ultrathin films, block copolymers and microemulsions is an important challenge.

4. Perspectives

We showed that the emergence and evolution with temperature of systems forming striped-like patterns can be conveniently described by an effective free energy which incorporates a nematic-like order parameter in a Landau expansion. This leads to a new perspective on the low temperature phases of two dimensional systems with competing isotropic interactions.

Many points remain to be clarified, we conclude by enumerating some of them:

- Is it possible to identify a *microscopic* origin for the phenomenological nematic term ? or in other words, what kind of microscopic Hamiltonians can give rise to the phenomenology described by the present Landau model ?
- What is the real importance of the other (infinite) terms in the expansion of quartic coupling function ? Can this question be answered by a RG analysis ?
- What are the effects of adding anisotropies on the low temperature phase of the systems ? What are the effects of confining or limiting lateral dimensions ?
- Are these equilibrium properties important from a dynamical perspective ? What is the nature of the relaxation dynamics of these pattern forming systems ?

We think that all these important questions are still to a great extent unresolved.

Acknowledgments

We wish to thank the *Conselho Nacional de Desenvolvimento Científico e Tecnológico* (CNPq) and the *"Fundação de Amparo à Pesquisa do Estado do Rio de Janeiro"* for financial support. We also thank the Abdus Salam International Centre of Theoretical Physics (ICTP) for Research Associateships.

References

[1] Portmann O, Vaterlaus A and Pescia D 2003 *Nature* **422** 701–704
[2] Angelescu D E, Harrison C K, Trawick M L, Register R A and Chaikin P M 2005 *Phys. Rev. Lett.* **95** 025702
[3] Imperio A and Reatto L 2006 *The Journal of Chemical Physics* **124** 164712
[4] Barci D G, Fradkin E, Kivelson S A and Oganesyan V 2002 *Phys. Rev. B* **65** 245319
[5] Kivelson S A, Fradkin E and Emery V J 1998 *Nature* **393** 550
[6] Harrison C, Chaikin P M, Huse D A, Register R A, Adamson D H, Daniel A, Huang E, Mansky P, Russell T P, Hawker C J, Egolf D A, Melnikov I V and Bodenschatz E 2000 *Macromolecules* **33** 857–865
[7] Widom B 1986 *The Journal of Chemical Physics* **84** 6943–6954
[8] Dawson K, Lipkin M and Widom B 1988 *The Journal of Chemical Physics* **88** 5149
[9] Chaikin P M and Lubensky T C 1995 *Principles of Condensed Matter Physics* (Cambridge: Cambridge University Press)
[10] Toner J and Nelson D R 1981 *Phys. Rev. B* **23** 316–334
[11] Abanov A, Kalatsky V, Pokrovsky V L and Saslow W M 1995 *Phys. Rev. B* **51** 1023–1038
[12] Brazovskii S A 1975 *Sov. Phys. JETP* **41** 85–89
[13] Fredrickson G H and Helfand E 1987 *The Journal of Chemical Physics* **87** 697–705
[14] Swift J and Hohenberg P C 1977 *Physical Review A* **15** 319
[15] Hohenberg P C and Swift J B 1995 *Phys. Rev. E* **52** 1828–1844
[16] Barci D G and Stariolo D A 2007 *Physical Review Letters* **98** 200604
[17] Barci D G and Stariolo D A 2009 *Physical Review B (Condensed Matter and Materials Physics)* **79** 075437
[18] Carubelli M, Billoni O V, Pighín S A, Cannas S A, Stariolo D A and Tamarit F A 2008 *Phys. Rev. B* **77** 134417
[19] Nicolao L and Stariolo D A 2007 *Phys. Rev. B* . **76** 054453
[20] Levin Y 2007 *Physical Review Letters* **99** 228903
[21] Barci D G and Stariolo D A 2007 *Physical Review Letters* **99** 228904
[22] Kosterlitz J M and Thouless D J 1973 *Journal of Physics C: Solid State Physics* **6** 1181–1203

Coarsening in the Potts model: out-of-equilibrium geometric properties

M P O Loureiro[1], J J Arenzon[1], L F Cugliandolo[2] and A Sicilia[3]

[1] Instituto de Física and INCT Sistemas Complexos, Universidade Federal do Rio Grande do Sul, CP 15051, 91501-970 Porto Alegre RS, Brazil
[2] Université Pierre et Marie Curie – Paris VI, LPTHE UMR 7589, 4 Place Jussieu, 75252 Paris Cedex 05, France
[3] Department of Chemistry, University of Cambridge, Lensfield Road, CB2 1EW, Cambridge, United Kingdom

E-mail: `arenzon@if.ufrgs.br`

Abstract. Geometric properties of polymixtures after a sudden quench in temperature are studied through the q-states Potts model on a square lattice, and their evolution with Monte Carlo simulations with non-conserved order parameter. We analyse the distribution of hull enclosed areas for different initial conditions and compare with exact and numerical findings for the $q = 2$ (Ising) case.

1. Introduction

Curvature-driven ordering processes, in non conserved scalar order parameter systems, underlie many interesting cellular growth processes and are of both theoretical and technological importance, applications including foams [1], cellular tissues [2], superconductors [3], magnetic domains [4, 5], liquid crystals [6], adsorbed atoms on surfaces, *etc.* These systems are spatially divided in many coexisting domains, in which one of the q possible spin orientations dominates, separated by interfaces whose local velocity is ruled by the Allen-Cahn equation, thus being proportional to the local curvature, $v = -(\lambda/2\pi)\kappa$, where λ is a temperature and q-dependent dimensional constant related with the surface tension and mobility of a domain wall and κ is the local curvature. The sign is such that the domain wall curvature is diminished along the evolution. In $d = 2$ the time dependence of the area contained within any finite domain interface (the hull) on a flat surface is obtained by integrating the velocity around the hull and using the Gauss-Bonnet theorem:

$$\frac{dA_n}{dt} = \begin{cases} -\lambda, & q = 2 \\ \dfrac{\lambda}{6}(n-6), & q > 2 \end{cases} \tag{1}$$

where n is the number of sides (or, equivalently, vertices). At each one of these vertices the tangent vector to the surface has a turning angle, that are absent for $q = 2$ (since in this case each domain is surrounded by a single neighbour). Notice that in both cases, the rate is size independent [7, 8], while for $q = 2$ it does not depend on the number of sides.

For $q = 2$, the growth law is homogeneous and all hull enclosed areas decrease with the same rate, what has led us to exactly obtain [8] the number of hull-enclosed areas per unit system area, in the interval $(A, A + dA)$. For example, if one takes as initial states $(t = 0)$, equilibrium configurations at T_c, whose distribution is exactly known for $q = 2$ [9], $n_h(A, 0) = c_h^{(2)}/A^2$ with $c_h^{(2)} = 8\pi\sqrt{3}$, the distribution at any $t > 0$ is

$$n_h(A,t) = \frac{c_h^{(2)}}{(A + \lambda_h t)^2}, \tag{2}$$

that compares extremely well with simulation data for the Ising model on a square lattice [8, 10, 11]. Moreover, the above equation scales as $n(A, t) = t^{-2} f(A/t)$, in accordance with the scaling hypothesis. Although this is usually only a hypothesized, but well-established feature, for the $q = 2$ model it can be obtained from first principles [8]. For a quench from infinite temperature, on the other hand, the initial distribution corresponds to the one of the critical random continuous percolation [8,10]. This observation is an essential ingredient to understand the fact that, and obtain the probability with which, the system attains a striped frozen state at zero temperature (see [12] and references therein). Interestingly, on a square lattice, the initial state does not correspond to the random percolation critical point but the coarsening evolution gets very close to it after one or two Monte Carlo steps. Therefore, although the analytical results of Ref. [8] were obtained with a continuous description, a priori not guaranteed to apply on a lattice, they do describe the coarsening dynamics of the discrete Ising model with remarkable accuracy.

For $q > 2$, whether a cell grows, shrinks or remains with constant area depends on its number of sides being, respectively, larger than, smaller than or equal to 6, what is known as the von Neumann's law. Therefore, one cannot write a simple relation to link the area distribution at time t to the one at the initial time $t = 0$ and the distribution might get scrambled in a non-trivial way during the coarsening process (for example, when a domain disappears, the number of sides of the neighboring domains changes, along with their growth rate) [13]. Nonetheless, the system again evolves to a scaling state in which the domain morphology is statistically the same at all times when lengths are measured in units of $R(t)$, a single characteristic growing length scale, in accordance with the scaling hypothesis.

In this paper we briefly compare the cases $q = 2$ and $q > 2$, in particular when the initial state was equilibrated at $T = T_c$. We first present the $d = 1$ case and then the much richer two dimensional one, reviewing some of the results already published in Refs. [8, 13]. Besides completeness, it is interesting to compare the uni and two dimensional cases as they strongly differ.

2. Coarsening in $d = 1$

The Hamiltonian in this case is

$$\mathcal{H} = -J \sum_i \delta_{s_i, s_{i+1}}, \tag{3}$$

where $s_i = 0, ..., q - 1$ and $J > 0$. We measure the domain length distribution at time t, $n(\ell, t)$ $(1 \leq \ell \leq L)$ after quenching the system from $T_0 \to \infty$ to the final, working temperature, $T = 0$. There is no finite static critical temperature and the system orders ferromagnetically only at $T = 0$. The completely disordered initial condition starts evolving in a coarsening regime in which regions of finite length order. Differently from the higher dimensional case where domain growth is driven by interfacial tension, in $d = 1$ coarsening is driven by the diffusion of domain walls and annihilation when they meet. At this point when a domain disappears, the two neighboring domains always coalesce if $q = 2$. This may not be the case for $q > 2$ and becomes less probable as q increases. Thus, the average length size of the domains, $\langle \ell \rangle$, grows faster the smaller the value of q. Indeed, the growth law is $\langle \ell \rangle = q\sqrt{\pi t}/(q - 1)$ [14, 15], where the $t^{1/2}$ behaviour is expected from the diffusive interface motion.

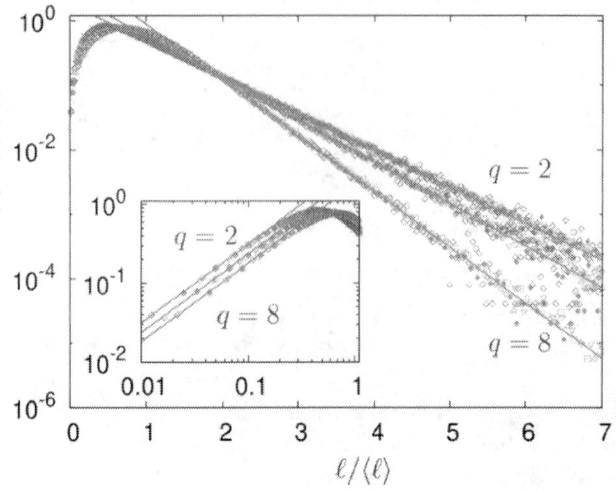

Figure 1. Cluster length distribution for the unidimensional Potts model with $q = 2$, 3 and 8 versus the rescaled length, $\ell/\langle \ell \rangle$. The qualitative behaviour is the same, linear for small lengths and exponential at large ones. The (red) lines are analytical predictions of Derrida and Zeitak [15], eq. (5) with (A_q, B_q) equal to $(1.306, 0.597)$, $(1.500, 0.963)$ and $(1.993, 1.876)$ for $q = 2$, 3 and 8, respectively. The system is 10^5 large and averages were taken over 1000 samples. Data corresponds to times from $t = 2^2$ to 2^9.

During the coarsening regime, the distribution of domain sizes obeys the scaling behaviour

$$n(\ell, t) = \langle \ell(t) \rangle^{-1} f\left(\frac{\ell}{\langle \ell(t) \rangle}\right). \tag{4}$$

At zero working temperature the universal function $f(x)$ is given by [14–17]

$$f(x) = \begin{cases} \dfrac{\pi q x}{2(q-1)} & x \ll 1\,, \\ \exp(-A_q x + B_q) & x \gg 1\,, \end{cases} \tag{5}$$

where the constants are known exactly [15]. In Fig. 1 we show, for $q = 2$, 3 and 8, that the above equations fully agree with numerical simulations (the $q = 2$ case was previously presented in Ref. [10]). In all cases the initial state has null correlations, all spins being fully uncorrelated, what seems to be the key ingredient for the exponential tail in the distribution. As we will see in the next section, the $q = 2$ case, starting from infinite temperature, is an exception to such behaviour.

3. Coarsening in $d = 2$

The detailed evolution of the system during the coarsening dynamics depends on the correlations already present in the initial state. Below we show the behaviour when such correlations are either absent ($T_0 \to \infty$) or long-ranged ($T_0 = T_c$ for $q \leq 4$). In the latter case, since the thermodynamic transition also corresponds to a percolation transition in $2d$, the initial state already presents one spanning cluster at $t = 0$. On the other hand, for short-range initial correlations, such spanning domains are either formed very fast (e.g. in the case $q = 2$ with $T_0 \to \infty$), or not formed at all. We should emphasize the exceptionality of the $q = 2$ case: having the highest concentration of a single species at $T_0 \to \infty$, the proximity with the percolation critical point strongly affects the system's evolution [8]. Indeed, the hull-enclosed area distribution (geometric domains present an analogous behaviour [10]) becomes power-law in a couple of steps, and the effective initial condition is given by $n_h(A,0) = 2c_h^{(2)}/A^2$ [9]. Power-laws are also the initial distributions for $T_0 = T_c$ for $q \leq 4$ and generally given by (the prefactor is different from the previous case)

$$n_h(A,0) = \frac{(q-1)c_h^{(q)}}{A^2}. \tag{6}$$

Notice that, unless for $q = 2$, the value of $c_h^{(q)}$ is not known exactly, although they are all close to each other.

For $2 \leq q \leq 4$, the Potts model presents a continuous transition, and one could imagine that their non equilibrium coarsening phenomenology would be similar as well. Indeed, the equal time correlation function seems to share the same universal scaling function and the related correlation length grows with the same power of time, $t^{1/2}$. There are, however, fundamental differences. Differently from the $q = 2$ case, that presents a percolating domain with probability almost one as early as $t = 2$ after the quench [8, 10], the $q > 2$, $T_0 \to \infty$ initial condition is sufficiently far from critical percolation that the system remains, at least in the time window of our simulations, distant from the percolation threshold (in spite of the largest domain steadily, but slowly, increasing with time). In this regard, the $q = 2$ case is the exception, while the $q = 3$ and 4 are similar. As the system evolves after the quench, the distribution keeps memory of the initial state, that corresponds to random percolation with occupation probability $p = 1/q$. And, by not getting close to a critical point, the distributions do not become critical and, as a consequence, do not develop a power law tail, as illustrated in Fig. 2 for $q = 3$ and $T_0 \to \infty$. There is, however, an A^{-2} envelope that is a direct consequence of dynamical scaling, present also for other values of q [13].

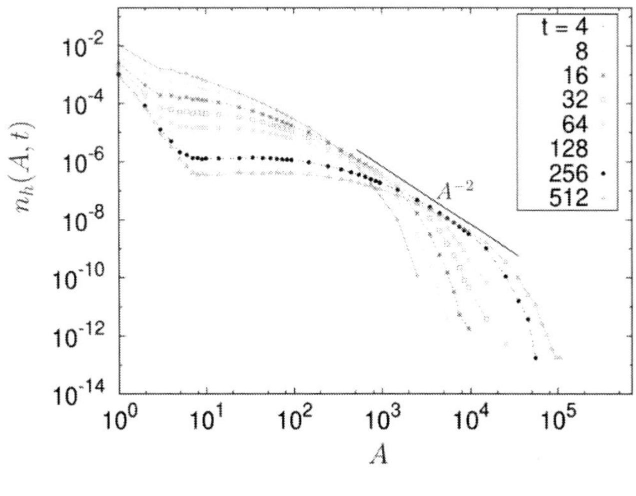

Figure 2. Hull-enclosed area distribution at several times (given in the key) after a quench from equilibrium at $T_0 \to \infty$ to $T_f = T_c/2$ in the $q = 3$ case. Analogous distributions are obtained for $q > 4$ and $T_0 \to \infty$ (not shown). The declivity of the envelope is -2 as a consequence of the scaling obeyed by the distribution. Figure from Ref. [13].

The collapsed distribution of hull-enclosed areas after a quench from $T_0 = T_c$ to $T_f = T_c/2$ for $q = 3$ is shown in Fig. 3. Although for $q = 2$ this distribution conserves its form during the coarsening regime due to an overall advection to the right as domains increase in size, for $q > 2$ there is a dependence on the number of sides in von Neumann's equation, some domains increase while other decrease in size, and thus there is no obvious reason for the distribution at $t > 0$ to keep its initial power-law form, eq. 6. Remarkably, the general behavior is similar to the $q = 2$ case, in which the collapse of curves for different times onto a single universal function demonstrates the existence of a single length scale that, moreover, follows the Allen-Cahn growth law, $R(t) \sim t^{1/2}$. We can make a mean-field-like approximation and replace the number of sides in the von Neumann equation (1) by a constant mean, $n \to \langle n \rangle$. Using Eq. (6) and the results in Refs. [8, 10], the hull-enclosed area distribution for $q \leq 4$ becomes

$$n_h(A,t) = \frac{(q-1)c_h^{(q)}}{\left(A + \lambda_h^{(q)} t\right)^2}, \tag{7}$$

that fits very accurately the data using $\lambda_h^{(3)} \simeq 1.4$. The deviations present at small values of A/t are due to thermal fluctuations that are visible in the inset (see [10] for details). In order to test the above approximation, we measured the average change in area, dA/dt, that is, the number of spins included or excluded in those domains that survived during a given time interval. The above rate, from von Neumann's equation, depends on $n - 6$ (both in absolute value and sign), and is different for domains with different numbers of neighbours (except for $q = 2$ where it is constant). However, upon average, there is an effective, constant $\lambda_{\rm eff}$. Similarly, only those that have $\lambda_{\rm eff} \leq 0$ present a power law distribution. Thus, there seems to be a net difference between cases that give a power law distribution function from the ones that do not.

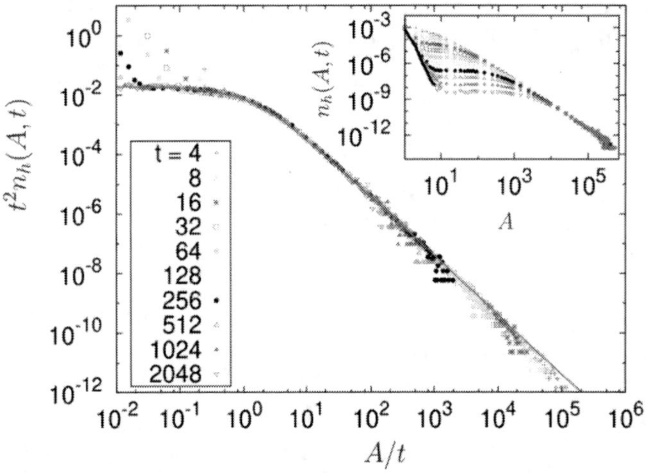

Figure 3. Collapsed hull-enclosed area distributions at several times after a quench from equilibrium at $T_0 = T_c$ to $T_f = T_c/2$, for $q = 3$. The line is Eq. (7) with $\lambda_h^{(3)} \simeq 1.4$. The points at $A/t \ll 1$ that deviate from the scaling function are due to thermal fluctuations. These fluctuations have been independently measured and are depicted as a continuous black line in the inset (notice that their distribution is time independent). Figure from Ref. [13].

4. Conclusions

We studied the Potts model during the coarsening dynamics after a sudden quench in temperature, emphasizing the cases $q = 2$ and 3 that present a continuous transition. Although the theory developed in Refs. [8, 10] for the Ising model ($q = 2$) cannot be easily extended to $q > 2$, our numerical results show the existence of similarities between both cases. Besides the equilibrium continuous transition, there is a dynamical length scale that grows with $t^{1/2}$ after a sudden quench in temperature and the same universal scaling function for the rescaled correlation function in both cases. Nonetheless, very different are the distributions of dynamic hull-enclosed areas (and geometric domains) evolved from initial states with zero correlation lengths, as those obtained in equilibrium at $T_0 \to \infty$. One expects that the scaling functions of the dynamic distributions will be reminiscent of the disordered state at the initial temperature, with an exponential tail. This is indeed what happens in $d = 1$ and, in $d = 2$, for $q > 2$. The remarkable exception is $q = 2$ (in $d = 2$): the proximity from the percolation critical point, and the subsequent flow to it after the temperature quench, develops a power-law tail in the distribution. For $q = 3$ the initial state is no longer close to any critical point and this does not occur. A further surprise occurs for the distributions after a quench from $T_0 = T_c$. Despite von Neumann's law showing that differently sided domains could either grow or shrink, scrambling the distribution, it still shows a power law, in analogy with the Ising case but without a complete explanation up to now besides a simple mean-field-like argument. Alongside, many questions are still unanswered for this long studied, classic problem and surely many surprises are still awaiting.

Acknowledgments

Work partially supported by CAPES/Cofecub grant 667/10. JJA and MPOL are partially supported by the Brazilian agency CNPq.

References

[1] Glazier J A, Anderson M P and Grest G S 1990 *Phil. Mag. B* **62** 615–645
[2] Mombach J C M, de Almeida R M C and Iglesias J R 1993 *Phys. Rev. E* **48** 598–602
[3] Prozorov R, Fidler A F, Hoberg J R and Canfield P C 2008 *Nature Physics* **4** 327–332
[4] Babcock K L, Seshadri R and Westervelt R M 1990 *Phys. Rev. A* **41** 1952–1962
[5] Jagla E A 2004 *Phys. Rev. E* **70** 046204
[6] Sicilia A, Arenzon J J, Dierking I, Bray A J, Cugliandolo L F, Martinez-Perdiguero J, Alonso I and Pintre I C 2008 *Phys. Rev. Lett.* **101** 197801
[7] Mullins W W 1956 *J. Appl. Phys.* **27** 900–904
[8] Arenzon J J, Bray A J, Cugliandolo L F and Sicilia A 2007 *Phys. Rev. Lett.* **98** 145701
[9] Cardy J and Ziff R M 2003 *J. Stat. Phys.* **110** 1–33
[10] Sicilia A, Arenzon J J, Bray A J and Cugliandolo L F 2007 *Phys. Rev. E* **76** 061116
[11] Sicilia A, Sarrazin Y, Arenzon J J, Bray A J and Cugliandolo L F 2009 *Phys. Rev. E* **80** 031121
[12] Barros K, Krapivsky P L and Redner S 2009 *Phys. Rev. E* **80** 040101(R)
[13] Loureiro M P O, Arenzon J J, Cugliandolo L F and Sicilia A 2010 *Phys. Rev. E* **81** 021129
[14] Amar J G and Family F 1990 *Phys. Rev. A* **41** 3258–3262
[15] Derrida B and Zeitak R 1996 *Phys. Rev. E* **54** 2513–2525
[16] Alemany P A and ben Avraham D 1995 *Phys. Lett. A* **206** 18–25
[17] Krapivsky P L and Ben-Naim E 1997 *Phys. Rev. E* **56** 3788–3798

Journal of Physics: Conference Series **246** (2010) 012023 doi:10.1088/1742-6596/246/1/012023

Loewner equation for Laplacian growth: A Schwarz-Christoffel-transformation approach

M. Durán and G. L. Vasconcelos

Laboratório de Física Teórica e Computacional, Departamento de Física, Universidade Federal de Pernambuco, 50670-901, Recife, Brazil.

E-mail: `giovani@df.ufpe.br`

Abstract. The problem of Laplacian growth is considered within the Loewner-equation framework. A new method of deriving the Loewner equation for a large class of growth problems in the half-plane is presented. The method is based on the Schwarz-Christoffel transformation between the so-called 'mathematical planes' at two infinitesimally separated times. Our method not only reproduces the correct Loewner evolution for the case of slit-like fingers but also can be extended to treat more general growth problems. In particular, the Loewner equation for the case of a bubble growing into the half-plane is presented.

1. Introduction

The Loewner equation [1] is an important result in the theory of univalent functions [2] that has found important applications in nonlinear dynamics, statistical physics, and conformal field theory [3, 4]. In its most basic formulation, the Loewner equation is a first-order differential equation for the conformal mapping $g_t(z)$ from a given 'physical domain,' consisting of a complex region $\mathbb{P}$ minus a curve Γ_t emanating from its boundary, onto a 'mathematical domain' represented by $\mathbb{P}$ itself. Usually, $\mathbb{P}$ is either the upper half-plane or the exterior of the unit circle, but recently the Loewner equation for the channel geometry was also considered [5]. The Loewner equation depends on a driving function, here called $a(t)$, that is the image of the growing tip under the mapping $g_t(z)$. An important development on the theory of the Loewner equation was the discovery by Schramm [6] that when the driving function $a(t)$ is a Brownian motion the resulting Loewner evolution describes the scaling limit of certain statistical mechanics models. This result spurred great interest in the so-called stochastic Loewner equation [4].

Recently, the deterministic Loewner equation was also used to study the problem of Laplacian fingered growth in both the half-plane and radial geometries [7, 8] as well as in the channel geometry [5]. In this case, the driving function $a(t)$ has to follow a specific time evolution in order to ensure that the finger tip grows along the gradient lines of the corresponding Laplacian field. The idea of using iterated conformal maps to generate aggregates was first deployed by Hastings and Levitov [9] in the context of stochastic growth models, such as diffusion limited aggregation; see, e.g., [10] for further developments along those lines. A deterministic version of the Hastings-Levitov model that is closely related to the Loewner-equation approach—albeit not using explicitly such a formalism—was studied in [11].

In this paper we consider the problem of Laplacian growth within the context of the Loewner evolution, and present a new method of deriving the corresponding Loewner equation for a broad class of growth models in the half-plane. Our method is based on the Schwarz-Christoffel (SC) transformation between the mathematical planes g_t and $g_{t+\tau}$, where τ is an infinitesimal time interval. More specifically, the method consists of expanding the integrand of the SC formula in powers of the appropriate infinitesimal quantity (related to τ) and then performing the integrals up to the leading-order term. Our method correctly yields the Loewner evolution for the case of slit-like fingers studied before [5, 8]. More importantly, the method is able to handle more general growth problems, so long as the growth rule can be specified (in the mathematical plane) in terms of a polygonal curve, in which case the Schwarz-Christoffel transformation can be used. An example is given for the case of a bubble growing from the real axis into the upper half-plane.

2. Loewner Equation for Slit-like Fingers

2.1. The Case of a Single Finger

In order to set the stage for the remainder of the paper, we wish to begin our discussion by considering the simplest Loewner evolution, namely, that in which a curve starts from the real axis at $t = 0$ and then grows into the upper half-z-plane $\mathbb{H}$, where

$$\mathbb{H} = \{z = x + iy \in \mathbb{C} : y > 0\}.$$

The curve at time t is denoted by Γ_t and its growing tip is labeled by $\gamma(t)$. Now let $g_t(z)$ be the conformal mapping that maps the 'physical domain,' corresponding to the upper half-z-plane minus the curve Γ_t, onto the upper half-plane of an auxiliary complex w-plane, called the 'mathematical plane,' i.e., we have $w = g_t(z)$, where

$$g_t : \mathbb{H} \backslash \Gamma_t \to \mathbb{H}, \tag{1}$$

with the curve tip $\gamma(t)$ being mapped to a point $a(t)$ on the real axis in the w-plane; see Fig. 1. Furthermore, we consider the growth process to be such that the accrued portion of the curve from t to $t + \tau$, where τ is an infinitesimal time interval, is mapped under $g_t(z)$ to a vertical slit in the mathematical w-plane; see Fig. 1. The mapping function $g_t(z)$ must also satisfy the initial condition

$$g_0(z) = z, \tag{2}$$

since we start with an empty upper half-plane. We also impose the so-called hydrodynamic normalization condition at infinity:

$$g_t(z) = z + O(\frac{1}{|z|}), \qquad z \to \infty. \tag{3}$$

These conditions specify uniquely the mapping function $g_t(z)$.

From a more physical viewpoint, the problem formulated above belong to the class of Laplacian growth models where an interface evolves between two phases driven by a scalar field $\phi(x, y; t)$, representing, for example, temperature, pressure, or concentration, depending on the physical problem at hand. In one phase, initially occupying the entire upper half-plane, the scalar field ϕ satisfies the Laplace equation

$$\nabla^2 \phi = 0, \tag{4}$$

whereas in the other phase one considers ϕ=const., say $\phi = 0$, with the curve Γ_t representing a finger-like advancing interface between the two phases. (Here the finger is assumed to be infinitesimally thin.) The complex potential for the problem can then be defined as $w(z, t) = \psi(x, y; t) + i\phi(x, y; t)$, where ψ is the function harmonically conjugated to ϕ. On the boundary of the physical domain, consisting here of the real axis together with the curve Γ_t,

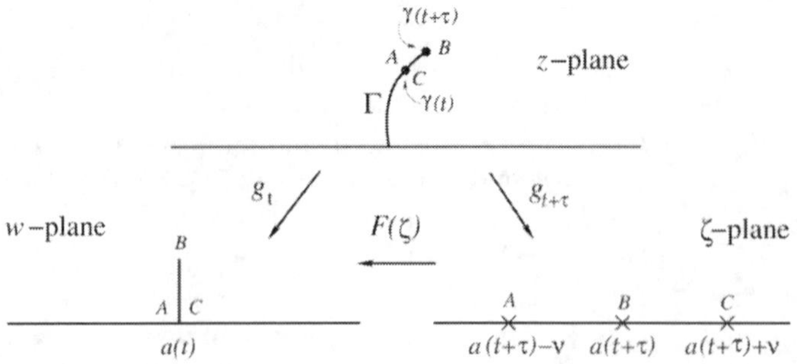

Figure 1. The physical z-plane and the mathematical w- and ζ-planes at times t and $t+\tau$, respectively, for a single finger in the upper half-plane. The mapping g_t maps the curve Γ (at time t) onto a segment of the real axis on the w-plane, whereas the accrued portion of the curve (during time interval τ) is mapped to a vertical slit. The mapping g_t is obtained as the composition of $g_{t+\tau}$ and the slit mapping F; see text.

we impose the condition $\phi = 0$, whereas at infinity we assume a uniform gradient field, $\vec{\nabla}\phi \approx \hat{y}$, or alternatively,

$$w(z,t) \approx z, \qquad z \to \infty. \tag{5}$$

From this point of view, the mapping function $g_t(z)$ introduced above corresponds precisely to the complex potential $w(z,t)$ of the problem. In particular, the fact that in the w-plane the curve grows along a vertical line implies that the finger tip grows along gradient lines in the z-plane. To specify completely a given physical model, one has also to prescribe the interface velocity, which is usually taken to be proportional to some power η of the gradient field: $v \sim |\vec{\nabla}\phi|^\eta$. For most of the problems considered here the specific velocity model is not relevant, in the sense that the finger shapes will be independent of the exponent η, which only affects the time scale of the problem. (However, there are situations, such as the case of competing asymmetrical fingers [5], where different η's may yield different patterns.)

For convenience of notation, we shall represent the mathematical plane at time $t+\tau$ as the complex ζ-plane and so we write $\zeta = g_{t+\tau}(z)$. Now consider the mapping $w = F(\zeta)$, from the upper half-ζ-plane onto the mathematical domain in the w-plane; see Fig. 1. The mapping function $g_{t+\tau}(z)$ can then be given in terms of $g_t(z)$ as

$$g_{t+\tau} = F^{-1} \circ g_t, \tag{6}$$

where F^{-1} is the inverse of $F(\zeta)$. The above relation governs the time evolution of the function $g_t(z)$ and naturally leads to the Loewner equation. A standard way of showing this is to construct the slit mapping $F(\zeta)$ explicitly, substitute its inverse in (6), and then take the limit $\tau \to 0$. One then finds the Loewner equation

$$\dot{g}_t(z) = \frac{d(t)}{g_t(z) - a(t)}, \tag{7}$$

where $d(t)$ is the so-called growth factor which is related to the tip velocity. One can show [5] that $d(t) = |f_t''(a(t))|^{-\eta/2-1}$, where $f_t(z)$ is the inverse of $g_t(z)$. Here, however, the specific form of $d(t)$ is not relevant since we can always rescale the time coordinate in (7) so as to set $d(t) = 1$. From symmetry, one also gets

$$\dot{a}(t) = 0, \tag{8}$$

so that $a(t) = a_0 = \text{const.}$, which implies that the tip $\gamma(t)$ simply traces out a vertical line in the z-plane.

In more general situations, such as the case of multiple fingers, the function $F(\zeta)$ can no longer be obtained in closed form and so one has to resort to alternative approaches to derive the Loewner evolution. Previous methods [5, 8] consider a series of compositions of the basic one-slit mapping. Here, however, we will apply a more direct method based on the Schwarz-Christoffel transformation to obtain the Loewner equation for slit-like fingers in the half-plane. In the next section, our method will be extended to treat the case of a growing bubble in the upper half-plane. To illustrate how the method works, let us first use it to derive the Loewner equation (7) for a single finger.

We begin by inverting (6) so as to write

$$g_t = F \circ g_{t+\tau}. \tag{9}$$

The mapping $F(\zeta)$ from the upper half-ζ-plane onto the upper half-w-plane with a vertical slit (see Fig. 1) is easily found by a direct application of the Schwarz-Christoffel formula [12]. One then finds

$$g_t = F(g_{t+\tau}) = \int_{\zeta_0}^{g_{t+\tau}} \frac{\zeta - a(t+\tau)}{\sqrt{(\zeta - a(t+\tau))^2 - \nu^2}} d\zeta + a(t), \tag{10}$$

where $\zeta_0 = a(t+\tau) + \nu$. Note from Fig. 1 that the parameter ν is related to the (infinitesimal) height of the slit in the w-plane. The above integral can be performed exactly, as already mentioned, but here we take an alternative approach, namely, we first expand the integrand in (10) in powers of ν and then compute the relevant integrals afterwards. To do that we first rewrite (10) in the form

$$g_t = \int_{\zeta_0}^{g_{t+\tau}} \frac{d\zeta}{\sqrt{1 - \nu^2[\zeta - a(t+\tau)]^{-2}}} + a(t). \tag{11}$$

After expanding the integrand in powers of ν^2 and performing the relevant integrals, one obtains that, up to order ν^2, equation (11) becomes

$$g_{t+\tau} - g_t = \frac{\nu^2/2}{g_{t+\tau} - a(t+\tau)} + a(t+\tau) - a(t). \tag{12}$$

Now expanding this equation up to the first order in τ, dividing by τ, and then taking $\tau \to 0$, one gets

$$\dot{g}_t = \frac{d(t)}{g_t(z) - a(t)} + \dot{a}, \tag{13}$$

where

$$d(t) = \lim_{\tau \to 0} \frac{\nu^2}{2\tau}. \tag{14}$$

Using the boundary condition $\lim_{g_t \to \infty} \dot{g}_t = 0$, which follows from (3), yields precisely the Loewner equation (7) together with the condition (8).

2.2. The Multifinger Case

Here we consider the case of multiples fingers Γ_i, $i = 1, ..., n$, growing from the real axis into the upper half-z-plane, as shown in Fig. 2. As before, the map $g_t(z)$ maps the physical domain in the z-plane onto the upper half-w-plane and the tips are required to grow along gradient lines, so that the accrued portions of the curves Γ_i during an infinitesimal time τ are mapped under $g_t(z)$ to vertical slits emanating from the real axis; see Fig 2.

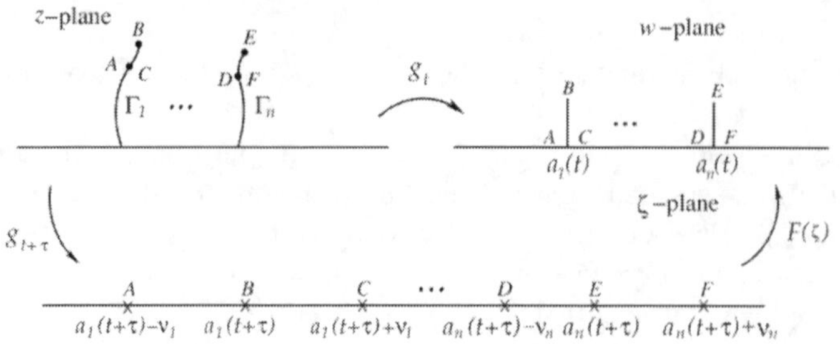

Figure 2. The physical and mathematical planes for the case of multiple fingers in the upper half-plane.

The mapping $w = F(\zeta)$, from the upper half-ζ-plane to the upper half-w-plane with n vertical slits, can again be easily obtained from the Schwarz-Christoffel transformation [12]:

$$g_t = F(g_{t+\tau}) = \int_{\zeta_0}^{g_{t+\tau}} \prod_{i=1}^{n} \frac{\zeta - a_i(t+\tau)}{\sqrt{(\zeta - a_i(t+\tau))^2 - \nu_i^2}} d\zeta + a_j(t), \tag{15}$$

where $\zeta_0 = a_j(t+\tau) + \nu_j$ for a given j. [We remark parenthetically that in writing (15) we have assumed, for simplicity, that the slits in the w-plane are mapped under $F^{-1}(w)$ onto symmetrical segments on the real ζ-axis; see Fig 2. Rigorously speaking, this is valid only in the limit that the slit heights become vanishingly small, that is, when $\tau \to 0$, which is the relevant limit for us here.]

The integral in (15) cannot be performed exactly for arbitrary n, hence in order to obtain the Loewner equation for this case we first need to expand the integrand in powers of the infinitesimal parameters ν_i and then proceed with the integration. Note, however, that each term in (15) is of the same form as that appearing in (10) for the case of a single finger. We can thus build upon our experience with that case to treat the present situation. In particular, we notice that the mixed terms involving different ν_i's in the expansion of the integrand in (15) are all of orders higher than ν^2 and hence need not be considered, for they do not contribute to the final result in the limit $\tau \to 0$. Thus, to the extent that the mixed terms can be neglected, we can rewrite (15) as

$$g_t \approx \sum_{i=1}^{n} \int_{\zeta_0}^{g_{t+\tau}} \frac{d\zeta}{\sqrt{1 - \nu_i^2[\zeta - a_i(t+\tau)]^2}} + a_j(t). \tag{16}$$

Now repeating the exactly same procedure used for the single finger case, see Eqs. (11)-(13), one readily obtains

$$\dot{g}_t = \sum_{i=1}^{n} \frac{d_i(t)}{g_t - a_i(t)} - \sum_{\substack{i=1 \\ i \neq j}}^{n} \frac{d_i(t)}{a_j(t) - a_i(t)} + \dot{a}_j(t), \tag{17}$$

where

$$d_i(t) = \frac{\nu_i^2}{2\tau}. \tag{18}$$

After using the condition $\lim_{g_t \to \infty} \dot{g}_t = 0$ in (17), we get the Loewner equation for multiple curves

$$\dot{g}_t = \sum_{i=1}^{n} \frac{d_i(t)}{g_t - a_i(t)}, \tag{19}$$

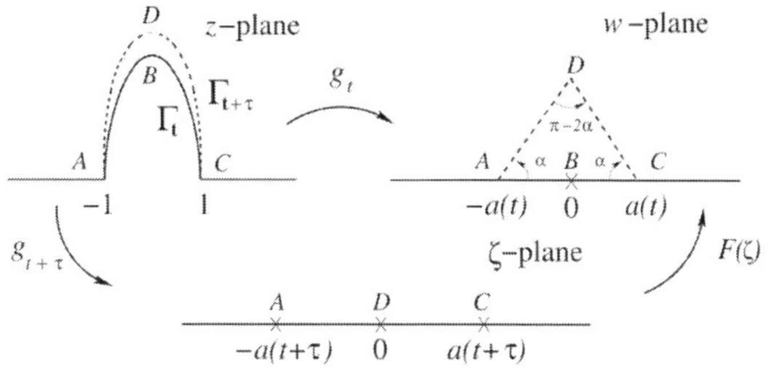

Figure 3. The physical and mathematical planes for the case of a growing bubble in the upper half-plane. Here the new interface at time $t+\tau$ is mapped under g_t to a tent-like shape; see text for details.

with the time evolution of the points $a_i(t)$ being given by the following system of ordinary differential equations

$$\dot{a}_i(t) = \sum_{\substack{j=1\\ j\neq i}}^{n} \frac{d_j(t)}{a_i(t) - a_j(t)}. \tag{20}$$

If the growth factors $d_i(t)$ are all the same, we can again rescale the time variable so as to set $d_i = 1$. In particular, in the case of two symmetrical fingers (i.e., $d_1 = d_2 = 1$), equation (19) can be integrated exactly to yield the mapping function $g_t(z)$, from which the finger shapes can be computed analytically [5]. A related exact solution for two fingers was obtained in [13]. An alternative derivation of (19) was given elsewhere [8] using a composition of n single-slit mappings. Our method, however, is somewhat more direct in the sense that it considers a single mapping with n slits as shown in Fig 2. In the next section we will extend our method to include the case of a growing bubble in the half-plane.

3. Loewner Equation for a Bubble in the Half-Plane

Here we consider the problem of an interface starting initially from a segment, say, $[-1,1]$, along the real axis and then growing into the upper half-z-plane, as indicated in Fig. 3. As before, we denote the interface at time t by Γ_t and consider the mapping $g_t(z)$ from the physical domain in the z-plane to the mathematical w-plane:

$$g_t : \mathbb{H}\backslash\Gamma_t \to \mathbb{H}.$$

The mapping function $g_t(z)$ must also satisfy the hydrodynamic normalization condition (3). The growth dynamics is specified by requiring that each point on the interface grow along gradient lines, but with a growth rate that is maximum at the center (i.e., at the bubble tip) and zero at both 'contact points,' $z = \pm 1$, in such a way that the interface at time $t+\tau$ is mapped under $g_t(z)$ to a tent-like shape, as shown in Fig. 3. The height of this tent, or alternatively, the angle α formed with the real axis, is linearly related to τ in the limit $\tau \to 0$, as we will see below. Here we consider only the case of symmetric bubbles, where the contact points, $z = \pm 1$, are mapped by $g_t(z)$ to the points $w = \pm a(t)$, respectively, and the tip (point B in Fig. 3) is mapped to the origin.

Since the domain in the w-plane has a polygonal shape, the mapping $w = F(\zeta)$ can once again be obtained from the Schwarz-Christoffel transformation. In this case we have

$$g_t = F(g_{t+\tau}) = \int_{a(t+\tau)}^{g_{t+\tau}} \frac{\zeta^{2\alpha/\pi}}{[\zeta^2 - a^2(t+\tau)]^{\alpha/\pi}}\, d\zeta + a(t), \tag{21}$$

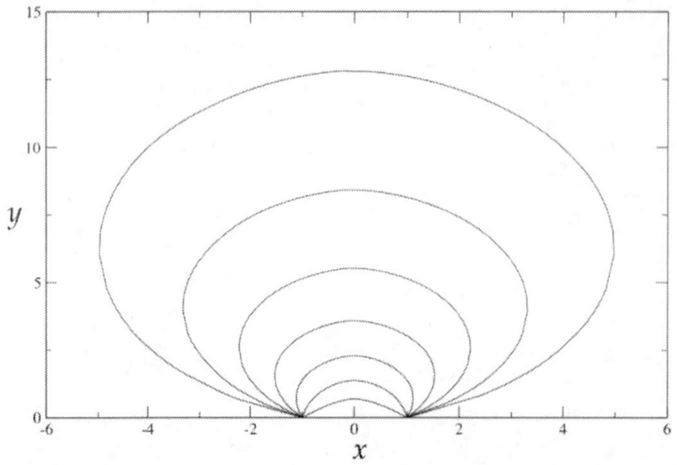

Figure 4. Loewner evolution for a bubble growing in the upper half-plane for $d(t) = 1$. The bubble starts at $t = 0$ from the interval $[-1, 1]$ on the real axis. The different curves represent the interface at time intervals of $\Delta t = 0.3$.

which up to the first order in the parameter α becomes

$$g_t \approx \int_{a(t+\tau)}^{g_{t+\tau}} \left\{1 + \frac{\alpha}{\pi}\left[2\ln\zeta - \ln(\zeta^2 - a^2(t+\tau))\right]\right\} d\zeta + a(t), \tag{22}$$

After performing the integral above and taking the limit $\tau \to 0$, one readily obtains the following generalized Loewner equation:

$$\dot{g}_t(z) = d(t)\left\{[g_t + a(t)]\ln[g_t + a(t)] + [g_t - a(t)]\ln[g_t - a(t)] - 2g_t \ln g_t\right\}, \tag{23}$$

together with the governing equation for the logarithmic singularity $a(t)$

$$\dot{a}(t) = (\ln 4)d(t)a(t), \tag{24}$$

where the growth factor $d(t)$ is now given by

$$d(t) = \lim_{\tau\to 0} \frac{\alpha}{\pi\tau}. \tag{25}$$

Equation (24) can be solved exactly to yield $a(t) = a_0 4^{\int_0^t d(t')t'}$. As explained before, we can set $d(t) = 1$ by rescaling the time coordinate, in which case we have $a(t) = a_0 4^t$. Recall also that the Loewner equation (23) must be supplemented with the initial condition $g_0(z) = z$.

In Fig. 4 we show numerical solutions of (23) with $a_0 = 1$ for various final times t, starting from $t = 0.3$ up $t = 2.1$ with a time separation of $\Delta t = 0.3$ between successive curves. To generate the curves shown in Fig. 4 we used the numerical scheme described in [13]. More specifically, we start with a 'terminal condition' $g_t = w$, for $w \in [-a(t), a(t)]$, and integrate the Loewner equation (23) backwards in time, using a Runge-Kutta method of second order, to get the corresponding point g_0 on the interface. (See also [14] for a recent review on numerical integration of the Loewner equation.) From Fig. 4 one sees that the bubble initially grows somewhat slowly and then rapidly expands and tends to occupy the whole plane for large times. In fact, one can show that (for $d = 1$) the tip velocity grows exponentially with time as $t \to \infty$. It is possible to modify the growth factor $d(t)$ so to have the tip velocity related to the gradient of the field $\phi(x, y)$, as discussed in Sec. 2, but this does not change the interface shapes and only alters the time scale of the bubble evolution.

4. Conclusions

We have presented a novel method to derive the Loewner equation for Laplacian growth problems. The method is based on the Schwarz-Christoffel (SC) transformation and consists of expanding the integrand of the SC formula in the appropriate infinitesimal parameter, performing the relevant integrals, and then taking the limit in which the infinitesimal parameter goes to zero. Our method is able to reproduce the Loewner evolution for the problem of slit-like fingers in both the half-plane (Sec. 2) and the channel geometry (not shown here). Furthermore, the method can be extended to treat more complicated growth problems, so long as the growth dynamics in the complex-potential plane can be specified in terms of a polygonal boundary, in which case the Schwarz-Christoffel transformation can be used. We note however that the requirement that the growth rule be formulated in terms of a polygonal curve is not as restrictive as it seems, for any simple curve can in principle be approximated by a piecewise linear function. [Such more general growth models are currently under investigation.] In particular, we obtained the Loewner equation for a novel situation in which a bubble grows from a segment of the real line into the upper half-plane. Although in this case we refer to the evolving interface as a growing 'bubble,' in contrast to the slit-like 'fingers' of Sec. 2, this terminology should not be taken literally. Depending on the physical problem at hand, such growing interface may represent, say, an expanding front in combustion experiments or in electrochemical deposition [5]. Of course, further work is necessary to relate more directly the growth models discussed here to experiments.

Acknowledgments

This work was supported in part by the Brazilian agencies FINEP, CNPq, and FACEPE and by the special programs PRONEX and CT-PETRO.

References

[1] Löwner K 1923 *Math. Ann.* **89** 103.
[2] Duren P L 1983 *Univalent Functions* (Springer, New York.)
[3] Gruzberg I A and Kadanoff L P 2004 *J. Stat. Phys.* **114** 1183.
[4] Bauer M and Bernard D 2006 *Phys. Rep.* **432** 115.
[5] Gubiec T and Szymczak P 2008 *Phys. Rev.* E **77** 041602.
[6] Schramm O 2000 *Israel J. Math.* **118** 221.
[7] Carleson L and Makarov N 2002 *J. Anal. Math.* **87**, 103.
[8] Selander G 1999 *Ph.D. thesis* (Royal Institute of Technology, Stockholm, Sweden) (unpublished).
[9] Hastings M B and Levitov L S 1998 *Physica* D **116** 244.
[10] Davidovitch B, Feigenbaum M J, Hentschel H G E, and Procaccia I 2000 *Phys. Rev.* E **62** 1706 (2000);
[11] Hastings M B 2001 *Phys. Rev.* E **64** 046104.
[12] Carrier G F, Krook M and Pearson C E 1983 *Functions of a complex variable: theory and technique* (Hod Books, Ithaca.)
[13] Kager W, Nienhuis B and Kadanoff L P 2004 *J. Stat. Phys.* **115** 805.
[14] Kennedy T 2009 *J. Stat. Phys.* **137** 839.

Journal of Physics: Conference Series **246** (2010) 012024 doi:10.1088/1742-6596/246/1/012024

Nonlinear response and dynamical transitions in a phase-field crystal model for adsorbed overlayers

J.A.P. Ramos [1,2], E. Granato [2,3], S.C. Ying [3], C.V. Achim [4,5], K.R. Elder [6], and T. Ala-Nissila [3,4]

[1] Departamento de Ciências Exatas, Universidade Estadual do Sudoeste da Bahia, 45000-000 Vitória da Conquista, BA, Brazil
[2]Laboratório Associado de Sensores e Materiais, Instituto Nacional de Pesquisas Espaciais,12245-970 São José dos Campos, SP, Brazil
[3]Department of Physics, P.O. Box 1843, Brown University, Providence, RI 02912-1843, USA
[4]Department of Applied Physics, Aalto University School of Science and Technology, P.O. Box 11000, FI-00076 Aalto, Espoo, Finland
[5] Institut für Theoretische Physik II: Weiche Materie, Heinrich-Heine-Universität Düsseldorf, Universitätsstraße 1, D-40225 Düsseldorf, Germany
[6]Department of Physics, Oakland University, Rochester, Michigan 48309-4487, USA

E-mail: `Jorge@las.inpe.br`

Abstract. The nonlinear response and sliding friction behavior of a phase-field crystal model for driven adsorbed atomic layers is determined numerically. The model describes the layer as a continuous density field coupled to the pinning potential of the substrate and under an external driving force. Dynamical equations which take into account both thermal fluctuations and inertial effects are used for numerical simulations of commensurate and incommensurate layers. At low temperatures, the velocity response of an initially commensurate layer shows hysteresis with dynamical melting and freezing transitions at different critical forces. The main features of the sliding friction behavior are similar to the results obtained previously from molecular dynamics simulations of particle models. However, the dynamical transitions correspond to nucleations of stripes rather than closed domains.

1. Introduction

The nonlinear response of an adsorbed atomic monolayer under an external force is a central problem for understanding friction phenomena [1, 2] in experiments on sliding friction between two surfaces with a lubricant [3] or between adsorbed overlayers on an oscillating substrate [4, 5]. A fundamental issue in modeling such systems is the origin of the hysteresis and the dynamical melting and freezing transitions associated with the different static and kinetic frictional forces when increasing the driving force from zero and decreasing from a large value, respectively. Various elastic and simplified particle models have been used to study the driven dynamical transitions and the sliding friction behavior [1, 2, 6, 7, 8]. However, it is still not fully understood which features are universal and what are the most relevant parameters in such models.

In this work, we present results for the nonlinear response and sliding friction behavior of an adsorbed layer using a phase-field crystal (PFC) model with pinning [9] in presence of thermal fluctuations and inertial effects [10]. The model describes the layer in terms of a continuous density field coupled to a periodic pinning potential and external driving force. At low temperatures, the velocity response of commensurate and incommensurate layers are qualitative different. For an initially commensurate layer, the nonlinear response shows hysteresis with dynamical melting and freezing transitions at different critical values of the force. The main features of the velocity response are similar to the results obtained previously with molecular dynamics simulations of particles. However, the dynamical melting and freezing mechanisms are significantly different, corresponding to nucleation of stripes rather than closed domains.

2. Model and simulation

In the PFC model with pinning [9, 11, 12], the potential energy contribution due to the interacting particles forming the layer and the pinning potential can be written in a dimensionless form as

$$H_{\rm pfc} = F_0 \int d\vec{x} \{ \frac{1}{2}\psi[r + (1+\nabla^2)^2]\psi + \frac{\psi^4}{4} + V\psi \}, \tag{1}$$

where $\psi(\vec{x})$ is a continuous field and $V(\vec{x})$ is the external pinning potential. The phase field $\psi(\vec{x})$ is a measure of deviations of the particle density $\rho(\vec{x})$ from a uniform reference value ρ_0, such that $\psi(\vec{x}) = (\rho(\vec{x}) - \rho_0)/\rho_0$. The overall constant F_0 sets the energy scale and r is a parameter. In Eq. (1), $\psi(\vec{x})$ is a conserved field and the energy minima depend on the average value $\bar{\psi}$ of $\psi(\vec{x})$. When $V(\vec{x}) = 0$, the minimum energy configuration consists of an hexagonal pattern of peaks in $\psi(\vec{x})$ with wave vector $k_h \approx 1$. This phase-field crystal of an hexagonal array of peaks [13] is taken here as a simple model of an atomic layer with the pinning potential representing the effects of a rigid substrate.

To take into account inertial effects, we also include the momentum density $\vec{g}(\vec{x}) = \rho(\vec{x})\vec{v}(x)$ as an additional dynamical variable. As obtained recently [10], the resulting coupled dynamical equations of motion can be written as

$$\begin{aligned} \frac{\partial\psi}{\partial t} &= -\nabla\cdot\vec{g}; \\ \frac{\partial g_i}{\partial t} &= -\nabla_i \frac{\delta H_{pfc}}{\delta\psi} + \psi f_i - \eta g_i + \nu_i(\vec{x},t); \\ \langle \nu_i(\vec{x},t)\nu_j(\vec{x}',t')\rangle &= 2k_B T\eta\delta(\vec{x}-\vec{x}')\delta(t-t')\delta_{i,j}, \end{aligned} \tag{2}$$

where $\vec{f}$ is the spatially uniform external force and the thermal noise $\nu_i(\vec{x},t)$ satisfies the fluctuation-dissipation relation corresponding to a temperature T and damping parameter η.

We have considered a pinning potential $V(\vec{x})$ representing a substrate with square symmetry

$$V(\vec{x}) = -V_0[\cos(k_0 x) + \cos(k_0 y)]. \tag{3}$$

The parameters of the PFC model were fixed to $r = -0.25$, $\bar{\psi} = -0.25$, with a lattice mismatch between the layer and the pinning potential $\delta_{\rm m} = (k_h - k_0)/k_h = -0.5$.

For numerical calculations, the phase field $\psi(\vec{x})$ and momentum density field $\vec{g}(\vec{x})$ are defined on a space square grid with $dx = dy = \pi/4$ in systems with linear sizes $L = 64$ to 128 and periodic boundary conditions. The Laplacians and gradients were evaluated using finite differences. The coupled equations (2) are solved numerically using time steps $dt = 0.001 - 0.005$.

3. Numerical results and discussion

The velocity response was studied for commensurate and incommensurate layers. Different pinning amplitudes V_0 with the external force set to zero, were chosen to obtain the equilibrium states [9, 11]. For a small pinning amplitude $V_0 = 0.05$, the ground state and low temperature configurations consist of an incommensurate hexagonal phase (inset of Fig. 1a) while for a larger pinning amplitude $V_0 = 0.15$, the configuration is a commensurate $c(2 \times 2)$ structure

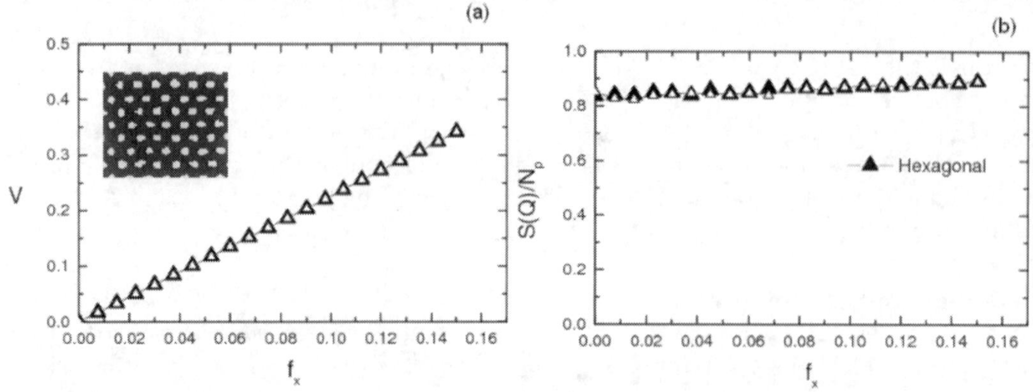

Figure 1. (a) Velocity as a function of applied force for an incommensurate layer. Inset: snapshot of the density field configuration at $f = 0$. (b) Scaled structure-factor peak $S(Q)/N_p$ as a function of applied force. Here $\vec{Q}$ stands for the primary reciprocal lattice vector of the hexagonal structure ($\vec{k}_h$). Filled and open symbols correspond to increasing and decreasing forces, respectively.

(inset of Fig. 2a), where every second site of the lattice of the pinning potential corresponds to a peak in the phase field $\psi(\vec{x})$. The solid commensurate configuration is stable against thermal fluctuations [11] below a melting transition temperature $T_c \approx 0.055$ and the incommensurate one below $T_h \approx 0.035$. We focus on equilibrium states which correspond to well ordered $c(2 \times 2)$ or incommensurate phases, at a temperature ($T = 0.01$) well below both transition temperatures (T_h, T_c) and a fixed damping parameter $\eta = 0.4$.

To obtain the driven response, a force along the x direction is increased from zero to a maximum value and then decreased back to zero. For each value of the force, 10^6 time steps were used to reach a steady state and an additional equal number were used to obtain time-averaged quantities. The velocity is measured by tracking the position of the peaks in the phase field $\psi(\vec{x})$ during the simulation [11, 12] and then the steady-state drift velocity $\vec{v}$ is obtained as a time average over all the peaks.

The main features of the velocity response to the driving force are shown in Fig. 1 for the incommensurate phase and in Fig. 2 for the commensurate phase. While for the incommensurate phase the response is nonzero even for small driving forces and shows no hysteresis behavior (Fig. 1a), for the commensurate phase the nonlinear response shows clear hysteresis behavior with two different critical force thresholds $f_a \approx 0.075$ for increasing forces and $f_b \approx 0.045$ for decreasing forces (Fig. 2a). At f_a, the velocity jumps abruptly from zero to a finite value whereas at f_b the velocity of the sliding layer drops abruptly to zero and the layer become pinned by the external potential to form an immobile commensurate state. These two threshold values correspond to the static frictional force and kinetic frictional force, respectively.

To characterize the dynamical transitions at f_a and f_b in more detail, we examine the behavior of the steady state structure factor $S(\vec{k})$ which is a measure of the translational order of the layer. It can be calculated from the the positions $\vec{R}_j$ of the peaks in the phase-field pattern as $S(\vec{k}) = \langle \sum_{j,j'=1}^{N_P} \exp(-i\vec{k} \cdot (\vec{R}_j - \vec{R}_{j'}))/N_p \rangle$, where, $\langle ... \rangle$ denotes a time average and N_p is the number of peaks. The structure factor evaluated at the dominant lattice wave vector Q of the layer is shown in Fig. 1b and 2b. Here $\vec{Q} = \vec{k}_c$ corresponds to the primary reciprocal lattice vector for a $c(2 \times 2)$ structure and $\vec{Q} = \vec{k}_h$ to the reciprocal lattice vector of a hexagonal structure. For the incommensurate case where there is no critical force, $S(k_h)$ remains almost unchanged with the applied force indicating that the sliding layer remains ordered as an hexagonal structure. However, for the commensurate case, on increasing the force beyond f_a, $S(k_c)$ drops abruptly to

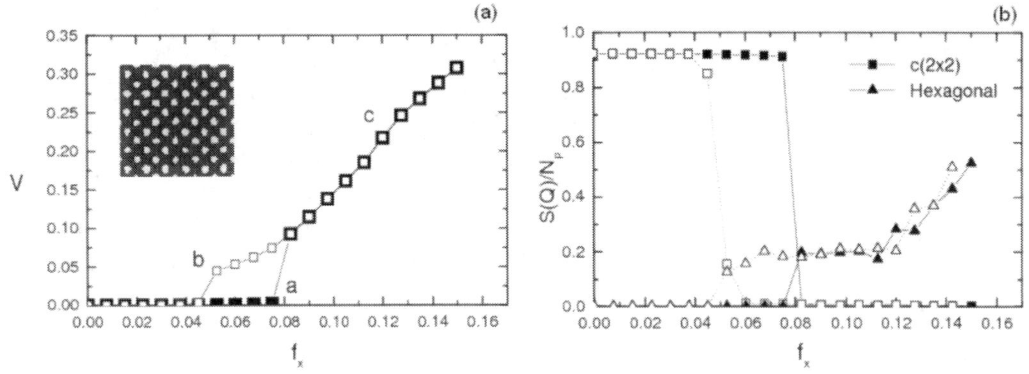

Figure 2. (a) Velocity as a function of applied force for a commensurate layer. The letters indicate the dynamical transitions at f_a, f_b and f_c. Inset: snapshot of the density field configuration at $f = 0$. (b) Scaled structure-factor peak $S(Q)/N_p$ as a function of applied force. Here $\vec{Q}$ stands for the primary reciprocal lattice vector for either the $c(2 \times 2)$ structure ($\vec{k}_c$) or the hexagonal structure ($\vec{k}_h$). Filled and open symbols correspond to increasing and decreasing forces, respectively.

zero consistent with the velocity response behavior in Fig. 2a. This is the onset of a force induced dynamical melting transition of the initial $c(2 \times 2)$ commensurate state. In fact, the behavior of $S(k_c)$ is analogous to the temperature induced melting transition in absence of a driving force [11]. On decreasing the force, the value of $S(k_c)$ stays vanishingly small until the force drops below the threshold f_b of a dynamical freezing transition, at which point $S(k_c)$ rapidly increases to a value corresponding to the commensurate pinned state. Further understanding of these dynamical transitions can be obtained from an analysis of the neighbors coordination numbers of the peaks in $\psi(x)$ using a Voronoi diagram construction [9, 10]. The qualitative behavior of the structure factor and the coordination number strongly suggest that the transitions at f_a and f_b can be regarded as a force induced dynamical melting and freezing transition respectively.

Finally, we examine snapshots of the phase field $\psi(x)$ obtained in the steady state. For driving forces above the dynamical melting threshold f_a (Fig. 3a) and the dynamical freezing threshold f_b (Fig. 3b), the configuration consists of stripes of commensurate $c(2 \times 2)$ phase separated by disordered domain walls. These domain walls are mobile liquid like regions as confirmed by a count of the fraction of various coordination numbers. The commensurate stripes have only 4-fold coordination numbers, while in the liquid like domain walls, there is a mixture of 5-fold, 6-fold and 7-fold coordination numbers. Starting from the steady state above f_b which still has a non-zero average sliding velocity, an examination of the time sequence of the freezing shows that the domain wall regions gradually shrink and eventually disappear. Above f_c the sliding layer shows an hexagonal structure with defects (Fig. 3c), indicating that at the crossover critical value f_c (Fig. 2a), there is another dynamic transition from a disordered phase into an incommensurate hexagonal phase corresponding to the increase of the structure factor $S(k_h)$ in Fig. 2b. This dynamical phase emerges as the average effect of the external pinning potential becomes less and less important at high sliding velocities and the steady state then corresponds effectively to the case of a weak pinning potential.

4. Conclusions

We have investigated the nonlinear response and sliding friction behavior of an adsorbed layer using a PFC model with pinning, where the adsorbed structure is described as a continuous density field allowing for elastic and plastic deformations. The main features of the velocity

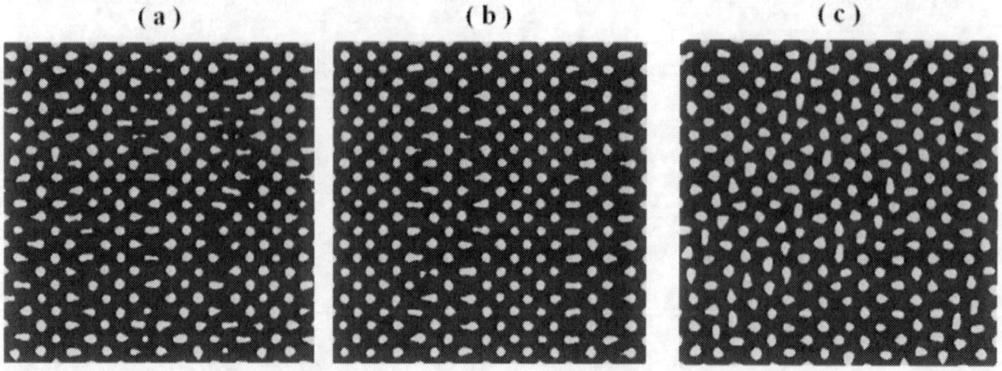

Figure 3. Snapshots of the density field at different values of the applied force in Fig. 2a : (a) $f_x = 0.0975$ (near f_a); (b) $f_x = 0.0525$ (near f_b) and (c) $f_x = 0.1425$ ($f_x > f_c$).

response, in particular the hysteresis loop separating the static friction and sliding friction thresholds for the commensurate layer, are similar to the results obtained previously with particle models with interacting Lennard-Jones potentials [1, 2, 7]. The similarity of results from these very different models demonstrates the universality of the hysteresis loop for models with inertial effects and the macroscopic consequence of stick and slip motion. However, the details of the dynamical melting and freezing mechanisms are significantly different. In the PFC model, they correspond to nucleation of stripes rather than closed domains found in particle models. The origin of this intriguing difference requires further investigation of both atomistic and PFC models.

5. Acknowledgments
This work was supported by Secretaria da Administração do Estado da Bahia - SAEB (J.A.P.R.), Fundação de Amparo à Pesquisa do Estado de São Paulo - FAPESP (Grant Nos. 07/08492-9) (E.G), a joint funding FAPESP (Grant No. 09/01942-4) (E.G. and S.C.Y.), the Academy of Finland through its COMP CoE grant,(T.A.N.), a joint funding EU STREP 016447 MagDot, NSF DMR Award No. 0502737, and a NSF fund under Grant No. DMR-0906676 (K.R.E.).

References
[1] Persson B N J 1998 *Sliding Friction: Physical Principles and Applications* (Springer, Heidelberg)
[2] Persson B N J 1993 *Phys. Rev. Lett.* **71** 1212
[3] Urbakh M Klafter J, Dourdon D and Israelachvili J 2004 *Nature* **430** 525
[4] Krim J, Solina D H and Chiarello R 1991 *Phys. Rev. Lett.* **66** 181
[5] Bruschi L, Fois G, Pontarollo A, Mistura G, Torre B, Buatier de Mongeot F, Boragno C, Buzio R and Valbusa U 2006 *Phys. Rev. Lett.* **96** 216101
[6] Braun O M, Dauxois T, Paliy M V and Peyrard M 1997 *Phys. Rev. Lett.* **78** 1295
[7] Granato E and Ying S C 2000 *Phys. Rev. Lett.* **85** 5368
[8] Granato E and Ying S C 2004 *Phys. Rev.* B **69** 125403
[9] Achim C V, Karttunen M, Elder K R, Granato E, Ala-Nissila T and Ying S C 2006 *Phys. Rev.* E **74** 021104
[10] Ramos J A P, Granato E , Ying S C, Achim C V, Elder K R and Ala-Nissila T 2010 *Phys. Rev.* E **81** 011121
[11] Ramos J A P, Granato E, Achim C V, Ying S C, Elder K R and Ala-Nissila T 2008 *Phys. Rev.* E **78** 031109
[12] Achim C V, Ramos J A P, Karttunen M, Elder K R, Granato E, Ala-Nissila T and Ying S C 2009 *Phys. Rev.* E **79** 011606
[13] Elder K R, Katakowski M, Haataja M and Grant M 2002 *Phys. Rev. Lett.* **88** 245701

Journal of Physics: Conference Series **246** (2010) 012025 doi:10.1088/1742-6596/246/1/012025

q-deformed statistical-mechanical structure in the dynamics of the Feigenbaum attractor

A. Robledo

Instituto de Física,
Universidad Nacional Autónoma de México,
Apartado Postal 20-364, México 01000 D.F., Mexico.

E-mail: `robledo@fisica.unam.mx`

Abstract. We show that the two complementary parts of the dynamics associated to the Feigenbaum attractor, inside and towards the attractor, form together a q-deformed statistical-mechanical structure. A time-dependent partition function produced by summing distances between neighboring positions of the attractor leads to a q-entropy that measures the ratio of ensemble trajectories still away at a given time from the attractor (and the repellor). The values of the q-indexes are given by the attractor's universal constants, while the thermodynamic framework is closely related to that first developed for multifractals.

The Feigenbaum attractor, an icon of the historical developments in the theory of nonlinear dynamics [1], is getting renewed attention [2]. This is because it offers a convenient model system to explore features that might reflect those of statistical mechanical systems under conditions of phase space mixing and ergodicity breakdown. It therefore offers insights on the limits of validity of ordinary statistical mechanics. Recently a thorough description with newly revealed features has been given [3] [4] of the intricate dynamics that takes place both inside and towards this famous multifractal attractor. Here we show that these two types of dynamics are related to each other in a statistical-mechanical fashion, i.e. the dynamics *at* the attractor provides the 'microscopic configurations' in a partition function while the *approach* to the attractor is economically described by an entropy obtained from it. As we show below, this property conforms to a q-deformation [5] [6] of the ordinary exponential (Boltzmann) weight statistics.

Furthermore, this novel statistical-mechanical property arises in many other cases, some very familiar, that involve multifractal attractors with vanishing Lyapunov exponents. We have already uncovered [7] the prerequisites for the quasiperiodic transition to chaos (specifically along the golden mean) to be another example of the same kind of q-statistics [8]. Features in problems that relate to the period-doubling or the quasiperiodic routes to chaos (with general nonlinearities differing from the usual quadratic or cubic) acquire thermodynamic q-deformed structures. These include, for instance, phase transitions in spin or plaquette models of systems with many degrees of freedom, when built-in scaling properties lead to iteration techniques, and their singular behavior appears linked to Mandelbrot-like sets [9]. The broader issue of incidence of q-statistical properties in the complex dynamics of Julia sets associated to neutral or indeferent fixed points is an interesting question to go in for.

Trajectories within the Feigenbaum attractor show self-similar temporal structures, they preserve memory of their previous locations and do not have the mixing property of chaotic trajectories [10]. The fluctuating sensitivity to initial conditions has the form of infinitely many interlaced q-exponential functions that fold into a single one with use of a two-time scaling property [3] [6] [7]. More precisely, there is a hierarchy of such families of interlaced q-exponentials; an intricate (and previously unknown) state of affairs that befits the rich scaling features of a multifractal attractor. Furthermore, the entire dynamics is made of a family of pairs of Mori's dynamical q-phase transitions [10] [3] [6] [7].

On the other hand, the process of convergence of trajectories into the Feigenbaum attractor is governed by another unlimited hierarchy feature built into the preimage structure of the attractor and its counterpart repellor [4]. The overall rate of approach of trajectories towards the attractor (and repellor) is conveniently measured by the fraction of (fine partition) bins W_{t_1} still occupied at time t_1 by an ensemble of trajectories with initial positions uniformly distributed over phase space [11] [4]. For the first few time steps the rate W_{t_1} remains constant, $W_{t_1} \simeq \Delta$, $1 \leq t_1 \leq t_0$, $t_0 = O(1)$ [12], after which a power-law decay with log-periodic modulation sets in, a signature of discrete-scale invariance [13]. This property of W_{t_1} is explained in terms of a sequential formation of gaps in phase space, and its self-similar features are seen to originate in the ladder feature of the preimage structure [4]. The rate W_{t_1} was originally presented in Ref. [11] where the power law exponent φ in

$$W_{t_1} \simeq \Delta\, h\left(\frac{\ln t}{\ln \Lambda}\right) t^{-\varphi},\ t = t_1 - t_0, \tag{1}$$

was obtained numerically. Above, $h(x)$ is a periodic function with $h(1) = 1$, and Λ is the scaling factor between the periods of two consecutive oscillations [14].

We proceed now to demonstrate the connection between the aforementioned dynamical properties. We recall [1] the definition of the interval lengths or diameters $d_{n,m}$ that measure the bifurcation forks that form the period-doubling cascade sequence in unimodal maps, here represented by the logistic map $f_\mu(x) = 1 - \mu x^2$, $-1 \leq x \leq 1$, $0 \leq \mu \leq 2$. These quantities are measured when considering the superstable periodic orbits of lengths 2^n, $n = 1, 2,...$; i.e. the 2^n-cycles that contain the point $x = 0$ at $\mu_n < \mu_\infty$, where $\mu_\infty = 1.401155189...$ is the value of the control parameter μ at the period-doubling accumulation point [15]. The positions of the limit 2^∞-cycle constitute the Feigenbaum attractor. The $d_{n,m}$ in these orbits are defined (here) as the (positive) distances of the elements x_m, $m = 0, 1, ..., 2^{n-1} - 1$, to their nearest neighbors $f_{\mu_n}^{(2^{n-1})}(x_m)$, i.e. $d_{n,m} \equiv \left| f_{\mu_n}^{(m+2^{n-1})}(0) - f_{\mu_n}^{(m)}(0) \right|$. For large n, $d_{n,0}/d_{n+1,0} \simeq \alpha$, where α is Feigenbaum's universal constant $\alpha \simeq 2.5091$.

Innermost to our arguments is the following comprehensive property: Time evolution at μ_∞ from $t = 0$ up to $t \to \infty$ traces the period-doubling cascade progression from $\mu = 0$ up to μ_∞. Not only is there a close resemblance between the two developments but also asymptotic quantitative agreement. For example, the trajectory inside the Feigenbaum attractor with initial condition $x_0 = 0$, takes positions x_t such that the distances between nearest neighbor pairs of them reproduce the diameters $d_{n,m}$ defined from the supercycle orbits with $\mu_n < \mu_\infty$. See Fig. 1. This property has been central to obtain rigorous results for the fluctuating sensitivity to initial conditions $\xi_t(x_0)$ within the Feigenbaum attractor, as separations at chosen times t of pairs of trajectories originating close to x_0 can be obtained as diameters $d_{n,m}$ [3] [6].

Further, the complex dynamical events that fix the decay rate W_{t_1} can be understood in terms of the correlation between time evolution at μ_∞ from $t = 0$ up to $t \to \infty$ and the 'static' period-doubling cascade progression from $\mu = 0$ up to μ_∞. As shown recently [16] each doubling of the period (obtained through the shift $\mu_n \to \mu_{n+1}$) introduces additional elements in the hierarchy of the preimage structure and in the family of sequentially-formed phase space gaps in the finite period cycles. The complexity of these added elements is similar to that of the total period 2^n system. Also, the shift $\mu_n \to \mu_{n+1}$ increases in one unit the number of undulations in the

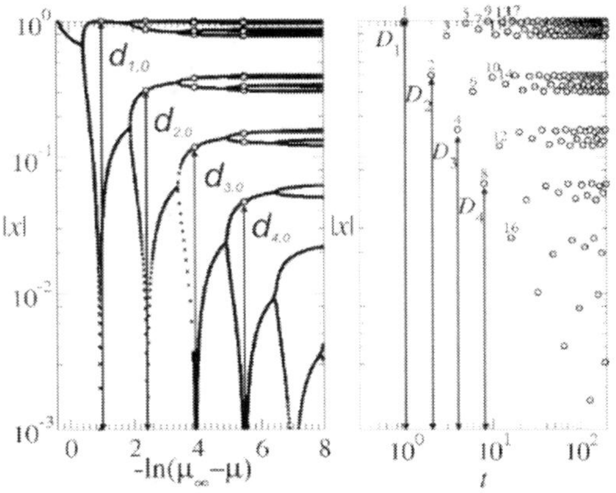

Figure 1. Left panel: Absolute value of attractor positions for the logistic map $f_\mu(x)$ in logarithmic scale as a function of $-\ln(\mu_\infty - \mu)$. Right panel: Absolute value of trajectory positions for $f_\mu(x)$ at μ_∞ with initial condition $x_0 = 0$ in logarithmic scale as a function of the logarithm of time t, also shown by the numbers close to the circles. The arrows indicate the equivalence between the diameters $d_{n,0}$ in the left panel, and position differences D_n with respect to $x_0 = 0$ in the right panel.

transitory log-periodic power law decay found for the corresponding rate W_{n,t_1} of approach to the 2^n-supercycle attractor [16]. As a consequence of this we have obtained detailed understanding of the mechanism by means of which the discrete scale invariance implied by the log-periodic property in $W_{t_1} \equiv \lim_{n\to\infty} W_{n,t_1}$ arises. What is more, the rate W_{t_1}, at the values of time for period doubling, can be obtained quantitatively from the supercycle diameters $d_{n,m}$, that is [4], $W_{t_1} = \Delta\ Z_t$, $t = t_1 - t_0$, and

$$Z_t = \sum_{m=0}^{2^{n-1}-1} d_{n,m},\ t = 2^{n-1},\ n = 1, 2, 3, ... \tag{2}$$

This is an explicit expression equivalent to the numerical procedure followed in Ref. [17] by the use of the triadic cantor set construction of the Feigenbaum attractor to evaluate the power law exponent φ, and from which the value for $\varphi \cong 0.800138194$ is reported.

Now, to reveal the aforesaid statistical-mechanical structure we identify the decay rate Z_t as a partition function. From this perspective the diameters $d_{n,m}$ are configurational terms and we go forward to determine their time dependence. The $d_{n,m}$ scale (asymptotically) with n for $m/2^n$ fixed as $d_{n,m} \simeq \alpha_y^{-n+1}$, n large, the α_y are universal constants obtained, for instance, from the finite jump discontinuities of Feigenbaum's trajectory scaling function $\sigma(y) = \lim_{n\to\infty} d_{n,m+1}/d_{n,m}$, $y = \lim_{n\to\infty} m/2^n$ [1]. The largest discontinuities of $\sigma(y)$ correspond to the most sparse and the most crowded regions of the multifractal attractor, and for these we have, respectively, $d_{n,0} \simeq \alpha^{-n+1}$ and $d_{n,1} \simeq \alpha^{-2(n-1)}$. (The 1st diameter $d_{0,0} = 1$ and the $d_{n,m}$ converge rapidly to α_y^{-n+1} as n increases). With use of the identity $A^{-n+1} \equiv (1+\beta)^{-\ln A/\ln 2}$, $\beta = 2^{n-1} - 1$, the power law $d_{n,m} \simeq \alpha_y^{-n+1}$ can be rewritten as a q-exponential ($\exp_q(x) \equiv [1-(q-1)x]^{-1/(q-1)}$), i.e., $d_{n,m} \simeq \exp_{q_y}(-\beta\nu_y)$,where $q_y = 1 + \nu_y^{-1}$, $\nu_y = \ln\alpha_y/\ln 2$, and $\beta = t - 1 = 2^{n-1} - 1$. Likewise, the partition function $Z_t \simeq t^{-\varphi}$ (or $Z_t \simeq \varepsilon^{-n+1}$), with $\varphi = \ln\varepsilon/\ln 2$ and $t = 2^{n-1}$, can be written as $Z_t \simeq \exp_Q(-\beta\varphi)$, $Q = 1 + \varphi^{-1}$ and again $\beta = t - 1 = 2^{n-1} - 1$.

Journal of Physics: Conference Series **246** (2010) 012025 doi:10.1088/1742-6596/246/1/012025

Our main point becomes apparent when the above q-exponential forms for $d_{n,m}$ and Z_t are used in Eq. (2), to yield

$$\exp_Q(-\beta\varphi) \simeq \sum_y \exp_{q_y}(-\beta\nu_y). \tag{3}$$

Eq. (3) resembles a basic statistical-mechanical expression where the quantities in it play the following roles: β an inverse temperature, φ a free energy (or the product $s = -\beta\varphi$ a Massieu thermodynamic potential, or entropy), and the ν_y configurational energies. However, the equality involves q-deformed exponentials in place of ordinary exponentials that would be recovered when $Q = q_y = 1$. It is worth noticing that there is a multiplicity of q-indexes associated to the configurational weights in Eq. (3), however their values form a well-defined family [6] determined by the discontinuities of Feigenbaum's function σ. To substantiate the usefulness and appropriateness of this identification we present in the remaining part of this paper: 1) A 'mean field' evaluation of Z_t and a thermodynamic interpretation of the time evolution process. 2) The relationship of Z_t with the familiar partition function developed for the description of multifractal geometry. 3) A crossover to $q = 1$ ordinary statistics.

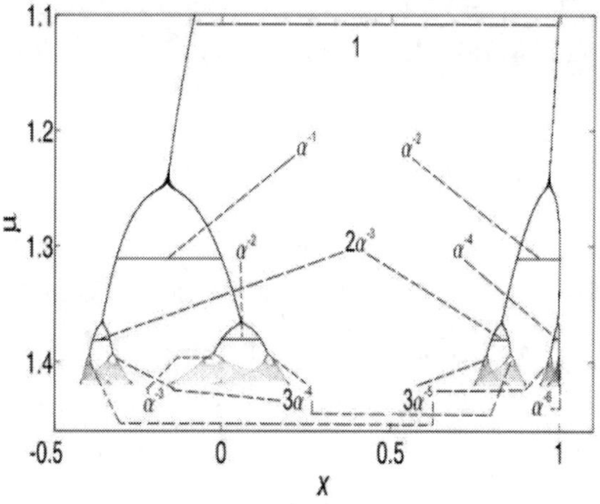

Figure 2. Fig. 2. Sector of the bifurcation tree for the logistic map $f_\mu(x)$ that shows the formation of a Pascal triangle of diameter lengths according to the scaling approximation expalined in the text, $\alpha \simeq 2.5091$ is the pertinent universal constant.

Akin to a mean field approximation we assume that, for a given value of n, e.g., $n = 3$, the diameters $d_{n,m}$ that are of comparable lengths have equal length and this is obtained from those of the shortest or longest diameters via a simple scale factor; e.g., $d_{3,3} = d_{3,2} = \alpha^{-1}d_{3,0} = \alpha d_{3,1}$. This introduces some degeneracy in the lengths that propagates across the bifurcation tree. See Fig. 2 [18]. Specifically, the $d_{n,m}$ scale now with increasing n according to a binomial combination of the scaling of those diameters that converge to the most crowded and most sparse regions of the multifractal attractor. This is to consider that the 2^{n-1} diameters at the n-th supercycle have lengths equal to $\alpha^{-(n-1-l)}\alpha^{-2l}$ and occur with multiplicities $\binom{n-1}{l}$, where $l = 0, 1, ..., n-1$. As seen in Fig. 2 the diameters form a Pascal triangle across the bifurcation cascade. The partition function can be immediately evaluated to yield

$$Z_t = \sum_{l=0}^{n-1} \binom{n-1}{l} \alpha^{-(n-1-l)}\alpha^{-2l} = \left(\alpha^{-1} + \alpha^{-2}\right)^{n-1}, \tag{4}$$

$t = 2^{n-1}$. We obtain $\varepsilon = (\alpha^{-2} + \alpha^{-1}) = 1.7883$, $\varphi = 0.8386$, and $Q = 2.1924$, a surprisingly good approximation when compared to the numerical estimates $\varphi = 0.8001$, and $Q = 2.2498$ of the exact values. Under this approximation all the indexes q_m in Eq. (3) are equal, $q_m = q = 1 + \nu^{-1}$, $\nu = \ln\alpha/\ln 2$, and Eq. (3) becomes

$$\exp_Q(-\beta\varphi) = \sum_{l=0}^{n-1} \Omega(n-1,l) \exp_q(-\beta\nu), \tag{5}$$

where$\Omega(n-1,l) = \alpha^{-l}\binom{n-1}{l}$. Thermodynamically, the approach to the attractor described by Eqs. (3) or (5) is a cooling process $\beta \to \infty$ in which the free energy (or energy) φ is fixed and therefore the entropy $s = -\beta\varphi$ is linear in β. It is illustrative to define an 'energy landscape' for the Feigenbaum attractor as being composed by an infinite number of 'wells' whose equal-valued minima at $\beta \to \infty$ coincide with the points of the attractor on the interval $[-\alpha^{-1}, 1]$ [14]. When $\beta = 2^{n-1} - 1$, n finite, the wells merge into 2^{n-1} intervals of widths equal to the diameters $d_{n,m}$.

As it is well known the geometric properties of multifractals conform to a statistical mechanical framework, the so-called thermodynamic formalism [15]. The partition function devised to study their properties, such as the spectrum of singularities $f(\tilde{\alpha})$ [15], is $Z(\tau, \mathsf{q}) \equiv \sum_m^M p_m^\tau l_m^{-\mathsf{q}}$, where the l_m (in one-dimensional systems) are M disjoint interval lengths that cover the multifractal set and the p_m are probabilities assigned to these intervals. The usual procedure consists of requiring that $Z(\tau, \mathsf{q})$ neither vanishes nor diverges in the limit $l_m \to 0$ for all m (and $M \to \infty$). In this case the exponents τ and q define a function $\tau(\mathsf{q})$ from which $f(\tilde{\alpha})$ is obtained via Legendre transformation [15]. When the multifractal is an attractor its elements become ordered dynamically, and for the Feigenbaum attractor the trajectory with initial condition $x_0 = 0$ generates sequentially the positions that form the diameters, producing all diameters $d_{n,m}$ for n fixed between times $t = 2^{n-1}$ and $t = 3 \cdot 2^{n-1}$. Since the diameters cover the attractor it is natural to choose the covering lengths at stage n to be $l_m^{(n)} = d_{n,m}$ and to assign to each of them the same probability $p_m^{(n)} = 1/2$. Within the two-scale approximation to the Feigenbaum multifractal [15], $l_k^{(n)} = \alpha^{-(n-1-k)}\alpha^{-2k}$, the condition $Z(\tau, \mathsf{q}) = 1$ reproduces Eq. (4) when $p_m^{(n)} = t^{-1} = 2^{-n+1}$, with $\tau = 1$ and $\mathsf{q} = -\varphi$. It should be kept in mind that the 'static' partition function $Z(\tau, \mathsf{q})$ is not meant to distinguish between chaotic and critical (vanishing Lyapunov exponent λ) multifractal attractors as we do here. It is the functional form of the link between the probabilities $p_m^{(n)}$ and actual time t that determines the nature of the statistical mechanical structure of the dynamical system.

The recursive method of backward iteration of chaotic maps provides a convenient way for reconstructing multifractal sets and obtaining their underlying statistical mechanics [19]. A chaotic unimodal map has a two-valued inverse and given a position $x = x_n$ a binary tree is formed under backward iteration, so there are 2^n initial conditions x_0 for trajectories that lead to x_n. Since now $\lambda > 0$, lengths expand under forward iteration according to $l \sim \exp(\lambda n)$ and contract under backward iteration as $l \sim \exp(-\lambda n)$. We can define, as above, a set of covering lengths $D_{n,m} = \delta_m \exp(-\lambda n)$, where m relates to the initial condition x_0 and use of them in a partition function like that in Eq. (2) gives $\exp(-\beta\varphi) = \sum_m \delta_m \exp(-\beta\lambda)$, where now $\beta = n$. Recalling Pesin's theorem, φ is clearly identified as the Kolmogorov-Sinai entropy. The crossover from q-deformed statistics to ordinary $q = 1$ statistics can be observed for control parameter values in the vicinity of the Feigenbaum attractor, $\mu \gtrsim \mu_\infty$, when the attractor consists of $2^{\overline{n}}$ bands, $\overline{n}$ large. The Lyapunov coefficient λ of the chaotic attractor decreases with$\Delta\ \mu = \mu - \mu_\infty$ as $\lambda \propto 2^{-\overline{n}} \sim \Delta\mu^\kappa$, $\kappa = \ln 2/\ln\delta_F(\varsigma)$, where δ_F is the Feigenbaum constant that measures the rate of development of the bifurcation tree in control parameter space [1]. The chaotic orbit consists of an interband periodic motion of period $2^{\overline{n}}$ and an intraband chaotic motion. The expansion rate $\sum_{i=0}^{t-1} \ln|df_\mu(x_i)/dx_i|$ fluctuates with increasing amplitude as $\ln t$ for $t < 2^{\overline{n}}$

but converges to a fixed number that grows linearly with t for $t \gg 2^{\overline{n}}$ [10]. This translates as dynamics with $q \neq 1$ for $t < 2^{\overline{n}}$ but ordinary dynamics with $q = 1$ for $t \gg 2^{\overline{n}}$.

We have shown that there is a statistical-mechanical property lying beneath the dynamics of an ensemble of trajectories en route to the Feigenbaum attractor. The fraction of phase space still occupied at time t is a partition function Z_t made up of q-exponential weighted configurations, while Z_t itself is the q-exponential of a thermodynamic potential function. This is a clear signature of q-deformation of ordinary statistical mechanics, and, to our knowledge, it is the first *bona fide* concrete instance (anticipated or not in the form presented here) where arguments can be made explicit and rigorous. There is a close parallel with the thermodynamic formalism for multifractal sets, but it should be emphasized that the departure from the usual exponential statistics is dynamical in origin, and due to the vanishing of the (only) Lyapunov exponent. Our results translate to other multifractal critical attractors.

Although there should not be any confusion that our discussion presented here refers to the dynamical properties of low-dimensional multifractal critical attractors, and explicitly only to the Feigenbaum attractor, we would like to remind the casual reader that our development is not intended to apply to thermal systems, Hamiltonian systems composed of many degrees of freedom, the customary subject of ordinary statistical mechanics. Rather, our analysis relates to the statistical-mechanical structure exhibited by static or dynamical properties of multifractal sets.

Acknowledgments

Support by DGAPA-UNAM and CONACyT (Mexican Agencies) is acknowledged.

References

[1] See, for example, H.G. Schuster 1988 *Deterministic Chaos. An Introduction* 2nd Revised Edition (Weinheim: VCH Publishers)
[2] A. Robledo 2005 *Europhys. News* **35** 214
[3] E. Mayoral and A. Robledo 2005 *Phys. Rev. E* **72** 026209
[4] A. Robledo and L.G. Moyano 2008 *Phys. Rev. E* **77** 036213 ; *2009 Braz. J. Phys.* **39** 364
[5] C. Tsallis 1988 *J. Stat. Phys.* **52** 479
[6] A. Robledo 2006 *Physica A* **370** 449
[7] H. Hernández-Saldaña and A. Robledo 2006 *Physica A* **370** 286
[8] Technically, translating properties from one route to chaos to the other entails little more than replacement of universal constants and period-doublings by Fibonacci number sequences.
[9] See, for example, N.S. Ananikian *et al* 1998 *Phys. Lett. A* **248** 381 ; N.S. Anakinian and R.G. Ghulghazaryan 2000 *Phys. Lett. A* **277** 249 and references therein.
[10] H. Mori, H. Hata, T. Horita and T. Kobayashi 1989 *Prog. Theor. Phys. Suppl.* **99** 1
[11] F.A.B.F. de Moura, U. Tirnakli and M.L. Lyra 2000 *Phys. Rev. E* **62** 6361
[12] The constant Δ is given by $\Delta = (1+\alpha^{-1})/2$, where α is Feigenbaum's universal number $\alpha \simeq 2.5091$. This follows from the fact that all initial conditions out of the interval $[-\alpha^{-1}, 1]$ take a value inside it in the 1st iteration. See Fig. (21) in Ref. [4].
[13] D. Sornette 1998 *Phys. Rep.* **297** 239
[14] Discrete scale invariance due to period doubling implies $\Lambda = 2$ but use of a finite number of trajectories N_{tr} in the numerical evaluation of W_{t_1} yieldsΛ < 2. See Ref. [4] for the convergenceΛ $\to 2$ as $N_{tr} \to \infty$.
[15] See, for example, C. Beck and F. Schlogl 1993 *Thermodynamics of Chaotic Systems* (UK: Cambridge University Press)
[16] L.G. Moyano, D. Silva and A. Robledo 2009 *Cent. Eur. J. Phys.* **7** 591
[17] P. Grassberger 2005 *Phys. Rev. Lett.* **95** 140601
[18] G. Linage, F. Montoya, A. Sarmiento, K. Showalter and P. Parmananda 2006 *Phys. Lett. A* **359** 638
[19] J.L. McCauley 1989 *Int. J. Mod. Phys.* **B3** 821

Journal of Physics: Conference Series **246** (2010) 012026 doi:10.1088/1742-6596/246/1/012026

Influence of the coupling mechanism on the stochastic resonance response in extended systems

Horacio S. Wio

Instituto de Física de Cantabria (UC and CSIC),
Avda. de los Castros, s/n, E-39005 Santander, Spain, EU

E-mail: wio@ifca.unican.es

Abstract. We analyze the stochastic resonance response in an extended system, considering different transport/coupling mechanisms: diffusion, KPZ, and also include the possibility of a non-local interaction. Our aim, since these mechanisms correspond to different forms of coupling of resonant units leading to an extended system, is to obtain information about the way to optimize the system's response to weak signals. To reach such a goal, we exploit the knowledge of the so called "non-equilibrium potential" for the above indicated situations.

Stochastic resonance (SR) has become a paradigm of the constructive effects of fluctuations on nonlinear systems [1, 2]. In a simplified picture, the phenomenon occurs whenever the Kramers' rate for the transition between attractors matches the typical frequency of a signal which is incapable by itself to trigger that transition (i.e. it is *subthreshold*).

In a recent series of papers we have studied the SR phenomenon for the transitions between two different patterns in extended systems, exploiting the concept of *nonequilibrium potential* (NEP). Some recent review-like papers [3, 4, 5], summarize the indicated results including several references to our work as well. Such NEP is a special Lyapunov's function of the associated deterministic system which, for non-equilibrium systems, plays a role similar to that of a thermodynamic potential in equilibrium thermodynamics [6, 7, 8, 9, 10]. It is closely related to the stationary solution of the system's Fokker–Planck equation, characterizing the global properties of the dynamics: attractors, linear and *relative stability* of these attractors, *height of the barriers* separating attraction basins. Moreover, it allows to evaluate the *transition rates* among the different attractors [6, 7, 8, 9, 10]. Regarding the problem of SR in extended systems, it was shown that the knowledge of the NEP allows to obtain a rather complete picture of the behavior of the output *signal-to-noise ratio* (SNR), and eventually other response measures. The novelty with non-equilibrium extended systems is that even point-like attractors in the medium's infinite-dimensional phase-space can be nontrivial field configurations (real-space *patterns*).

In this contribution we exploit previous results, and new ones as well, in order to discuss the role of the coupling mechanism on the SR response of a bistable extended system. We will consider three different situations: a simple reaction-diffusion mechanism [3, 4, 5, 11], a KPZ-like contribution [12, 13, 14], and finally the existence of a non-local coupling [15].

We start briefly sketching some relevant points related to the NEP concept. We refer to [3, 4] for details. If we have a kinetic equation defined by

$$\partial_t \phi(x,t) = \mathcal{L}\phi(x,t) + f(\phi(x,t)) + \xi(x,t),$$

where $\mathcal{L}$ is the coupling or transport operator; $\xi(x,t)$ is a Gaussian white noise [of zero mean, correlation $\langle\xi(x,t)\xi(x',t')\rangle = 2\gamma\delta(x-x')\delta(t-t')$, and intensity γ]; and $f(\phi(x)) = -\partial V(\phi(x))/\partial\phi$, [here we adopt $V(\phi(x)) = \frac{\alpha}{4}\phi(x)^4 - \frac{\beta}{2}\phi(x)^2$]. We denote by $\Phi[\phi]$ the NEP of such a system, whose first variation gives $\frac{\partial}{\partial t}\phi(x,t) = -\frac{\delta\Phi[\phi]}{\delta\phi(x,t)} + \xi(x,t)$, and that in the deterministic case fulfills the Lyapunov condition $\frac{d}{dt}\Phi[\phi] = -\left(\frac{\delta\Phi[\phi]}{\delta\phi(x,t)}\right)^2 \leq 0$.

As indicated in [3, 4], for a scalar bistable system and within the "two-state approximation" [1] (that implies an adiabatic assumption), the SNR is given by SNR $\approx \frac{\pi}{4\gamma^2}\frac{\mu_l\mu_r}{\mu_l+\mu_r}$, where $\mu_i \sim \exp\{\Delta\Phi[\phi_i]/\gamma\}$. Here $\Delta\Phi[\phi_i] = \Phi[\phi_o] - \Phi[\phi_i]$, with $\phi_i(x)$ $(i = l, r)$ one of the two **stable** stationary states, and $\phi_o(x)$ corresponds to the (**unstable**) pattern at the saddle separating both stable attractors. We will restrict the system to the interval $0 \leq x \leq \pi$ with Dirichlet b.c.

The first case we discuss here corresponds to a reaction-diffusion system described by

$$\partial_t\varphi(x,t) \;=\; \nu\partial_x^2\varphi(x,t) + f(\varphi(x,t)) + \xi(x,t). \tag{1}$$

The associated NEP reads [3, 4, 11]

$$\Phi[\varphi] = \int dx\left[\frac{\nu}{2}(\partial_x\varphi)^2 + V(\varphi(x))\right]. \tag{2}$$

As is known [3, 4, 11], for the above indicated form of $V(\varphi(x))$ and Dirichlet b.c., this system has two stable and one unstable solutions. We can readily use the expression for the SNR indicated above and obtain the result indicated by the continuous line on Fig. 1.

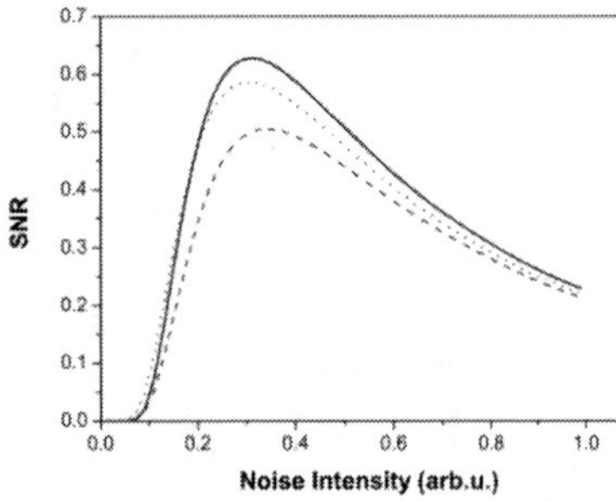

Figure 1. SNR vs noise intensity (γ). For the reaction-diffusion system in Eq. (1) we have the solid line on the left panel. Also, for the KPZ case we have $\lambda = 0$ (solid line), $\lambda = 0.25$ (dotted line) and $\lambda = 0.5$ (dashed line). The rest of parameters are $\nu = 2$, $\alpha = \beta = 30$.

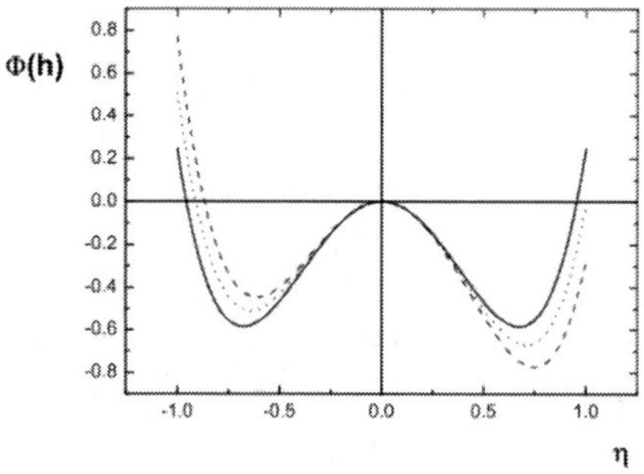

Figure 2. Nonequilibrium potential $\Phi(\eta)$ vs η, for the KPZ case, where η indicates a projection along the line joining the attractors through the saddle. We have $\lambda = 0$ (solid line), $\lambda = 0.25$ (dotted line) and $\lambda = 0.5$ (dashed line). The rest of parameters are $\nu = 2$, $\alpha = \beta = 30$.

Now we analyze the case of KPZ equation. For the 1-d case it reads

$$\partial_t h(x,t) \;=\; \nu\partial_x^2 h(x,t) + \frac{\lambda}{2}(\partial_x h(x,t))^2 + f(h(x,t)) + \xi(x,t). \tag{3}$$

This nonlinear stochastic partial differential equation describes the height of a fluctuating growing interface with surface tension ν, λ being proportional to the average growth velocity and arises because in this growth process the surface slope is transported in parallel. In Ref. [14] it was shown that the functional

$$\Phi[h] = \int dx \left[\frac{\nu}{2} (\partial_x h)^2 - \frac{\lambda}{2} \int_{h_{ref}}^{h(x)} d\psi \, (\partial_x \psi)^2 + V(h(x)) \right], \tag{4}$$

is the NEP for KPZ. The system is bounded by the potential V indicated before. In this case $h_i(x)$ $(i = l, r)$ are the two stable (nonhomogeneous) stationary states, while $h_o(x)$ corresponds to the flat (**unstable**) pattern at the saddle separating both stable attractors. We will work with the NEP's approximated form $\Phi[h_i(x)] \approx \int dx \left[\left(\frac{\nu}{2} - \frac{\lambda}{2} h_i(x) \right) (\partial_x h_i(x))^2 + V(h(x)) \right]$.

Figure 1 plots the SNR as a function of the noise intensity, for different values of λ. The trend is that for increasing λ, the system's response diminishes. This can be understood within the NEP picture, since the KPZ term resembles an effective (h dependent) diffusivity [16, 17], indicating a variation of the effective diffusivity with the spatial variable. This originates a breaking of the NEP's symmetry as λ is varied, yielding a SNR reduction [5]. Figure 2 depicts a projection of that NEP (along the line joining the attractors and crossing the saddle), where the rupture of symmetry due to the effect of λ is apparent.

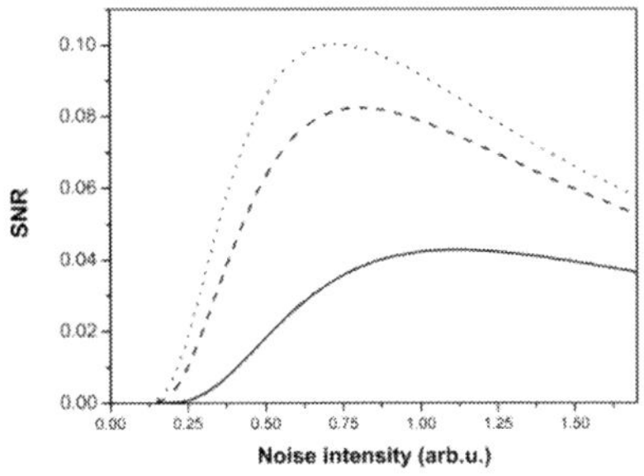

Figure 3. SNR vs noise intensity (γ) for cases with non-local coupling. The different curves corresponds to $l = 0.05$ (dotted line), $l = 0.5$ (dashed line) and $l = 1.$ (solid line). The rest of parameters are $\lambda = 0.1$, $\vartheta = 0.01$, $\nu = 2$, $\alpha = \beta = 30$.

The last example corresponds to the inclusion of a non-local coupling [1, 5]. In this case the equation has the form

$$\partial_t \phi(x,t) = \mathcal{L}\phi(x,t) + f(\phi) - \vartheta \int_0^{\pi} \mathcal{K}(x,x') \, \phi(x',t) \, dx' + \xi(x,t), \tag{5}$$

where $\mathcal{K}(x,x')$ is the non-local coupling [3, 4] that, in order to simplify the analysis, we choose as $\mathcal{K}(x,x') = \frac{1}{2l}$ if $|x - x'| \leq l$, and $= 0$ if $|x - x'| > l$ [15]. Hence, we can do the analysis just varying the interaction range $2l$. Such a contribution could arise from the adiabatic elimination of an inhibitor-like field in an activator-inhibitor system [3, 4]. The associated NEP has the form

$$\mathcal{F}[\phi] \;=\; \int_0^{\pi} dx \left\{ \frac{1}{2} (\partial_x \phi)^2 + V(\phi(x)) + \frac{\vartheta}{2} \int_0^{\pi} dx' \, \phi(x') \, \mathcal{K}(x,x') \, \phi(x) \right\}. \tag{6}$$

This NEP contains both, local and nonlocal couplings corresponding to the diffusive and the nonlocal contribution, respectively. The last one contains the nonlocal kernel given above, with a variable range $2l$, corresponding to the field interaction at points x' belonging to the region $[x - l, x + l]$. However, such points will contribute if and only if are inside the domain $[0, \pi]$. With this form of NEP we can directly use the expression for the SNR indicated above and obtain the result indicated on Fig. 3.

When comparing the results for the three cases there are some apparent trends. Firstly, for the 2nd case we see a reduction of the SNR when including a KPZ-like contribution [14]. Secondly, when considering nonlocal terms, at least for moderate values of ϑ and the range $2l$, there is an increase of the system's response. However, a deeper analysis of this problem requires to make a detailed comparison with Monte Carlo simulations, that will be the subject of a future work.

When searching for enhancement mechanisms, an interesting possibility involves analyzing the extension of the previous studies to systems with *selective coupling* as studied in [16, 17]. Other possibility could be to exploit the coupling with an auxiliary field as in [18], that seems to present differences respect to the activator-inhibitor-like systems studied in [3, 4, 5]. Also, to study the effect of a coupling based on a Swift-Hohenberg mechanism [19]. Such studies will be the subject of future work.

0.1. Acknowledgments

The author acknowledge discussions with R.Deza, G.Izús, J.A.Revelli, M.A.Rodríguez, C. Escudero. He also acknowledges financial support from MEC, Spain, project CGL2007-64387/CLI, and from AECID, Spain, projects A/013666/07 and A/018685/08.

References

[1] Gammaitoni L, Hänggi P, Jung P and Marchesoni F 1998 Rev. Mod. Phys. **70**, 223
[2] Wellens T, Shatokin V and Buchleitner A 2004 Rep. Prog. Phys. **67**, 45
[3] Wio H S, Bouzat S and von Haeften B 2002 Physica A **306C**, 140, Proc. 21st IUPAP Int. Conf. on Statistical Physics
[4] Wio H S and Deza R R 2007 Europ. Phys. J-Special Topics **146**, 111
[5] Wio H S, Revelli J A, Rodriguez M A, Deza R R and Izúz G G 2009 Proc. *Stochastic Resonance 2008*, Gammaitoni L, Hänggi P and Marchesoni F, Eds., Europ. Phys. J. B **69**, 71
[6] Graham R and Tél T 1990 Phys. Rev. A **42**, 4661
[7] Graham R and Tél T 1991 in *Instabilities and Nonequilibrium Structures III*, edited by E. Tirapegui, W. Zeller (Kluwert), p. 229
[8] Descalzi O and Graham R 1992 Phys. Lett. A **170**, 84
[9] Descalzi O and Graham R 1994 Z. Phys. B **93**, 509
[10] Wio H S 1997 in *4th Granada Seminar in Computational Physics*, edited by Garrido P and Marro J (Springer-Verlag, Berlin), p. 135
[11] Castelpoggi F and Wio H S 1997 Europhys. Lett. **38**, 91
[12] Kardar M, Parisi G and Zhang Y 1986 Phys. Rev. Lett. **56**, 889
[13] Medina E, Hwa T, Kardar M and Zhang Y 1989 Phys. Rev. A **39**, 3053
[14] Wio H S 2009 Int. J. Bif. Chaos **19**, 2813
[15] von Haeften B and Wio H S 2007 Physica A **376**, 199
[16] von Haeften B, Deza R R and Wio H S 2000 Phys. Rev. Lett. **84**, 404
[17] Tessone C and Wio H S 2007 Physica A **374**, 46
[18] Hildebrand M, Kuperman M, Wio H S, Mikhailov A S and Ertl G 1999 Phys. Rev. Lett. **83**, 1475
[19] Cross M C and Hohenberg P C 1993 Rev. Mod. Phys. **65**, 851

XI Latin American Workshop on Nonlinear Phenomena IOP Publishing
Journal of Physics: Conference Series **246** (2010) 012027 doi:10.1088/1742-6596/246/1/012027

Effective large-scale description of lateral diffusions on fluctuating interfaces

Claire Chevalier[1] **and Fabrice Debbasch**[2]

[1] Université Pierre et Marie Curie-Paris6, UMR 7600, LPTMC, 4 place Jussieu, 75255 Paris, France. [2] Université Pierre et Marie Curie-Paris6, UMR 8112, ERGA-LERMA, 3 rue Galilée, 94200 Ivry, France

E-mail: [1] `claire.chevalier26@gmail.com` [2] `fabrice.debbasch@gmail.com`

Abstract. The non-linear coupling between lateral diffusions on an interface and the geometry of this interface is studied numerically on a simple 1D model which presents the same kind of non-linearities as fully 2D ones. It is found that small-scale geometry fluctuations influence the diffusion at all scales, including scales much larger than those of the fluctuations. At large enough scales, the diffusion can be approximately described by introducing an effective or renormalized diffusion coefficient. Several theoretical estimates for this coefficient are introduced and compared to the value derived from numerical simulations.

1. Introduction

Lateral diffusions on interfaces are of great biological importance [1, 2, 3, 4, 5, 6]. They are usually modelled by 2D diffusions on 'simple' surfaces like a plane, a sphere or an ellipsoid. Yet, the geometry of real biological interfaces fluctuates in time and space [7, 5]; these fluctuations are coupled non-linearly to diffusions and, thus, influence transport in a non-trivial manner. The problem has been addressed recently in [8, 9, 10, 11]. This contribution focuses on a simple 1D model of diffusion which presents the same kind of non-linearities as a fully 2D Brownian motion on a fluctuating surface. The main conclusion is that small scale geometry fluctuations profoundly influence diffusions at all scales, including scales much larger than the typical scales of the fluctuations. In particular, the effective large-scale aspects of diffusions cannot be generically described through a renormalized diffusion coefficient and, even when they can, this coefficient generically depends on time and space in a non trivial manner.

2. General framework

2.1. Basics

A curved surface is a 2D differentiable manifold [12] $\mathcal{M}$ equipped with a Riemannian metric g. The metric codes for all local geometric properties of $\mathcal{M}$. Given a coordinate system $x = (x^1, x^2)$ on $\mathcal{M}$, the metric g is represented by its components g_{ij}, which generally depend on the time t and on both coordinates. The metric g defines a line element dl^2 and a surface element dS_g on $\mathcal{M}$. In coordinates, $dl^2 = g_{ij}dx^i dx^j$ and $dS_g = \sqrt{\det g_{kl}(t,x)}d^2x$.

Suppose now that the geometry of $\mathcal{M}$ can be naturally interpreted as the superposition of a mean (or reference) geometry, represented by a mean (or reference) metric $\bar{g}$ and some fluctuations. The fluctuations are represented by the difference $h = g - \bar{g}$. Consider a Brownian

Journal of Physics: Conference Series **246** (2010) 012027 doi:10.1088/1742-6596/246/1/012027

motion [13, 14, 15, 16] B_g in g which starts at time $t = 0$ from some given point O (with coordinates $x = 0$). This stochastic process is fully characterized in law by its density $n_g(t, x)$ with respect to dS_g, which obeys the diffusion equation:

$$\frac{\partial}{\partial t}\left(\sqrt{\det g_{kl}(t,x)}n_g(t,x)\right) = \frac{1}{2}\frac{\partial}{\partial x^i}\left(\sqrt{\det g_{kl}(t,x)}g^{ij}(t,x)\frac{\partial}{\partial x^j}n_g(t,x)\right) \tag{1}$$

with initial condition $n_g(0, x) = \delta(x)$. The same process can be alternately characterized by its density N_g with respect to $dS_{\bar{g}} = \sqrt{\det \bar{g}_{kl}(t,x)}d^2x$; the identity $n_g dS_g = N_g dS_{\bar{g}}$ gives

$$n_g(t,x) = \frac{\sqrt{\det \bar{g}_{kl}(t,x)}}{\sqrt{\det g_{kl}(t,x)}} N_g(t,x). \tag{2}$$

Inserting the above relation into eq. (1) leads to the diffusion equation obeyed by N_g:

$$\frac{\partial}{\partial t}\left(\sqrt{\det \bar{g}_{kl}(t,x)}N_g(t,x)\right) = \frac{1}{2}\frac{\partial}{\partial x^i}\left(\sqrt{\det g_{kl}(t,x)}g^{ij}(t,x)\frac{\partial}{\partial x^j}\left(\frac{\sqrt{\det \bar{g}_{kl}(t,x)}}{\sqrt{\det g_{kl}(t,x)}}N_g(t,x)\right)\right). \tag{3}$$

Consider now a Brownian motion $B_{\bar{g}}$ in $\bar{g}$ which also starts at point O. This process is fully characterized in law by its density $N_{\bar{g}}$ with respect to $dS_{\bar{g}}$. The two Brownian motions B_g and $B_{\bar{g}}$ are compared with each other by comparing their respective densities N_g and $N_{\bar{g}}$ with respect to the reference surface element $dS_{\bar{g}}$.

2.2. The 1D problem

Consider in the 2D (x, y)-Euclidean plane a curve defined by the equation $y = H(x)$. The squared line element along the rope, $dl^2 = dx^2 + dy^2 = \left(1 + H'^2(x)\right) dx^2$, defines a 1D metric g of single coordinate component $g_{xx} = 1 + H'^2(x)$. In 1D, this component coincides with $\det g_{kl}$. The inverse metric g^{-1} has $g^{xx} = 1/g_{xx}$ as single coordinate component. Since the problem is 1D, the line element dl also plays the role of the surface element dS_g. Thus, the Brownian motion along the rope starting at $x = 0$ is characterized by its density $n_g(t, x)$ with respect to dl. The differential equation obeyed by $n_g(t, x)$ is obtained by inserting the above expressions for g^{xx} and $\sqrt{\det g_{kl}}$ into eq. (1); this equation reads:

$$\frac{\partial}{\partial t}\left(\sqrt{1+H'^2(x)}n_g(t,x)\right) = \frac{1}{2}\frac{\partial}{\partial x}\left(\frac{1}{\sqrt{1+H'^2(x)}}\frac{\partial}{\partial x}n_g(t,x)\right), \tag{4}$$

with initial condition $n_g(0, x) = \delta(x)$.

We retain for H the following simple form:

$$H(x) = a\left(\cos\left(2\pi(K+\Delta K)x\right) - \cos\left(2\pi K x\right)\right), \tag{5}$$

and choose $a = 0.6$, $K = 10^{-1}$, $\Delta K = 10^{-2}$. The rope then presents 'small scale' oscillations of wave numbers 10^{-1} and $1.1 \cdot 10^{-1}$ around the straight line ('plane') with metric $\bar{g} = 1$; the difference $\Delta K = 10^{-2}$ between the wave numbers of these oscillations defines the 'large scale' $1/10^{-2} = 100$. The associated 'geometry' is thus 'flat' (straight line, equivalent to the plane) with oscillations of amplitude $\alpha = 4\pi^2 a^2$. The metric component g_{xx} reads:

$$\begin{aligned} g_{xx} =& 1 + \alpha\left(K^2 + K\Delta K + \frac{1}{2}\Delta K^2\right) - \alpha K(K+\Delta K)\cos(2\pi\Delta K x) \\ &+ \alpha K(K+\Delta K)\cos\left(2\pi(2K+\Delta K)x\right) - \alpha\frac{K^2}{2}\cos(4\pi K x) \\ &- \alpha\frac{(K+\Delta K)^2}{2}\cos\left(4\pi(K+\Delta K)x\right); \end{aligned} \tag{6}$$

the fluctuations thus generate in the metric both an offset in the lower 0 mode and contributions at wave-numbers $2\pi\Delta K$, $2 \times 2\pi K$, $2\pi(2K + \Delta K)$ and $2 \times 2\pi(K + \Delta K)$. As far as the inverse metric g^{xx} or the volume factor $\sqrt{\det g_{xx}}$ are concerned, the fluctuations thus also generate an offset in the 0 mode, together with an infinite number of contributions at wave-numbers, larger or equal to ΔK. One thus expects the fluctuations to influence diffusions, not only on 'small' scales comparable to K^{-1}, but also on 'large' scales, including scales of order ΔK^{-1} and larger ones. The form (5) has no direct physical relevance but is the simplest one which makes all interesting dynamical effects manifest: the form (5) is thus best suited for a detailed analysis of all significant physical effects.

Note that both the metric g and the initial condition $n_g(0, x)$ are symmetric with respect to $x = 0$ and that the density n_g has therefore this same symmetry property at all times. Plots below are thus given only for $x \geq 0$.

Let us replace x by another coordinate, say X, defined by:

$$dX = \sqrt{1 + H'^2(x)}dx, \tag{7}$$

and $X(0) = 0$. By definition of X, the metric component g_{XX} equals unity and the X-coordinate expression of the 'surface' element dS_g is simply dX. The density of a process with respect to dS_g is a scalar and is independent of the choice of coordinates. If one uses the X-coordinate, the process is thus represented by the function $\tilde{n}_g(t, X) = n_g(t, x(X))$, which obeys (1) with $g_{XX} = 1$ *i.e.* the standard diffusion equation $\partial_t \tilde{n}_g = (1/2) \times \partial_{XX} \tilde{n}_g$. Since the initial condition is $\tilde{n}_g(0, X) = \delta(X)$, one obtains immediately the standard expression $\tilde{n}_g(t, X) = (1/\sqrt{2\pi t}) \exp\left(-X^2/(2t)\right)$ for the density of the Brownian motion with respect to $dS_g = dX$, expressed as a function of t and X. This in turn furnishes the density $n_g(t, x)$ or, equivalently, $N_g(t, x)$, provided one knows X as a function of x. Obtaining this function directly from (7) involves computing a so-called oscillatory integral. It is therefore advisable to proceed indirectly by first expanding the integrand in powers of a around $a = 0$, thus obtaining a simple trigonometric series, and, then, to compute the integral of this series. This method works as long as the series converges. For the retained values of K and ΔK, the series converges for all (positive) $a < a^*$ where a^* is slightly superior to 0.6. The presented results thus correspond to a fluctuation amplitude close to the amplitude for which the retained numerical method breaks down and computational accuracy suffers. In practice, the expansion in power series of a has been pushed to order 30 and the relative amplitude of the corresponding neglected rest is typically 10^{-7}.

3. Results

3.1. Description

The unperturbed Brownian motion starting at point $x = 0$ associates the time $\tau(k) = 2\ln 2/(2\pi k)^2$ to any spatial scale k^{-1}; $\tau(k)$ is the time at which the half-width of the Gaussian density coincides with the scale k^{-1} and represents physically the time before which the diffusion does not 'see' the scale k^{-1}. The effects of small-scale geometry fluctuations on the large-scale ΔK^{-1} are thus best studied after time $\tau(\Delta K) \sim 351$. One also expects effects to be cumulative. All results below are thus presented at time $t_0 = 10^4 \sim 30\,\tau(\Delta K)$, which is large enough for large-scale effects to be fully developed. The complete time evolution of the diffusion will be presented in full detail in a future article.

The density $N_g(10^4, x)$ and the relative density contrast $\Delta N/N_{\bar{g}}$ at time $t = 10^4$ are plotted against the x-coordinate x in Fig. 1 and Fig. 2.

The length $1/\Delta K = 100$ is clearly much larger than the metric fluctuation amplitude $a = 0.6$; yet, the relative density contrast is not small compared to unity and neglecting the geometry fluctuations would lead a typical relative error of at least 20%. The density $N_g(10^4, x)$ also

Journal of Physics: Conference Series **246** (2010) 012027 doi:10.1088/1742-6596/246/1/012027

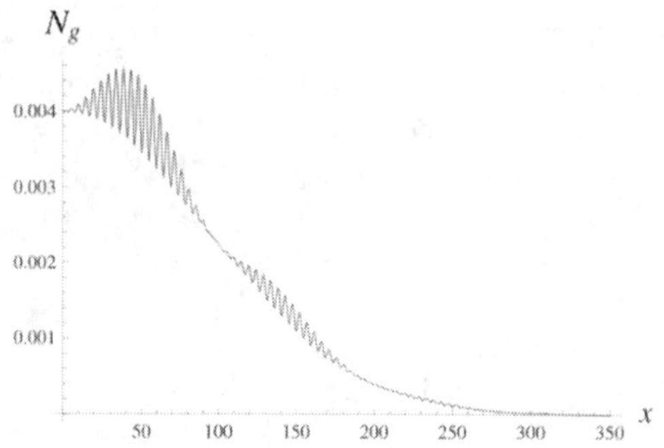

Figure 1. Plot of the density N_g with respect to x at fixed t ($t = 10^4$).

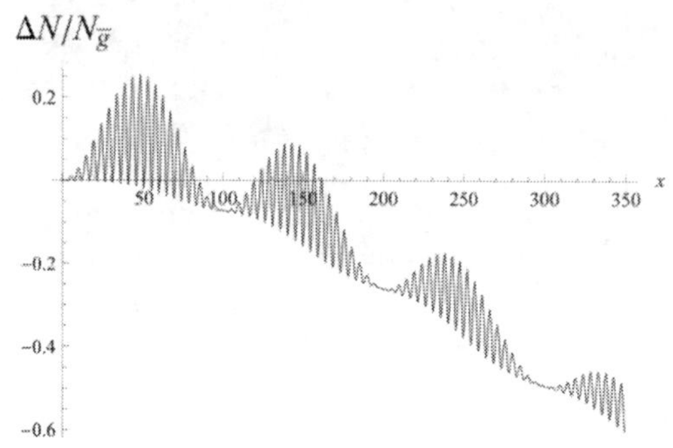

Figure 2. Plot of the relative density contrast $(N_g - N_{\bar{g}})/N_{\bar{g}}$ with respect to x at fixed t ($t = 10^4$).

displays small-scale oscillations around a function that resembles a Gaussian. This qualitative assessment is confirmed by inspecting the spatial Fourier transform of N_g at time $t = 10^4$, which is plotted in Fig. 3.

For $k \leq K_m = 7 \cdot 10^{-3}$ ($\log K_m \simeq -2.3$), the Fourier transform resembles a parabole. But the spectrum has also several small-scale pics, including ones around the base wave-number $K = 10^{-1}$ and one at the wave-number $\Delta K = 10^{-2}$. Note also that there is no pic of wave number smaller than ΔK. In other words, $1/\Delta K$ is the largest 'excited' scale in the system. On the other hand, there is apparently no smallest excited scale. In particular, the small scale ($\log k \rightarrow 0$) pics are not numerical artefacts and correspond to scales which are naturally excited by the non-linearity of the problem. Observing the diffusion on large scales comes down to filtering the density N_g and put to zero the contributions of all wave-numbers larger than a certain cut-off, say K^*. Let $N_g^<(K^*)$ be the corresponding filtered density. If K^* is larger than K_m, the pic at ΔK is not entirely filtered and the filtered density $N_g^<(K^*)$ does not resemble a Gaussian; an example of this situation is plotted in Fig. 4 for K^* larger, but of the same order than K_m. The larger K^*, more the filtered density $N_g^<$ displays oscillations and resembles the unfiltered density N_g.

On the other hand, the filtered density $N_g^<(K^*)$ does resemble a Gaussian if the cut-off K^* is smaller than K_m, as shown in the example plotted on Fig. 5.

An effective, large-scale diffusion coefficient $D^<(K^*)$ can be associated to $N_g^<(K^*)$ if the usual diffusion equation $\partial_t N_g^<(K^*) = \partial_x(D^<(K^*)\partial_x N_g^<)$ can be solved for $D^<(K^*)$. A straightforward analysis shows that no coefficient $D^<(K^*)$ can be associated to $N_g^<(K^*)$ if K^* is larger than K_m. Technically, this is due to the fact that $N_g^<(K^*)$ has then more than one extremum. On the other hand, an effective diffusion coefficient can be associated to any $N_g^<(K^*)$ with K^* smaller

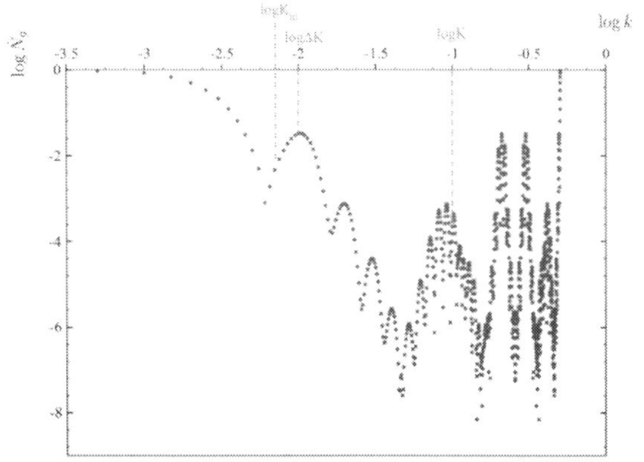

Figure 3. Plot of the decimal logarithm of the Fourier transform $\hat{N}_g$ of the density N_g with respect to the decimal logarithm of the wave number k, at fixed time $t = 10^4$.

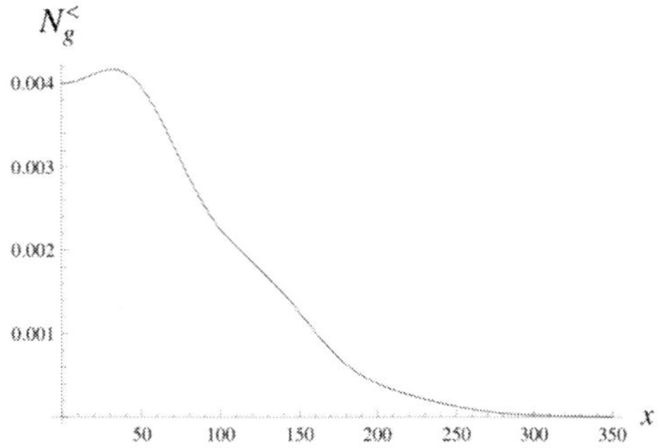

Figure 4. Plot of $N_g^<(K^*)$ with respect to x at fixed t ($t = 10^4$) for $K^* = 1.6 \times 10^{-2} > \Delta K = 10^{-2}$.

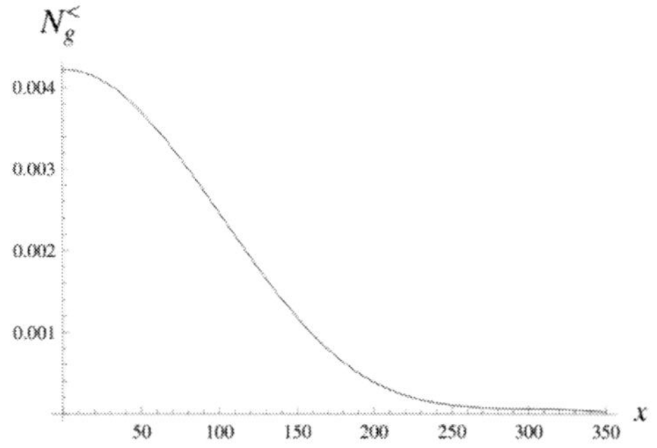

Figure 5. Plot of $N_g^<(K^*)$ with respect to x at fixed t ($t = 10^4$) for $K^* = 4.0 \times 10^{-3} < \Delta K = 10^{-2}$.

than K_m. Figure 6 shows the relative discrepancy for $K^* = 4.0 \times 10^{-3}$ between $D^<(K^*)$ and the real diffusion coefficient $D_0 = 1/2$ given by equation (4).

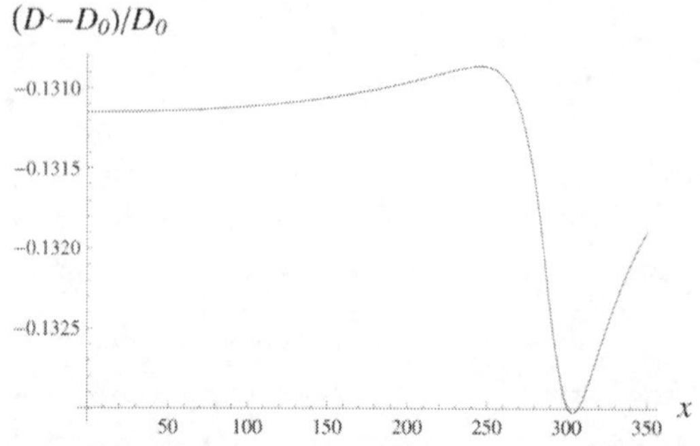

Figure 6. Plot of the ratio $(D^< - D_0)/D_0$ with respect to x at fixed t ($t = 10^4$).

This coefficient depends on x and is approximately 13% lower than the original, true coefficient $D_0 = 1/2$.

Though the precise shape of the function $D^<$ naturally depends on the choice of cutoff K^*, the order of magnitude of $D^<$ is found, perhaps surprisingly, to be actually cutoff-independent and the relative difference between real and effective diffusion coefficients remains around 13% for all reasonnable cutoffs *i.e.* cutoffs which generate a smoothed density $N_g^<$ which looks like a Gaussian. It is therefore natural to wonder where this rather stable percentage comes from. The next Section is devoted to this problem.

3.2. Interpretation

Equation (4) can be rewritten as

$$\frac{\partial n_g}{\partial t} = \frac{1}{2g}\left[\frac{\partial^2 n_g}{\partial x^2} - \frac{1}{2g}\frac{\partial g}{\partial x}\frac{\partial n_g}{\partial x}\right] \tag{8}$$

where $g(x) = g_{xx}(x) = 1 + H'^2(x)$. Since g is oscillating at scales much smaller than the typical extension $L \sim 150$ of the density (at the chose observation time $t = 10^4$), the second term on the right-hand side of (8) essentially 'averages' to zero and the above form of the transport equation suggests that the effective diffusion coefficient is essentially the 'average' of $1/g$, or alternately, that the inverse effective diffusion coefficient is the average of g. Of course, the above statements make sense and, thus, can be validated or invalidated only after a precise definition of averages has been given. Two simple alternative definitions come to mind.

The first possible definition of averages is simply:

$$< f >_1 = \lim_{x\to+\infty} \frac{1}{2X}\int_{-X}^{+X} f(x)dx \tag{9}$$

and the second one is:

$$< f >_2 (t, K^*) = \int_{-\infty}^{+\infty} N_g^<(t, x) f(x)dx. \tag{10}$$

Note that averages of the second type depend on both the transport (and thus on time) and the cutoff K^*.

Both types of averages have been evaluated for $f = 1/g$ and $f = g$, $t = 10^4$ and $K^* = 4\times 10^{-3}$. One finds $< 1/g >_1 \sim 0.882$ and $< 1/g >_2 (10^4, 4\times 10^{-3}) \sim 0.873$, corresponding respectively to $(D^< - D_0)/D_0 \sim -0.118$ and $(D^< - D_0)/D_0 \sim -0.127$, and $< g >_1 \sim 1.157$ and

$< g >_2 (10^4, 4 \times 10^{-3}) \sim 0.170$, corresponding respectively to $(D^< - D_0)/D_0 \sim -0.136$ and $(D^< - D_0)/D_0 \sim -0.145$.

The average relative importance of the $\partial_x n$ term in (8) is measured by the product $L < \dfrac{1}{2g}\dfrac{\partial g}{\partial x} >_i$ where $i = 1$ or 2 and $L \sim 150$ is the typical spatial variation scale of $N_g^<$. Both averages lead to a product of order 10^{-15}; this is actually smaller than the evaluated precision (10^{-11}) of the discrete Fourier transforms used in this contribution.

These results indicate that the second contribution to the right-hand side of (8) is indeed probably negligible on the average and, thus, that the effective large-scale dynamics should be describable by introducing an effective diffusion coefficient. They also indicate that the relative difference between the effective diffusion coefficient and the true diffusion coefficient should lie between -11% and -15%, which is the correct order of magnitude. But none of these results really coincides with numerically observed relative differences $(D^< - D_0)/D_0$, which typically vary (with cutoff and position) between -0.131 and -0.133. Nevertheless, both $< 1/g >_2$ and $< g >_1$ deliver rather good approximations for the effective diffusion coefficients.

4. Conclusion

We have presented numerical computations of a 1D diffusion which capture some essential properties of lateral diffusions on nearly flat interfaces. Our results show that small scale geometry fluctuations can influence diffusions at all scales, including scales much larger that the caracteristic scales of these fluctuations. The large scale aspects of diffusion cannot always be described by simply introducing a renormalized diffusion coefficient and, even when they do, this coefficient depends on both space and time and is substantially different from the true diffusion coefficient. The order of magnitude of the correction seems to be predictable by averaging some functionals of the metric.

Let us now conclude by mentioning some interesting, yet still open problems. The geometry fluctuations studied in this Letter are time-independent and have a discrete spatial Fourier spectrum; one wonders if time-dependent fluctuations, and/or fluctuations with a richer spectrum (for example, a continuous spectrum) have a comparable impact on diffusions. Studying fluctuations around a compact shape like an ellipsoid (or a circle) is also mandatory if one wants to take into account finite size effects in lateral diffusions on cell membranes; in this vein the influence of fluctuations on first-passage times in certainly worth studying. Finally, determining when and how the large-scale average diffusion can be described by introducing an effective diffusion coefficient is a very interesting and still open homogenization problem. Further numerical computations should help in indicating which theorems could be proved about effective transport in fluctuating geometries.

References

[1] Cowan A E, Koppel D E, Vargas L and Hunnicutt G 2001 *Developmental Biology* **236** 502
[2] Borgdorff A and Choquet D 2002 *Nature* **417** 649
[3] Collingridge G L, Isaac J and Wang Y 2004 *Nat. Rev. NeuroSci.* **5** 952
[4] Ashby M, Maier S, Nishimune A and Henley J 2006 *J. NeuroSci.* **26** 7046
[5] Sbalzarini I, Hayer A, Helenius A and Koumoutsakos P 2006 *Biophys. J.* **90** 878
[6] Gadella B 2008 *Animal Reproduction Science* **107** 229
[7] Christensen M 2004 *J. Comput. Phys.* **201** 421
[8] Chevalier C and Debbasch F 2007 *Eur. Phys. Lett.* **77** 20005
[9] Chevalier C and Debbasch F 2008 *J. Stat. Phys.* **131** 717
[10] Chevalier C and Debbasch F 2008 *International Journal of Numerical Analysis and Modeling* **5** 36
[11] Chevalier C and Debbasch F 2010 *accepted for publication in Eur. Phys. Lett.*
[12] Dubrovin B, Novikov S and Fomenko A 1984 *Modern geometry - Methods and applications* (New York: Springer-Verlag)
[13] Itô K 1950 *Nagoya Math. J.* **1** 35

[14] Itô K 1953 *MK* **28** 82
[15] Emery M 1989 *Stochastic calculus in manifolds* (Springer-Verlag)
[16] Øksendal B 1998 *Stochastic Differential Equations, 5th Edition* (Springer-Verlag)

Causal random geometry from stochastic quantization

J Ambjørn[1,2], R Loll[2], W Westra[3] and S Zohren[4,5]

[1] Niels Bohr Institute, Copenhagen, Denmark
[2] Institute for Theoretical Physics, Utrecht University, The Netherlands
[3] Department of Mathematics, University of Iceland, Iceland
[4] Mathematical Institute, Leiden University, The Netherlands
[5] Blackett Laboratory, Imperial College London, UK

E-mail: ambjorn@nbi.dk, loll@phys.uu.nl, wwestra@raunvis.hi.is, zohren@math.leidenuniv.nl

Abstract. In this short note we review a recently found formulation of two-dimensional causal quantum gravity defined through Causal Dynamical Triangulations and stochastic quantization. This procedure enables one to extract the nonperturbative quantum Hamiltonian of the random surface model including the sum over topologies. Interestingly, the generally fictitious stochastic time corresponds to proper time on the geometries.

1. Introduction

Two-dimensional random geometry can either be viewed as a simple toy model for a string theory of fundamental particles or as a playground for four-dimensional quantum gravity. The simplification from the string theory point of view is that the two-dimensional random surfaces we discuss are not embedded in an ambient space, only the intrinsic two-dimensional geometry of the string is considered. Although this removes many features of a full blown string theory, it allows us to focus on the signature of the metric on the surface, also called worldsheet, of the propagating string. In traditional approaches to two-dimensional random geometry, matrix models and Liouville theory, the signature of the metric is considered to be Euclidean whereas genuine string theory is presumed to possess Lorentzian signature worldsheets. Causal Dynamical Triangulations (CDT) describes the random geometry of two-dimensional surfaces with Lorentzian signature without using coordinates, hence avoiding complications due to gauge invariance [1]. The essential difference between causal surfaces and Euclidean surfaces is that the time dependent process where a string splits into two strings is nonsingular in the Euclidean case and singular in the Lorentzian case. This leads to a different assignment of coupling constants and a markedly different continuum theory [2]. In this short note we review a recent observation [3] that the time on the Lorentzian surfaces can be identified with the "fifth time" variable of the stochastic quantization method introduced by Parisi and Wu [4]. Remarkably the splitting and merging dynamics of the string with respect to this stochastic time can be described by an *exactly solvable* Schrödinger like equation [3]! In other words the method of stochastic quantization provides a natural nonperturbative definition of the quantum dynamics of causal random surfaces of arbitrary topology.

2. Stochastic quantization

Stochastic quantization was introduced by Parisi and Wu [4] in 1981 as a quantization scheme for Euclidean field theories (see [5] for a standard review and [6] for an interesting more recent account). The basic idea is to start off with a statistical physics system which evolves according to a, in general, fictitious time t and to observe that there exists a correspondence between certain correlation functions in the equilibrium statistical physics model at $t \to \infty$ and propagators in the Euclidean field theory. We will outline this correspondence in the simplest case of zero-dimensional Euclidean field theory which is sufficient for our purposes.

Consider a time dependent random variable X_t with initial value $X_0 = x_0$ whose time evolution is given by the Langevin equation

$$\frac{dX_t}{dt} = -f(X_t) + \sqrt{2\Omega}\,\nu(t), \quad f(x) = \frac{\partial S(x)}{\partial x}, \tag{1}$$

where $\nu(t)$ is Gaussian white noise of width one and $f(x)$ is a conservative drift. Let us now consider the probability density for $X_t = x$ given that $X_0 = x_0$,

$$P(x, x_0, t) = E_\nu\{\delta(x - X_t)|X_0 = x_0\}, \tag{2}$$

where E_ν refers to the expectation value with respect to the Gaussian white noise $\nu(t)$. From the Langevin equation (1) one can then derive the so-called Fokker-Planck equation for the probability density

$$\frac{\partial}{\partial t}P(x, x_0, t) = \frac{\partial}{\partial x}\left(\frac{\partial S}{\partial x} + \Omega\frac{\partial}{\partial x}\right)P(x, x_0, t). \tag{3}$$

From this one can directly verify that for $t \to \infty$,

$$\lim_{t\to\infty} P(x, x_0, t) = P^{eq}(x) = \frac{1}{Z}e^{-\frac{1}{\Omega}S(x)}, \quad Z = \int dx e^{-\frac{1}{\Omega}S(x)} \tag{4}$$

which is the correct measure arising from the Euclidean path integral with classical action $S(x)$ and thus points towards the correspondence with Euclidean field theory. Here Ω can be see as the analog of $\hbar$ in the quantum mechanical setting.

In principle one can now use (3) to extract the effective quantum Hamiltonian of the model with classical action $S(x)$. In the following we use this procedure to deduce the effective quantum Hamiltonian of a model of two-dimensional causal quantum gravity which we introduce in the forthcoming section.

3. Two-dimensional causal quantum gravity

The Causal Dynamical Triangulation (CDT) approach to quantum gravity aims to define two-dimensional Lorentzian quantum gravity through a nonperturbative path integral over geometries $[g_{\mu\nu}]$

$$Z(\kappa, \lambda) = \int \mathcal{D}[g_{\mu\nu}]\, \mathrm{e}^{-S[g_{\mu\nu}]}, \tag{5}$$

where the (Euclidean) Einstein-Hilbert action is given by

$$S[g_{\mu\nu}] = -\frac{1}{2\pi G_N}\int\sqrt{\det g_{\mu\nu}}\, R + \lambda\int\sqrt{\det g_{\mu\nu}}. \tag{6}$$

Here λ denotes the cosmological constant, G_N the Newton's constant and $\kappa = \mathrm{e}^{-1/G_N}$ is the dimensionless string coupling.

In the lattice regularization of CDT the path integral is defined as a continuum limit of a sum over triangulated surfaces (see [1] or the recent review [7]). This is done still with Lorentzian

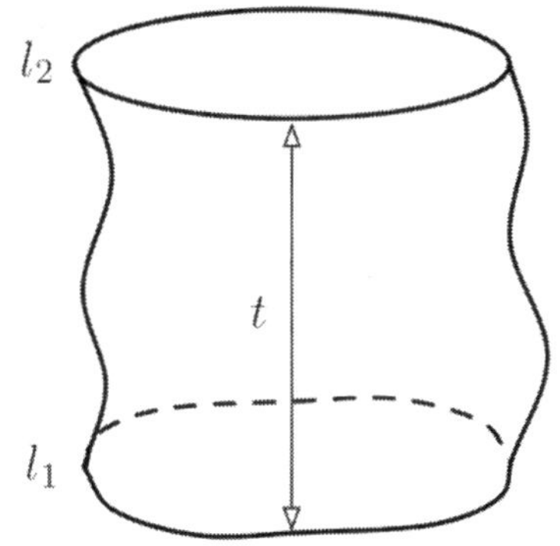

Figure 1. Illustration of the propagator $G_0(l_1, l_2; t)$, describing the amplitude from a fixed spatial boundary of length l_1 to a boundary of length l_2 in time t.

Figure 2. Snapshot from a Monte Carlo simulation of a two-dimensional CDT universe.

signature and a Wick rotation to Euclidean signature is performed at the level of the individual triangulations. In contrast to the Euclidean model defined through Dynamical Triangulations (DT), the Euclidean path integral of the Lorentzian model defined through CDT includes only geometries with a fixed time-sliced structure disallowing for spatial topology changes.

Let us denote the (genus zero) propagator from an initial marked spatial slice of length l_1 to a final unmarked spatial slice of length l_2 in proper time t by $G_0(l_1, l_2; t)$ (see Fig. 1). Let us for convenience also introduce the Laplace transformed propagator

$$G_0(x_1, x_2; t) = \int_0^\infty \mathrm{d}l_1 \int_0^\infty \mathrm{d}l_2 \; \mathrm{e}^{-x_1 l_1 - x_2 l_2} G_0(l_1, l_2; t). \tag{7}$$

Using the CDT program one can derive the following differential equation for the Laplace transformed propagator

$$\frac{\partial}{\partial t} G_0(x_1, x_2; t) = \frac{\partial}{\partial x_1} \left(\left(\lambda - x_1^2\right) G_0(x_1, x_2; t) \right). \tag{8}$$

In terms of the length of the spatial boundary the propagator can then also be written as

$$G_0(l_1, l_2; t) = \langle l_2 | \, e^{-t\mathcal{H}_0} \, | l_1 \rangle, \quad \mathcal{H}_0 = -l \frac{\partial^2}{\partial l^2} + \lambda l. \tag{9}$$

Here the effective quantum Hamiltonian describes the time evolution of the length of the spatial slice. In particular, one obtains that the average length of a spatial slice as well as its fluctuations are both of the same order $1/\sqrt{\lambda}$ as is also illustrated in Fig. 2.

4. Causal string field theory and stochastic quantization

Viewing each spatial slice as a string one can interpret the two-dimensional fluctuating surfaces considered in the previous section as so-called worldsheets of propagating strings. In principle one can then also allow strings to split and join, thus creating surfaces of arbitrary topology.

Such a so-called string field theory for CDT was introduced in [2, 8] where each splitting and joining of a string is weighted with a factor κ according to the action (6). Using this framework it was observed in [9, 10] that the partition function (5) can then be written as the following simple integral

$$Z(\kappa,\lambda) = \int dx \exp\left[-\frac{1}{g_s}\left(\lambda x - \frac{1}{3}x^3\right)\right], \quad \kappa = \frac{g_s}{\lambda^{3/2}}, \tag{10}$$

where the integral is formal in the sense that it should be understood as a suitable expansion in powers of the dimensionless string coupling $\kappa = g_s/\lambda^{3/2}$.

In view of (3) and (4) this suggests that the effective quantum Hamiltonian can be derived from a stochastic quantization procedure defined by the following Langevin equation [3]

$$\frac{dX_t}{dt} = -\left.\frac{\partial S(x)}{\partial x}\right|_{x=X_t} + \sqrt{2g_s}\,\nu(t), \quad S(x) = \lambda x - \frac{1}{3}x^3. \tag{11}$$

The corresponding Fokker-Planck equation reads

$$\frac{\partial}{\partial t}P(x,x_0,t) = \frac{\partial}{\partial x}\left(\lambda - x^2 + g_s\frac{\partial}{\partial x}\right)P(x,x_0,t). \tag{12}$$

One can now directly identify $P(x,x_0,t)$ with the propagator $G(x,x_0,t)$ of the string field theory. By performing an inverse Laplace transformation of (12) one obtains the propagator and Hamiltonian in the "length representation". These describe the transition amplitude of a string to propagate from a specific initial length l_1 to a certain final length l_2,

$$G(l_1,l_2;t) = \langle l_2|\, e^{-t\mathcal{H}}\, |l_1\rangle, \quad \mathcal{H} = -l\frac{\partial^2}{\partial l^2} + \lambda l - g_s l^2. \tag{13}$$

One observes that Eqs. (12) and (13) exactly yield (8) and (9) for $g_s = 0$. Hence we see that for $g_s = 0$ we exactly obtain the CDT propagator without any spatial or space-time topology changes. Furthermore, the stochastic time t which is in general a fictitious time variable exactly corresponds to the proper time on the worldsheets.

For $g_s > 0$ the effective quantum Hamiltonian (13) describes fluctuating surfaces where spatial slices are allowed to split and joint each weighted by a factor g_s. The corresponding amplitudes thus include a summation over all genera! It is interesting that (13) can be used to explicitly give solutions for some amplitudes. The simplest amplitude to look at is the disc function with is a propagator with one boundary shrunken to zero and integrated over time, i.e.

$$W(l) = \int_0^\infty \mathrm{d}t\, G(l, l_2 = 0; t) \tag{14}$$

From (12) one can derive that $W(l)$ is determined by

$$\mathcal{H}\, W(l) = 0, \tag{15}$$

where $\mathcal{H}$ is given in (13). This equation can be interpreted as a Wheeler-deWitt equation for the spatial slices. The explicit solution to this equation can be given in terms of Airy functions $\mathrm{Bi}(\cdot)$ and reads

$$W(l) = \frac{\mathrm{Bi}\left(\frac{\lambda}{g_s^{2/3}} - g_s^{1/3} l\right)}{\mathrm{Bi}\left(\frac{\lambda}{g_s^{2/3}}\right)}. \tag{16}$$

This expression corresponds to the following genus expansion in the dimensionless string coupling $\kappa = g_s/\lambda^{3/2}$:

$$W(l) \quad = $$

$$+\,\mathcal{O}(\kappa^3). \tag{17}$$

5. Conclusion

We presented a recently found stochastic quantization formulation of a "causal" random surface model defined through CDT. This formulation is reminiscent of an analogous formulation for Euclidean quantum gravity defined through DT [11, 12]. The stochastic quantization procedure enabled us to determine the effective quantum Hamiltonian (13) of the nonperturbative dynamics including the sum over topologies. A detailed physical interpretation of this Hamiltonian including a discussion on the problem of unboundedness can be found in [3].

In our model the external stochastic time is identified with proper time on the string worldsheets. This special role of an external time variable and unitarity which is inherent to the CDT formulation are very similar to the setup of Hořava-Lifshitz gravity [13] for which the concept of stochastic quantization might also be relevant. This points towards a possible relation between these two models. In fact, it was observed very recently that CDT might not only serve as a lattice regularization of the renormalization group approach to quantum gravity, but also for Hořava-Lifshitz gravity [14].

Acknowledgments

JA, RL, WW acknowledge support by ENRAGE (European Network on Random Geometry), a Marie Curie Research Training Network, contract MRTN-CT-2004-005616 in the European Community's Sixth Framework Programme. RL acknowledges support by the Netherlands Organisation for Scientific Research (NWO) under their VICI program. SZ would like to thank the Department of Statistics at São Paulo University (IME-USP) as well as the Institute for Pure and Applied Mathematics (IMPA) for kind hospitality and the ISAC program (Erasmus Mundus) for financial support.

References

[1] Ambjørn J and Loll R 1998 *Nucl. Phys. B* 536 407-434 arXiv:hep-th/9805108
[2] Ambjørn J, Loll R, Westra W and Zohren S 2007 *JHEP* **0712** 017 arXiv:0709.2784 [gr-qc]
[3] Ambjørn J, Loll R, Westra W and Zohren S 2009 *Phys. Lett. B* **680** 359 arXiv:0908.4224 [hep-th]
[4] Parisi G and Wu Y 1981 *Sci. Sin.* **24**, 483
[5] Damgaard P H and Huffel H 1987 *Phys. Rept.* **152** 227
[6] Dijkgraaf R, Orlando D and Reffert S 2010 *Nucl. Phys.* B **824** 365 arXiv:0903.0732 [hep-th]
[7] Ambjørn J, Loll R, Watabiki Y, Westra W and Zohren S arXiv:0911.4208 [hep-th]
[8] Ambjørn J, Loll R, Watabiki Y, Westra W and Zohren S 2008 *JHEP* **0805** 032 arXiv:0802.0719 [hep-th]
[9] Ambjørn J, Loll R, Watabiki Y, Westra W and Zohren S 2008 *Phys. Lett. B* **665** 252 arXiv:0804.0252 [hep-th]
[10] Ambjørn J, Loll R, Westra W and Zohren S 2009 *Phys. Lett. B* **678** 227 arXiv:0905.2108 [hep-th]
[11] Ikehara M, Ishibashi N, Kawai H, Mogami T, Nakayama R and Sasakura N 1995 *Prog. Theor. Phys. Suppl.* **118** 241 arXiv:hep-th/9409101
[12] Jevicki A and Rodrigues J P 1994 *Nucl. Phys. B* **421** 278 arXiv:hep-th/9312118
[13] Hořava P 2009 *Phys. Rev. D* **79** 084008 arXiv:0901.3775 [hep-th]
[14] Ambjørn J, Görlich A, Jordan S, Jurkiewicz J and Loll R arXiv:1002.3298 [hep-th]

Journal of Physics: Conference Series **246** (2010) 012029 doi:10.1088/1742-6596/246/1/012029

Nonlinear effects in the dynamics governed by non-Markovian stochastic Langevin-like equations

R. L. S. Farias[1], R. O. Ramos[2,3] and L.A. da Silva[2]

[1]Departamento de Ciências Naturais, Universidade Federal de São João del Rei, São João del Rei, MG 36301-160, Brazil

[2]Departamento de Física Teórica, Universidade do Estado do Rio de Janeiro, Rio de Janeiro, RJ 20559-900, Brazil

[3]School of Physics and Astronomy, University of Edinburgh, Edinburgh, EH9 3JZ, United Kingdom

E-mail: [1]ricardofarias@ufsj.edu.br, [2]rudnei@uerj.br and [2]las.leandro@gmail.com

Abstract.
The influence of nonlinear effects in stochastic equations of motion with both additive and multiplicative noises is studied. Non-Markovian stochastic dynamics are compared with their corresponding Markovian (local approximations). Non-Markovian effects are implemented through Ornstein-Uhlenbeck and exponential damped harmonic dissipative kernels.

1. Introduction

In nature, almost as a rule, most systems behave as open ones, interacting with an environment (e.g., a thermal bath). Interactions with an environment lead, as a consequence, to dissipative as well-stochastic effects. Nonlinear stochastic dynamical systems are of interest in many different areas of investigation with applications in diverse fields, like in the study of double quantum dot systems (1; 2; 3), in fundamental aspects of biological systems (4; 5; 6), in studies concerning domain growth processes (7) and in the context of field theory models (8; 9). In double quantum dot systems, for instance, the importance resides in their possible applications in nanotechnology and quantum information processing (10; 11). Fluctuation and dissipation appear in these systems due to quantum operations and measurements and these can change the physics of the system's coherent evolution. Recently, many investigations have been performed using different forms of noises and interactions that can change the coherent evolution of electrons in the double dot quantum system.

Another area of interest is the application of stochastic models in biological systems. In this case, for example, one of the aspects studied is the issue of how cooperative behavior may emerge. The competition among different growth and death processes and the inclusion of external mechanisms can influence the global properties of these systems. Many efforts have been dedicated to provide a theoretical framework for understanding the cooperative behavior. Another example that has been a focus of attention in recent years is the analysis of tumor growth. In this case, a population of proliferating cells can be considered as a stochastic dynamical system far from equilibrium. To obtain a better understanding of the behavior of this type of phenomenon is necessary to include stochastic effects and, ideally, spatial correlations (12).

Nonlinear stochastic dynamical systems also appear in the context of field theory models (8; 9). It has been shown in that case how generalized equations of motion can be derived microscopically, resulting is a complicated structure depending on the form of the couplings involved between the fields, like between some background configuration we are interested in the dynamics and other fields and degrees of freedom taken as an environment or (quantum or thermal) bath fields.

One common way for describing all the above mentioned forms of evolution is by means of generalized stochastic Langevin-like equations of motion. These non-deterministic equations of motion are used in many applications of interest, including, e.g., for the study the system mentioned above. These equations, for example, can simulate Brownian motion in (classical and quantum) statistical mechanics and in other areas of physical interest (13; 14).

A classical example of a system whose dynamics is modeled by a Langevin equation of motion is the one that describes the Brownian motion of a classical particle described by a coordinate q, unitary mass and subjected to a potential $V(q)$ (as usual dots mean derivative with respect to time and $V'[q(t)] \equiv dV/dq$),

$$\ddot{q}(t) + \eta\dot{q}(t) + V'[q(t)] = \xi(t) \ , \tag{1}$$

where η is a Markovian (local) dissipation term and $\xi(t)$ is a stochastic term describing a white noise with Gaussian properties, satisfying (throughout this work we consider the Boltzmann constant $k_B = 1$)

$$\langle \xi(t) \rangle = 0 \ , \quad \langle \xi(t)\xi(t') \rangle = 2T\eta\delta(t-t') \ . \tag{2}$$

Approaches with Langevin equations like Eq. (1) and their generalizations, are used in different contexts, e.g. in classical statistical mechanics to study problems with dissipation and noise, to determine how order parameters equilibrate and in the studies like dynamic scaling and dynamic critical phenomena (15; 16). Though extensively used, equations of the form of Eq. (1) and with noise properties as given by Eq. (2), can only be considered phenomenologically. This is because it implicitly assumes that the environment interacts instantaneously with the system. This is a physically unacceptable situation that violates causality, since the environment bath has no memory time.

In a microscopic description, the effects of the environment on some select variable, taken as the system, dissipation and stochastic noise terms are expected to originate from scattering events, thus giving origin to finite interaction times that reflect in the system's equation of motion as being nonlocal, *i.e.*, they have non-Markovian terms with finite memory. The simplest archetype of this is the description of the system-environment as being modeled by linearly coupled harmonic oscillators (17) (for a general review, see, e.g., Ref. (18)), which also become to be known as Caldeira-Leggett type of models (19; 20; 21).

In this work our objective is to study the nonlinear effects on the dynamics governed by a non-Markovian generalized Langevin equation when compared to the approximated Markovian form. We will consider some of the most common forms for the dissipation kernels. These forms for the dissipation kernels include, for example, the one that describes an Ornstein-Uhlenbeck (OU) process (22) and the exponential damped harmonic (EDH) kernel (23; 24). We study both the cases of additive and multiplicative noises, including system dependent dissipation terms, according to the fluctuation dissipation theorem (for details see (25)). A detailed numerical analysis is made when the various regimes of non-linearity are considered.

2. Non-Markovian Generalized Langevin Equation (GLE)

Here we study a GLE describing the interaction of a system, denoted by a variable ϕ (which can be e.g. the coordinate of a particle) interacting with a thermal bath. The GLE studied here has the generic form

$$\ddot{\phi}(t) + \phi^n(t) \int_{t_0}^{t} dt' \phi^n(t') K(t-t') \dot{\phi}(t') + V'(\phi) = \phi^n(t)\, \xi(t) , \tag{3}$$

where $n = 0, 1$, with $n = 0$ giving the standard GLE of additive noise, while for $n = 1$ gives a multiplicative form of GLE. The inclusion of multiplicative noise and system dependent dissipation is motivated from field theory calculations (for details, see e.g. Refs. (8; 9)). The potential in Eq. (3) is considered to be one with quadratic and quartic terms, given by

$$V(\phi) = \frac{m^2}{2}\phi^2 + \frac{\lambda}{4}\phi^4 , \tag{4}$$

where m^2 and λ are parameters depending on the details of the system under study. The parameter m can be associated with the system's frequency and λ characterizes the degree of nonlinearity.

Non-Markovian and nonlinear GLEs of the form of Eq. (3) are very difficult to solve in general. Analytical methods can only be used in linear regimes (26), like in the additive noise case and when the quartic term in the potential can be neglected. In this case, the equation is in the form of a convolution, thus easily solvable through Laplace transform for instance (26). Otherwise, in more general cases, we must resort to numerical methods. This is the approach we follow in this paper in order to analyze the dynamics obtained from the full nonlinear Eq. (3). Though there are some specific numerical methods using e.g. Fourier transform that may apply for equations with non-Markovian kernels of generic form (27), we still would like to be able to solve equations like Eq. (3) through standard methods, which are less numerically expensive than other alternatives. Recently, the authors (26) have demonstrated the reliability of using a fourth-order Runge-Kutta method when solving GLE of the OU and EDH forms. The way this can be done stems from the fact that non-Markovian equations with kernels of those forms can be replaced by a system of completely local first-order differential equations, as described in details in (25; 26; 28).

The local version of the GLE that we will use to contrast the Markovian and the non-Markovian dynamics of the system is obtained by the following approximation (29; 30)

$$\begin{aligned} &\phi^n(t) \int_{t_0}^{t} dt' D(t-t') \phi^n(t')\, \dot{\phi}(t') \simeq \\ &\phi^{2n}(t)\, \dot{\phi}(t) \int_{-\infty}^{t} dt' D(t-t') \to \eta\, \phi^{2n}(t)\, \dot{\phi}(t) , \end{aligned} \tag{5}$$

where η is the local dissipation coefficient. Our local version of the GLE is then given by

$$\ddot{\phi}(t) + \eta\, \phi^{2n}(t)\, \dot{\phi}(t) + V'(\phi) = \phi^n(t)\, \xi(t) . \tag{6}$$

In this work we concentrate our study in equations like Eq. (3) with non-Markovian kernels of either the OU type (22),

$$D_{OU}(t-t') = \eta\, \gamma e^{-\gamma(t-t')} , \tag{7}$$

or with the EDH type (23; 24),

$$D_H(t-t') = \eta\, e^{-\gamma(t-t')} \frac{\Omega_0^2}{2\gamma} \left\{ \cos[\Omega_1(t-t')] + \frac{\gamma}{\Omega_1} \sin[\Omega_1(t-t')] \right\} , \tag{8}$$

where in the equations above, η sets the magnitude of the dissipation, γ sets the relaxation time for the bath kernels, $\tau = 1/\gamma$, and Ω_0 gives the oscillation time scale in the case of the EDH kernel. In Eq. (8), $\Omega_1^2 = \Omega_0^2 - \gamma^2$, and so, in the EDH case the values of γ and Ω_0 are restricted such that $\Omega_1^2 \geq 0$.

The GLE equations with kernels of those forms can be replaced by a system of completely local first order differential equations (25; 26; 28). For the OU case,

$$\begin{aligned}
\dot{\phi} &= y\,, \\
\dot{y} &= -V'(\phi) + \phi^n w_{OU} + \phi^n \xi_{OU}\,, \\
\dot{w}_{OU} &= -\gamma w_{OU} - D_{OU}(0)\phi^n y\,, \\
\dot{\xi}_{OU} &= -\gamma\left[\xi_{OU} - \sqrt{2T\eta}\,\zeta\right]\,,
\end{aligned} \tag{9}$$

while for the EDH case we obtain

$$\begin{aligned}
\dot{\phi} &= y\,, \\
\dot{y} &= -V'(\phi) + \phi^n w_H + \phi^n \xi_H\,, \\
\dot{w}_H &= u - 2\gamma w_H - D_H(0)\,\phi^n y\,, \\
\dot{u} &= -\Omega_0^2 w_H + \dot{D}_H(0)\,\phi^n y - 2\gamma D_H(0)\,\phi^n y\,, \\
\dot{\xi}_H &= z\,, \\
\dot{z} &= -2\gamma z - \Omega_0^2 \xi_H + \Omega_0^2\sqrt{2T\eta}\,\zeta\,.
\end{aligned} \tag{10}$$

In the above equations, $D_{OU}(0) = \eta\gamma$, $D_H(0) = \eta\,\Omega_0^2/(2\gamma)$ and $\dot{D}_H(0) = 0$, which follow from Eqs. (7) and (8). As before, the additive noise case is when $n = 0$ is used in Eqs. (9) and (10), while $n = 1$ is for the multiplicative noise case. ζ is a white Gaussian noise.

We take the initial time as $t_0 = 0$ and define new variables $u(t)$ and $w(t)$ by (25; 26):

$$w_{OU,H}(t) = -\int_0^t dt' \phi^n(t') D_{OU,H}(t-t')\dot{\phi}(t')\,, \tag{11}$$

and

$$u(t) = \int_0^t dt' \left[\frac{dk_H(t-t')}{dt'} - \gamma D_H(t-t')\right]\frac{d\phi(t')}{dt'}\,. \tag{12}$$

In Ref. (25) we have studied how the variation of the various parameters of the noise and dissipation kernels affect the dynamics. Here, we study how the variation of the non-linearity parameter λ affects overall system dynamics and its effects in the applicability or not of the simpler Markovian approximation. Of course, we expect that both dynamics, the non-Markovian and the Markovian ones to have the same asyntoptic state because, by definition, the system reaches the equilibrium, but the question we want to address here is, given a set of model parameters representing the system and the thermal bath to which it is coupled to, for how long can we expect the memory effects due to the non-Markovian terms to be important and if the discrepancy between the Markovian and non-Markovian dynamics changes if we vary the λ parameter ? Since the representation of the dynamics in a local form as given by Eq. (6) represents a considerable simplification, for both a numerical point of view, or for analytical analysis (when it is possible), when compared, e.g., with the full nonlocal, nonlinear integro-differential stochastic equation (3), these then become important questions to be accessed for most practical studies that make use of nonlinear stochastic equations of motion.

In the following we will study how the non-linearity (when the quartic constant term in the potential is varied) change the equilibration and thermalization of the stochastic system.

$\Delta\phi_{\text{OU add}}$	$\Delta\phi_{\text{OU mult}}$	$\Delta\phi_{\text{EDH add}}$	$\Delta\phi_{\text{EDH mult}}$	
0.0	20	19	44	35
5.0	15	14	32	38
15.0	11	10	23	75

Table 1. The time scale (in units of $1/m$),using $\eta = 1.0$, for the non-Markovian dynamics to approach the Markovian one, within a precision of 10^{-3} for the differences defined in Eq. (13).

3. Numerical Results

Let us now turn to our numerical results for the Markovian and non-Markovian dynamics of the system. Numerical simulations of the system of differential first-order equations, Eqs. (9) and (10), for the non-Markovian GLE with OU and EDH kernels, respectively, are compared with those obtained through the local approximation given by Eq. (6). All our simulations were performed with $300,000$ realizations over the noise and we have integrated all differential equations using a standard fourth-order Runge-Kutta method with a time stepsize of $\delta t = 0.001$, which was found to be more than enough for both numerical stability and also for numerical precision (as determined in Ref. (26), these values already assure an overall numerical error always smaller than about one percent, which suffices for our comparison purposes set here). In all our simulations we have also used the initial conditions $\phi(0) = 1$ and $\dot{\phi}(0) = 0$. The time in all our evolutions is in units of the (inverse of the) frequency for the system (which is equivalent to consider $m = 1$ throughout). Comparisons between the Markovian and non-Markovian dynamics are made changing the degree of non-linearity of the system's potential. The other model parameters are kept fixed.

3.1. The dynamics of the system

Let us first consider the analysis of the dynamics for ϕ. In Fig. 1 we plot side by side, using $\eta = 1$, our results for the difference between the Markovian and non-Markovian dynamics, which can be estimated by defining the quantities below:

$$\Delta\varphi = \langle\phi\rangle_{\text{non-Markovian}} - \langle\phi\rangle_{\text{Markovian}} \,. \tag{13}$$

In Eq. 13, $\langle\phi\rangle$ is the ensemble averaged macroscopic system variable $\phi(t)$, where the average is over the noise realizations. In Fig. 2, we perform the same numerical computation, but using $\eta = 10.0$. The results from the plots shown in Figs. 1 and 2 are useful to determine within which time scale the Markovian and non-Markovian dynamics become sufficiently close (within to some given precision) and how the non-linearity of the system influences this time scale. The results for these time scales for the different simulations we have performed, for the Markovian dynamics and for the non-Markovian dynamics with the two types of memory kernels, are given in tables 1 and 2.

The effect of changing λ is clear: The larger is λ, the more pronounced are the memory effects, resulting in a strong difference with respect to the local approximation. As expected, at some sufficient long time, the two dynamics, Markovian and non-Markovian approximate each other. This can also be seen in the case where we have plotted the correlation $\langle\phi^2(t)\rangle$ for both OU and EDH cases. There is another interesting aspect of our results. If we look at Fig. 2, we can see that the larger is λ, the faster the difference between Markovian and non-Markovian dynamics seems to tend to zero. It occurs more explicitly in the EDH cases (top panels). However, note that the behavior shown in Fig. 1 is exactly the opposite: the larger is λ, the slower the difference between Markovian and non-Markovian dynamics reduces.

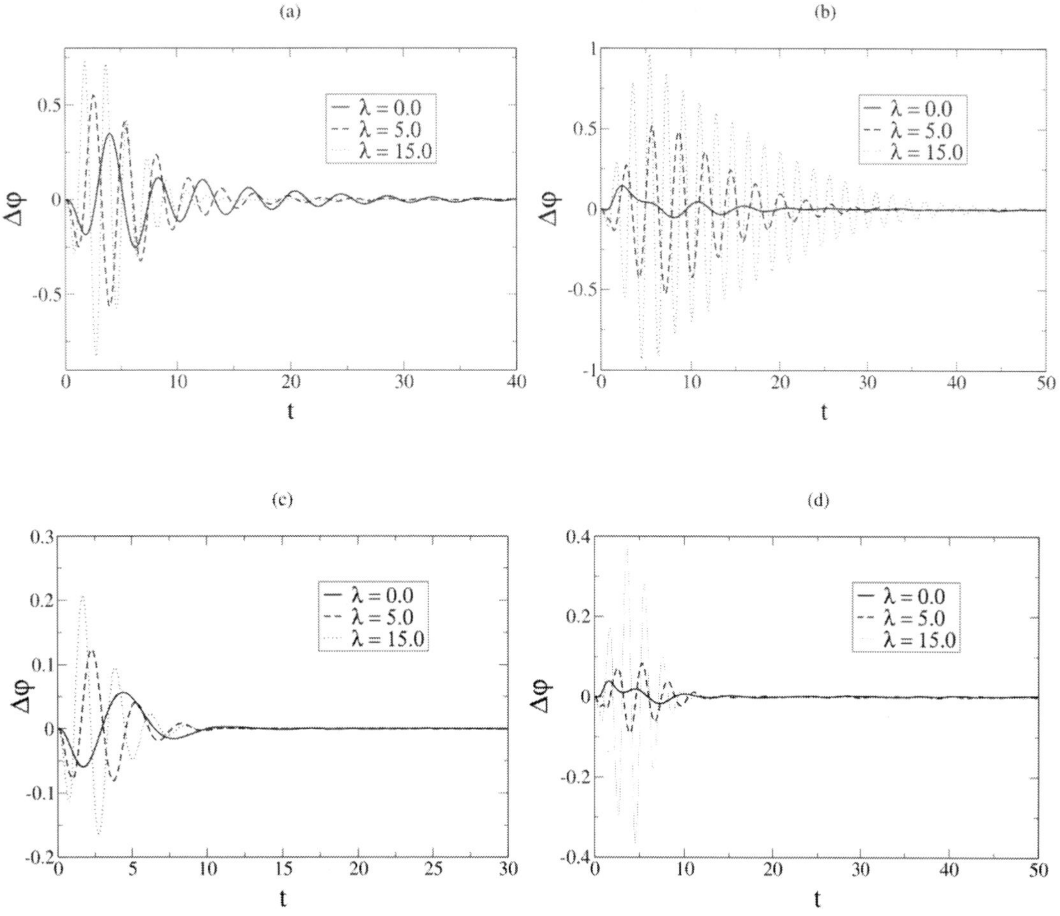

Figure 1. The time evolution for $\Delta\phi = \langle\phi\rangle_{\text{non-Markovian}} - \langle\phi\rangle_{\text{Markovian}}$ fixing $\eta = 1.0$ and varying the λ parameter for (a) EDH additive noise case, (b) EDH multiplicative noise case, (c) OU additive noise case and (d) OU multiplicative noise case. The other parameters used are: $T = 1.0$, $\Omega_0 = 1.0$, $m^2 = 1.0$ and $\gamma = 0.5$ (EDH case) or $\gamma = 5.0$ (OU case).

$\Delta\phi_{\text{OU add}}$	$\Delta\phi_{\text{OU mult}}$	$\Delta\phi_{\text{EDH add}}$	$\Delta\phi_{\text{EDH mult}}$	
0.0	58	23	92	150
5.0	20	16	29	14
15.0	11	10	22	13

Table 2. The time scale (in units of $1/m$),using $\eta = 10.0$, for the non-Markovian dynamics to approach the Markovian one, within a precision of 10^{-3} for the differences defined in Eq. (13).

3.2. The thermalization and nonlinear effects

It is also useful to determine how the parameter λ influences the thermalization time for the system when put in contact with the thermal bath at some temperature T. For this, let us define an effective time dependent temperature for the system according to the equipartition of kinetic energy:

$$T_{\text{eff}}(t) = \langle\dot{\phi}^2(t)\rangle \ . \tag{14}$$

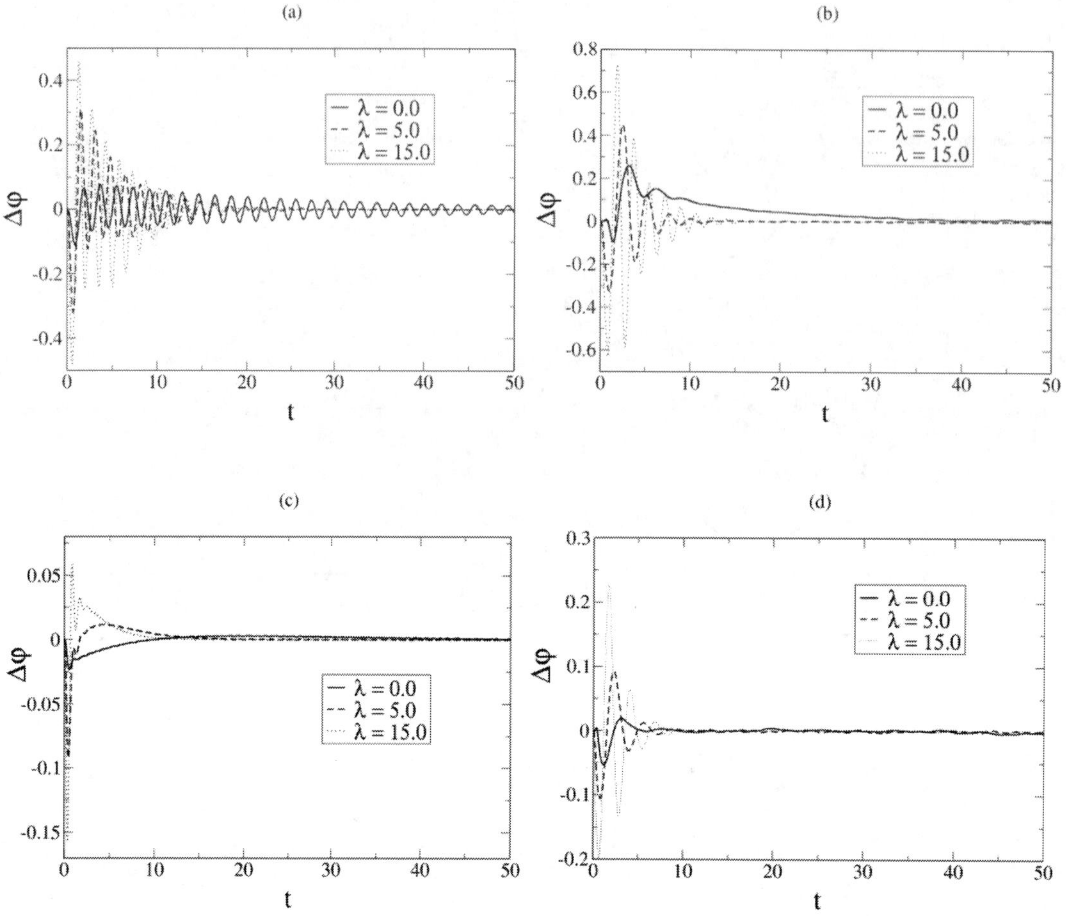

Figure 2. The time evolution for $\Delta\phi$ fixing $\eta = 10.0$ and varying the λ parameter for (a) EDH additive noise case, (b) EDH multiplicative noise case, (c) OU additive noise case and (d) OU multiplicative noise case. The other parameters used are: $T = 1.0$, $\Omega_0 = 1.0$, $m^2 = 1.0$ and $\gamma = 0.5$ (EDH case) or $\gamma = 5.0$ (OU case).

From the plots shown in both Figs. 3 and 4 we can see how the non-linearity of the system reflects in the thermalization of the system. Larger values of λ lead to a longer time for the system to thermalize. Typically, for comparable λ values, in the additive case the system tends to thermalize faster than in the multiplicative case. Also, we can see that the variation of λ produces more striking changes in the thermalization time of the EDH case, in both additive and multiplicative noise cases. In Tables 3 and 4 we give the approximate time (in units of $1/m$) for thermalization for all the cases studied above. For comparative purposes, we show in Fig. 5 the additive and multiplicative Markovian dynamics for the effective temperature $T_{\rm eff}$ for both $\eta = 1.0$ and $\eta = 10.0$.

4. Conclusions

In this work we have studied how the variation of the nonlinear parameter λ of the system's quartic potential term affects the discrepancy between the dynamics of a GLE, given by Eq. (3), and the dynamics of its Markovian or local approximated form. We here have concentrated in two forms for the non-Markovian memory kernel, the OU and EDH cases, and we have analyzed

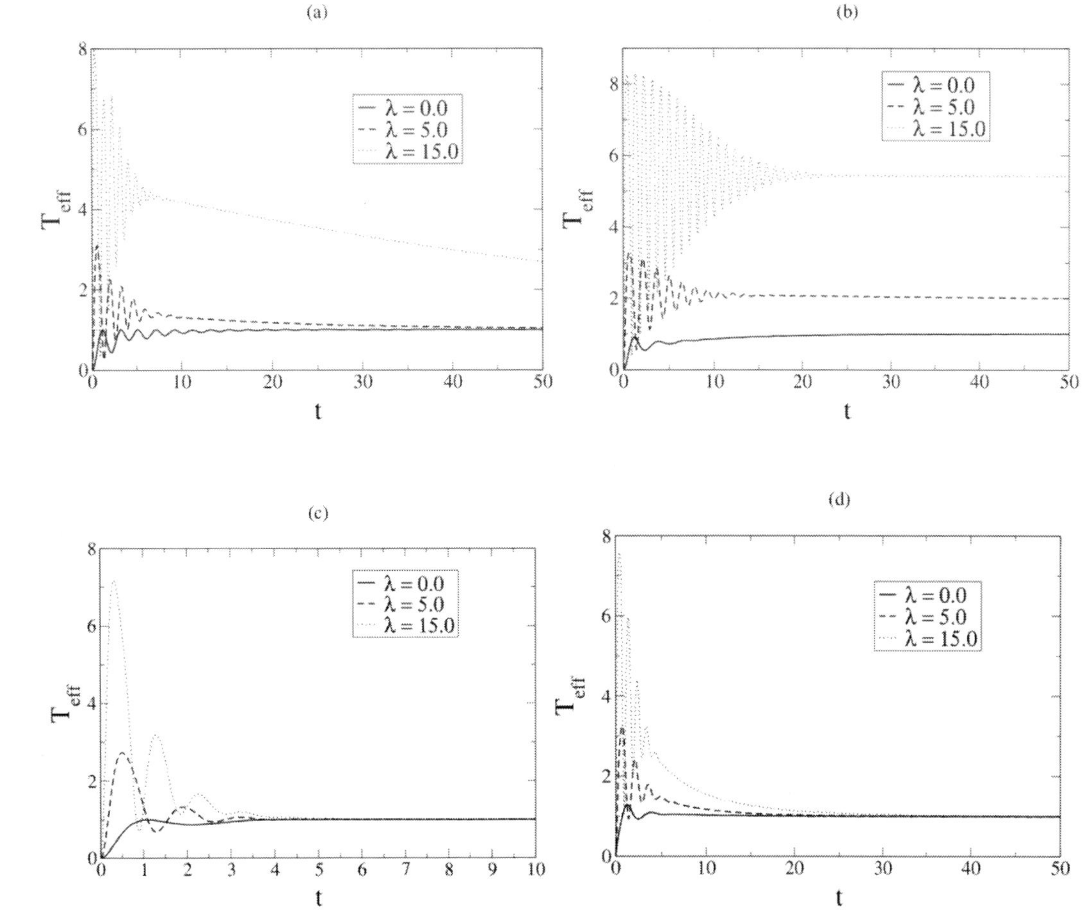

Figure 3. Time evolution for T_{eff}. (a) EDH additive noise case, (b) EDH multiplicative noise case, (c) OU additive noise case and (d) OU multiplicative noise case. The other parameters used are: $\eta = 1.0$, $T = 1.0$, $\Omega_0 = 1.0$ and $\gamma = 0.5$ (EDH case) or $\gamma = 5.0$ (OU case).

$\tau_{\text{markov add}}$	$\tau_{\text{OU add}}$	$\tau_{\text{EDH add}}$	$\tau_{\text{markov mult}}$	$\tau_{\text{OU mult}}$	$\tau_{\text{EDH mult}}$	
0.0	7	8	146	32	47	200
5.0	7	17	180	40	54	650
15.0	8	36	500	70	115	2500

Table 3. The approximate time for thermalization for $\eta = 1.0$, in units of $1/m$, for the Markovian and non-Markovian dynamics, determined when (14) approaches the temperature of the thermal bath.

the cases of additive and multiplicative noises in both cases.

Our results show that the Markovian approximation becomes less applicable as the value of λ increases. Obviously, since these are all dissipative systems, we expect the two dynamics to tend to each other asymptotically. We have then analyzed how long it takes for the discrepancy between the non-Markovian and Markovian dynamics to tend to vanish. We have analyzed the time for the difference $\Delta\varphi$ and also the thermalization time for each of the dynamics by studying the behavior of the correlation function $\langle\dot{\phi}^2\rangle$, which is associated to the temperature of equilibration through the equipartition theorem. We have fixed the all other parameters and

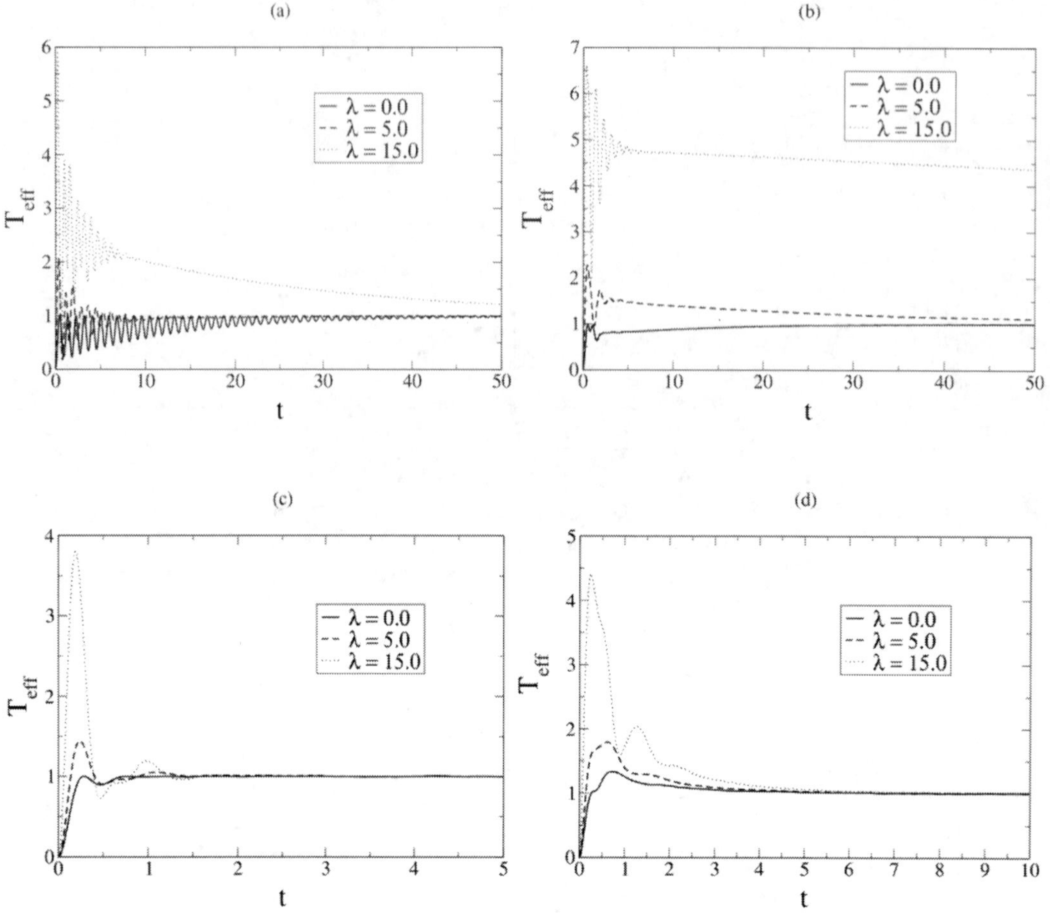

Figure 4. Time evolution for T_{eff}. (a) EDH additive noise case, (b) EDH multiplicative noise case, (c) OU additive noise case and (d) OU multiplicative noise case. The parameters used are: $\eta = 10.0$, $T = 1.0$, $\Omega_0 = 1.0$ and $\gamma = 0.5$ (EDH case) or $\gamma = 5.0$ (OU case).

$\tau_{\text{markov add}}$	$\tau_{\text{OU add}}$	$\tau_{\text{EDH add}}$	$\tau_{\text{markov mult}}$	$\tau_{\text{OU mult}}$	$\tau_{\text{EDH mult}}$	
0.0	6	10	120	9	27	150
5.0	5	10	125	10	28	320
15.0	6	10	420	13	25	2300

Table 4. The approximate time for thermalization for $\eta = 1.0$, in units of $1/m$, for the Markovian and non-Markovian dynamics, determined when (14) approaches the temperature of the thermal bath.

then analyzed the system dynamics varying λ for two cases of dissipation coefficient η. By increasing λ, we have seen that in general, the larger is the λ parameter, the more important are the memory effects, resulting in a strong difference with respect to the local approximation. This difference is seen most notably at short times, obviously, but can last for considerable longer times.

Another interesting aspect of our results is the antagonistic behavior of the system dynamics face the variation of the magnitude of the dissipation parameter, η: looking at Fig. 2, we can see that the larger is λ, the faster the difference between Markovian and non-Markovian dynamics

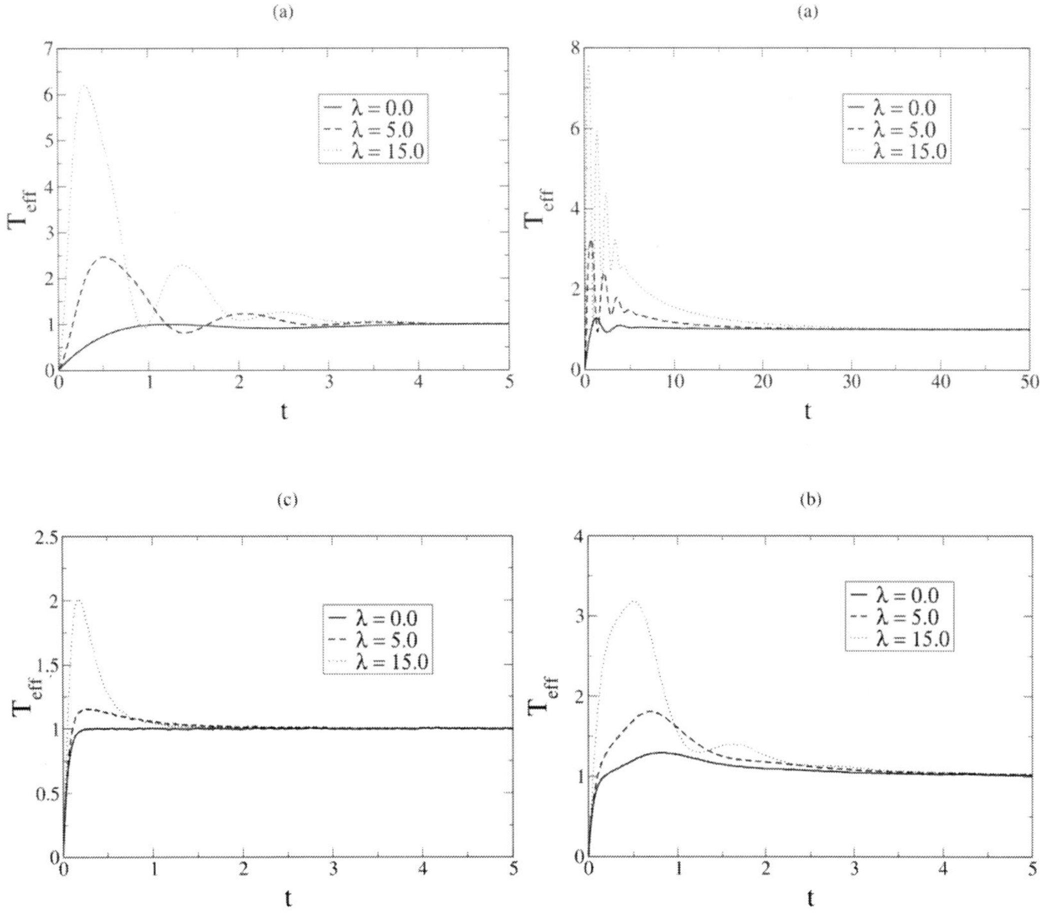

Figure 5. Markovian time evolution for T_{eff} varying the λ parameter. (a) Additive noise case and (b) Multiplicative noise case for $\eta = 1.0$. (c) Additive noise case and (d) Multiplicative noise case for $\eta = 10.0$. The other parameters used are: $T = 1.0$ and $m^2 = 1.0$.

seems to tend to zero. The case we have shown in Fig. 1 is exactly the opposite. In general, the Markovian dynamics tend more and more to overestimate the time for thermalization as the value of λ increases. We can see that this behavior is more accentuated in both additive and multiplicative EDH cases. The results we have obtained here show that, in many cases, the local approximation is not a reliable description of the true non-Markovian dynamics and this becomes more accentuated as the dynamics becomes more nonlinear.

References

[1] Tu M W Y and Zhang W M 2008 *Phys. Rev. B* **78** 235311
[2] M Genkin E W and Lindroth E 2008 *Phys. Rev. B* **79** 245310
[3] P Zedler G Schaller G K C E and Brandes T 2009 *Phys. Rev. B* **80** 045309
[4] Murray J D 1993 *Mathematical Biology* (Berlin: Springer)
[5] M Marusiae Z Bajzer S V P and J P Feyer B 1994 *Math. Biol.* **56** 617
[6] Brenner N and Shokef Y 2007 *Phys. Rev. Lett.* **99** 138102
[7] M Ibañes J García-Ojalvo R T and Sancho J M 2000 *Eur. Phys. J. B* **18** 663

Journal of Physics: Conference Series **246** (2010) 012029 doi:10.1088/1742-6596/246/1/012029

[8] Gleiser M and Ramos R O 1994 *Phys. Rev. D.* **50** 2441

[9] A Berera I G M and Ramos R O 2009 *Rep. Prog. Phys.* **72** 026901

[10] T Fujisawa T H and Sasaki S 2006 *Rep. Prog. Phys.* **69** 759

[11] R Hanson L P Kouwenhoven J R P S T and Vandersypen L M K 2007 *Rev. Mod. Phys.* **79** 1217

[12] García-Ojalvo J and Sancho J M 1999 *Noise in Spatially Extended Systems* (New York: Springer)

[13] van Kampem N G 1992 *Stochastic processes in physics and chemistry* (Amsterdam: North-Holland)

[14] E S Fraga G Krein L F P *arXiv:0910.4369*

[15] Kawasaki K 1976 *Phase Transitions and Critical Phenomena* (New York: Academic)

[16] Hohenberg P C and Halperin B I 1977 *Rev. Mod. Phys.* **49** 435

[17] G W Ford M K and Mazur P 1965 *J. Math. Phys.* **6** 504

[18] Weiss U 1999 *Quantum Dissipative Systems* (Singapore: World Scientific)

[19] Caldeira A O and Leggett A J 1981 *Phys. Rev. Lett.* **46** 211

[20] Caldeira A O and Leggett A J 1983 *Ann. Phys.* **149** 374

[21] Caldeira A O and Leggett A J 1984 *Ann. Phys.* **153** 445

[22] Uhlenbeck G E and Ornstein L S 1930 *Phys. Rev.* **36** 823

[23] Schimansky-Geier L and Zülicke C 1990 *Z. Phys. B: Condens. Matter* **79** 451

[24] R Bartussek P Hanggi B L and Geier L S 1997 *Physica D Matter* **109** 17

[25] R L S Farias R O R and da Silva L A 2009 *Phys. Rev.* E **80** 031143

[26] R L S Farias R O R and da Silva L A 2009 *Comp. Phys. Comm.* **180** 574

[27] Lu K and Bao J D 2005 *Phys. Rev. E* **72** 067701

[28] R L S Farias R O R and da Silva L A 2008 *Braz. J. Phys.* **38** 499

[29] Ingold G L 1998 *Quantum Transport and Dissipation, edited by T. Dittrich, P. Hänggi, G.-L. Ingold, B. Kramer, G. Schön, and W. Zwerger* (Wiley-VCH, Weinheim)

[30] Hänggi P and Jung P 1995 *Adv. Chem. Phys.* **89** 239

Diverging Fano factors

F. Pennini[1,2], **A. Plastino**[1]

[1]Instituto de Física La Plata–CCT-CONICET, Universidad Nacional de La Plata, C.C. 727, 1900 La Plata, Argentina

[2]Departamento de Física, Universidad Católica del Norte, Av. Angamos 0610, Antofagasta, Chile

E-mail: plastino@fisica.unlp.edu.ar

Abstract. We illustrate, with reference to the so-called Fano factor, that dramatic differences between the two most popular phase-space distributions (Wigner's and Husimi's) become evident at the level of the harmonic oscillator treatment.

1. Introduction

In statistics, the Fano factor [1] is a measure of the dispersion of a probability distribution, defined as

$$F = \frac{\sigma_U^2}{\mu_U}, \tag{1}$$

where σ_U^2 is the variance and μ_U is the mean of a random process in some time-window U. The Fano factor can be viewed as a kind of noise-to-signal ratio, being a measure of the reliability with which the random variable can be estimated from a time window that on average contains several random events. For a Poisson process, the variance in the count equals the mean count, so that $F = 1$. If the time window is chosen to be infinity, the Fano factor is similar to the variance-to-mean ratio (VMR) which in statistics is also known as the "index of dispersion", dispersion index or coefficient of dispersion. This is a normalized measure of the dispersion of a probability distribution. In other words, it tells us whether a set of observed occurrences are clustered or dispersed compared to a standard statistical model. As just stated, the Poisson distribution has $F = 1$, but both the geometric and the negative binomial distributions have $F > 1$, while in the binomial instance we find $F < 1$. A constant "random variable" has $F = 0$. The Fano factor turns out to be a convenient noise-indicator of a non-classical field. In the case of a photon distribution (N photons) it reads [2]

$$F \equiv \sigma = \frac{(\Delta N)^2}{\langle \hat{N} \rangle}, \tag{2}$$

and is in this sense related to the so-called Mandel parameter Q [3]

$$Q = \frac{(\Delta N)^2}{\langle \hat{N} \rangle} - 1 \equiv F - 1, \tag{3}$$

In consonance with the above remarks, for $F < 1$ ($Q \leq 0$), emitted light is referred to as sub-Poissonian since it has photo-count noise smaller than that of coherent (ideal laser) light with the same intensity ($F = 1$; $Q = 0$), whereas for $F > 1$, ($Q > 0$) the light is called super-Poissonian, exhibiting photo-count noise higher than the coherent-light noise. Of course, one wishes to minimize the Fano factor.

For a coherent state the Mandel parameter vanishes, i.e., $Q = 0$ and $F = 1$. It is important to note that a field in a coherent state is considered to be the closest possible one to a classical field, since it saturates the Heisenberg uncertainty relation and has the same uncertainty in each quadrature component. Therefore, $Q = 0$ or $F = 1$ define a boundary between a classical and a quantum field. It is clear then that both Q and F can function as indicators on non-classicality. Indeed, for a thermal state one has $Q > 0$ and $F > 1$, corresponding to a photon distribution broader than the Poissonian. For $Q < 0$, ($F < 1$) the photon distribution becomes narrower than that of a Poisson-PDF and the corresponding state is non-classical. The most elementary examples of non-classical states are number states. Since they are eigenstates of the photon number operator $\hat{N}$ the fluctuations in $\hat{N}$ vanish and the Mandel parameter reads $Q = -1$ ($F = 0$) [4].

Here we will concern ourselves with "quantal" phase-space distributions and are forced thus to speak of the celebrated Wigner quasi-probability distribution $D_W(x,p)$ (also called the Wigner function or the Wigner-Ville distribution), a special type of quasi-probability distribution. It was introduced by Eugene Wigner in 1932 [5] to study quantum corrections to classical statistical mechanics. In trying to approximate it in some fashion one is prone to fall into the semiclassical domain. Wigner's goal was to supplant the wave-function that appears in Schröedinger's equation with a probability distribution D_W in phase space. This D should function as a generating function for all spatial autocorrelation functions of a given quantum-mechanical wave-function $\psi(x)$. Thus, in the map between real phase-space functions and Hermitian operators introduced by Hermann Weyl in 1927 [6], D_W maps on the quantum density matrix [6]. One speaks of the Weyl-Wigner transform of the density matrix. In 1949 Moyal [7], who had also re-derived it independently, recognized D_W as the quantum moment-generating functional, *i.e.*, as the basis of an elegant encoding of all quantum expectation values, and hence quantum mechanics in phase space (*Weyl quantization*). It has applications in statistical mechanics, quantum chemistry, quantum optics, classical optics and signal analysis in diverse fields such as electrical engineering, seismology, biology, speech processing, and engine design [8]. Thus, Wigner's is the most elaborate phase-space (PS) formulation of quantum mechanics [5, 9, 10]. A rival phase space distribution is the one developed by Husimi (see Sect. 4 below) [11, 12, 13]. Although both the Wigner and the Husimi distributions carry complete information regarding a quantum state, they exhibit different features. The Wigner function displays large oscillations and may adopt negative values which make it a quasi-distribution rather than a classical probability density. On a compact phase space of area A it is able to reveal fine structures on a sub- Planck scale of order $\hbar^2/A$ [14], structures that can be traced to quantum interferences from distant localized objects [14, 15, 16, 17], that in turn enhance the state's sensitivity to perturbations [14, 15, 16, 17]. Instead, the Husimi distribution is known to be a Gaussian smearing of the Wigner function on an area of size $\hbar$ that washes out the negative part and hence it is suitable as a probability density [18]. However, such smoothing may hide significant important attributes or aspects of the Wigner function [19]. Summing up, while the Wigner function exhibits high resolution, it is not free of long range quantum interferences. The Husimi distribution washes out quantum interferences at the price of hiding important semiclassical structures [19]. We will here illustrate Wigner-Husimi differences in dramatic fashion by recourse to the Fano factor for the harmonic oscillator (HO).

In this work we will compare the Wigner and Husimi distributions in the presence of noise as described by the Fano factor and see that remarkable oddities emerge.

2. Fisher information

The last years have viewed a great deal of effort revolving around physical applications of Fisher's information measure (FIM) [20]. FIM is the source of a powerful variational principle, the extreme physical information one, that yields most of the canonical Lagrangians of theoretical physics [20], characterizing also in quite a proper fashion an "arrow of time", alternative to the one associated with Boltzmann's entropy [21, 22]. The classical Fisher information associated with translations of a one-dimensional observable x with corresponding probability density $\rho(x)$ is [23]

$$I_x = \int \mathrm{d}x\, \rho(x) \left(\frac{\partial \ln \rho(x)}{\partial x} \right)^2, \tag{4}$$

which obeys the so-called Cramer-Rao inequality

$$(\Delta x)^2 \geq I_x^{-1} \tag{5}$$

involving the variance of the stochastic variable x [23]

$$(\Delta x)^2 = \langle x^2 \rangle - \langle x \rangle^2 = \int \mathrm{d}x\, \rho(x)\, x^2 - \left(\int \mathrm{d}x\, \rho(x)\, x \right)^2. \tag{6}$$

3. Thermal Wigner distribution

It is well known that the Wigner distribution for a general density matrix is given by [24]

$$f_w(x,p) = \int_{-\infty}^{\infty} \mathrm{d}s\, e^{ips/\hbar} \left\langle x - \frac{s}{2} \middle| \hat{\rho} \middle| x + \frac{s}{2} \right\rangle, \tag{7}$$

which is normalized according to $\int (\mathrm{d}^2 z/\pi) f_w(x,p) = 1$, with $\mathrm{d}^2 z = \mathrm{d}x\mathrm{d}p/2\pi\hbar$. The pertinent analytic expression for the harmonic-oscillator thermal density is thus [24]

$$f_w(x,p) = 2\tanh(\beta\hbar\omega/2)\, e^{-2\tanh(\frac{\beta\hbar\omega}{2})|z|^2}, \tag{8}$$

where $|z|^2 = x^2/4\sigma_x^2 + p^2/4\sigma_p^2$, with $\sigma_x^2 = \hbar/2m\omega$ and $\sigma_p^2 = \hbar m\omega/2$.
The HO-Fisher Information measure associated to the thermal Wigner distribution is then [13]

$$I_w = \frac{1}{4} \int \frac{\mathrm{d}^2 z}{\pi}\, f_w(z) \left(\frac{\partial \ln f_w(z)}{\partial |z|} \right)^2, \tag{9}$$

and integrating over phase space we immediately get [25, 26]

$$I_w = 2\tanh(\beta\hbar\omega/2). \tag{10}$$

(Note that $0 \leq I_w^{sc} \leq 2$.) Thus, we realize that the HO-thermal Wigner function can be expressed entirely in terms of the Fisher measure (10) [13], i.e.,

$$f_w(x,p) = I_w\, e^{-I_w |z|^2}. \tag{11}$$

We know [27] that, in phase space terms, the HO-Hamiltonian reads

$$\hat{H} = \hbar\omega |z|^2, \tag{12}$$

so that the mean energy becomes

$$E_{Wigner} = \hbar\omega \int \frac{\mathrm{d}^2 z}{\pi} |z|^2 f_w(|z|) = \frac{\hbar\omega}{2\tanh(\beta\hbar\omega/2)} = \frac{\hbar\omega}{I_w}. \tag{13}$$

4. Husimi distribution

Although the oldest and most elaborate phase space (PS) formulation of quantum mechanics is the one discussed above [5, 10], the ensuing distribution function can, as has been noted (regrettably enough), assume negative values so that a probabilistic interpretation becomes questionable. Such limitation was overcome, among others, by Husimi [11].

The Husimi probability distributions $\mu(x,p)$ can be regarded as a "smoothed Wigner distributions" [9]. Indeed, $\mu(x,p)$ is a Wigner-distribution smeared over an $\hbar$ sized region of phase-space [27]. Thus, it is an approximation to the "true" distribution that, specifically, neglects effects of order $\hbar^2$ [9]. As a consequence, it should not be surprising to find out that a Husimi treatment yields results different from the exact ones.

In terms of the concomitant Husimi probability distributions, quantum mechanics can be completely reformulated [12, 28, 29, 30]. This phase space distribution has the appearance

$$\mu(x,p) \equiv \mu(z) = \langle z|\hat{\rho}|z\rangle, \tag{14}$$

where $\hat{\rho}$ is the density operator of the system and $|z\rangle$ are the coherent states (see, for instance, [31] and references therein). The function $\mu(x,p)$ is normalized in the fashion

$$\int \frac{\mathrm{d}^2 z}{\pi} \mu(x,p) = 1. \tag{15}$$

For a thermal equilibrium case $\hat{\rho} = Z^{-1} e^{-\beta\hat{H}}$, $Z = \mathrm{Tr}(e^{-\beta\hat{H}})$ is the partition function and $\beta = 1/T$. Specializing things for the HO of frequency ω, with eigenstates $|n\rangle$ associated to the eigenenergies $E_n = \hbar\omega\,(n+1/2)$, one has [13]

$$\langle z|\hat{\rho}|z\rangle = \frac{1}{Z}\sum_n e^{-\beta\hat{H}} |\langle z|n\rangle|^2, \tag{16}$$

and the normalized Husimi probability distribution is [13]

$$\mu(z) = (1 - e^{-\beta\hbar\omega})\, e^{-(1-e^{-\beta\hbar\omega})|z|^2}. \tag{17}$$

The Fisher information associated to an Husimi PDF has been derived in Ref. [27]. One starts with

$$I_h = \frac{1}{4}\int \frac{\mathrm{d}^2 z}{\pi}\, \mu(z) \left(\frac{\partial \ln \mu(z)}{\partial |z|}\right)^2, \tag{18}$$

so that inserting the μ–expression into (18) we straightforwardly find for the HO-case the analytic form [27]

$$I_h = 1 - e^{-\beta\hbar\omega}, \tag{19}$$

leading to the following limits: for $T \to 0$ one has $I_h = 1$ and for $T \to \infty$ one has $I_h = 0$ as it should be expected (null information at infinite temperature). On the other hand, if we evaluate the mean energy with the Husimi distribution we obtain [27]

$$E_{Husimi} = \hbar\omega \int \frac{\mathrm{d}^2 z}{\pi} |z|^2 \mu(|z|) = \frac{\hbar\omega}{1 - e^{-\beta\hbar\omega}} = \frac{\hbar\omega}{I_h}. \tag{20}$$

Remember that on a compact phase space of area A the Wigner function reveals fine structures on a sub- Planck scale of order $\hbar^2/A$ [14], while the Husimi distribution is known to be a Gaussian smearing of the Wigner function on an area of size $\hbar$ that washes out the negative Wigner parts and hence it is suitable as a probability density [18]. However, such smoothing may hide significant important attributes or aspects of the Wigner function [19].

5. HO-Canonical ensemble: energy fluctuation

It is of the essence for our present purposes to revisit the HO-canonical ensemble machinery [32], which yields correct statistical mechanical results. For practical purposes we here from use the abbreviated notation

$$\Omega = \hbar\omega; \;\; b = \beta\Omega; \;\; e_n = \Omega(n+1/2); \;\; \text{and} \;\; I_h^{-1} = \frac{1}{1-e^{-b}}. \tag{21}$$

Thus, the canonical partition function reads [32]

$$Z = \sum_n \exp\left[-b\,(n+1/2)\right] = e^{(-b/2)}\, I_h^{-1}, \tag{22}$$

and the mean energy is [32]

$$\overline{E} = -\frac{d}{d\beta}(\ln Z) = \frac{\Omega}{2} + \frac{\Omega}{e^b - 1} = \frac{\Omega(2-I_h)}{2I_h}. \tag{23}$$

We arrive now at the vital, decisive stage of our communication. The crux of the matter we are concerned with here is that of the energy fluctuation [32],

$$(\Delta H)^2 = \overline{(E^2)} - (\overline{E})^2 = -\frac{\partial \overline{E}}{\partial \beta}, \tag{24}$$

that implies

$$(\Delta H)^2 = \frac{\Omega^2 e^b}{(e^b-1)^2} = \frac{\Omega^2(1-I_h)}{I_h^2}. \tag{25}$$

It is now evident that the canonical HO distribution adopts the appearance

$$f(\epsilon_n) = \frac{e^{-b(n+1/2)}}{Z} = I_h\, \exp\left(-b\, n\right), \tag{26}$$

which does not coincide, of course, neither with the Husimi (17) nor with the Wigner (11) distributions. Also, from Eqs. (23) and (25) we see that the correct statistical mechanics' HO energy-dispersion is

$$(\Delta H)^2 = (\overline{E})^2 - \Omega^2/4, \tag{27}$$

while, for the energy mean-value one gets

$$\overline{E} = \frac{\Omega}{2} + \frac{\Omega}{e^b-1} = \frac{\Omega(e^b+1)}{2(e^b-1)} = \frac{\Omega}{2}\frac{1+e^{-b}}{1-e^{-b}} = \frac{\Omega}{2I_h}\left(1+e^{-b}\right) = \frac{\Omega}{I_w}, \tag{28}$$

which, as should be expected, coincides with the, also exact, Wigner energy (13). We thus find the nice (and original, as far as we know) information theory relation:

$$I_w = \frac{2I_h}{1+e^{-b}}. \tag{29}$$

Now comes the crucial point, regarding the Fano factor F, intimately linked to the energy fluctuation because the Hamiltonian equals $\Omega(N+1/2)$. We have

$$F_c = \frac{(\Delta N)^2}{\overline{N}} = \frac{(\Delta H)^2}{\Omega\left[\overline{E} - \Omega/2\right]}, \tag{30}$$

and, in more detailed fashion

$$F_c = \frac{(\Delta N)^2}{\overline{N}} == \frac{\exp(b)}{\exp(b) - 1} = \frac{1}{1 - \exp(-b)} = I_h^{-1}. \tag{31}$$

Accordingly,

$$\lim_{\beta \to \infty} F_c = 1, \text{ coherent state's value} \tag{32}$$

is the correct limiting quantum statistical result for the Fano factor.

6. The Husimi and Wigner fluctuation-treatments

We see from Eqs. (11) and (17) that we can recast the general distribution in a common form [33]

$$f_{h,w}(z) = \lambda_{h,w}\, e^{-\lambda_{h,w}|z|^2}, \tag{33}$$

with $\lambda_h = I_h$ for the Husimi case or $\lambda_w = I_w$ for the Wigner instance. Accordingly, the mean energy is

$$\overline{E}_{h,w} = \frac{\Omega}{\lambda_{h,w}}, \tag{34}$$

and

$$\overline{E^2}_{h,w} = 2\frac{\Omega^2}{\lambda_{h,w}^2}. \tag{35}$$

Therefore, the energy fluctuation adopts the appearance

$$(\Delta H)^2 = \overline{E^2}_{h,w} - (\overline{E}_{h,w})^2 = \frac{\Omega^2}{\lambda_{h,w}^2}, \tag{36}$$

and we ascertain that this result disagrees with the correct canonical one of the preceding Section. Then, the Fano factor becomes now

$$F_{h,w} = \frac{\lambda_{h,w}^{-1}}{1 - \frac{\lambda_{h,w}}{2}}. \tag{37}$$

We illustrate in Figure 1 the thermal Fano factor as a function of T.
Since $I_w = 2$ for $T = 0$, we see that the HO-Wigner Fano factor diverges at this temperature (infinite noise), which is of course wrong. We obtain three different Fano factor-forms, with the canonical the correct one. That the Wigner instance exhibits a non-physical divergence is our main result.

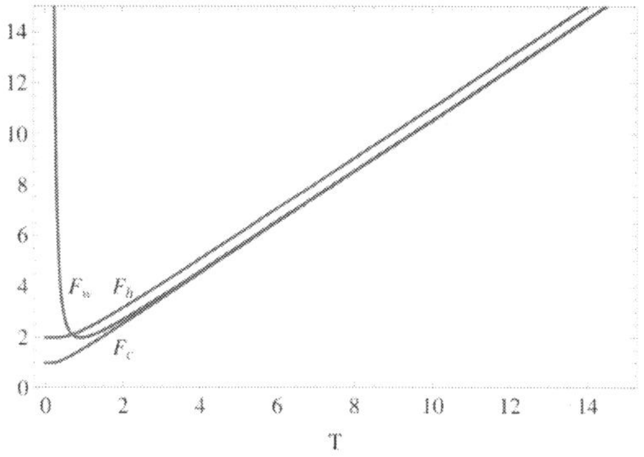

Figure 1. Fano factor as a function of T for canonical, Husimi and Wigner treatments.

7. Conclusions

It is known that there are some cases for which the Weyl-Wigner procedure, that maps Hermitian operators $\hat{A}$ to phase-space functions $A_W(x,p)$, does not give the the correct correspondence between classical and quantum operators [34]. Among these cases one encounters the square of the Hamiltonian [34], an instance in which the Wigner function yields wrong results [34]. Thus, a priori one could have anticipated problems with the Wigner-expectation value of this operator. What is new here is the dramatic nature of the errors in connection with the Fano factor.

Indeed, we have vividly illustrated this effect for the case of the HO-Fano factor, which diverges when evaluated à la Wigner. Instead, the Husimi treatment (HT) is no affected by the Weyl-failure. Since the HT consists in a smearing of the Wigner function over areas of the order of $\hbar$, we gather, as a further result, that the Wigner defect is a sub-$\hbar$ (sub-Planck) effect, of order $\hbar^2$, that disappears after smearing procedures and becomes weaker and weaker with increasing temperatures. Of course, the canonical treatment of statistical mechanics gives always the correct result.

Note also the important role played by the Fisher information measure in these proceedings.

Acknowledgments

F. Pennini would like to thank for partial financial support FONDECYT, grant 1080487.

[1] Fano U 1947 *Physical Review* **72** 26
[2] Bajer J and Miranowicz A 2000 *J. Opt. B* **2** L10
[3] Mandel L and Wolf E 1995 *Optical coherence and quantum optics* (Cambridge: University Press)
[4] Haug F, Freyberger M, Vogel K and Schleich W P 1993 *Quantum Optics* **5** 65
[5] Wigner E P 1932 *Phys. Rev.* **40** 749
[6] Weyl H 1927 *Z. Phys.* **46** 1
[7] Moyal J E 1949 *Proc. of the Cambridge Phil. Soc.* **45** 99
[8] Zachos C, Fairlie D and Curtright T 2005 *Quantum Mechanics in Phase Space* (Singapore: World Scientific)
[9] Lee H W 1995 *Phys. Rep.* **259** 147
[10] Wlodarz J J 2003 *Int. J. Theor. Phys.* **42** 1075
[11] Husimi K 1940 *Proc. Phys. Math. Soc. Japan* **22** 264
[12] Mizrahi S S 1984 *Physica A* **127** 241
[13] Pennini F, Ferri G L and Plastino A 2009 *Entropy* **11** 972
[14] Zurek W H 2001 *Nature* **412** 712
[15] Karkuszewski Z P, Jarzynski C and Zurek W H 2002 *Phys. Rev. Lett.* **89** 170405
[16] Wisniacki D A 2003 *Phys. Rev. E* **67** 016205
[17] Rivas A M F, Vergini E G and Wisniacki D A 2005 *Eur. Phys. J. D* **32** 355
[18] Manfredi G and Feix M R 2000 *Phys. Rev. E* **62** 4665
[19] Rivas A M F and Ozorio de Almeida A M 2002 *Nonlinearity* **15** 681

[20] Frieden B R 1998 *Physics from Fisher information*, (England: Cambridge University Press, Cambridge); *Science from Fisher information* 2004 (England: Cambridge University Press, Cambridge)
[21] Plastino A R and Plastino A 1996 *Phys. Rev. E* **54** 4423
[22] Plastino A, Plastino A R and Miller H G 1997 *Phys. Lett. A* **235** 129
[23] Hall M J W 2000 *Phys. Rev. A* **62** 012107
[24] Tannor D J 2007 *Introduction to Quantum Mechanics: Time-Dependent Perspective* (USA: University Science Books)
[25] Pennini F, Ferri G L and Plastino A 2009 *Central European Journal of Physics* **7** 624
[26] Pennini F and Plastino A 2010 accepted for publication in *Phys. Lett. A*
[27] Pennini F and Plastino A 2004 *Phys. Rev. E* **69** 057101
[28] O' Connel R F and Wigner E P 1981 *Phys. Lett. A* **85** 121
[29] Mizrahi S S 1986 *Physica A* **135** 237
[30] Mizrahi S S 1988 *Physica A* **150** 541
[31] Pennini F and Plastino A 2004 *Phys. Lett. A* **326** 20
[32] Reif F 1965 *Statistical and thermal physics* (NY: McGraw-Hill)
[33] Olivares F, Pennini F and Plastino A 2010 *Physica A* in press
[34] Reichl L E 1991 *A modern course in statistical physics* (London: Hodder & Stughton)

Do Arnold Tongues really constitute a Fractal Set?

Anderson Luis Gama
Mario Sergio Teixeira de Freitas
Paraná Federal University of Technology: Av. Sete de Setembro, 3165 - Rebouças - CEP 80230-901 - Curitiba - Paraná - Brazil

E-mail: gamaanderson@uol.com.br
msergio58@yahoo.com.br

Abstract. We review and analyze the main features of fractal sets, as well as argue that the geometric property called fractality is not a well defined concept in the literature. As an example, we consider the mode locking phenomenon exhibited by the Sine Circle Map, concerning those sets in parameter space called Arnold Tongues and Devil's Staircase (more exactly, their complements). It is shown that well-known results for the fat fractal exponent of the ergodic region turn out to be valid only in a tiny region of parameter space where $\rho \in [0, a], a \to 0$, being ρ the winding number. A careful geometric analysis shows that a misleading simplification has led to a loss of information that hindered relevant conclusions. We propose an alternative, broader and more rigorous approach in that it selects a generic interval from the whole domain which excludes the mentioned restricted region. Our results reveal that the measure shows a different dependence on tongue widths, so we argue that such a discrepancy is only possible if we assume that the set in question does not satisfy a scale-free property. Since there is no invariant exponent for the complementary set of Arnold Tongues (and consequently for that of Devil's Staircase), we conclude that the exponent found by Ecke *et al.* [4] does not characterize those sets as true fractals. We also consider those sets as an example that statistical criteria are insufficient to characterize any set as being a fractal; they should always follow an analysis of the topological process responsible for generating the set.

Keywords: Arnold Tongues, Fat fractal, Exponent, Ergodic Region.

1. Introduction

Geometrical features of phase space are a powerful tool for representing the behavior of dynamical systems, mainly since the concept of fractal sets was introduced by Mandelbrot in 1975 [1]. Though fractal geometry is strongly connected to nonlinear sciences, and power laws are ubiquitously related to its statistical properties, up to now there is no criterion in the bibliography, that could be considered as a formal definition for categorizing a geometrical set as being or not a fractal and, consequently, that leads to the risk of abusive fractal characterization of every apparently irregular set [2]. In this article, we provide an example of such a critical situation through a careful analysis of the particular case of Arnold Tongues.

The name 'Arnold Tongues' is normally applied to the parameter space of the dynamical system concerning the mode locking phenomenon, which occurs when a nonlinear oscillator is forced by another one; as the nonlinear parameter increases, periodic behavior can also be generated by some incommensurable frequencies [3]. The most complete bibliographic source on quantitative relations of the forementioned system dates from 1989 [4], where the authors have concluded that the complement of Arnold Tongues that is present in the Sine Map, namely, the ergodic region, is a fat fractal set. Such conclusion have also appeared later in some books [3], [5], and papers [6], [7], [8], [9].

Our main goal in this paper is to emphasize the need of a more careful analysis prior to labeling a set as a fractal, taking the Arnold Tongues as an example of a set which, we believe, is inappropriately characterized as a fractal. Falconer has already pointed out to such lack of criterion that concerns not only 'fractals', but also the term 'fractal dimension' [2]. It is clear that nonlinear sciences ought to have a more rigorous definition of the concept of fractal in order to properly develop. In our previous works, we have already investigated the recursive geometrical processes that are involved in the formation of some particular fractal sets [10], [11]. We suggest that a possible criterion for characterizing fractals could be based on the nature of such processes. However, such question must be extensively discussed, and it would be unrealistic to expect it to be solved in one single article.

In section 2, we review some fundamental concepts and reproduce some well-known statistical results, in order to single out some features of the Arnold Tongues that are unexpected for a typical fractal set [1]. Then, in section 3, we reanalyze [4], and we find out that a subtle misleading analytical procedure made it possible to obtain the non-representative 2/3 exponent. Our results and conclusions are discussed in section 4.

2. Brief review about fat fractal sets and Arnold Tongues

Instead of introducing general basic notions about fractals, in this section we discuss key concepts that normally are not paid enough attention in the bibliography, regarding whether some geometrical set must be considered or not as a fractal.

Fractal is a recent discovered class of geometrical sets that violates some axioms and theorems of Euclidean Geometry [1]. The lack of a formal definition of fractals lead them to be analyzed through a comparison to some classic sets that are postulated as fractals, based on their geometric structure and also on some statistical properties, most of them involving power laws.

One of the most traditional fractals is the Cantor set [1], which is a recursive process in which one starts from one unit length segment and, upon successive middle-interval subtractions, ends up with a limit set of null measure. The fat fractal version of the Cantor set [2] has a positive Lebesgue measure and an integer dimension. In order to generate a fat Cantor set, one should remove the same sequence of intervals from a set that is otherwise larger than unity; as a result, the length of the final set will be the difference between the two initial intervals [12]. The intervals size, relatively to the remaining length in each iteration, will vanish much faster than that of the standard Cantor Set. Figure 1a shows the Cantor set formation, and Figure 1b its fat fractal counterpart.

It also is appropriate to briefly review what defines the multifractal version of the Cantor set, which can be either thin or fat: the removed intervals are not necessarily selected from the middle of the former iterated set; so, the two remaining intervals are not equal, although their size still vanish exponentially with the number of iterations, just like a standard Cantor set (Figure 1c).

We stress here that, along the iterations, both the decrease of size and increase of number of intervals follow an exponential law, although the Capacity Dimension, that is used to characterize thin fractals, is based on a simple power law [2], [1]

$$N(\varepsilon) = C\varepsilon^{-d}. \tag{1}$$

In equation 1, $N(\varepsilon)$ is the smallest number of intervals smaller than ε needed to cover the set,

[1] Unfortunately the names Arnold Tongues and Devil's Staircase [1] may also refer to those fractals that represent the Arnold Tongues and Devil's Staircase obtained from physical systems. However, their geometry is completely different, and for that reason, they are true fractals.

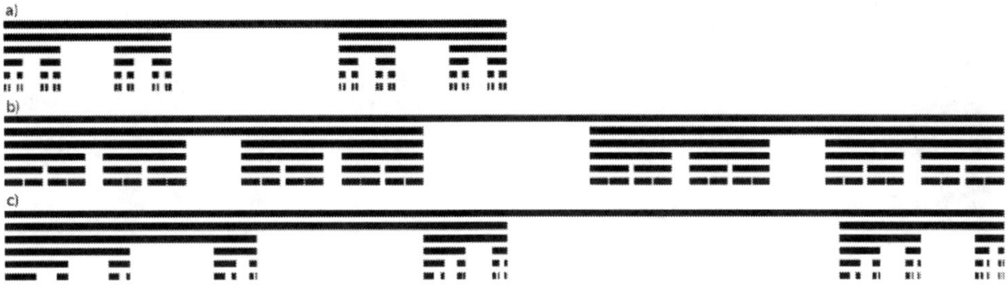

Figure 1. The recursive formation processes of the Cantor set (a), of the fat Cantor set (b), and of the multifractal Cantor set (c). In all cases, we start from a straight segment.

the exponent d is the Capacity Dimension (in the case of the Cantor set we have $d = \log_3 2$), and C is a scale-dependent positive constant.

Concerning the Capacity Dimension, it provides little relevant information about the fractal itself, since it is equals 1 for all one-dimensional fat fractals. For that reason, other kinds of measure that are valid for fat fractals have been established, and in this paper we will choose the coarse-grained measure [4]

$$\mu(\varepsilon) = \mu(0) + A\varepsilon^{\beta}, \quad (2)$$

where $\mu(\varepsilon)$ is the length of the set in which all discontinuities smaller than ε are neglected; β is the fat fractal exponent, and A is a scale-depending positive constant. For thin fractals as well as for the original Cantor set, $\mu(0) = 0$, and it is easily demonstrated that

$$\mu(\varepsilon) = A\varepsilon^{D-d}. \quad (3)$$

Next, we review the structure of Arnold Tongues, stressing that, up to now, there have been no exact criterion for characterizing a fractal set, and most characteristics are not universal [2]. Following this short review, we reproduce and discuss some results obtained by Ecke [4], that will be relevant in next section.

Circle maps, generated by the Poincaré section of two coupled oscillators through a nonlinear term [4], [3], can be encountered in several areas, ranging from simple fixed points to highly chaotic motion. The particular Sine Map is given by

$$X_{t+1} = X_t - (k/2\pi)\sin(2\pi X_t) + \Omega, \quad (4)$$

where k is the nonlinearity parameter and Ω is the frequency of the first oscillator, being the second frequency unitary. For $0 \le k \le 1$, the ubiquitous non-chaotic motion is characterized by the winding number

$$\rho = \lim_{t\to\infty} \frac{X_t - X_0}{t}. \quad (5)$$

For periodic behavior $\rho = p/q$, where p and q are the smallest integers that respect

$$X_{t+q} = X_t + p. \quad (6)$$

Each Arnold tongue is the set of all pairs (Ω, k) sharing the same pair (p, q). Figure 2 depicts the set of the largest tongues in $\Omega \le 1/2$. The Devil's Staircase may help to visualize the whole parameter space.

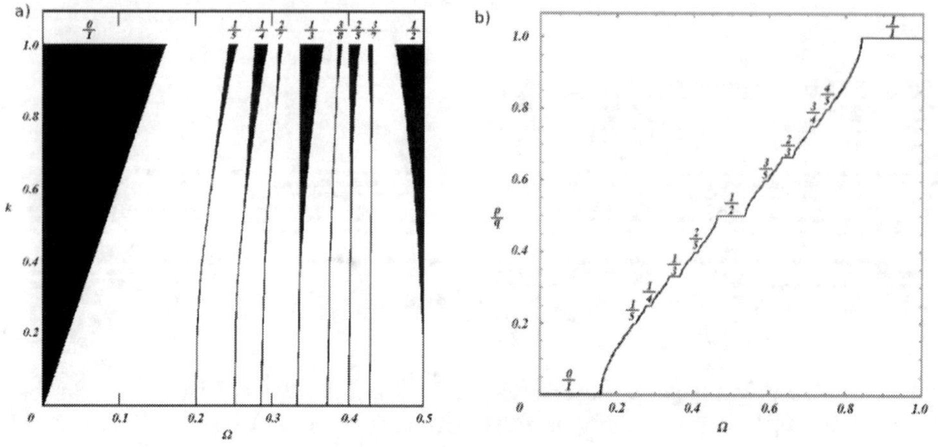

Figure 2. a) The set of the largest Arnold tongues of the Sine Map is shown in black and the Complement, that is the Ergodic Region, in white. In higher resolution there would appear tiny black tongues inside every white region, but otherwise no tiny white region inside black ones. On the top of each tongue, the corresponding winding number p/q is indicated. b) The complete Devil's Staircase represents the graph of the winding number p/q in function of Ω for a specific k, in this case $k = 1$. What seems to be a continuous line is in fact several smaller and smaller steps [5].

Based on extensive numerical simulations, Ecke *et al.* have found some power laws concerning dependency of tongues' widths ($\Delta\Omega$) on periods (p and q) for several values of nonlinearity parameter (k), the most important being [4]

$$\lim_{k\to 1}\lim_{q\to\infty} \Delta\Omega = C \cdot q^{-3}, \tag{7}$$

where C is a proportionality constant that varies as a rather complicated and non-monotonic decreasing way with p [4]. After some approximations, the authors have established the *ad hoc* assumption, which is supposed to be valid for large values of p [2] and q

$$\Delta\Omega = Aq^{-3}p^{-\alpha}, \tag{8}$$

where A and α are positive constants, and $\alpha \gg 3$, based on data evidence. However, Ecke [4] could not ensure that an invariant α would be established for large values of q. Nevertheless, the statistical tool $p^{-\alpha}$ will be essential in determining the fat fractal exponent. It is also interesting to point out that the Arnold Tongues are ruled by power laws, differently from fractals, which display exponential laws [1].

In order to evidence some measuring problems that Ecke and his collaborators [4] have faced regarding geometrical features of the Arnold Tongues, we reproduce in Figures 3 and 4, with the authors' caption, two of their numerical results, where data do exhibit strong fluctuations. The fitting to black circles, in the second figure, provides an exponent that has been estimated to be 0.68 ± 0.05, contrasting with $2/3$, that was analytically obtained for $0.4 < k < 0.9$.

In next section, we develop a broader analytical method to investigate the validity of a fat fractal exponent to the complement of Arnold Tongues.

[2] Originally, Ecke [4] has adopted the symbol $\bar{p}$ instead of p, representing the sequence of largest $\Delta\Omega$. But it is particularly important that the differences between $\bar{p}$ and p should not be significant, which is the case for most values of k (more precisely, for $k < 0.9$); so we decided to keep only p.

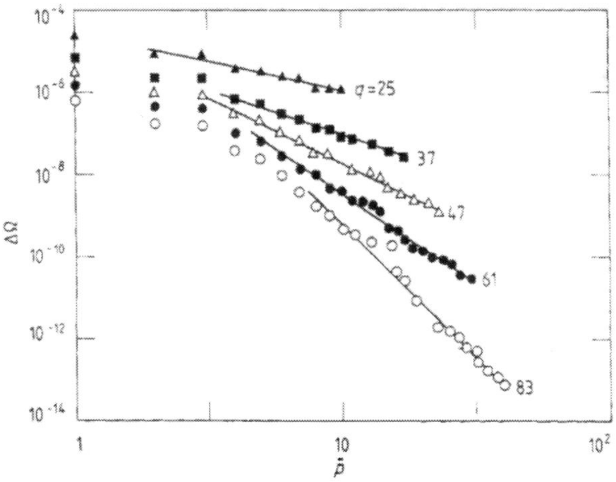

Figure 3. The distribution of interval sizes at fixed period q. At fixed $k = 0.8$ we compute all the periodic intervals with a given q, and sort them in order according to their size. The order is labeled by an index $\bar{p} = 1, 2, \ldots, \bar{p}_{MAX}$. $\Delta\Omega$ is plotted against $\bar{p}$ for several different values of q (extracted from [4]).

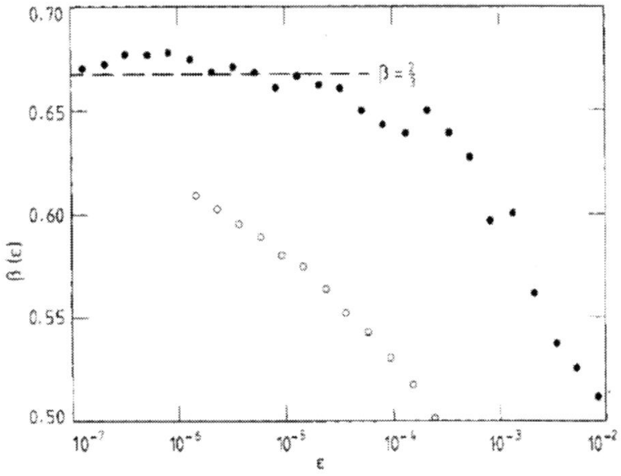

Figure 4. An illustration of the convergence of the computed value of the fat-fractal scaling exponent β with the scale of resolution ε. The numerical data of figure 3 (not shown in this paper) are numerically differentiated. The resulting slopes are plotted against the corresponding values of $\log \varepsilon$, to give an idea of the rate of convergence. We show two cases. At $k = 0.80(\bullet)$ the convergence is within experimental error at the smallest values of ε. At $k = 0.95(\circ)$, however, due to crossover phenomena there is no convergence, even at the smallest value of ε that we were able to compute (extracted from [4]).

3. Algebraic Procedures for the Fat Fractal Exponent

It is convenient here, before presenting our results, to review how Ecke and his collaborators [4] have obtained the value 2/3 for the fat fractal exponent. Though all approximate procedures seem to be reasonable, we have detected a paradox: while we were developing a broader and more exact model, we obtained our unexpected result for a fractal. That was due to some characteristics of a very strict interval in parameter space, which prevail over those of the whole set and masks the true hierarchic structure of the tongues; we will show that, if any generic segment of the set is sampled, an exponent different from 2/3 is obtained. As a result, we conclude that the authors' approximate method has hindered a reliable diagnosis, in other words, we will demonstrate that the ergodic region does not scale as a fat fractal exponent.

Ecke *et al.* [4] started considering the length of the ergodic region. For a particular value of k, such length equals the total interval of parameter Ω, minus the summation of the widths of all tongues; so, they conclude that

$$\mu(\varepsilon) - \mu(0) = \sum_{\Delta\Omega<\varepsilon} \Delta\Omega(q,p) = \sum_{\Delta\Omega<\varepsilon} Aq^{-3}p^{-\alpha}. \tag{9}$$

For methodological reasons, the authors separate out the sum of equation 9 in two parts: in Figure 5, values of p inside region I hold for values of q that are greater than the critical curve $Q(p)$ in which $\Delta\Omega = \varepsilon$; in region II, otherwise, values of p hold for every q that is greater than p. The boundary value of p between regions I and II is called p'.

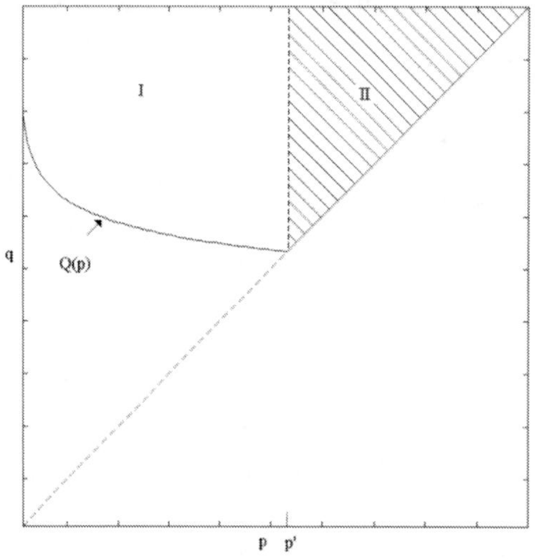

Figure 5. Sketch of the pairs (p, q) corresponding to the summation in equation 9 [4]. $Q(p)$ represents the critical curve where $\Delta\Omega = \varepsilon$, and p' is the value for which $Q(p') = p'$. Just for a better qualitative visualization, the authors chose the value $\alpha = 0.1$.

Summation I is limited by lines $q = Q(p)$ and $p = p'$. The scalings of $Q(p)$ and of p' are obtained from equation 8

$$\Delta\Omega(Q(p), p) = \varepsilon \Leftrightarrow Q(p) = (\varepsilon \cdot p^{\alpha}/A)^{-1/3}, \tag{10}$$

$$Q(p') = p' \Leftrightarrow p' = (\varepsilon/A)^{-1/(3+\alpha)}. \tag{11}$$

Summing over each region, equation 9 becomes like

$$\mu(\varepsilon) - \mu(0) = \sum_{\substack{1\leq p<p' \\ q>Q(p)}} Aq^{-3}p^{-\alpha} + \sum_{\substack{p'\leq p<q \\ q>p'}} Aq^{-3}p^{-\alpha}. \tag{12}$$

In order to solve the above equation, variables p and q are interpreted as being continuous, so the summations are written as integrals

$$\mu(\varepsilon) - \mu(0) = \int_1^{p'} \int_{Q(p)}^{\infty} Aq^{-3}p^{-\alpha}\, dq\, dp + \int_{p'}^{\infty} \int_{p}^{\infty} Aq^{-3}p^{-\alpha}\, dq\, dp. \tag{13}$$

Using equation 10 for $Q(p)$ and equation 11 for p', the integrals are straightforward, resulting

$$\mu(\varepsilon) - \mu(0) = C\epsilon^{2/3} + \bar{C}\varepsilon^{(\alpha+1)/(\alpha+3)}, \tag{14}$$

where C and $\bar{C}$ are constants. In fact, the approximation of equation 12 by 13 represents a lower bound and the upper bound can be found replacing p by $p-1$ and q by $q-1$ in equation 13. Ecke shows that this implies the same equation 14 with only different constants [4]. As a result, he argues $\mu(\varepsilon) - \mu(0)$ follows the same scaling of equation 14.

The key point is that, since $\alpha \gg 3$, we have $\lim_{\varepsilon\to 0} \mu(\varepsilon) - \mu(0) = C\epsilon^{2/3}$; from that, Ecke and his collaborators have concluded that the ergodic region scales as a power law with a fat fractal exponent that equals 2/3 for the subcritical regime [4].

We now present our approach. Although the methodology is similar, an important difference lies in the restriction of p/q by the variables a and b.

Although Eckes result seems a reasonable one, a more careful investigation reveals that the measure value is not scale invariant. In order to make evident the misleading step, we follow all of the steps that have been described previously, adopting a more general analysis for obtaining the coarse-grained measure: instead of summing the tongue widths directly along the whole domain of the winding number $0 < (p/q) < 1$, we take initially an arbitrary interval $0 \le a < (p/q) < b \le 1$, that belongs to the same domain

$$\mu(\varepsilon) - \mu(0) = \sum_{\substack{\Delta\Omega<\varepsilon \\ a<(p/q)<b}} \Delta\Omega(q,p) = \sum_{\substack{\Delta\Omega<\varepsilon \\ a<(p/q)<b}} Aq^{-3}p^{-\alpha}. \tag{15}$$

In what follows, we apply the methodological procedure of separating equation 15 into two summations, corresponding to regions III and IV in Figure 6, which is analogous to the qualitative graph of Figure 5, originally presented by Ecke *et al.* [4]. Figure 6 is our original result, and it will lead us to the main conclusion.

Summation III is limited by $p = p''$ and by $p = p'''$. According to equation 8, p'' and p''' scale as

$$p'' = \frac{\varepsilon}{a^3 A}^{-1/(3+\alpha)}, \tag{16}$$

$$p''' = \frac{\varepsilon}{b^3 A}^{-1/(3+\alpha)}. \tag{17}$$

So, analogously to equations 12 and 13 we have

$$\mu(\varepsilon) - \mu(0) = \sum_{\substack{1\le p''<p<p''' \\ Q(p)<q<p/a}} Aq^{-3}p^{-\alpha} + \sum_{\substack{p'''\le p \\ p/b<q<p/a}} Aq^{-3}p^{-\alpha}, \tag{18}$$

$$\mu(\varepsilon) - \mu(0) = \int_{p''}^{p'''} \int_{Q(p)}^{p/a} Aq^{-3}p^{-\alpha}\, dq\, dp + \int_{p'''}^{\infty} \int_{p/b}^{p/a} Aq^{-3}p^{-\alpha}\, dq\, dp. \tag{19}$$

Solving the above integrals, we obtain

$$\mu(\varepsilon) - \mu(0) = C\epsilon^{(\alpha+1)/(\alpha+3)}, \tag{20}$$

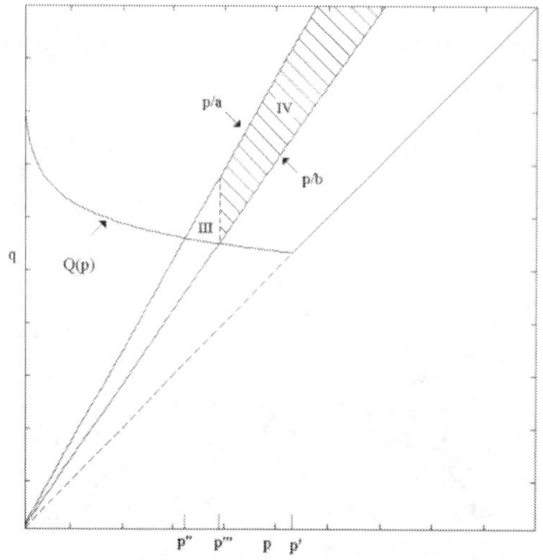

Figure 6. Representation of the pairs (p, q) that are summed in equation 15. $Q(p)$ represents the critical curve for which $\Delta\Omega = \varepsilon$, and p', p'' and p''' represent the values for which $Q(p') = p'$, $Q(p'') = p''/a$ and $Q(p''') = p'''/b$. Differently from the other figures, this one is original from our work.

where C is some positive constant. Also here the approximation of equation 18 by 19 is a lower bound and since $\Delta\Omega$ is monotonic an upper bound can be found replacing p by $p-1$ and q by $q-1$ in equation 18. By solving the resulting integral the same equation 20 is found, up to a constant. As a result we conclude that $\mu(\varepsilon) - \mu(0)$ scales as $\epsilon^{(\alpha+1)/(\alpha+3)}$ for $a < (p/q) < b$. Note that in our modified method the unitary lower limit does not appear in the first integral of the right hand side of equation 13. Solving the integral allows us to conclude that this particular change led the first term of the right hand side of equation 14 to vanish. As a result we find a different fat fractal exponent $\beta = (\alpha+1)/(\alpha+3)$.

Furthermore, equations 14 and 20 do evidence a paradox: the complement of the Arnold Tongues in the interval $\rho \in]a, b[\subset]0, 1[$ scales as the fat fractal exponent $(\alpha+1)/(\alpha+3) > 2/3$, while in the particular case of $\rho \in]0, 1[$ the fat fractal exponent equals 2/3. We make clear that our method is somewhat limited: as a approaches zero, the method fails for $p'' < 1$. As a result, we conclude that, close enough to zero, the intervals $]a, b[$ also have a fat fractal exponent equal to 2/3. Therefore, we also conclude that the 2/3 fat fractal exponent obtained by Ecke *et al.* [4] results from the tongues only with small winding numbers, being the remaining tongues too narrow, so that they were not considered. But, by removing the tongues with small winding numbers, our method shows a different fat fractal exponent.

This kind of behavior is not only unexpected for a fractal, it is not desirable either. Nevertheless, that is not accidental but rather a direct consequence from the rules that describe the system, mainly equation 8. Ecke himself stress that as the exponent 2/3 is not dependent on α, only the tongues with small[3] p are relevant for the geometry of the set, for the other ones are just too narrow to affect the exponent [4]. But we remark that such argument involves a geometric consequence: all tongues with small p are located on the left side of the set, having $\rho \to 0$. Though we agree that multifractals also have a concentration of large discontinuities on one of the set's edge, they also have smaller concentration on other points, unlike the ergodic region we are analyzing. Equation 8 predicts that the size of discontinuities generally decreases

[3] here small has an absolute meaning, not relative to q

from left to right, with no smaller concentration, and such a structure resembles a logarithmic accumulation, rather than a fractal [13], [14].

A possible solution for this situation would be to consider that the ergodic region excluding the neighborhood of $\rho = 0$ can be a fractal set with exponent equal to $(\alpha + 1)/(\alpha + 3)$. However, Ecke and his collaborators themselves stress that the obtained exponent shows no dependence on α [4]. On the other hand, our result is strongly influenced by the $\Delta\Omega(p)$ dependence. Moreover, numerical data that we reproduce in Figure 3 show that not only α increases with q, but also that such increase is uniform. As a result, we argue that it would not be prudent to suppose $(\alpha + 1)/(\alpha + 3)$ as the exponent's value, before proving its own validity.

In the next section, we discuss the previous results, and we conclude that the complement of Arnold Tongues, and consequently the complement of the Devil's Staircase, have no reliable relationship to any traditional fractal set, unless a new procedure for such an analysis leads to an invariant value of its exponent.

4. Discussion

In this article, we developed a careful algebraic analysis of properties of the parameter space of the Sine Map and showed that some current approximate approaches have led to a misinterpretation of the set known as Arnold Tongues as being a true fractal. We have shown that, although some well-known results are already widely mentioned in the literature concerning Arnold Tongues, such results are, unfortunately, not reliable.

Our main contribution is that, according to the analytical results that we have obtained through a broader algebraic method, the value of 2/3 for the fat fractal exponent corresponds strictly to the region where the winding number $\rho \to 0$, not being representative for the whole set. Furthermore, such exponent is not an invariant one.

We have also established, through geometrical arguments, that the accumulated structure near the edge that corresponds to small values of p/q in parameter space does not display the behavior that would be expected from a fractal set.

Consequently, we do not wish to replace the exponent of 2/3 by the remaining exponent, that is $\beta = (\alpha + 1)/(\alpha + 3)$ in equation 20, as some kind of correction since such exponent turns out to be based on a heuristic assumption present in the literature, namely the α in $\Delta\Omega = Aq^{-3}p^{-\alpha}$. Such assumption was supposedly proved through some doubtful statistical evidences. We argue that, since it has not been possible to determine the value of α and the only fact we know it that $\alpha > 3$.

In summary, since the complement of Arnold Tongues persists up to now without a valid exponent that characterizes it, we do not have a clear proof that it has any relationship to traditional fractals and multifractals, despite its apparent irregularity.

A thorough study of intricate geometrical sets deserves more rigorous criteria than arbitrarily comparing them to some so-called irregular objects or applying some indicative statistical measurements. The conclusion we have reached at this writing, concerning the lack of criterion for characterizing the Arnold Tongues as a fractal, refers to just one of the examples we are investigating [10], [11]. We expect that our methodology and results may contribute to stimulate more intensive research on fractal geometry, aiming at the establishment of a formal, rigorous definition of what a fractal is. We argue that such eventual definition should take into account the topological process that generates the set, rather than indicative statistical properties, such as the several different exponents called Fractal Dimensions, including, among others, Box-Counting Dimension, Correlation Dimension, Lyapunov Dimension, and Coarse-Grained Measure [2], [3], [5], [4].

Acknowledgments
The authors are grateful to Dr. Pedro Zambianchi from Department of Physics of UTFPR for his careful revison of our text.

References
[1] Mandelbrot B B 1983 *The Fractal Geometry of Nature* (New York: W.H. Freeman and Company)
[2] Falconer K 1990 *Fractal Geometry: Mathematical Foundations and Applications* (Chichester: John Wiley)
[3] Lichtenberg A J and Lieberman M A 1992 *Regular and Chaotic Dynamics* (Berlin: Springer)
[4] Ecke R E, Farmer J D and Umberger D K Scaling of the Arnold tongues 1989 *Nonlinearity* **2** 175-196
[5] Ott E 1993 *Chaos in Dynamical Systems* (Cambridge: Cambridge University Press)
[6] Paar V and Pavin N Intermingled fractal Arnold tongues 1998 *Phys. Rev.* E **57** 1544-1549
[7] Paar V, Pavin N and Rosandi M Link between Truncated Fractals and Coupled Oscillators in Biological Systems 2001 *Journal of Theoretical Biology* **218** 47-56
[8] Kennedy M P, Krieg K R and Chua L O The Devil's Staircase: The Electrical Engineer's Fractal 1989 *Circuits and Systems, IEEE Transactions on* **36** 1113-1139
[9] Wenzel W and Biham O Periodic orbits in the dissipative standard map *Phys. Rev.* A **43** 6550-6557
[10] Gama A L and de Freitas M S T Fractal Formation in the Circular Chaos Game: a Topological Investigation 2010 *Internation Journal of Bifurcation and Chaos* **20** 703-713
[11] Gama A L and de Freitas M S T Fractal Formation Process in the Mandelbrot Set, submited for the Journal *Fractals*
[12] Vicsek T 1992 *Fractal Growth Phenomena* (Singapure: World Scientific)
[13] McDonald S W, Grebogi C, Ott E and Yorke J A Fractal basin boundaries 1985 *Phys.* D **17** 125-153
[14] de Freitas M S T, Viana R L and Grebogi C Multistability, basin boundary structure, and chaotic behavior in a suspension bridge model 2004 *Internation Journal of Bifurcation and Chaos* **14** 927-950

Journal of Physics: Conference Series **246** (2010) 012032 doi:10.1088/1742-6596/246/1/012032

Efficiency of financial markets and algorithmic complexity

R. Giglio, S. da Silva
Department of Economics, Federal University of Santa Catarina, 88049970 Florianopolis SC, Brazil

Iram Gleria
Institute of Physics, Federal University of Alagoas,
57072-970 Maceió AL, Brazil

E-mail: iram@if.ufal.br

A. Ranciaro
Department of Economics, Federal University of Alagoas,
57072-970 Maceió AL, Brazil

R. Matsushita, A. Figueiredo
Institute of Physics, University of Brasilia, 70910900 Brasilia DF, Brazil

Abstract. In this work we are interested in the concept of market efficiency and its relationship with the algorithmic complexity theory. We employ a methodology based on the Lempel-Ziv index to analyze the relative efficiency of high-frequency data coming from the Brazilian stock market.

1. Introduction

When the information in possession of market participants is instantaneously incorporated into the prices of assets, the market is said to be information efficient. In an efficient market the prices must fluctuate in a random way [3]. In pioneer works *e.g*, [1] and [2], statistical analysis suggested a Brownian motion. In an efficient market there is no room for arbitrage because the prices move instantly so as to prevent informed traders to take advantage of their information. The uninformed traders can then infer from an observed price increase that some traders in the market have favorable information about the asset [4]-[5]. This classic definition of efficiency refers to an absolute concept of efficiency. In a recent article [6], some of the authors used the algorithmic complexity theory to analyze the relative efficiency of a given market. In fact, Campbell and coauthors [8] pointed that the relative notion of efficiency may be a more useful approach. They observed that, in real markets prices reflect only the information for which the acquisition costs does not outweigh the benefits (including transaction costs). Furthermore, information is not homogeneously distributed and upon the arrival of new information market participants may diverge from one another in how they think it will impact prices (heterogeneous expectations). Thus, residual inefficiencies are always present in actual markets and can introduce artificial patterns and redundant information in real-world data. Within this framework, an efficiency index is expected to measure the distance of one given market from the idealized efficient market.

In [6] some of the authors applied a technique based on the Lempel-Ziv approach to sort stock exchanges according to their relative efficiency. The De Bruijn sequence was considered to be a useful approximation for the ideal efficient market; real markets having their efficiency measured against it. In this work, we use the same method to analyze high frequency data; however, we also implement some changes in our approach in order to achieve more robust results.

The paper is organized as follows. In the next section we review our perspective based on the Lempel-Ziv ("deterministic") complexity index. Then, we present data and perform analysis.

2. Lempel-Ziv algorithmic complexity

According to Kolmogorov, the complexity of a string is given by the length of the shortest computer program that can reproduce it. Since this shortest algorithm cannot be computable, several alternative calculations of the complexity have been suggested in literature. Lempel and Ziv [18] put forward a useful measure and Kaspar and Schuster [19] provided an algorithm to calculate the Lempel-Ziv (LZ) index. Consider, for example, the binary string $S = s_1, \ldots, s_n$. The Kaspar and Schuster procedure reproduces the string using two operations: insertion of a new digit into S or copy of the digit. In the $r - th$ step we have that digit s_r cannot belong to the subset $s_1, \ldots, s_{r-1}$. This is so because s_r could be copied from this subset if it belonged to the substring $s_1, \ldots, s_{r-1}$. To check whether the last digits of S can be reconstructed by either the procedure of copying or inserting new digits we take s_{r+1}. We check whether s_{r+1} belongs to one of the substrings of S , in which case we simply copy it from S. The routine goes on until the appearance of a new digit (which needs to be inserted). The complexity c of S is then obtained as the number of newly inserted digits plus one (if the last copy step is not followed by inserting a digit). This technique is similar to those used in most of the compressing algorithms, such as WinZip and WinRar.

Obviously, the complexity of a string is a function of the size of S. A truly random string asymptotically approaches its maximum complexity r as its length $n \to \infty$ according to $lim_{n\to\infty} c = r = n/log_2 n$ [19]. One may thus define a normalized index $LZ \equiv c/r$, which provides a measure of the complexity of a string relative to that of a random one. As the size of the string grows, $LZ \to 1$. Very complex series obtained from experimental data usually have an LZ a little bit above one. It was shown [19] that the index is suitable for comparing strings of distinct lengths as long as the lengths are ≥ 1000.

If we think on efficiency as the lack of statistical dependence, the LZ complexity can also capture it. In a sense, the LZ index provides a times-series statistic more basic than those produced by autoregressive processes of any order. If a time series exhibits statistical dependence, we expect it to present more patterns and accordingly to have lower complexity than an independent one. That the LZ complexity measure can also capture this feature can be seen in the example of an unfair coin, whose probability distribution of the current value is dependent on the previous one with probability p. If $p = 1/2$ the process collapses to that of the fair coin with no memory (that is, an independent process); if either $p = 0$ or $p = 1$ the current value is perfectly predictable from past observation, that is, the process is totally dependent. Figure 1 shows the LZ index of the sequences generated by the unfair coin for different probabilities. In the parabola-shaped curve, the index reaches its maximum when $p = 1/2$, and approaches its minimum when there is total dependency ($p = 0$ or $p = 1$).

Some applications of the LZ index include the following. Use of the LZ index for DNA sequences to reconstruct the phylogenetic tree of several species of placental mammals [20], and analysis of the complexity of the heart rate variability signals to identify intrauterine growth

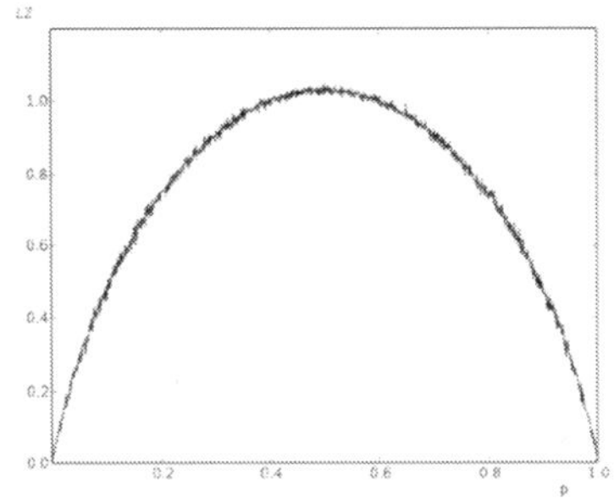

Figure 1. Experiment with an unfair coin. Maximum complexity is reached if the probability $p = 1/2$, and lack of complexity obtains as the probability is either 0 or 1.

restricted fetuses [17]. Shmulevich and Povel [21] suggested that the LZ index is able to identify temporal complexity in musical rhythms despite the fact that one needs long sequences in order to measure complexity.

As in [6], here we consider sliding time windows rather than calculating one single index for the entire string. We calculate the index for several subsets (windows), and then get the average. For instance, for a time series of N data points, a chosen time window of $n < N$ observations, and a chosen step $s < n$, we first compute the LZ index of the window from 1 to n, then we calculate the LZ from s to $n + s$, and so on, up to the index of the last window. Then we take the average of the indices.

3. Data and analysis

We collected high frequency data from selected stocks listed on Bovespa, the Brazilian stock exchange. Those are tick-by-tick data for the year 2007, ranging from 100,000 to 500,000 data points. Analysis was performed with returns (that is, the differences) of the raw series. We coded a return series according the following rule. First, we define a stability region (SR) through the parameter b, which assumes real values. Given a return observation ρ_t, the data point d_t is equal to $d_t = 0$ if $\rho_t < -b$, $d_t = 1$ if $\rho_t > b$ and $d_t = 2$ if $-b < \rho_t < b$. Thus, we coded the series as a ternary string if $b \neq 0$, and a binary string if $b = 0$.

In [6] it was found that data coming from daily returns are very complex. They look more like the genuinely random series than the totally redundant, perfectly predictable series. Inspired by the experiment we decided to consider $LZ = 1$ as a threshold in order to assess the relative efficiency of the series (Figure 1). The number of occurrences where the LZ index was caught above one was considered as a measure of relative efficiency. For example, in a computer generated pseudo-random series the LZ was surpassed $LZ = 1$ 98.8% of the times; thus this series is 98.8% efficient. For high frequency data this approach is not suitable. In fact, tick-by-tick data are usually correlated and we thus expect to obtain $LZ < 1$. Although the data are short range correlated, there is a correlation "time" (number of transactions, not physical time) of roughly 10 units (Figure 2). Then, the procedure described in [6] would give 0% efficiency. An alternative approach is to take the average LZ value as a good measure of complexity. Instead of presenting a ranking similar to the one presented in [6], we performed an analysis of the sensibility of the LZ index with the stability region SR (parameterized by b), the window size, and the size of the steps s. In Figure 3 we present for the BRTP3 and TMAR5 stocks the value of LZ versus SR for three different windows and a fixed jump s. In Figure 4 we plot LZ versus SR for a fixed window size and several steps. The LZ value is somewhat robust for changes of the window size and the step size. Nevertheless, we observe that the LZ rises when $b \to 0$,

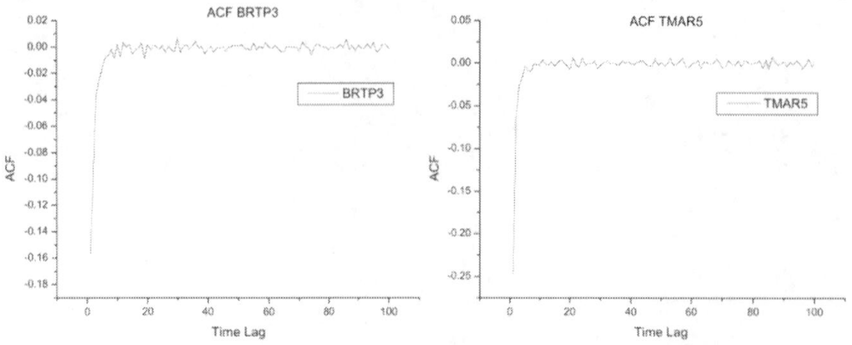

Figure 2. Aucorrelation function for the tick-by-tick data.

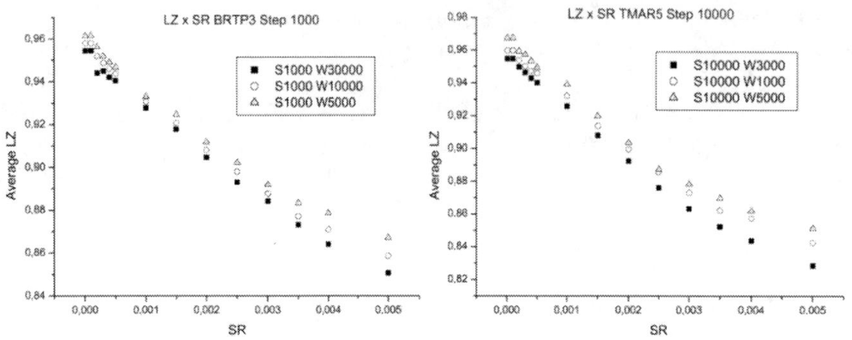

Figure 3. The value of LZ versus SR for 3 different windows and a fixed step s.

reaching a maximum for $b = 0$ with maximum complexity ($LZ \approx 0.96$). In tick-by-tick data, since the variations of the stock price from one trade to another are usually small, raising b has the effect of introducing repetitive patterns of symbols, lowering the complexity of the series. Thus, we expect maximum complexity for $b = 0$. This is in stark contrast with the daily data analyzed in [6], where the results obtained were robust to changes in b.

In reference [6] some of the authors compared the methodology based on the algorithm complexity and the Hurst exponent reckoned by detrended fluctuation analysis (DFA) [23]. The DFA outperforms the traditional Hurst exponents and other methods designed to track long range memory [23]. Considering the percentage of times the DFA exponent values were caught inside a given interval, a rank similar to the one based on the LZ index was presented. Although the stock indices and exchange rates of developed markets were top ranked, the matching between the LZ indices and the DFA exponents were not that perfect.

More recently, a methodology based on complexity-entropy causality plane was presented [24] to distinguish the stage of stock market development and reveal the presence of correlations and some degree of order in emerging markets, a result in line with the LZ methodology. Other alternative approaches can be found in references [25]-[29].

4. Conclusions

We used high-frequency data from the Bovespa stock exchange to assess the relative efficiency of its stocks with the interpretation provided by the algorithmic complexity theory. We adapted the previous analysis in Giglio et al. to the high-frequency data. Tick-by-tick data were more correlated than daily data, thus rendering $LZ < 1$. The correlation time was found to be of around 10 units. We also carried out an analysis of the sensibility of the LZ index to the

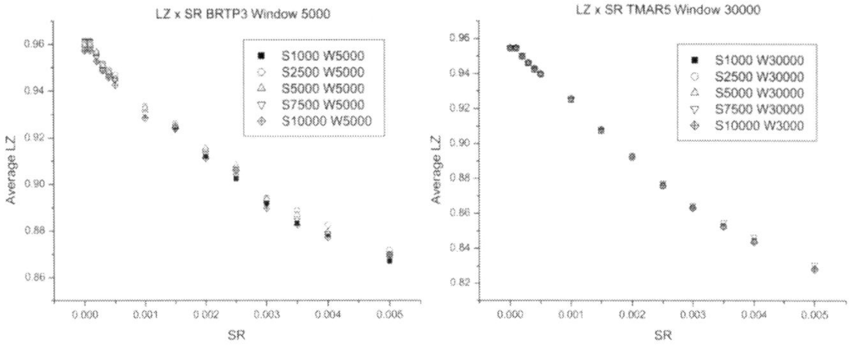

Figure 4. LZ versus SR for a fixed window size and several steps.

stability region SR, the window size, and the size of the steps s. In the tick-by-tick data, since the variations of the stock price from one trade to another are usually small, raising b has the effect of introducing repetitive patterns of symbols, thus lowering the complexity of the series. We then expect maximum complexity for $b = 0$. This is to be compared with the analysis for daily data, where the results obtained did not depend on b.

5. Acknowledgments

We acknowledge financial support from the Brazilian agencies CNPq and CAPES-Procad. We also thanks the Bovespa stock exchange staff for the data.

References

[1] M. L. Bachelier 1900 *Theorie de la Speculation* (Gauthier-Villars, Paris) .
[2] E. F. Fama 1965 *Journal of Business* **38** 34
[3] P. A. Samuelson 1965 *Industrial Management Review* **6** 41
[4] F. A. Hayek 1945 *American Economic Review* **35** 519
[5] S. J. Grossman 1976 *Journal of Finance* **31** 573
[6] R. Giglio, R. Matsushita, A. Figueiredo, I. Gleria and S. DaSilva 2008 *Europhysics Letters* **84** 48005
[7] E. F. Fama 1970 *Journal of Finance* **25** 383
[8] J. Y. Campbell, A. W. Lo and A. C. Mackinlay 1970 *The Econometrics of Financial Markets* (Princeton University Press, Princeton)
[9] R. N. Mantegna and E. Stanley 2000 *An Introduction to Econophysics: Correlations and Complexity in Finance* (Cambridge University Press, Cambridge)
[10] K. Matia, Y. Ashkenazy and H. E. Stanley 2003 *Europhysics Letters* **61** 422
[11] C. M. Nascimento, H. B. N. Junior, H. D. Jennings, M. Serva, I. Gleria and G. M. Visvanathan 2008 *Europhysics Letters* **81** 18002
[12] M. Beechey, D. Gruen and J. Vickery 2000 *Reserve Bank of Australia Research Discussion Paper* 2000-01
[13] A. Shmilovivi, Y. Alon-Brimer and S. Hauser 2003 *Computational Economics* **22** 273
[14] O. Gabjin, S. Kim and C. Eom 2007 *Physica A* **382** 209
[15] S. Pincus 1991 *Proceedings of the National Academy of Sciences of the United States of America* **88** 2297
[16] J. S. Richman and J. R. Moormam 2000 *American Journal of Physiology: Heart and Circulatory Physiology* **278** 2039
[17] M. Ferrario, M. Signorini and G. Magenes 2005 *Computers in Cardiology* **32** 989
[18] A. Lempel and J. Ziv 1976 *IEEE Transactions on Information Theory* **22** 75
[19] F. Kaspar and H. G. Schuster 1987 *Physical Review A* **36** 842
[20] B. Li, Y. B. Li and H. B. He 2005 *Genomics, Proteomics & Bioinformatics* **3** 206
[21] I. Shmulevich and D. J. Povel 2000 *Journal of New Music Research* **29** 61
[22] R. Matsushita, A. Figueiredo, I. Gleria and S. DaSilva 2007 *Physica A* **378** 427
[23] L. Xu, P. C. Ivanov, K. Hu, Z. Chen, A. Carbone and H. E. Stanley 2005 *Physical Review E* **71** 1
[24] L. Zunino, M. Zanin, B. M. Tabak, D. G. Perez and O. A. Rosso 2010 *Physica A* **389** 1891
[25] L. Zunino, B. M. Tabak, A. Figliola, D. G. Perez, M. Garavaglia and O. A. Rosso 2008 *Physica A* **387** 6558

[26] L. Zunino, M. Zanin, B. M. Tabak, D. G. Perez and O. A. Rosso 2009 *Physica A* **388** 2854
[27] K. Matia, M. Pal, H. Salunkay and H. E. Stanley 2004 *Europhysics Letters* **66** 909
[28] G. Oh, S. Kim and C. Eom 2007 *Physica A* **382** 209
[29] B. L. Lan and Y. O. Tan 2007 *Physica A* **375** 605

Analysing and controlling the tax evasion dynamics via majority-vote model*

F.W.S. Lima

Departamento de Física, Universidade Federal do Piauí, 64049-550, Teresina - PI, Brazil.

E-mail: fwslima@gmail.com, wel@ufpi.edu.br

⋆ This paper is dedicated to Dietrich Stauffer

Abstract. Within the context of agent-based Monte-Carlo simulations, we study the well-known majority-vote model (MVM) with noise applied to tax evasion on simple square lattices, Voronoi-Delaunay random lattices, Barabasi-Albert networks, and Erdös-Rényi random graphs. In the order to analyse and to control the fluctuations for tax evasion in the economics model proposed by Zaklan, MVM is applied in the neighborhod of the noise critical q_c to evolve the Zaklan model. The Zaklan model had been studied recently using the equilibrium Ising model. Here we show that the Zaklan model is robust because this can be studied using equilibrium dynamics of Ising model also through the nonequilibrium MVM and on various topologies cited above giving the same behavior regardless of dynamic or topology used here.
Keywords: Opinion dynamics, Sociophysics, Majority vote, Nonequilibrium.

1. Introduction

The Ising model [1, 2] has been used during long time as a "toy model" for diverses objectives, as to test and to improve new algorithms and methods of high precision for calculation of critical exponents in Equilibrium Statistical Mechanics using the Monte Carlo method as Metropolis [3], Swendsen-Wang [4], Wang-Landau [5] algorithms, Single histogram [6] and Broad histogram [7] methods. The Ising model also has become a excellent tool for to study other models of social application. The Ising model was already applied decades ago to explain how a school of fish aligns into one direction for swimming [8] or how workers decide whether or not to go on strike [9]. In the Latané model of Social Impact [10] the Ising model has been used to give a consensus, a fragmentation into many different opinions, or a leadership effect when a few people change the opinion of lots of others. To some extent the voter model of Liggett [11] is an Ising-type model: opinions follow the majority of the neighbourhood, similar to Schelling [12], all these cited models and others can be found out in [13]. Already Föllmer (1974) [14] applied the Ising model to economics.

Realistic models of tax evasion appear to be necessary because tax evasion remain to be a major predicament facing governments [15, 16, 17, 18]. Experimental evidence provided by Gächter [19] indeed suggests that tax payers tend to condition their decision regarding whether to pay taxes or not on the tax evasion decision of the members of their group. Frey and Torgler [20] also provide empirical evidence on the relevance of conditional cooperation for tax morale. Following the same context, recently, Zaklan et al. [21, 22] developed an economics model to study the problem of tax evasion dynamics using the Ising model through Monte-Carlo simulations with the Glauber and heatbath algorithms (that obey detailed-balance equilibrium) to study the proposed model. I have introduced for the first time the use of local majority rules in social systems. I also include a review paper on all my contributions to the field of sociophysics. Another one shows that a unifying paper on all discrete opinion models. I hope you will find these papers of interest.

Grinstein et al. [23] have argued that nonequilibrium stochastic spin systems on regular square lattices with up-down symmetry fall into the universality class of the equilibrium Ising model [23]. This conjecture was confirmed for various Archimedean lattices and in several models that do not obey detailed balance [24, 25, 26, 27, 28]. The majority-vote model (MVM) is a nonequilibrium model proposed by M.J. Oliveira in 1992 and defined by stochastic dynamics with local rules and with up-down symmetry on a regular lattice shows a second-order phase transition with critical exponents β, γ, ν which characterize the system in the vicinity of the phase transition identical [27, 29] with those of the equilibrim Ising model [1] for regular lattices. Lima et al. [30] studied MVM on Voronoi-Delaunay random lattices with periodic boundary conditions. These lattices posses natural quenched disorder in their connections. They showed that presence of quenched connectivity disorder is enough to alter the exponents β/ν and γ/ν from the pure model and therefore that is a relevant term to such non-equilibrium phase-transition with disagree with the arguments of Grinstein et al. [23]. Recently, simulations on both *undirected* and *directed* scale-free networks [31, 32, 34, 35, 36, 37, 38], random graphs [47, 48] and social networks [49, 50, 51], have attracted interest of researchers from various areas. These complex networks have been studied extensively by Lima et al. in the context of magnetism (MVM, Ising, and Potts model) [52, 53, 41, 54, 55, 56], econophysics models [22, 57] and sociophysics model [58].

In the present work, we study the behavior of the tax evasion on two-dimensional square lattices, Voronoi-Delaunay random lattices, Barabási-Albert networks, and Erdös-Rényi random graphs using the dynamics of MVM, furthermore add a policy makers's tax enforcement mechanism consisting of two components: a probability of an audit each person is subject to in every period and a length of time detected tax evaders remain honest. We aim here is to extend the study of Zaklan et al. [21, 22], which illustrates how different levels of enforcement affect the tax evasion over time, to dynamics of MVM as an alternative model of nonequilibrium to the Ising model that is capable of reproduce the same results for analysis and control of the tax evasion fluctuations. Then, we show that the Zaklan model is very robust for equilibruim and nonequilibrium models and also for various topologies used here. We show that the choice of using the Ising (equilibrium dynamics) or MVM (nonequilibrium dynamics) used to evolve the Zaklan model is irrelevant, because the results obtained in this work are about the same for both Ising and MVM. The Zaklan model also is robust, because it works on square lattice, scale free network, Voronoi-Delaunay random lattices, Barabási-Albert networks, and Erdös-Rényi random graphs. We show that for different topologies the Zaklan model reaches our objective, that is, to control the tax evasion of a country (Germany and others). This does not occur with other models as Axelrod-Ross model for evolution of ethnocentrism [58], because the results are different depending of the topology of the network. The Ising model also is not robust, because on directed Barabási-Albert network this no phase transition present as also on directed square lattice, $3D$, $4D$ and directed hypercubics lattices [38]. As described above, the MVM was proposed by M.J. Oliveira in 1992 [27] in order to improve the criterion of Grinstein et al. [23], initially described above. In the order to achieve his goal he used 4^4 (square lattice) Archimedean lattice. However, also with the aim of improve this criterion other researchers studied MVM on several other topologies that are not Archimedeans [39, 30, 53, 40, 53, 42, 43]. To their surprise all results obtained for the critical exponents are different from results obtained by M.J. Oliveira, and are also different for each topology used. Pereira et al. [40] then concluded that MVM has different universality classes which depend only on the topology used, and that all have one thing in common that is their $2\beta/\nu + \gamma/\nu = D_{eff}$ effective dimension, obtained by critical exponents for each topology used, equals $D_{eff} = 1$. Here, we show that the Zaklan model behavior is identical for all topologies or dynamics studied here. Therefore, we believe that this model is very robust, different the other models cited above. Galam [44, 45, 46] introduced for the first time local majority rules in social systems to the field of sociophysics using discrete opinion models. Here, we also hope to introduce for the first time the use of MVM to the field of sociophysics or econophysics using discrete opinions as in the Zaklan model. Therefore, we do not live in a social equilibrium, any rumor or gossip can lead to a government or market chaos and we believe that nothing is better than a nonequilibrium model (MVM) to explain events of nonequilibrium. stock market generalized to market, in order to include currency exchange.

The remainder of our paper is organised as follows. In section 2, we present the Zaklan model evolving with dynamics of MVM. In section 3 we make an analysis of tax evasion dynamics with the Zaklan model on two-dimensional square lattices using MVM for their temporal evolution under different enforcement regimes; we discuss the results obtained. In section 4 we show that MVM also is capable to control the different levels of the tax evasion analysed in section 3, as it was made by Zaklan et al. [22] using Ising models. We use the enforcement mechanism cited above on various structures: square lattice, Voronoi-Delaunay random lattice, Barabási-Albert network and Erdös-Rényi graph; we discuss the resulting tax evasion dynamics. Finally in section 5 we present our conclusions about the study of the Zaklan model using MVM.

2. Zaklan model

On a square lattice each site of the lattice is inhabited, at each time step, by an agent with "voters" or spin variables σ taking the values $+1$ representing an honest tax payer, or -1 trying to at least partially escape her tax duty. Here is assumed that initially everybody is honest. Each period individuals can rethink their behavior and have the opportunity to become the opposite type of agent they were in previous period. In each time period the system evolves by a single spin-flip dynamics with a probability w_i given by

$$w_i(\sigma) = \frac{1}{2}\left[1 - (1-2q)\sigma_i S\left(\sum_{\delta=1}^{k_i} \sigma_{i+\delta}\right)\right], \tag{1}$$

where $S(x)$ is the sign ± 1 of x if $x \neq 0$, $S(x) = 0$ if $x = 0$, and the summation runs over all k_i nearest-neighbour sites $\sigma_{i+\delta}$ of σ_i. In this model an agent assumes the value ± 1 depending on the opinion of the majority of its neighbors. The control noise parameter q plays the role of the temperature in equilibrium systems and measures the probability of aligning antiparallel to the majority of neighbors. Then various degrees of homogeneity regarding either position are possible. An extremely homogenous group is entirely made of honest people or tax evaders, depending of the sign $S(x)$ of the majority of neighbhors. If $S(x)$ of the neighbors is zero the agent σ_i will be honest or evader in the next time period with probability $1/2$. We further introduce a probability of an efficient audit (p). Therefore, if tax evasion is detected, the agent must remain honest for a number k of time steps. Here, one time step is one sweep through the entire lattice.

3. Analysing tax evasion dynamics

Here as Zaklan et al [21] we also use square lattices with lattice size $L = 1000$ ($L \times L = 10^6$ sites) and simulate tax evasion dynamics for various values of q and various punishments (k). When k is zero, no punishment is present and the model describes the baseline majority-vote model. Several degrees of punishment are introduced for different noise (q) or "Social Temperature", by setting k consecutively equal to 10 and 50 periods for all considered levels of the noise. The probability of an audit is sequentially increased, in steps, from 0 to 100 percent. above remark suggests that within one continuous simulation you increase k;actually you use only a few k values, presumably in separate runs. For a given probability of an audit the dynamics of the tax evasion (measured as portion of the entire population) is shown over 300 time steps. Here, we have observed the behavior of tax evasion dynamics on a square lattice for high $q = 0.70$ and low $q = 0.050$ noise values of the critical value for square lattices ($q_c = 0.075$) MVM . For both punishment times, $k = 50$ and 10, we also use two probability of an audit is either at a realistic level ($p = 0.05$) or at a rather high level ($p = 0.9$) values in this model. The tax evaders have a greater influence to turn honest citizens into tax evaders if they contitute a majority in given neighbourhood. If the majority evades, one is likely to also evade. On other way, if the majority of agents in the vicinity are honest, the respective individual is likely to become a honest citizen even though it has been a tax evader before. Therefore, Figs.1-3 show that on square lattices, MVM (nonequilibrium model) is capable to reproduce the same behavior and results for analysis of tax evasion levels, as Zaklan et al. [21] made using the Ising model (equilibrium model). How strong the influence from the neighbourhood is can be controlled by adjusting the noise parameter, q, next to (q_c).

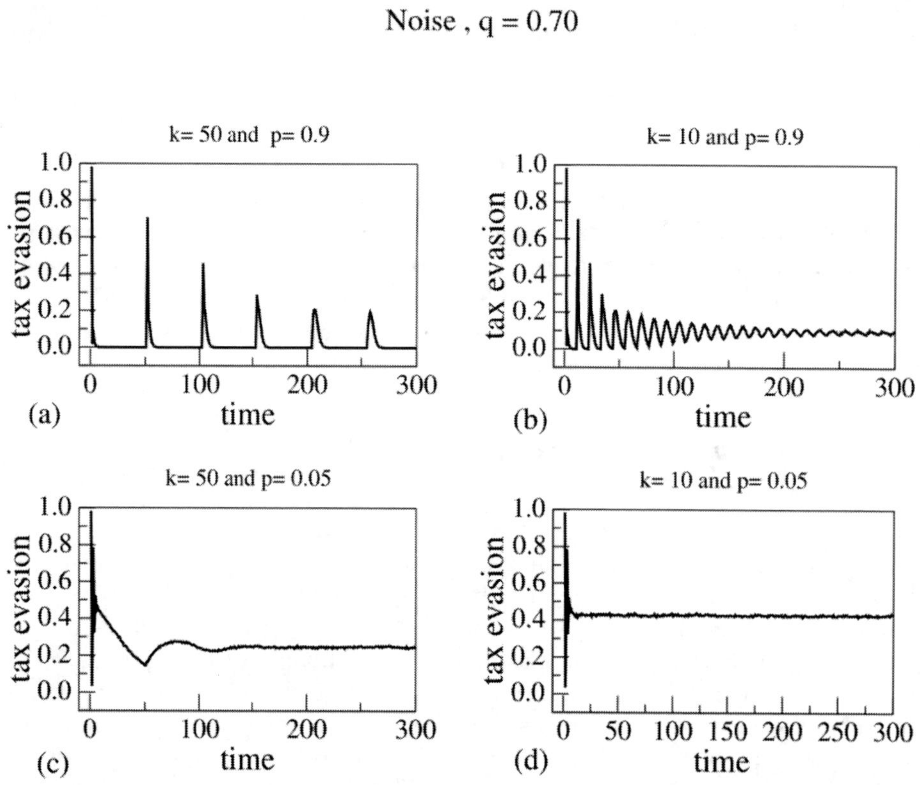

Figure 1. Tax evasion dynamics for high "social temperature" $q = 0.70$ ($q_c = 0.075$) and 300 time steps, $L = 1000$ (10^6 sites), with length of punishment at either $k = 50$ or 10 periods and two different probabilites of an audit, i.e. $p = 0.05$ and $p = 0.9$.

In figure 1 we plot tax evasion for high noise $q = 0.70$, for punishment periods $k = 50$ and 10. If the penalty is high enough, $k = 50$, tax evasion can be reduced to 0% in the short run, given that the probability of an audit p is sufficiently high. In the case of $k = 50$ and $p = 90\%$, picture

(a) of Fig. 1, within only a few time steps each individual eventually is forced to remain honest. This happens, because the spins flip more often at this noise, which is too high compared to the critical noise of this model in the square lattice ($q_c = 0.075$) [27], where q_c is the critical noise for phase transition of MVM on square lattice. In this picture (a) we observe peaks of different levels of the tax evasion, that result from the fact that 90% of the initial large number of tax evaders gets caught and after k iterations becomes free to decide whether to evade or not. Moreover consecutive peaks in non-compliance diminish less over time and it takes longer until perfect non-compliance is established, further out on the time scale at which the evolution of tax evasion is considered. Therefore, in our simulation using MVM the perfect compliance is not attained in agreement with the results observed by Zaklan et al. [21] using the Ising model where an equilibrium level of about 2% non-compliance is attained.

If punishment is set equal to $k = 10$ periods (picture (b) of Fig. 1) tax evasion reaches 0% in the first ten time steps and finally comes to rest at a level of 10%. When the probability of an audit $p = 0.05$ is much lower the tax evasion level reaches values for non-compliance of about 24% and 43% in pictures (c) and (d), respectively, of Fig. 1. These two pictures show well that punishment is a suitable enforcement mechanism when the probability of an audit is set to a realistic level.

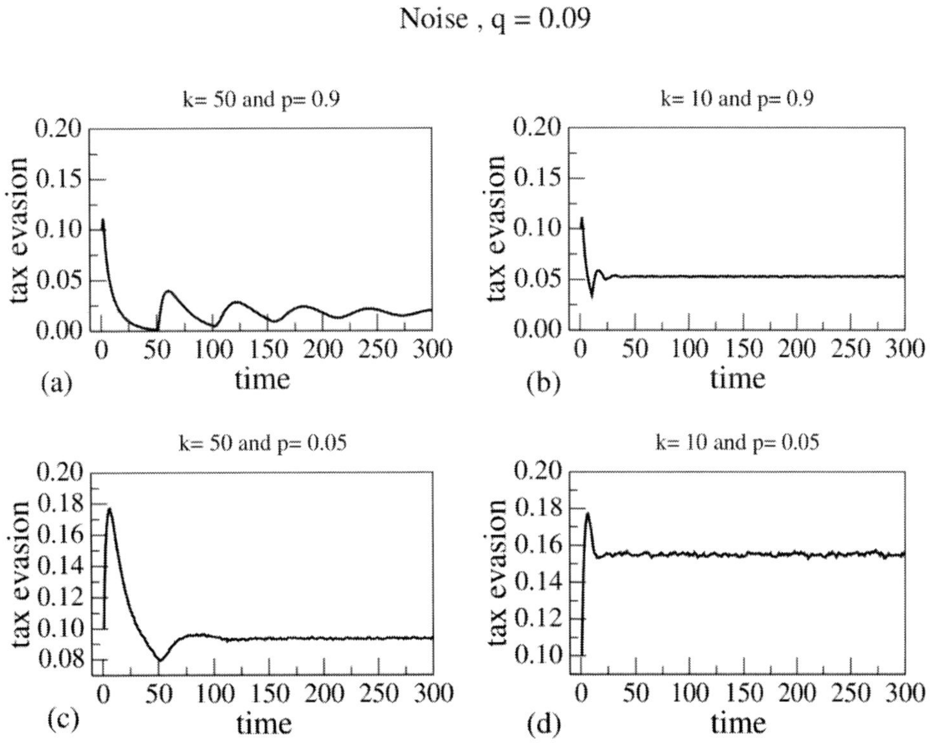

Figure 2. Tax evasion dynamics for "social temperature" $q = 0.090$ ($q_c = 0.075$). The same simulation design as in figure 1.

We also consider two other noises above q_c, $q = 0.090$, and below it, $q = 0.050$. At $q = 0.090$ the individual becomes tax evader more slowly, because the tax evasion problem is less pronounced already from the beginning compared to higher noises, because spins flip less frequently at lower noises, the same enforcement mechanisms may work less efficiently in

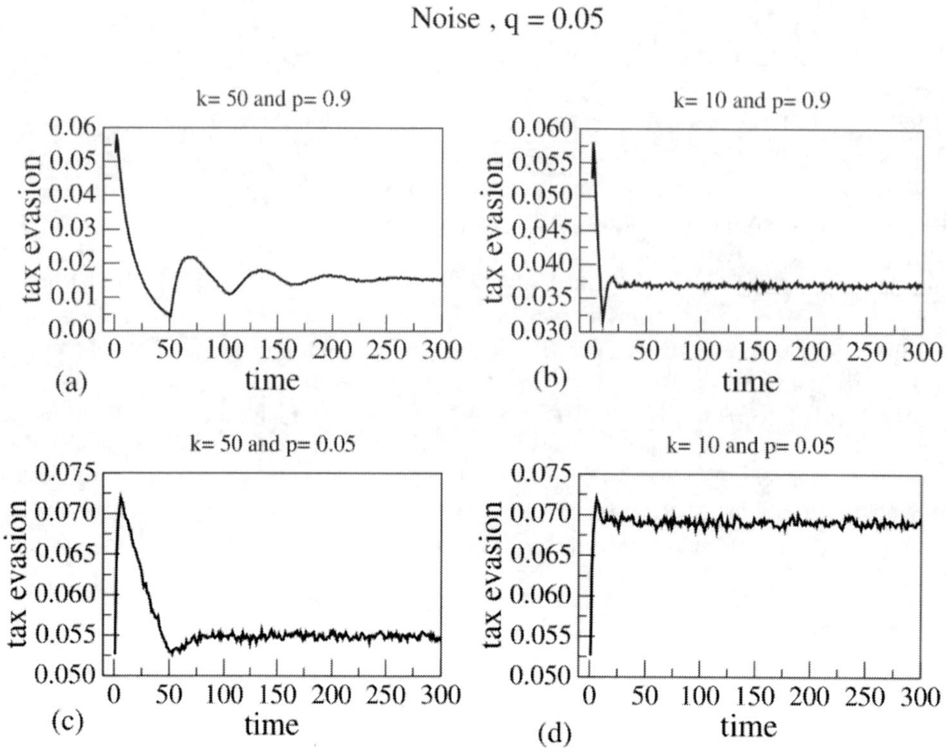

Figure 3. Tax evasion dynamics for "social temprature" $q = 0.050$. The same simulation design as in figure 1.

the short run than at higher noise. Therefore as shown in Fig. 2 at $q = 0.090$ with $k = 50$ and $p = 0.9$ one clearly sees that evasion cannot be reduced to zero percent anymore. At $q = 0.050$ the shape of tax evasion is similar to $q = 0.090$, but the level of tax evasion is smaller, because spins flip less frequently at lower noises $q = 0.050$ as shown in Fig. 3, again enforcement mechanisms cannot work efficiently in the short run as opposed to higher noise ($q = 0.70$). At such low noise individuals seldomly decide to become non-compliant because their vicinity which is mostly compliant on average exerts strong influence on them to be honest as well. Therefore the equilibrium level of tax evasion is smaller than at higher noises.

4. Controlling the tax evasion dynamics

Here, we first will present the baseline case $k = 0$, i.e., no use of enforcement, for different network structure. We use for square lattice $q = 0.95q_c$, for Voronoi-Delaunay lattice at $q = q_c = 0.117$ (critical noise for MVM on Voronoi-Delaunay lattice), for Barabási-Albert $q = 0.9q_c = 0.306$ and Erdös-Rényi $q = 0.9q_c = 0.181$. All simulation are performed over $20,000$ time steps, as shown in Fig. 4.

For very low noises the part of autonomous decisions almost completely disappears. The individuals then base their decision solely on what most of their neighbours do. A rising noise has the opposite effect. Individuals then decide more autonomously. For MVM it is known that for $q > q_c$, half of the people are honest and other half cheat, while for $q < q_c$ states dominated by cheating or by correlated changed into dominated; you always have correlations compliance prevail for most of the time. Because this behavior we set some values close to q_c, where the case that agents distribute in equal proportions onto the two alternatives is excluded. Then having set the noise parameter, q, close to ($q_c = 0.075$) on the square lattice, as suggested in section 3, we vary the degrees of punishment ($k = 1$, 10 and 50) and audit probability rate ($p = 0.5\%$, 10% and 90%). Therefore, if tax evasion is detected, the enforcement mechanism (p) and the period time of punishment k are triggered in order of to control the tax evasion level. The individual remain honests for a certain number of periods, as explained before in section 2 and 3. We also extend our study to other networks as the Voronoi-Delaunay lattices, Barabási-Albert networks and Erdös-Rényi graphs with $N = 400$ sites. As before the initial configurations is with all honest agents ($\sigma_i = +1$) at fixed "Social Temperature" (q). Here, have been performed simulations of 10,000 to 20,000 time steps.

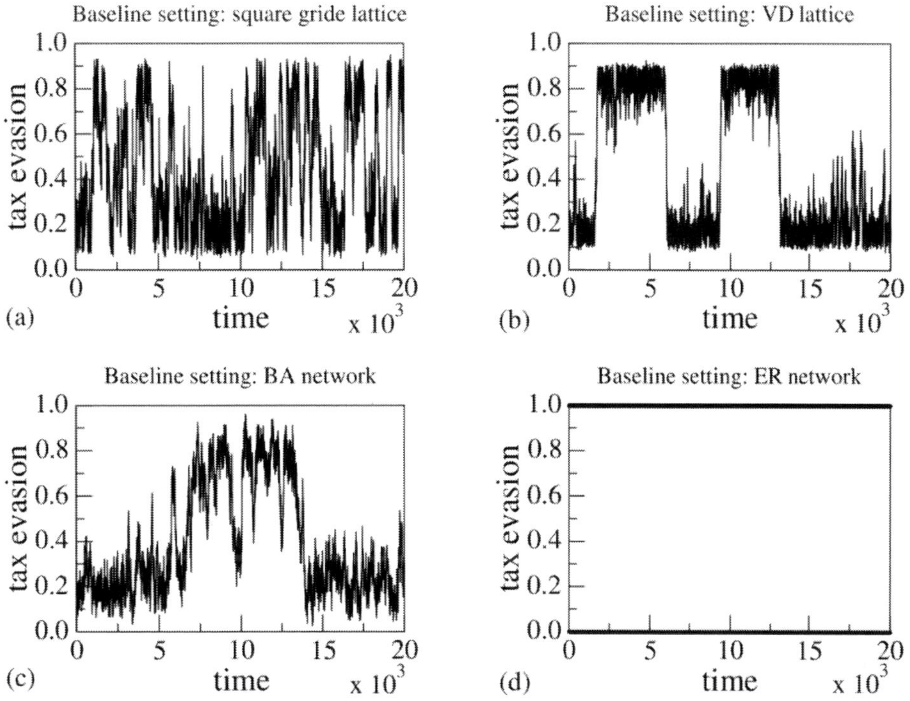

Figure 4. Baseline case for different network structure. Where we use for square lattice $q = 0.95q_c$, for Voronoi-Delaunay lattice at $q = q_c = 0.117$ (critical noise for MVM on Voronoi-Delaunay lattice), for Barabási-Albert $q = 0.9q_c = 0.306$ and Erdös-Rényi $q = 0.9q_c = 0.181$. All simulation are performed over 20,000 time steps.

In Fig. 4 we plot the baseline case $k = 0$, i.e., no use of enforcement, for the square lattice (a), Voronoi-Delaunay (b), Barabási-Albert (c), and Erdös-Rényi (d) for dynamics of the tax evasion over 20,000 time steps. Although everybody is honest initially, it is impossible to predict which level of tax compliance will be reached at some time step in the future.

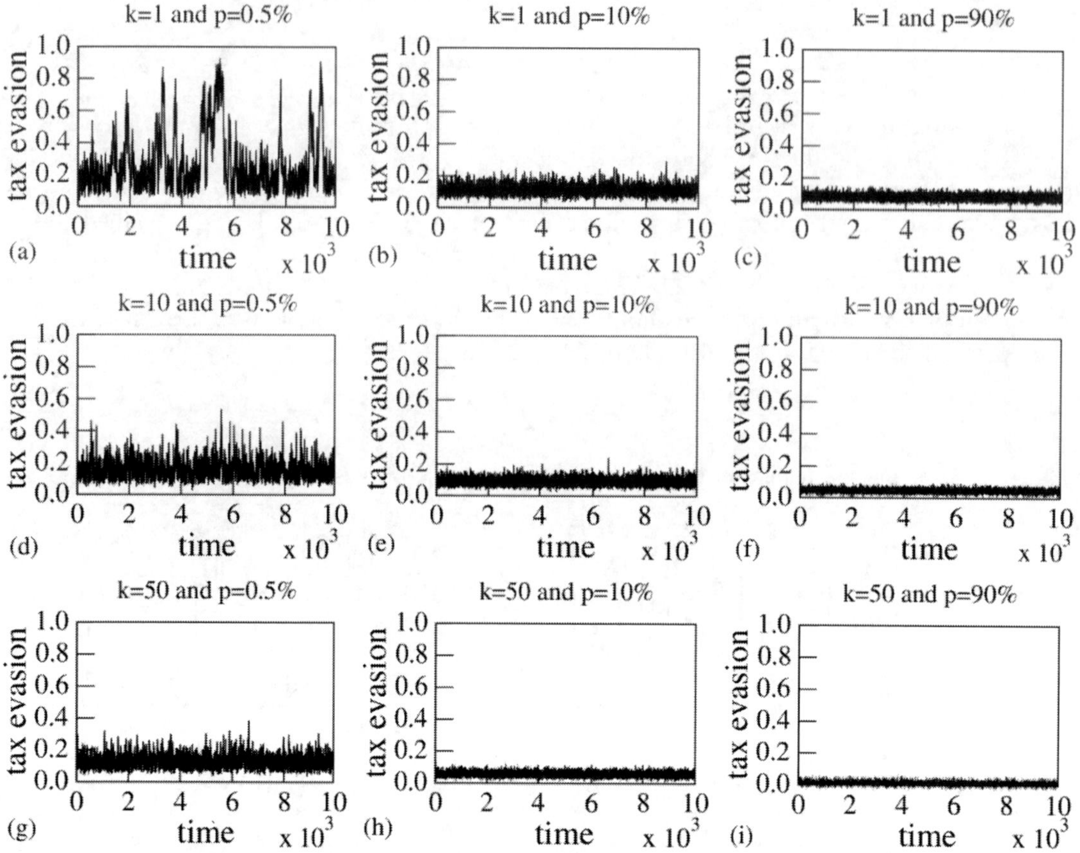

Figure 5. The square lattice model of tax evasion with various degrees p of enforcement. $q = 0.95q_c$ and 20,000 time steps.

Figure 5 illustrates different simulation settings for square lattice, for each considered combination of degree of punishment ($k = 1$, 10 and 50) and audit probability rate ($p = 0.5\%$, 10% and 90%), where the tax evasion is plotted over 10,000 time steps. Here we show that even a very small level the enforcement ($p = 0.5\%$ and $k = 1$) suffices to reduce fluctuations in tax evasion and to establish mainly compliance. Both a rise in audit probability (greater p) and higher penalty (greater k) work to flatten the time series of tax evasion and to shift the band of possible non-compliance values towards more compliance. However, the simulations show that even extreme enforcement measures ($p = 90\%$ and $k = 50$) cannot fully solve the problem of tax evasion.

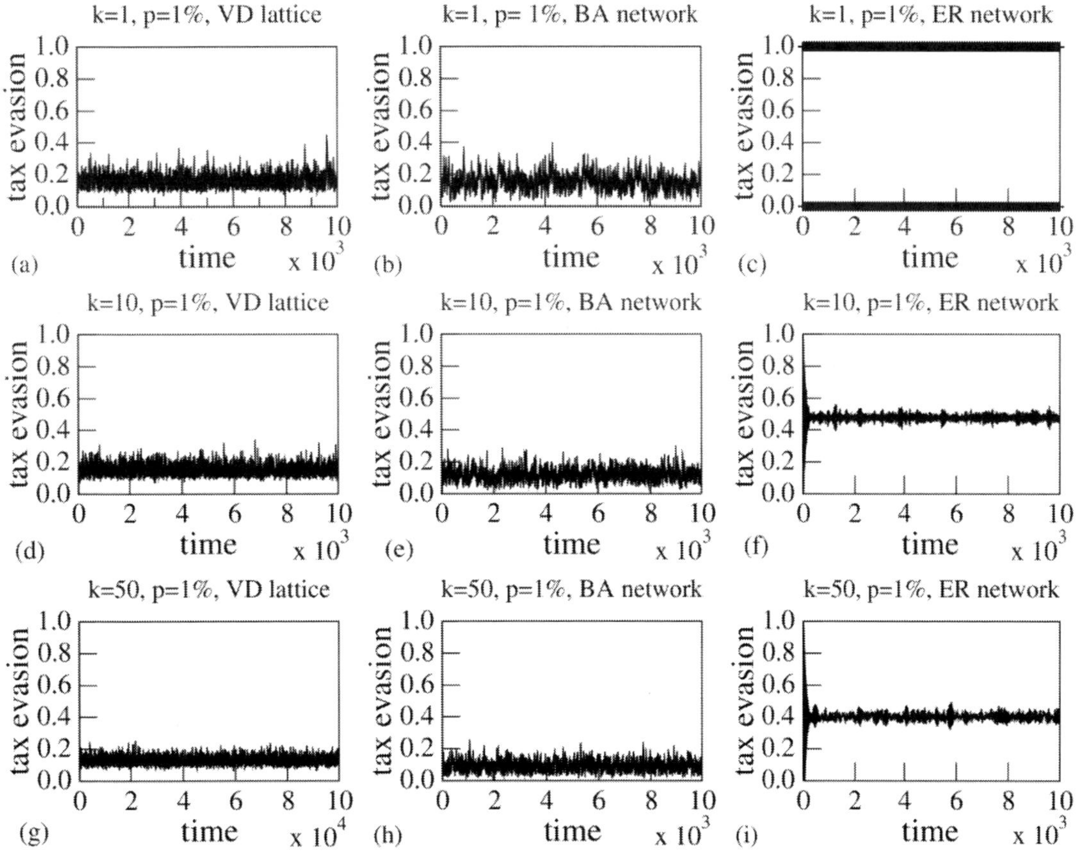

Figure 6. The first column displays tax evasion for different enforcement regimes for Voronoi-Delaunay network ((a), (d), and (g)). The next two column depict the tax evasion for Barabási-Albert network ((b), (e), and (h)) and Erdös-Rényi graph((c), (f), and (i)). Again, we use 10,000 time steps.

In Fig. 6 we display tax evasion for Voronoi-Delaunay random lattices, Barabási-Albert networks, and Erdös-Rényi random graphs for different enforcement for $k = 1$, 10, and 50 with the same audit probability $p = 1\%$. We observe for the Voronoi-Delaunay random lattice and Barabási-Albert network that the tax evasion level decreases with increasing time periods k of punishment. For the case of the Erdös-Rényi random graph the tax evasion fluctuations for $k = 1$ and $p = 1\%$, case (c), is identical to the baseline case for this random graph, only for $k = 10$ and 50 at $p = 1\%$ the tax evasion level decreases to about 48% (f) and 40% (i), showing that the control for tax evasion for small audit probability $p = 1\%$ is more difficult to reach in these Erdös-Rényi random graphs.

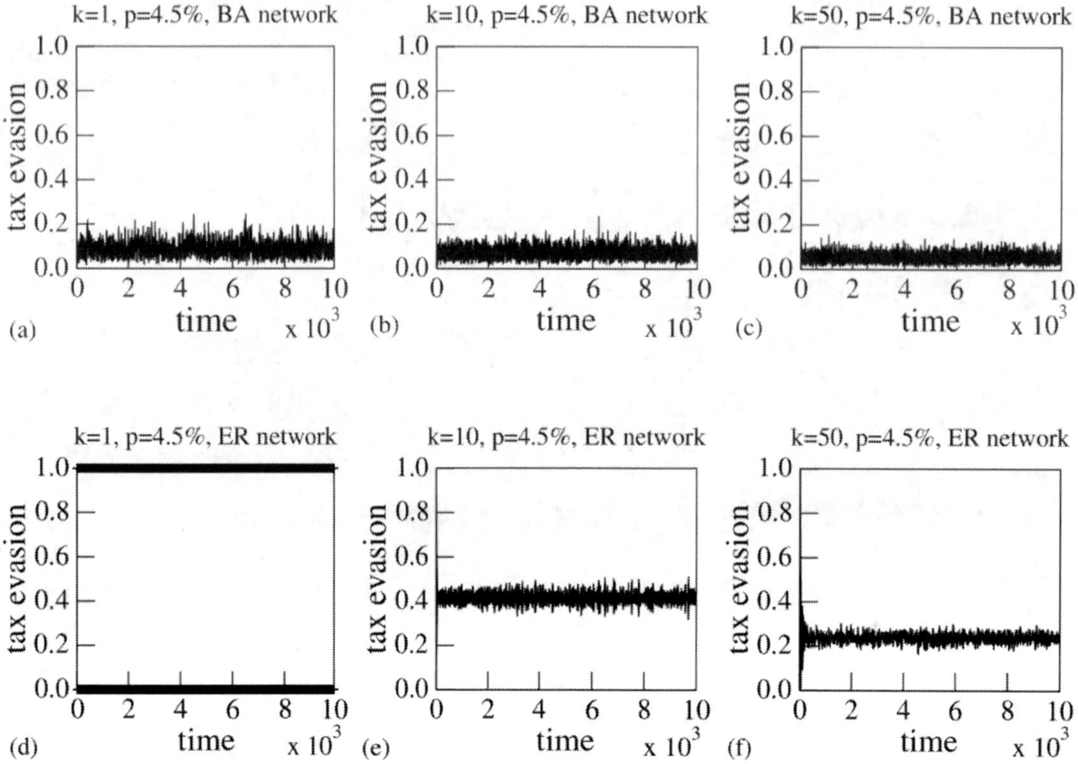

Figure 7. The first line display the resulting tax evasion for different enforcement regimes for Barabási-Albert networks ((a), (b), and (c)) for degrees of punishment $k = 1$, 10 and 50 and audit probability $p = 4.5\%$,. The second line depicts tax evasion for Erdös-Rényi graphs((d), (e), and (f)) for the same values the case as line 1.

In Fig. 7 we plot tax evasion for Barabási-Albert networks and Erdös-Rényi random graphs, again for different enforcement $k = 1$, 10, and 50, but now with audit probability $p = 4.5\%$. For Barabási-Albert networks the tax evasion level decreases with increasing audit probability p showing that an increase of the audit probability favors the control of tax evasion. Again, for $k = 1$ and $p = 4.5\%$ (d) the tax evasion for the Erdös-Rényi graphs is about the same for case baseline $k = 0$, i.e., no use of enforcement and the case (c) of Fig. 6, showing that for these random graphs the time period k of punishment is important to control tax evasion.

5. Conclusion

In summary, tax evasion can vary widely across nations, reaching extremely high values in some developing countries. Wintrobe and Gërxhani [59] explains the observed higher level of tax evasion in generally less developed countries with a lower amount of trust that people have in governmental institutions. To study this problem Zaklan et al. [21, 22] proposed a model, called here call the Zaklan model, using Monte Carlo simulations and a equilibrium dynamics (Ising model) on square lattices. Their results are good agreement with analytical and experimental results obtained by [14, 15, 16, 17, 18, 19, 20, 59]. In this work we show that the Zaklan model is very robust for analysis and control of tax evasion, because we use a nonequilibrium dynamics (MVM) to simulate the Zaklan model, that is the opposite of the study done by [21, 22] equilibrium dynamics (Ising model), and also on various topologies used here. Our results are qualitatively and quantitatively identical the results obtained by Zaklan et al. [21, 22] giving the same behavior regardless of dynamic or topology. Here, we also hope to have introduced for the first time the use of MVM to the field of sociophysics and econophysics using discrete opinion model as Zaklan model. As we do not live in a social equilibrium and any rumor or gossip can lead to a government or market chaos, we believe that nothing is better than a nonequilibrium model (MVM) to explain events of nonequilibrium. Therefore, as the Zaklan model is a sociophysics and econophysics model, we also believe that the best topology used for simulations of this model are social networks of Albert-Barabási type.

5.1. Acknowledgments

The author thanks D. Stauffer for many suggestion and fruitful discussions during the development this work and also for the reading this paper. We also acknowledge the Brazilian agency FAPEPI (Teresina-Piauí-Brasil) for its financial support. This work also was supported the system SGI Altix 1350 the computational park CENAPAD.UNICAMP-USP, SP-BRAZIL.

[1] Lenz W 1921 *Z. Phys.* **21** 613 , Ising E 1925 *Z. Phys.* **31** 235
[2] Baxter R J 1982 *Exactly solved models in statistical mechanics*, London, Academic Press
[3] Metropolis N, Rosenbluth A W, Rosenbluth M N, Teller A H and Teller E 1953 *J. Chem. Phys.* **21** 1087
[4] Swendsen R H and Wang J S 1987 *Phys. Rev. Lett.* **58** 86
[5] Wang F and Landau D P 2001 *Phys. Rev. Lett.* bf 86 2050
[6] Ferrenberg A M and Swendsen R H 1988 *Phys. Rev. Lett.* **61** 2635
[7] de Oliveira P M C, Penna T J P and Herrmann H J 1996 *Braz J. Phys.* **26** 677
[8] Callen E and Shapero D 1974 *Physics Today* **27** 23
[9] Galam S, Gefen Y and Shapir Y 1982 *J. Mathematical Sociology* **9** 1
[10] Latané B 1981 *Am. Psychologist* bf 36 343
[11] Liggett T M 1985 *Interacting Particles Systems*, Springer, New York
[12] Schelling T C 1971 *J. Mathematical Sociology* **1** 143
[13] Stauffer D, de Oliveira S. Moss, de Oliveira P M C and Sá Martins J S 2006 *Biology, Sociology, Geology by Computational Physicists* Elsevier Amsterdam ; Castellano C, Fortunato S and Loreto V 2009 *Rev. Mod. Phys.* **81** 591
[14] Föllmer H 1974 *Journal of Mathematical Economics* **1** 51
[15] Bloomquist K 2006 *Social Science Computer Review* **24** 411
[16] Andreoni J, Erard B and Feinstein J 1998 *Journal of Economic Literature* **36** 818
[17] Lederman L 2003 *Public Law Research paper No. 50*
[18] Slemrod J 2007 *Journal of Economic Perspective* **21** 25
[19] Gächter S 2006 *Discussion Papers 2006-03* CeDEx, University of Nottingham
[20] Frey B S and Togler B 2006 *IEW-Working papers* 286 Institute for Empirical Research in Economics, University of Zurich
[21] Zaklan G, Westerhoff F and Stauffer D 2008 *arXiv:0801.2980.*, Zaklan G, Westerhoff F and Stauffer D 2008 *J. Econ. Interact. Coordination*
[22] Zaklan G, Lima F W S and Westerhoff F 2008 *Physica A* **387** 5857
[23] Grinstein G, Jayaprakash C and He Y 1985 *Phys. Rev. Lett.* **55** 2527
[24] Bennett C H 1985 *Phys. Rev. Lett.* **55** 657
[25] Wang J S and Lebowitz J L 1988 *J. Stat. Phys.* **51** 893
[26] Marques M C 1990 *Phys. Lett. A* **145** 379
[27] Oliveira M J 1992 *J. Stat. Phys.* **66** 273
[28] Lima F W S 2006 *Int. J. Mod. Phys. C* **17** 1273
[29] Santos M A and Teixeira S 1995 *J. Stat. Phys.* **78** 963
[30] Lima F W S, Fulco U L and Costa Filho R N 2005 *Phys. Rev. E* **71** 036105
[31] Newman M E J, Strogatz S H and Watts D J 2001 *Phys. Rev. E* **64** 026118
[32] Sanchez A D, Lopez J M and Rodriguez M A 2002 *Phys. Rev. Lett.* **88** 048701
[33] Palla G, Farkas I J, Pollner P, Derenyi I and Vicsek T 2007 *New J. Phys.* **9** 186
[34] Barabási A -L, Albert R 1999 *Science* **286** 509
[35] Aleksiejuk A, Holyst J A and Stauffer D 2002 *Physica A* **310** 269
[36] Sumour M A and Shabat M M 2005 *Int. J. Mod. Phys. C* **16** 585
[37] Sumour M A, Shabat M M and Stauffer D 2006 *Islamic University Journal (Gaza)* **14** 209
[38] Lima F W S and Stauffer D 2006 *Physica A* **359** 423
[39] Campos P R, Oliveira V M and Moreira F G B 2003 *Phys. Rev. E* **67** 026104
[40] Pereira L F C and Moreira F G B, Phys. Rev. E 71 (2005) 016123.

Journal of Physics: Conference Series **246** (2010) 012033 doi:10.1088/1742-6596/246/1/012033

[41] Lima F W S 2007 *Commun. Comput. Phys.* **2** 358
[42] Edina M S Luz and Lima F W S 2007 *Internat J. Modern Phys. C* **18** 1251
[43] Lima F W S , Sousa A O and Sumuor M A 2008 *Physica A* **387** 3503
[44] Galam S, Chopard B, Masselot A and Droz M 1998 *Eur. Phys. J. B* **4** 529
[45] Galam S 2005 *European Physicl Society* **70** 705
[46] Galam S 2008 *Int. J. Mod. Phys. C* **19** 409
[47] Erdös P and Rényi A 1959 *Publ. Math Debrecen* **6** 290 , Erdös P and Rényi A 1960 *Publ. Math. Inst. Hung. Acad. Sci.* **5** 17 , Erdös P and Rényi A 1961 *Bull. Inst. INt. Stat.* **38** 343
[48] Balobás B 1985 *Random Graphs* 1985 Academic Press (New York)
[49] Watts D J and Strogatz S H 1998 *Nature* **393** 440
[50] Wasseman S, Faust K and Balobás B 1994 *Social Networks Analysis* Cambridge University Press. (Cambridge)
[51] Stauffer D, Hohnisch M and Pittnauer S 2006 *Physica A* **370** 734
[52] Lima F W S 2006 *Int. J. Mod. Phys. C* **17** 1267
[53] Lima F W S 2006 *Int. J. Mod. Phys. C* **17** 1257
[54] Lima F W S 2007 *Commun. Comput Phys.* **2** 522
[55] Lima F W S 2008 *Int. J. Mod. Phys. C* **19** 1063
[56] Fernandes F P and Lima F W S 2008 Int. J. Mod. Phys. C *19* 1777
[57] Lima F W S and Zaklan G 2008 *Int. J. Mod Phys. C* **19** 1797
[58] Lima F W S , Hadzibeganovic T and Stauffer D 2009 *Physica A* **388** 4999.
[59] Wintrobe R and Gërxhani K 2004 *Proceedings of the Anuual Meeting of the European Public Choice Society.*

Correlation and multifractality in climatological time series

I T Pedron
Paraná Western State University
Campus de Marechal Cândido Rondon, Paraná, 85960-000, Brazil
E-mail: itpedron@yahoo.com.br

Abstract. Climate can be described by statistical analysis of mean values of atmospheric variables over a period. It is possible to detect correlations in climatological time series and to classify its behavior. In this work the Hurst exponent, which can characterize correlation and persistence in time series, is obtained by using the Detrended Fluctuation Analysis (DFA) method. Data series of temperature, precipitation, humidity, solar radiation, wind speed, maximum squall, atmospheric pressure and randomic series are studied. Furthermore, the multifractality of such series is analized applying the Multifractal Detrended Fluctuation Analysis (MF-DFA) method. The results indicate presence of correlation (persistent character) in all climatological series and multifractality as well. A larger set of data, and longer, could provide better results indicating the universality of the exponents.

1. Introduction

The behavior of the climatological variables affect directly crucial aspects of the people's daily lives. Typically measures of these variables are a sequence of values that constitute a time series and time series analysis tools can contribute effectively to the study of climatological variables. This work is focused on the investigation of correlations and memory effects in such series and to capture underlying multifractality. It is applied the Detrended Fluctuation Analysis method (DFA), a well-established method for the detection of long-range correlations. Usually trends may mask the effect of correlations. DFA can systematically eliminate polynomial of different order. This method was proposed in [1] and has successfully been applied to many different fields, and particularly in the study of variables associated with the weather and climate. The DFA method gives the Hurst exponent (H) and estimating such exponent from the given data is an effective way to determine the nature of correlation in it. Values of H in the range (0, 0.5) characterize anti-persistence, whereas those in the range (0.5,1) characterize persistent long-range correlations. The value $H=0.5$ is associated with uncorrelated noise. However, when the series points to the existence of more than one exponent for its characterization we are dealing with multifractal behavior. Multifractal signals are far more complex than monofractal signals and require more exponents (theoretically infinite) to characterize their scaling properties. In this work multifractality in time series data of climatological variables are studied using the Multifractal Detrended Fluctuation Analysis (MF-DFA) proposed in [2]. It is a modified version of DFA to detect multifractal properties of time series and provides a systematic means to identify and quantify the multiple scaling exponents in the data.

2. Data set and method

The data series were obtained in Cascavel meteorological station (24°53' S, 53°33' W, 719 m) located in the South of Brazil. The following series of (daily) data were studied: average temperature, precipitation, relative humidity of air, solar radiation, wind speed, maximum squall, atmospheric pressure, and randomic series. All data set is within the period from Sep-1998 to Feb-2009.

The MF-DFA method is a generalization of the standard DFA, being based on identification of the scaling of the *mth*-order moments of the time series which may be non-stationary. The modified MF-DFA procedure consists of a sequence of steps and detailed information about computation can be found in [2]. The first steps are essentially identical to the conventional DFA procedure. First we construct the profile $X(i)$ as $X(i)=\sum_{k}^{i} x'_k$. The index i counts the data points in the record, $i=1,2,..., N$. For eliminating the periodic seasonal trends was obtained daily differences $x'_i = x_i - \bar{x}_i$ where $\bar{x}_i$ represents the average value for each calendar date i. The profile is then divided into $N_s = int(N/s)$ non-overlapping segments of length s. To accommodate the fact that some of data points may be left out, the procedure is repeated from other end of the data set [2]. The local trend is determined by using the least-squared fit to each segment ν and we obtain the detrended time series $X_s(i) = X(i) - p_\nu(i)$ where $p_\nu(i)$ is the polynomial fit to the νth segment. In this work it is used linear fit (DFA1). The variance of the detrended time series is calculated for each segment as

$$F^2(\nu,s) = \frac{1}{s}\left\{\sum_{i=1}^{s} X_s^2[(\nu-1)s+i]\right\}. \qquad (1)$$

Averaging over all segments the *mth* order fluctuation can be obtained and

$$F_m(s) = \left\{\frac{1}{2N_s}\sum_{\nu=1}^{2N_s}[\,F^2(\nu,s)\,]^{m/2}\right\}^{1/m}. \qquad (2)$$

For $m=2$ the standard DFA procedure is retrieved. If the original series are long-range power-law correlated the fluctuation function will vary as $F_m(s) \propto s^{h(m)}$ (for large values of s). Note that $h(m)$ is the generalized Hurst exponent and $h(2)$ is the usual exponent H previously mentioned. A multifractal description can also be obtained from considering partitions functions $Z_m(s) = \sum_{\nu=1}^{N_s}\left|X_{\nu s} - x_{(\nu-1)s}\right|^m \propto s^{\tau(m)}$ where $\tau(m)$ is the Renyi exponent [3]. A linear scaling of $\tau(m)$ with m is characteristic of a monofractal data set, whereas a nonlinear scaling is indicative of multifractal behavior. The exponent $h(m)$ is related to the Renyi exponent $\tau(m)$ by

$$\tau(m) = m\,h(m) - 1. \qquad (3)$$

It is also possible to verify the multifractality degree defining the ratio $H_</H_>$ where $H_<$ is the slope of the function $\tau(m)$ *versus* m for $m<0$ and $H_>$ is the equivalent for $m>0$. By definition such relation is equal to unity in monofractal signals and deviation from this value indicate multifractal properties.

3. Results and discussion

The H exponent for the studied data series are presented in Table 1. Temperature and precipitation values agree with values presented in some works [4, 5] despite some references claim that the scaling exponent is not universal for the temperature data [6, 7]. Contrary to results presented in the studies cited, the precipitation showed here a more pronounced multifractality compared with temperature.

Remembering that $H > 0.5$ is related with correlation and persistence the high value for it in the relative humidity shows a stronger persistence in the relative humidity fluctuations, results also obtained in [8, 9]. This feature can be associated with the water behavior and probably has a local characterization. For wind speed the H exponent is less than that presented in [10,11] but is closest to those found in [12,13]. One can ensure the universality of the correlations but its exponents are related to local circulation patterns. These conclusions can also be extended to the multifractal behaviour. The other variable present relatively weak correlation, a not surprising result. In contrast with our result, solar radiation present high degree of anti-persistence in [14]. All these variables are strongly correlated but apparently the results for the scaling exponents of their series is not related in a simple way. In Table 1 it is also possible to observe the ratio $H_</H_>$ which indicates the deviation of the monofractal behavior. Monofractal signals are characterized by the unity in this relation. In this sense, Figure 1 shows the behavior of the exponent $\tau(m)$ with different values of m. Non-linearity in these curves indicates multifractal properties as shown in Table 1. It is necessary to apply the procedure in a larger number of stations to obtain more conclusive results and to characterize a particular site through their exponents. Forthcoming works will be addressed in this task.

Table 1. Hurst exponent H for the climatological series from Cascavel – PR meteorological station and the ratio $H_</H_>$ indicating the multifractality degree (value 1 means monofractality in the data series).

variable	H	$H_</H_>$
humidity	0.722	1.099
wind speed	0.628	1.373
temperature	0,625	1.157
solar radiation	0.593	1.036
maximum squall	0.560	0.962
precipitation	0.557	1.651
atmospheric pressure	0.543	1.417
randomic series	0.495	1.018

4. Conclusion

In this work the DFA method was applied to detect long range correlation in several climatological series of one station. Results indicate that climatological series present correlation in some way. Long memory or persistent effects imply the possibility to predict conditions in the future based on historical records. Furthermore, to analize their scaling properties it was applied the MF-DFA method to the data set. Results indicate that the climatological series are multifractal, but in different degrees. These characteristics mean that the processes may be governed by more than one scaling exponent to capture the complex dynamics inherent in the data and show the non-stationary character of these series. It should be noted here that it is worthy to consider the problem taking into account a larger set of stations and longer series of data to generalize the new results and recover others in the literature. Both methods are powerful to analize the statistical behavior of the series and to obtain important information on the climate from its historical data.

Journal of Physics: Conference Series **246** (2010) 012034 doi:10.1088/1742-6596/246/1/012034

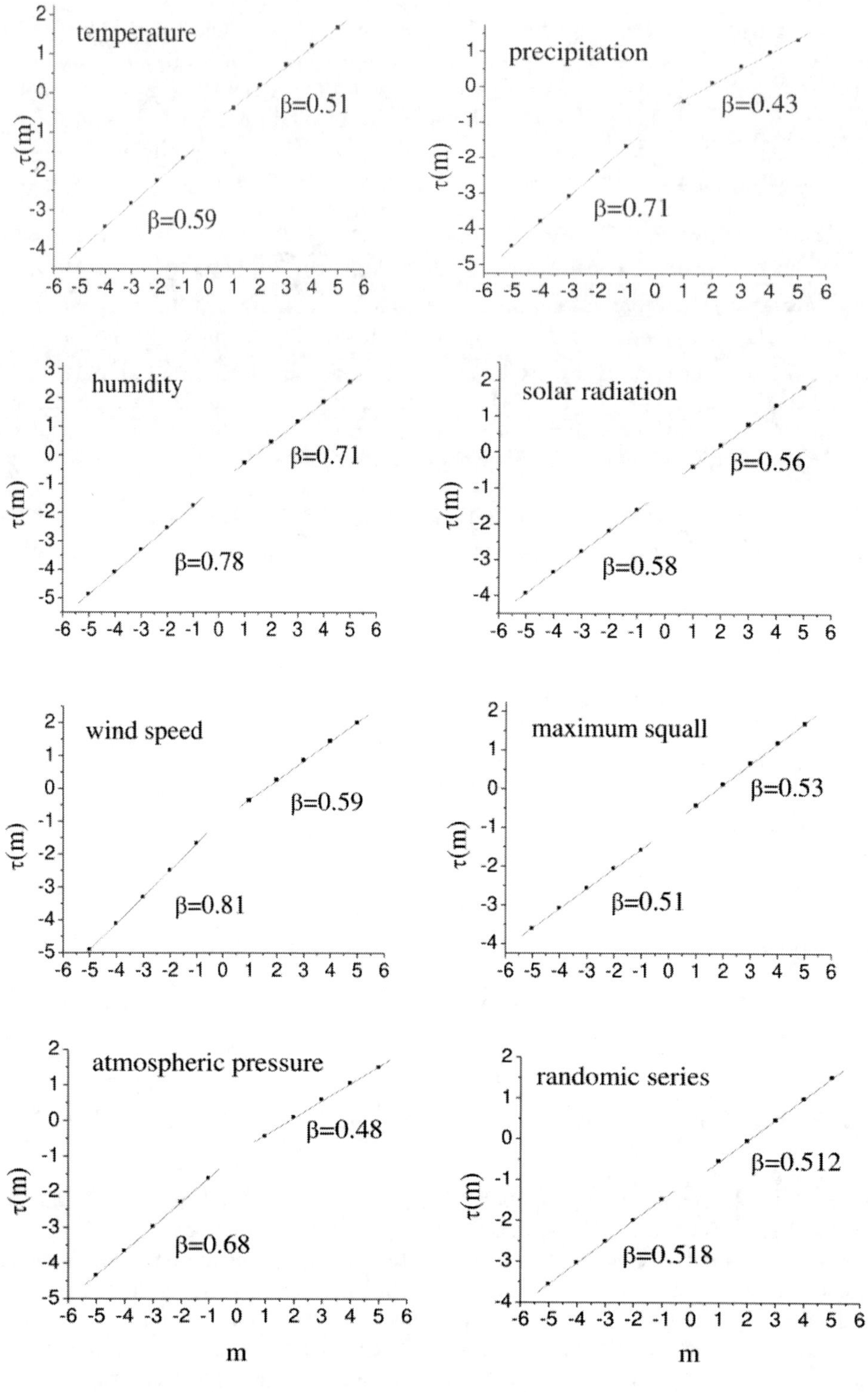

Figure 1: The multifractal exponent $\tau(m)$ versus several moments m. Different slopes indicate multifractal characterization for the data series.

5. References

[1] Peng C-K, Buldyrev S V, Havlin S, Simons M, Stanley H E, Goldberger A L 1994 *Phys. Rev.* **49** 1685
[2] Kantelhardt J W, Zschiegner S A, Koscielny-Bunde E, Bunde A, Havlin S, Stanley H E 2002 *Physica A* **316** 87
[3] Barabasi A and Vicsek T 1991 *Phys. Rev.* A 44 2730
[4] Koscielny-Bunde E, Bunde A, Havlin S, Roman H E, Goldreich Y, Schellnhuber HJ 1998 *Phys. Rev. Lett.* **81** 729
[5] Bunde A, Havlin 2002 *Physica A* **314** 15
[6] Király A, Jánosi I M 2005 *Meteorol. Atmos. Phys.* **88** 119
[7] Rybski D, Bunde A, von Storch H 2008 *J. Geophys. Res.* **113** D02106
[8] Chen X, Guangxing L, Zuntao F 2007 *Geophys. Res. Lett.* **34** L07804
[9] Guangxing L, Chen X, Zuntao F 2007 *Physica A* **383** 585
[10] Govindan RB, Kantz H 2004 *Europhys. Lett.* **68** 184
[11] Koçak K 2009 *Energy* **34** 1980
[12] Kavasseri R G, Nagarajan R 2005 *Chaos, Solitons & Fractals* **24** 165
[13] Feng T, Fu ZT, Deng X, Mao J 2009 *Physics Letters A* **373** 4134
[14] Harrouni S, Guessoum A 2009 *Chaos, Solitons & Fractals* **41** 1520

Catastrophic shifts in ecosystems: spatial early warnings and management procedures (Inspired in the physics of phase transitions)

H Fort[1], **N Mazzeo**[2], **M Scheffer**[3] **and E van Nes**[3]

[1] Complex Systems Group, Instituto de Física, Facultad de Ciencias, Universidad de la República, Iguá 4225, 11400 Montevideo, Uruguay

[2] Depto. de Ecología, Facultad de Ciencias, Universidad de la República, Iguá 4225, 11400 Montevideo, Uruguay

[3] Wageningen Agricultural University, Aquatic Ecology and Water Quality Management Group, PO Box 47, 6700 AA Wageningen, The Netherlands

E-mail: hugo@fisica.edu.uy

Abstract. Ecosystems are complex systems which can respond to gradual changes of their conditions by a sudden shift to a contrasting regime or alternative stable state (ASS). Predicting such critical points before they are reached is extremely difficult and providing early warnings is fundamental to design management protocols for ecosystems. Here we study different spatial versions of popular ecological models which are known to exhibit ASS. The spatial heterogeneity is introduced by a local parameter varying from cell to cell in a regular lattice. Transport of biomass among cells occurs by simple diffusion. We investigate whether different quantities from statistical mechanics -like the variance, the two-point correlation function and the patchiness-may serve as early warnings of catastrophic phase transitions between the ASS. In particular, we find that the patch-size distribution follows a power law when the system is close to the catastrophic transition. We also provide links between spatial and temporal indicators and analyze how the interplay between diffusion and spatial heterogeneity may affect the earliness of each of the observables. Finally, we comment on similarities and differences between these catastrophic shifts and paradigmatic thermodynamic phase transitions like the liquid-vapor change of state for a fluid like water.

1. Catastrophic Shifts in Ecosystems

Several ecosystems are known to display sudden catastrophic regime shifts when exposed to gradual change in external conditions such as climate, inputs of nutrients, toxic chemicals, etc. Recent examples illustrating such changes are the shift in Caribbean coral reefs [1, 2], shallow lakes that become overgrown by floating plants [3], savannahs that have been suddenly encroached by bushes [4, 5] and lakes that shift from clear to turbid [6, 7]. Such drastic shifts occur because the ecosystem has alternative stable states (ASS) [8, 9]. In other words, under the same external conditions the system can be in two or more stable states. Hence, when subjected to a slowly changing external factor, an ecosystem may show little change until at a critical point where a sudden transition between two different ASS occurs. The presence of ASS implies hysteresis, *i.e.* once a system has gone through a state shift, it tends to remain in the new state until the control variable is changed back to a much lower or higher level.

1.1. The Search of Spatial Early Warnings

The simplest models to describe alternative states in ecosystems are *mean-field* (MF) models. Neglecting all spatial heterogeneities, these models describe the change over time of some population that characterizes the state of the ecosystem. These models are easy to analyse and in cases without significant heterogeneity their predictions are not very different from those of spatial models. However, in other cases the presence of a spatial dimension profoundly alters population dynamics or opportunities for coexistence in the real world [10]. In fact, the oversimplification of MF models casts doubt on whether the occurrence of an alternative stable state could be an artefact. Moreover, verifications and predictive power with respect to catastrophic responses to a changing environmental conditions are still scarce for spatially extensive ecosystems. Analysis of spatially explicit models are relevant, for example, to understand phenomena like clumping and spatial segregation in plant communities [11]. It was shown that vegetation patches, which have been extensively studied for arid lands [12], can be approached as a pattern formation phenomenon [13]-[14]. It has been hypothesized that vegetation patchiness could be used as a signature of imminent catastrophic shifts between alternative states [15]. Evidences that the patch-size distribution of vegetation follows a power law were later found in arid Mediterranean ecosystems [16]. This implies that vegetation patches were present over a wide range of size scales, thus displaying scale invariance. It was also found that with increasing grazing pressure, the field data revealed deviations from power laws. Hence, the authors proposed that this power law behaviour may be a warning signal for the onset of desertification. These spatial *early warnings* complement temporal ones like the variance of time series introduced to detect lake eutrophication [17]. [Eutrophication is an increase in nutrients leading to an enhanced growth of aquatic vegetation or phytoplankton and further effects including lack of oxygen and severe reductions in water quality, fish, and other animal populations.]

1.2. Preventive Actions and Management Protocols

Once you have early warnings of an undesired catastrophic shifts, the next step is to implement preventive actions to avoid such transitions.

For instance, an example of catastrophic shift is the desertification of arid lands produced by over-grazing. Management of grazing pressure (either by stock or native species or pests) is crucial to prevent environmental degradation and to combat desertification. We will illustrate this by considering the effects of a simple remedial action in a *grazing model.*

Often the interests of different ecosystem users are in conflict. This is illustrated with lake management of aquatic vegetation. Nature conservationists prefer a dense vegetation because it promotes biodiversity and avoids potentially toxic agents like cyanobacteria. On the other hand, recreational users (boaters, surfers, swimmers) are hindered by aquatic plants. To consider the interests of all users, harvesting strategies that keep an intermediate level of submerged biomass might be a possible solution. Hence we will address the problem of lakes that shift among different primary producers dominance (phytoplankton, submerged or floating plants) through a another model, a *harvesting model* for submerged plants.

2. Spatial Ecological Models

2.1. Mean-Field Models

In this work we will consider the spatial version of some popular ecological models in terms of just a single-species biomass density which grows logistically and whose consumption, loss or removal (either by grazing, predation or harvesting) is represented by a saturation curve of Holling type II or III [18].

Both these models, in terms of two parameters, are known to have ASS.

2.1.1. First Example: Grazing Model A population model initially introduced to describe grazing systems [19], and later used in general for several ecosystems [20] and in particular for the case of the spruce budworm [21, 22], involves a Holling type III consumption term. The corresponding dynamical equation, in terms of all non-dimensional quantities, can be written as:

$$\frac{dX}{dt} = X\left(1 - \frac{X}{K}\right) - c\frac{X^2}{1+X^2}, \tag{1}$$

where X corresponds to the biomass density, K to the carrying capacity, or the number of individuals which can be supported in a given area within natural resource limits, and c to the maximum consumption rate.

The r.h.s. of (1) may be thought as the gradient of a potential V_g :

$$V_g = -\int dX \left[X\left(1 - \frac{X}{K}\right) - c\frac{X^2}{1+X^2}\right] = -\frac{X^2}{2} + \frac{X^3}{3K} + c\,(X - \arctan X)\,, \tag{2}$$

so the equilibria or attractors correspond to the roots of the first derivative of V_g.

2.1.2. Second Example: Harvesting Model for Submerged Plants The harvesting of submerged plants in lakes can be modelled by a Holling type II consumption term [23]. So, again in terms of non-dimensional quantities, X evolves according to

$$\frac{dX}{dt} = X\left(1 - \frac{X}{K}\right) - c\frac{X}{1+X}. \tag{3}$$

In complete analogy with the grazing model, the associated potential is given by: $V_h = -\frac{X^2}{2} + \frac{X^3}{3K} + c\,(X - \ln(1+X))$.

2.2. Cellular Automata

In order to take into account the spatial heterogeneity of the landscape one of the two parameters, the *local* parameter, is taken as dependent on the position. The other parameter, the *global* parameter, is taken uniform in all the system.

A two dimensional spatial version of the previous mean-field models is given by:

$$\frac{dX(x,y;t)}{dt} = X\left(1 - \frac{X(x,y;t)}{K(x,y)}\right) - c\frac{X(x,y;t)^q}{1+X(x,y;t)^q} + D\nabla^2 X(x,y;t) \tag{4}$$

where the carrying capacity $K(x,y)$ is a spatial heterogeneous parameter that varies from point to point (while the parameter c is taken as uniform), the exponent q is equal to 1 (2) for the harvesting (grazing) model and D is the diffusion coefficient measuring dispersion of X in space. We simulated these models in a $L \times L$ regular square lattice, so each cell, centred at integer coordinates (i,j), can be associated with a patch of the ecosystem. Each cell is connected to its four nearest neighbours *i.e.* the von Neumann neighbourhood is used. In other words, we get a cellular automaton whose update rule is given by:

$$\begin{aligned} X(i,j;t+1) = X(i,j,t) + X(i,j;t)\left(1 - \frac{X(i,j;t)}{K(i,j)}\right) - c\frac{X(i,j;t)^q}{1+X(i,j;t)^q} + \\ d[X(i+1,j;t) + X(i-1,j;t) + X(i,j+1;t) + X(i,j-1;t) - 4X(i,j;t)], \end{aligned} \tag{5}$$

where d is a reduced diffusion coefficient related with D and the lattice spacing a by $d = 4D/a^2$. Periodic boundary conditions (PBC) were used and L ranged from 100 to 800 (in fact, for different values of L in this range, no important differences were found). The number of time steps is typically 1000. Depending on the ecosystem, each time step could correspond to a day, or a month, or a year, etc.

The range of values for the model parameters that we use are chosen to contain the region of alternative stable states determined by the MF equations: the carrying capacity $K(i,j)$ varies randomly from cell to cell around a fixed spatial mean $\langle K\rangle$ (a typical value for this average is 7.5) in the interval $[-\delta_K, \delta_K]$ where $\delta_K = 1.0 - 2.5$. Typical values for the consumption rate c are between 1 and 3 and for for d are between 0.1 and 5.

2.3. Observables

Several quantities can be measured:

- The *spatial mean* $\langle X\rangle$:

$$\langle X\rangle(t) = \frac{1}{L^2}\sum_{i,j} X(i,j,t) \tag{6}$$

  (i and j locate each cell of the array).
- The *spatial variance* σ_X^2:

$$\sigma_X^2 = \langle X^2\rangle - \langle X\rangle^2 \tag{7}$$

- The *temporal variance* σ_t^2 computed from mean values of X at different times, $\bar{X}(t)$, (here we take $\bar{X} \equiv \langle X\rangle(t)$) which is defined as:

$$\sigma_t^2 = \frac{1}{\tau}\sum_{t'=t-\tau}^{t} \bar{X}(t')^2 - \left(\frac{1}{\tau}\sum_{t'=t-\tau}^{t} \bar{X}(t')\right)^2 \tag{8}$$

  for temporal bins of size τ (typical values for τ are from 50 to 150).
- The *patchiness* or *cluster structure*. Clusters of high (low) X are defined as connected regions of cells with $X(i,j,t) > X_m$ ($X(i,j,t) < X_m$) where X_m is a threshold value. There are different criteria to define X_m, one of which is stated in section 3.1.
- The *two-point correlation function* for pairs of cells at (i_1, j_1) and (i_2, j_2), separated a given distance R, which is given by:

$$G_2(R) = \langle X(i_1,j_1)X(i_2,j_2)\rangle - \langle X(i_1,i_1)\rangle\langle X(i_2,j_2)\rangle \tag{9}$$

3. Early Warning Signals

We will discuss first the results for the spatial grazing model (q=2 Holling type III consumption) and in a following subsection we do the same for the spatial harvesting model (q=1 Holling type II consumption).

For the grazing model the global parameter c is taken as the *control* parameter, varying with time, representing a changing grazing pressure. For simplicity, is assumed that the the local parameter $K(i,j)$ doesn't change *i.e.* $\langle K\rangle$ = constant = 7.5.

On the other hand, with the harvesting model we illustrate a different situation in which the environmental conditions are changing much more quickly than the consumption rate. For example, sudden droughts or floods can produce drastic modifications of the carrying capacity of a lake. Therefore, we take the global parameter c constant and the local carrying capacity $K(i,j)$ varying with time *i.e.* $\langle K\rangle = \langle K\rangle(t)$.

3.1. Grazing Model

3.1.1. Hysteresis Loops and Spatial Variance Let us start by studying the effect of gradually varying the stress on the system, increasing c from 1 to 3 in 1000 steps (we checked that reducing the number of steps to, for instance, 100 produces only slight quantitative changes). figure 1 shows hysteresis cycles and the spatial variance σ_X^2 for different values of d and $\langle K \rangle = 7.5$ ($X(i,j)$ takes an initial random value in the interval $[0, \langle K \rangle]$ for all cell). Notice that the peak of σ_X^2 occurs always at $c_m \simeq 2.08$ clearly announcing the coming shift. We observe two additional remarkable facts. First, the peak in σ_X^2 is always narrower for the backward transition than in the forward transition. Second, the width of the hysteresis loop and the height of the peak of the variance both decrease with d, this is expected since diffusion tends to mitigate heterogeneities

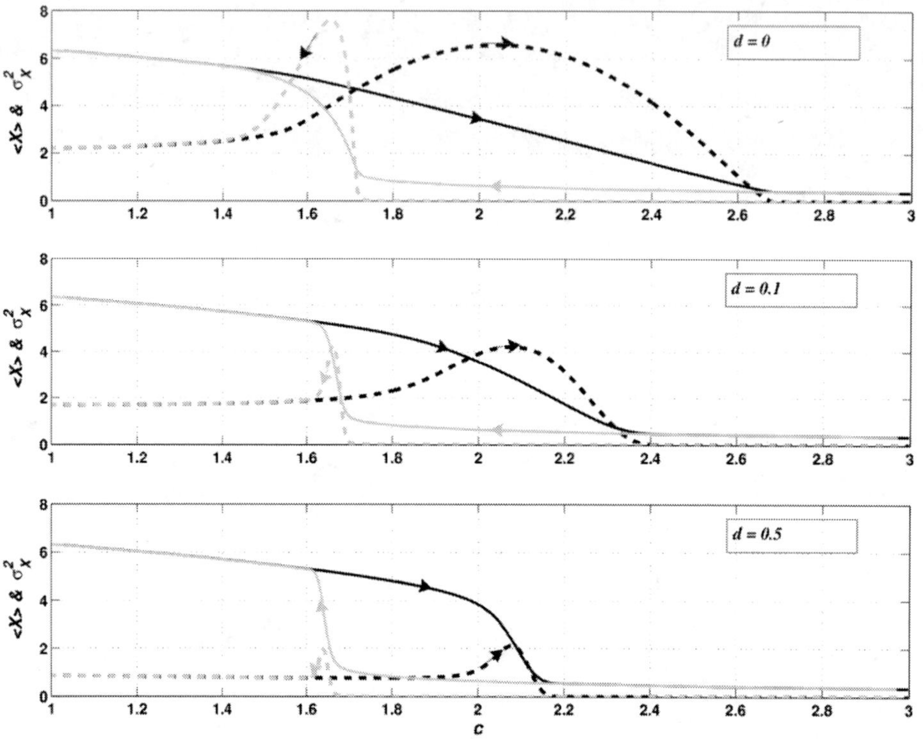

Figure 1. Grazing model: $\langle X \rangle$ (filled curves) and σ_X^2 (dashed curves) for $\langle K \rangle = 7.5$ & $\delta_K = 2.5$, computed for forward (black) and backward (gray) changes of the control parameter c. Results for d=0 (above), d=0.1 (middle) and d=0.5 (below).

3.1.2. Patchiness: Cluster structure In the upper row of figure 2 we show colour maps illustrating the state of the system just at $c = c_m = 2.08$ for d=0.1 and d=0.5. Red (blue) cells correspond to high (low) density of vegetation. As expected, the domain structure becomes more clear as d grows allowing greater segregation.

In order to study the cluster structure we must define a threshold X_m as a reference for the grid values $X(i,j)$. This X_m can be either local, as for example the value of the unstable root that separates the two attractors at each cell, or global e.g. the spatial average of these unstable roots or even the arithmetic mean of both attractors (we checked that these alternative choices do not introduce substantial changes). Thus, cells with X above (below) X_m belong to

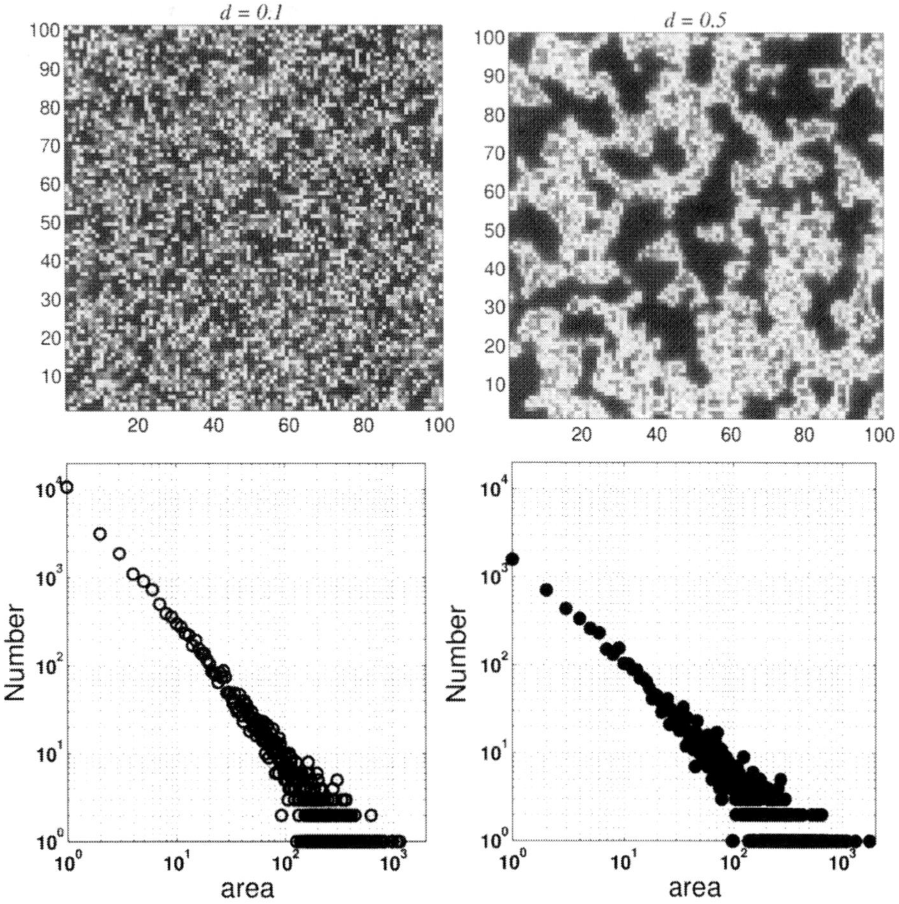

Figure 2. Grazing model: Above: A snapshot at $c = c_m = 2.08$ showing a portion of 100×100 cells, from the original 800×800 lattice, for $\langle K \rangle = 7.5$. Left (right) panel correspond to $d = 0.1$ ($d = 0.5$). Below: The corresponding patch-size distributions measured on the entire 800×800 lattice.

clusters of high (low) concentration of biomass. The value of $\langle X \rangle$ corresponding to $c_m \approx 2.08$ is $\langle X \rangle_{c_m} \approx 2.84$ and we will take it as the threshold. In the lower row of figure 2 we show the corresponding patch-size distributions for d=0.1 and d=0.5. Notice that they follow a power law $N(s){\sim}s^{-\gamma}$ over more than two decades, which disappears for smaller or greater values of c e.g. for $c = c_m \pm 0.1$ (not shown). Therefore this particular distribution may be considered as a signature of an upcoming catastrophic shift in the system. It is remarkable that for d=0.5, $\gamma \simeq 1.19$ and thus it is precisely in the range found by Kéfi *et al* for arid Mediterranean ecosystems [16]. Namely between γ=1.06 (for Greece) and $\gamma = 1.23$ (for Spain). For d=0.1 the exponent becomes a little larger, around 1.34.

3.2. Harvesting Model

Variations of the average carrying capacity $\langle K \rangle$ produce hysteresis cycles shown in figure 3 together with the spatial variance σ^2_X for $d = 0.1$ and c=2.4.

In figure 4 we show, also for $d = 0.1$ and c=2.4, $G_2(R)$ measured over the forward variation

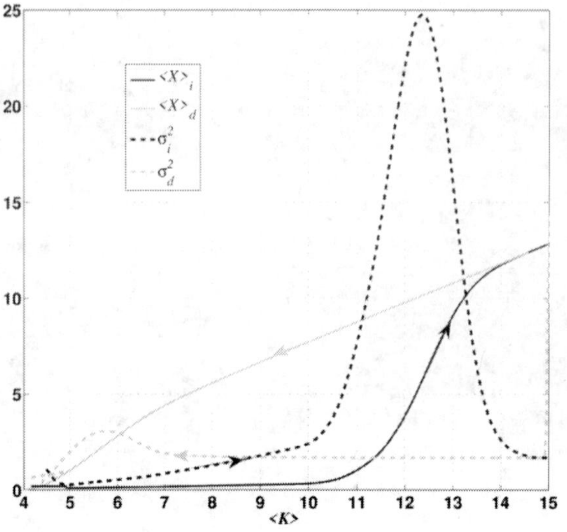

Figure 3. Harvesting model: $\langle X\rangle$ (filled curves) and σ_X^2 (dashed curves) for c =2.4 and d=0.1, computed for forward (black) and backward (gray) changes of the control parameter c.

of $\langle K\rangle$ (from 4.5 to 15) at different values of $\langle K\rangle$. Notice that the correlation increases as the average carrying capacity increases unil the correlation length, ξ, becomes maximum when $\langle K\rangle_m$ = 12.35 (*i.e.* it coincides with the maximum of σ_X^2). It is remarkable that for a wide range of values of $\langle K\rangle$ (relatively far from $\langle K\rangle_m$), the correlation length is greater than 1 lattice spacing (ξ=1.4 for $\langle K\rangle$=10.35).

4. Usefulness of the spatial early warnings

To determine the usefulness of the warning indicators presented in the previous section it is necessary first to assess their practicality and second if they really allow the implementation of corrective actions to avoid the catastrophic shift.

Concerning the question of the practicity of measurements: Calculating variances over grids consisting in a large number of sites (e.g. 400×400 or 800×800) is easy on a computer but involve a formidable task of field data acquisition. So, in order to assess the practical difficulty of estimating σ_X^2, we have performed calculations over sample grids of different sizes $L_s < L$. In figure 5 we show for the grazing model that the signal does not depend qualitatively on the number of points on the grid that are considered to estimate σ_X^2. In fact, even for a very small sample of 9 points, σ_X^2 still exhibits a noticeable peak. Of course, the quality of the signal improves with the size of the sample.

In figure 6 we show spatial variances for the harvesting model computed on 5×5 lattices for forward and backward shifts. Again the signal remains quite clear.

Additionally, since the data from real ecosystems may be very noisy, it is worth considering how the presence of noise alters results. So we assume some level of noise by adding to c a random value belonging to some interval $[-\delta_c, \delta_c]$. In figure 7 we show $\langle X\rangle$ and σ_X^2 for $\delta_c = 0.5$. The rise of σ_X^2 and the anticipation to the temporal variance are still observed.

In relation to possible remedial actions, we will study the consequences of a simple, at least theoretically, corrective measure consisting in immediately stopping the increase of the control parameter after it reaches some threshold value c^*. In figure 8 we show the effect of keeping c

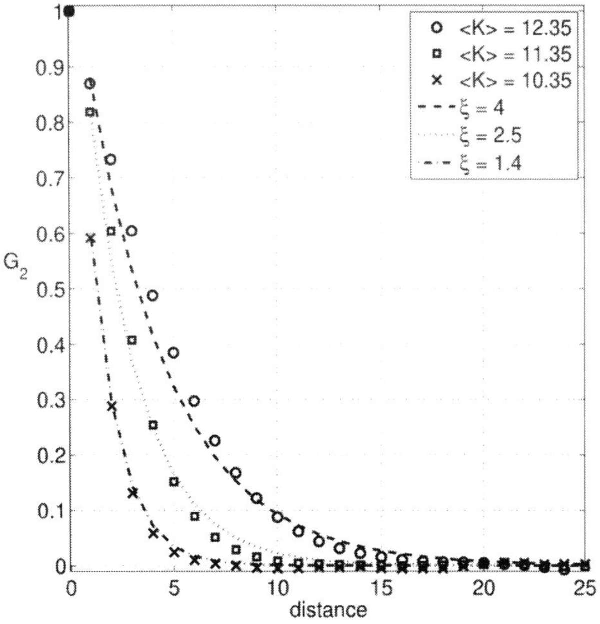

Figure 4. Harvesting model: G_2 vs. the distance R for c =2.4 and d=0.1, computed for forward changes of $\langle K \rangle$ and their corresponding exponential fits $\sim e^{-\xi/R}$.

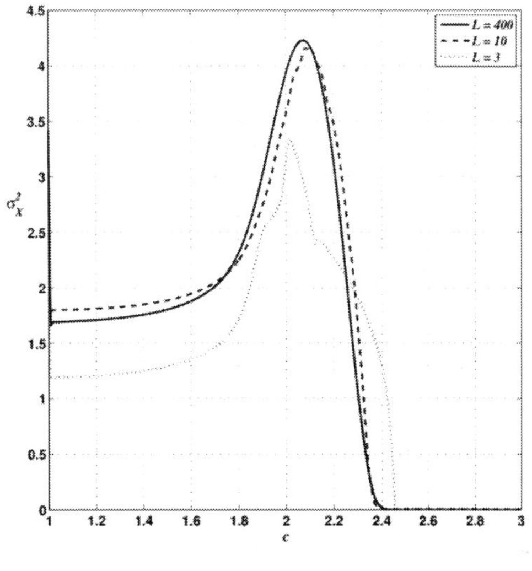

Figure 5. Harvesting model: σ_X^2 for $d = 0.1$, $\langle K \rangle = 7.5$, $\delta_K = 2.5$ calculated on lattices of size L_s=3 (dotted line), L_s=10 and L_s=400 (the entire lattice).

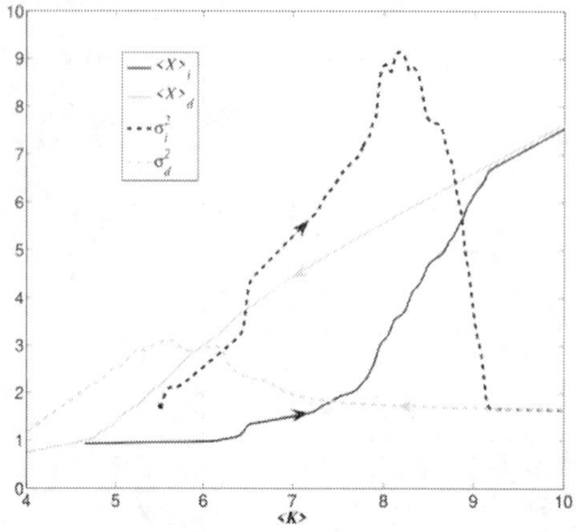

Figure 6. Harvesting model: $\langle X \rangle$ (filled curves) and σ^2_X (dashed curves) for c =2.4 and d=0.1, computed for forward (black) and backward (gray) changes of the control parameter c for a 5×5 sample lattice.

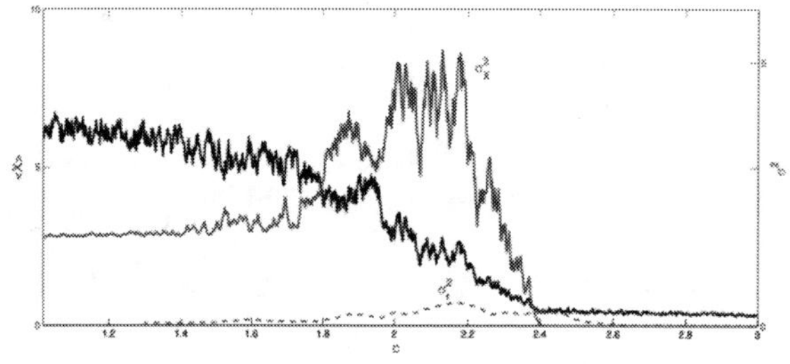

Figure 7. Grazing model: $\langle X \rangle$ (black) and σ^2_X (blue) for a random noise of amplitude $\delta c = 0.5$.

constant to c^* for different values of c^* and d. For instance, if the measure is applied at the very position of the peak of σ^2_X, $c^* = c_m \simeq 2.08$ ($\langle K \rangle = 7.5$), its usefulness depends on the value of d. For d small ($d = 0.1$) the decay in $\langle X(t) \rangle$ stabilizes soon to a value above 2 *i.e.* the system remains in a mixed state. On the other hand, for larger values of d ($d = 0.5$) the decay in $\langle X(t) \rangle$ continues and the ecosystem passes to the alternative state with low biomass, $\langle X(t) \rangle < 1$. This figure also shows that, for $d = 0.5$, the remedial measure is effective when applied before σ^2_X reaches its maximum at c_m, for $c^* = 1.9$. We checked that, for moderate or high diffusion (d $\gtrsim$ 0.5), this recipe of management works if c^* is taken between the line corresponding to $\mathcal{S}_M$ and the right fold line of $\mathcal{S}_B$ (closer to the first than to the second one). So a possible criterion to choose c^* is as the points belonging to $\mathcal{S}_M$.

We conclude this section discussing examples of alternative intervention strategies that have been carried out in practice to avoid ASS shifts in different ecosystems as well as potential measures to drive an ecological system back to the original desired ASS.

Fishing pressure is the prime reason that many fish stocks around the world have been strongly reduced [24]. Management has often failed to achieve sustainability. This failure is primarily due to continually increasing harvest rates and the intrinsic uncertainty in predicting the harvest that will cause population collapse. Models suggest that the mechanism of exploitation can by

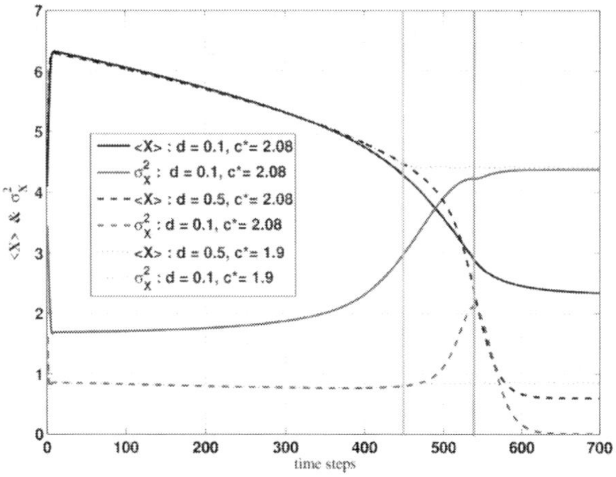

Figure 8. $\langle X \rangle$ (black) and σ_X^2 (blue) for $\langle K \rangle = 7.5$ in the case of a remedial action consisting in keeping constant the control parameter after it reaches some threshold value c^*. The red line indicates a threshold c^* coinciding with the peak of σ_X^2, $c^* = c_m \simeq 2.08$. Full (dashed) curves correspond to d=0.1 (d=0.5). The green line points a value of c^* before c_m, $c^* = 1.9$; in this case we get, for d=0.1, the dotted curves.

itself cause the overexploited state to be an ASS [25]. In that sense, cod populations represent a well-known example of how an economically important fish stock can crash in a dramatic way. However, the lack of recovery decades after the closure of fisheries on Newfoundland cod has raised the question of whether other mechanisms may keep collapsed stocks from recovering. For example, it has been hypothesized that a threshold population of adult stock is needed to control potential predators and competitors of their offspring [26]. Indeed, size structure predator prey interactions may well lead to Allee effects that make it difficult to recover once the population passes this critical threshold [27].

It is interesting that a similar situation occurs for the case of overgrazing in the dry-lands which could cause a stable overexploited state. Some degraded lands did not seem to recover upon removal of the animals, and it was found that an additional positive feedback (between soil condition and vegetation presence) can probably keep the system irreversibly trapped in the degraded state [28].

Another well studied problem is the restoration of non-vegetated turbid shallow lakes to a clear vegetated state (something which sounds good in theory but can be very difficult to achieve in practice). The pristine state of shallow lakes is in general one of clear water and a rich aquatic vegetation, which contribute to preserve this state. Nutrient loading, mainly due to the input of fertilizers from surrounding lands, used for agriculture or farming, has changed this situation in many cases. The increase of nutrients loading rate plays a similar role to the one played by the increase in the consumption rate in harvesting and fisheries, promoting a shift to an undesired ASS. As lakes shift from clear to turbid, submerged plants disappear. A later reduction of the nutrient loading may have little effect in restoring the clear water sate since, during the *eutrophication* enrichment period, a large amount of phosphorus has been absorbed by the sediments. Thus a reduction of the external loading is often compensated by 'internal loading', delaying the response of the lake water concentration to the reduction of external loading. Therefore, in many cases, nutrient reduction alone may be insufficient to restore the clear state in shallow lakes. Additional "indirect" measures such as removal of part of the fish stock and alteration of the water level have been successfully used as a way to break the feedback that keeps such lakes turbid.

The three examples mentioned demonstrate that the restoration of the original conditions demand strong efforts and often the probability of success can be quite low. An increase of the resilience capacity of ecosystems is the easiest and cheaper way to sustain their critical services like food or water.

5. Analyzing Ecosystem Shifts by using Catastrophe Theory

A general formalism for treating these catastrophic regime shifts in MF is the *Elementary Catastrophe Theory* (ECT) developed by R. Thom [29]. However, ECT works for static and homogeneous (MF) systems, where there is no time or spatial dependence of the potential. To discuss dynamics or local properties, ECT must be extended by incorporating some external assumptions. A change of the control parameter, reflecting changes of the external conditions, modifies the form of the potential. Therefore, as the shape of the potential changes, an original global minimum in which the system sits may become a metastable local minimum because other minimum assumes a lower value, or it even may disappear. In this case the system must jump from the original global minimum to the new one. ECT does not tell us when, and to which minimum, the jump occurs. A criterion to elucidate this is called, in the Catastrophe Theory parlance, a *convention*. The grazing model will serve us to illustrate the effects of the heterogeneity , diffusion and a varying control parameter (c) in modifying the predicitons of ECT.

5.1. Delay vs Maxwell Conventions

The MF 'phase diagram' for the grazing model in the K-c plane is given by the so-called *bifurcation set* $\mathcal{S}_B$ [30]. This is the set of points at which equilibria are either created or destroyed. Therefore it divides the phase space into two regions corresponding either to single stability (one attractor) or bistability (two competing attractors or ASS) (see figure 9). For the (c, K) points on this curve the second derivative of the potential (2) vanishes, so the bifurcation set is given in its parametric form by:

$$c = \frac{(x_1^2+1)^2}{2x_1^3}, \quad K = \frac{2x_1^3}{x_1^2-1} \qquad \text{for } x_1 > 1. \tag{10}$$

$\mathcal{S}_B$ has a cusp point at $c = 8/3^{3/2}$, $K_c = 3^{3/2} \simeq 5.196$. The attractor to the left (right) of $\mathcal{S}_B$ corresponds to high (low) biomass.

Before discussing conventions we need to introduce another important set of points in parameter space which control structural changes of the potential. This second set of points is called the *Maxwell set* $\mathcal{S}_M$ [30]. On $\mathcal{S}_M$ the values of the potential V at two or more stable equilibria are equal. In our case it is defined by:

$$\left(\frac{dV_g}{dX}\right)_{x_1,x_2} = 0 \tag{11}$$

$$V_g(x_1) = V_g(x_2), \tag{12}$$

and is the dashed curve in figure 9.

$\mathcal{S}_B$ and $\mathcal{S}_M$ are connected to two commonly applied criteria or conventions. Systems which remain in the equilibrium that they are in until it disappears are said to obey the *delay* convention. On the other hand, systems which always seek a global minimum of V are said to obey the *Maxwell* convention. Indeed these two conventions correspond to two extremes in a continuum of possibilities. Furthermore, real systems may obey either of these two conventions depending on the rate of change of the control parameters or on other external conditions (see next subsection).

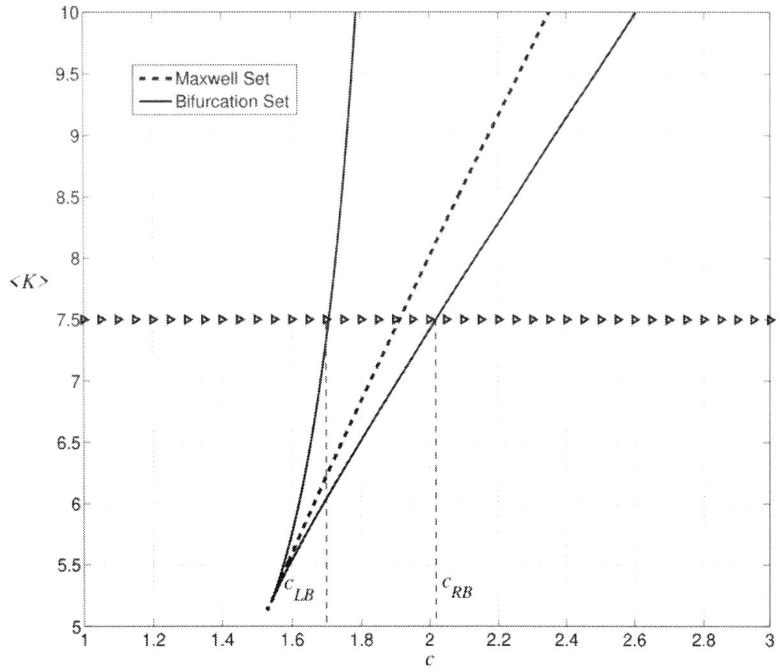

Figure 9. Bifurcation set (solid line) and Maxwell set (dashed) for the grazing model. The line of triangles denote a path at constant $\langle K \rangle = 7.5$, it intersects $\mathcal{S}_B$ at two points c_{LB}=1.7079 and c_{RB}=2.0195.

5.2. The effects of Heterogeneity, Diffusion and Varying Consumption

In our simulations we start from c=1, which is at the left of $\mathcal{S}_B$ *i.e.* with the ecosystem in the upper biomass attractor. Then, increasing the consumption rate c along an horizontal line of constant $\langle K \rangle$ (above K_c) in 1000 time steps $\mathcal{S}_B$ is traversed until c=3, at the right of $\mathcal{S}_B$, where the ecosystem is in the lower biomass attractor.

We found that within the bistability region, under the above conditions, the ecosystem remains always in the upper attractor or very close to it. The transition to the lower attractor occurs for values of c at the right of $\mathcal{S}_B$ and not just over it right component (where the upper attractor disappears). This 'delay' occurs because, since c is increasing, the state of the ecosystem at time t reflects smaller (earlier) values of the control parameter. In other words, in the case of constant c, within the region delimited by $\mathcal{S}_B$ the ecosystem should be in a mixed state and immediately at the right component of $\mathcal{S}_B$ all the ecosystem should be in the lower attractor. In fact we found this is true only when δ_K and d are equal to zero: the spatial heterogeneity together with diffusion modify this.

In order to dissect the effects of the three combined factors -spatial heterogeneity, diffusion and varying consumption rate- we analyze them by separate. That is, we proceed by steps: in the first step we consider the simplest situation, *i.e.* an homogeneous non diffusive and static system, and we add at the other two steps, one by one, the remaining factors. So, starting from a random configuration of $X(i,j)$, depending on the situations listed below, the outcomes within the region delimited by $\mathcal{S}_B$ are as follows.

- *Homogeneous non diffusive and static system i.e.* δ_K=0 (*i.e.* K=7.5 for all cells), d=0 and $dc/dt = 0$: *Bistable state.* In this simplest situation the ecosystems displays a two-colour state. That is, depending if the initial value of X for each cell is above or below the unstable root separating both attractors for the given value of c, this cell adopts respectively the upper or lower attractors. For instance, over the right component of $\mathcal{S}_B$, $c = c_{RB}$, X takes only two values for all the cells: $X_u = 3.455$ (upper attractor) or $X_l = 0.637$ (lower attractor), as shown in figure 10-A. Just at the right of this curve, $c = c^+_{RB}$, the whole ecosystem collapses to its lower attractor.

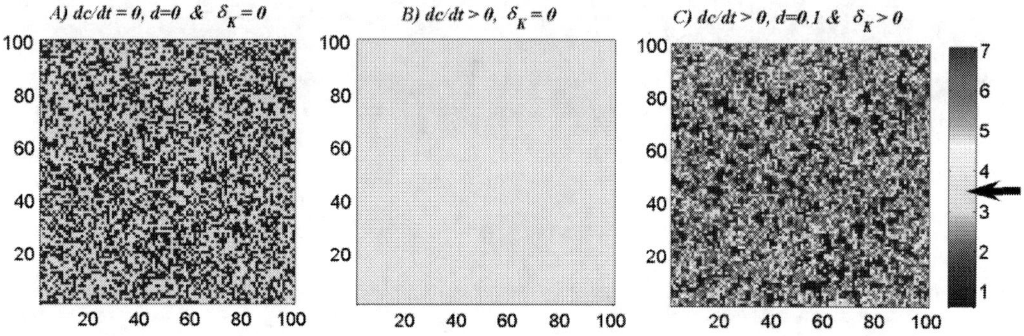

Figure 10. Snapshots at $c = c_{RB}$ showing a portion of 100×100 cells, from the original 800×800 lattice, for the three different situations listed in the text. The black arrow points to the value of the upper attractor at $c = c_{RB}$: X_u=3.455.

- *Dynamic but homogeneous system i.e.* δ_K=0, either with d=0 or $d > 0$, and $dc/dt > 0$: *Homogeneous state (upper attractor).* Now, the point representing the system in figure 9 starts at the single upper attractor region $c < c_{LB}$. As c increases, the ecosystem quickly adopts an homogeneous state coinciding with the corresponding upper attractor. When it enters the region delimited by $\mathcal{S}_B$ it "keeps memory" and remains in the upper attractor state. In figure 10-B it is shown this for $c = c_{RB}$ for which the whole ecosystem adopted a state of $X(i,j) \simeq 3.816\ \forall(i,j)$ which is slightly higher than X_u. This small difference is explained by the reaction delay mentioned above (*i.e.* if c is freezed to c_{RB} all the cells converge to X_u). The homogeneity of the state persists even after the ecosystem traversed the entire region delimited by $\mathcal{S}_B$ and the transition to the lower attractor occurs for $c > c_{RB}$ in a quite gradual and soft way.

- *Heterogeneous diffusive dynamic system i.e.* $\delta_K > 0$, $d > 0$, $dc/dt > 0$: *Quasi upper attractor.* In this ordinary situation, the underlying heterogeneity of the environment leads to an heterogeneous state. However, at least on average, this state is much more closer to the upper attractor than to the lower one: for $c = c_{RB}$ $\langle X(i,j)\rangle \simeq 3.41$ which is slightly lower than X_u, the value of the corresponding upper attractor (see figure 10-C).

5.3. Comparison between Shifts in Ecosystems and the Phenomenology of Thermodynamic Phase Transitions

Some of the characteristic fingerprints or 'wave flags' for catastrophes are: *modality, sudden jumps, hysteresis* and a *large* or *anomalous variance* [30]. These are precisely the signals we found for the considered spatial heterogeneous ecological models representing a species or set of species subject to exploitation (either grazing or harvesting).

It is interesting to analyze similarities and differences with the liquid-vapour transition in a fluid, like water. Therefore, the biomass density X would correspond to the fluid density, the liquid to the high biomass density attractor and the vapour to the low biomass density attractor. Let us compare the above catastrophe flags for the fluid vs. the ecosystem:

- Modality: The fluid is bimodal in the neighbourhood of the liquid-gas coexistence curve, having well defined liquid and gas states (bubbles in liquid or droplets in vapour). So this is similar in both systems.

- Sudden Jumps: In the case of the fluid it is certainly true that sudden jumps occur, since there is an abrupt increase in volume when a liquid transforms into vapour. However, this large change in volume occurs when a slight change in the temperature and pressure moves the fluid from one side of the coexistence curve to the other. Hence, the liquid-vapour coexistence curve can be identified with $\mathcal{S}_M$ and the water changes of state obey in general the Maxwell convention.
  On the other hand, the shift in the considered model always obeys the delay convention: the ecosystem remains in the higher attractor (higher values of X) until the bifurcation set is completely traversed. However, when perturbations are big enough they can allow the switching between equilibria on different stability branches, the systems may follow the Maxwell convention. We have checked that the effect of a sudden perturbation of the environment, represented for example by a sharp decrease of the average carrying capacity $\langle K\rangle$ followed by a slow recovery, produces a change of convention: from delay to Maxwell [31].

- Hysteresis: In everyday situations one does not observe hysteresis in the liquid-gas phase transition of water: the liquid usually boils at the same temperature at which the vapour condenses. In other words, water changes of state obey in general the Maxwell convention. Nevertheless, a careful experimentalist can obtain an hysteresis cycle by first raising the temperature and superheating the liquid, and after evaporation, cooling the gas below the condensation point. Indeed the coexistence curve is surrounded by two *spinodal lines* which determine the limits to superheating and supersaturation. These spinodal or fold lines can then be identified with $\mathcal{S}_B$.

- Anomalous Variance: When a fluid condenses (boils) from its gas (liquid) to its liquid (gas) state, small droplets (bubbles) are formed. As a consequence, the variance of the volume may become large, similarly to what happens for the ecosystem.

6. Acknowledgments

H. F. and N. M. acknowledge support from Project PDT 63/13 and PEDECIBA-Uruguay and from ANII.

References

[1] L J McCook. Macroalgae, nutrients and phase shifts on coral reefs: scientific issues and management consequences for the great barrier reef. *Coral Reef*, 18:357, 1999.

[2] M Nystrom et al. Coral reef disturbance and resilience in a human-dominated environment. *Trends Ecol. Evol.*, 15:413, 2000.

[3] M Scheffer et al. Floating plant dominance as a stable state. *Proc. Natl. Acad. Sci. USA*, 100:4040, 2003.

[4] D Ludwig et al. Sustainability, stability, and resilience. *Conserv. Ecol.*, 1:http://www.consecol.org/vol1/iss1/art7, 1997.

[5] B H Walker. Rangeland ecology: understanding and managing change. *Ambio*, 22:2, 1993.

[6] M Scheffer. *Ecology of Shallow Lakes*. Chapman & Hall, 1998.

[7] S R Carpenter et al. Management of eutrophication for lakes subject to potentially irreversible change. *Ecol. Appl.*, 9:751, 1999.

[8] M Scheffer et al. Catastrophic shifts in ecosystems. *Nature*, 413:591, 2001.

[9] S Carpenter. Alternate states of ecosystems: evidence and some implications. In Huntly N. Press, M. C. and S. Levin, editors, *Ecology: achievement and challenge*, pages 357–381. Blackwell, London, UK, 2001.

[10] E K Steinberg and P Kareiva. Challenges and opportunities for empirical evaluation of spatial theory. In D. Tilman and P. Kareiva, editors, *Ecology: achievement and challenge*, pages 318–332. Princeton University, 1997.

[11] S A Levin and S W Pacala. Theories of simplification and scaling of spatially distributed processes. In D. Tilman and P. Kareiva, editors, *Ecology: achievement and challenge*, pages 271–296. Princeton University, 1997.

[12] M R Aguiar and O E Sala. Patch structure, dynamics and implications for the functioning of arid ecosystems. *Tree*, 14:273, 1999.

[13] C A Klausmeier. Regular and irregular patterns in semiarid vegetation. *Science*, 284:1826, 1999.

[14] J. von andd others Hardenberg. Diversity of vegetation patterns and desertification. *Phys. Rev. Lett.*, 87:1981011, 2001.

[15] M Rietkerk et al. Self-organized patchiness and catastrophic shifts in ecosystems. *Science*, 305:1926, 2004.

[16] S Kéfi et al. Spatial vegetation patterns and imminent desertification in mediterranean arid ecosystems. *Nature*, 449:213, 2007.

[17] S R Carpenter and W A Brock. Rising variance: a leading indicator of ecological transition. *Ecology Letters*, 9:311, 2006.

[18] C S Holling. The components of predation as revealed by a study of small mammal predation of the european pine sawfly. *Canadian Entomol.*, 91:293, 1959.

[19] I Noy-Meir. Stability of grazing systems: an application of predator-prey graphs. *Jour. of Ecology*, 63:459, 1975.

[20] R M May. Thresholds and breakpoints in ecosystems with a multiplicity of stable states. *Nature*, 269:471, 1977.

[21] D Ludwig, D D Jones, and C S Holling. Qualitative analysis of insect outbreak systems: the spruce budworm and forest. *Jour. Anim. Ecology*, 47:315, 1978.

[22] J D Murray. *Mathematical Biology*. Springer-Verlag, 1993.

[23] E van Nes, van den Bergb M S Scheffer, M, and H Coops. Aquatic macrophytes: restore, eradicate or is there a compromise? *Aquatic Botany*, 72:387, 2002.

[24] L W Botsford et al. The management of fisheries and marine ecosystems. *Science*, 277:509, 1997.

[25] J H Steele and Henderson E W. Modeling long-term fluctuations in fish stocks. *Science*, 224:985, 1984.

[26] C J Walters and J F Kitchell. Cultivation/depensation effects on juvenile survival and recruitment. *Aquat. Sci.*, 58:39, 2002.

[27] M. Scheffer. *Critical Transitions in Nature and Society*. Princeton University Press, 2009.

[28] M. Rietkerk and J Van de Koppel. Alternate stable states and threshold effects in semi-arid grazing systems. *Oikos*, 79:69, 1997.

[29] R Thom. *Structural Stability and Morphogenesis*. Reading: Benjamin, 1975.

[30] R Gilmore. *Catastrophe Theory for Scientists and Engineers*. Dover, 1981.

[31] A Fernandez and H Fort. Early warnings of catastrophic shifts in ecosystems: Comparison between spatial and temporal indicators. *J. Stat. Mech.*, P09014:1–17, 2009.

Imbalance of community structures in epilepsy

Guillermo J. Ortega[1], Iván Herrera Peco[1], Rafael García de Sola[1] and Jesús Pastor[2]

[1]Neurosurgery and [2]Clinical Neurophysiology Services, Hospital Universitario de la Princesa, Diego de León 62, 28006, Madrid, Spain.

E-mail: gjortega.hlpr@salud.madrid.org

Abstract. Epilepsy is commonly associated with synchronous activity in the form of spikes and also in developed seizures. Desynchronised activity seems to play an important role also in the seizure process, favouring the initiation of seizures. The aim of the present work is to explore synchronization activity in the inner areas in the temporal lobe of epileptic patients by a novel approach. Two temporal lobe epilepsy (TLE) patients' records have been analyzed through a cluster analysis. Electrical activity in the inner part of the temporal has been recorded by using Foramen Ovale Electrodes (FOE), a semi-invasive technique frequently used in drug resistant epileptic patients. Instead of tracking synchronized activity, we give here special attention to desynchronized activity, mainly those areas which are not included in synchronization clusters. Our results show that electrical activity in the epileptic side behaves in a less cohesive fashion than the contra-lateral side. There exists a clear tendency in the epileptic side to be organized as isolated clusters of electrical activity as compared with the contra-lateral side, which is organized in the form of large clusters of synchronous activity. In particular, we shall give special attention to the cluster desynchronization during the seizures. As we shall show, our results can help in understand several characteristics of the seizures dynamics.

1. Introduction

Traditional methods used to assess correlated and synchronized activity in a system of interacting members has come to be revisited and generalized in recent years in order to cope with nonlinear and chaotic phenomena. In this sense new concepts, as for example weak, strong and generalized synchronization [1][2] are now widely used to describe the collective behavior of interacting systems and in particular, systems composed of neurons at several scales. More recently, the complex network theory [3][4] has provided new insights to approach the study of several kind of networks, in particular, brain networks, allowing to extract the main and distinctive properties of large assemblies of interacting neurons. In particular, detection of community structures [5][6] allows classify and organize different degrees of synchronized activity in a very comprehensible way.

Neural dynamics nowadays is an active field covered from several different points of views since the pure theoretical basis to the most clinical and practical perspective. Epilepsy, which traditionally have been associated to a synchronization disorder [7], or even more, with a hyper-synchronization disease [8][9], it is currently revisited and reinterpreted under the light of the new desynchronization point of view [10][11][12].

In the classical view [13][14] the epileptic activity in the form of epileptiform spikes displayed in the ElectroEncephaloGram (EEG) is caused by the hyper-synchronous activity in thousand of neurons typically located in the hyppocampal and parahippocampal areas. This spiking activity remains confined to small regions, but sometimes, at the beginning of a seizure, the neurons' spiking activity overcome the surrounding inhibitions and then this synchronous activity begins to spread, activating synchronously large populations of neurons at distant areas [14], corresponding with the clinical manifestations of the epileptic seizures, commonly characterized as ‘hypersynchronous states”. Several works [9][10][11][12][15] however has apparently challenged this traditional view demonstrating that desynchronous activity is essential for the initiation and maintenance of epileptic seizures. Far from contradicting the classical knowledge, recent desynchronization results shed new light in the fragmented understanding we have today of the complex process which spark and sustain a developed and extended hypersynchronization activity during the clinical seizure. It is now fairly evident that in temporal lobe epilepsy (TLE) the whole process of seizure generation is associated with dramatic changes in neuronal synchronization at several, both temporal and spatial, scales mainly in the mesial region, amygdala, hippocampus and inner cortex of the brain, the major structures affected in this pathology.

In this paper we will introduce a new methodology to study synchronization activity in two TLE patients. Firstly, we will classify the community structures of the electrical activity in the inner side of the temporal lobe, recorded by FOE. We will identify desynchronized areas from the rest of the synchronization clusters by constructing a hierarchical tree of interactions, which we will call desynchronized electrodes (DE). We are able to identify the epileptic side, especially during the seizure by tracking the number of these DE along the entire FOE record. We will explain the proposed method in two TLE patients.

A

B

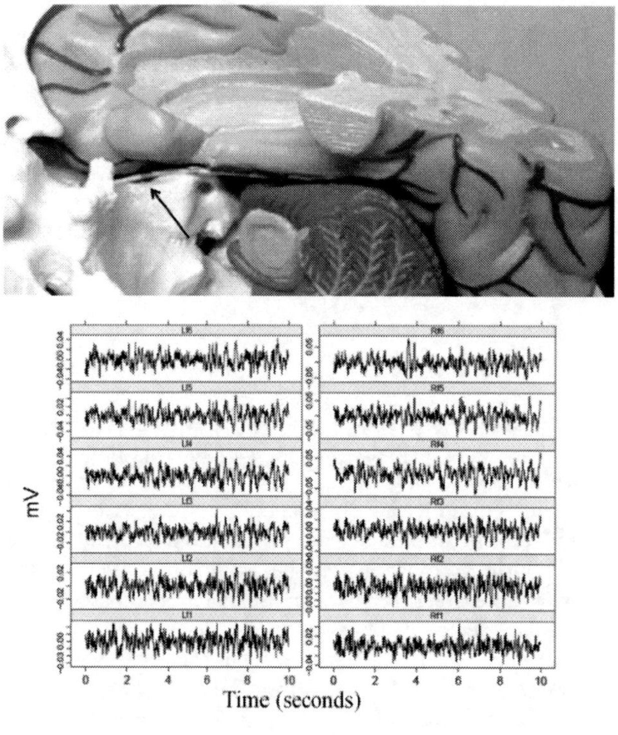

Figure 1: (A) Mesial aspect of the right temporal lobe and the right cerebellar hemisphere in a model of brain and skull. A foramen ovale electrode (FOE) was introduced through the right foramen ovale in order to demonstrate its localization within the mesial structures. The arrow indicate one of the six FOE locations. (B) Representative segments of FOE time series. Rf1-Rf6 stands for right FOE's #1 to #6 (right panel) and Lf1-Lf6 stands for left FOE's #1 to #6 (left panel).

Methods

1.1. Neurophysiological time series

Semi-invasive FOE technique has been developed [16] with the aim to record electrical activity at the inner part of the brain's temporal lobe, the main structure affected in temporal lobe epilepsy. Each FOE is a thin platinum wire, which is inserted through the ovale hole at the base of the skull. In this way, it allows to access deep structures in the head, especially the inner side of the brain, but without the need to go through into the brain. It records mainly the electrical activity at the surface of the inner side of the temporal lobe. Each wire consist of typically six contact electrodes [17] and during the placement procedure, two of these FOE are inserted bilaterally, one in each side of the brain. In figure 1 is displayed one FOE along its placement in between the inner part of the temporal lobe and other deep structures. The arrow in figure 1(A) shows one of the six electrodes in that FOE.

Digital FOE data, acquired at 500 Hz, filtered at 0.5–60 Hz and exported to ASCII format (XLTEK, Canada) at 200 Hz were used. Artifact free epochs lasting around 60 minutes were selected for further numerical analysis. All derivations were referenced to the scalp (Fz+Cz+Pz)/3 (standard 10-20 system). In figure 1(B) it is displayed ten seconds of a typical FOE record. Data were post-processed using Fortran and R programs. FOE will be named Lf1-Lf6 for the left electrodes and accordingly Rf1-Rf6 for the right electrodes.

Two neuro-physiological records have been analyzed in this work and both of them include a clinical seizure, characterized by a sudden change in the statistical properties of the electrical records which are translated in the typical epileptic symptoms. In figure 2 we show the standard deviation of all the scalp and FOE electrodes as a function of time. The clinical seizure onset, as indicated by the expert physician is around the minute 13.4. Clearly it can be seen the high and synchronous amplitude of several FOE, mainly in the left side. In this particular case, the patient suffers from a left TLE, which means that clinical seizures begin at this particular side of the temporal lobe, where the focus is supposedly located.

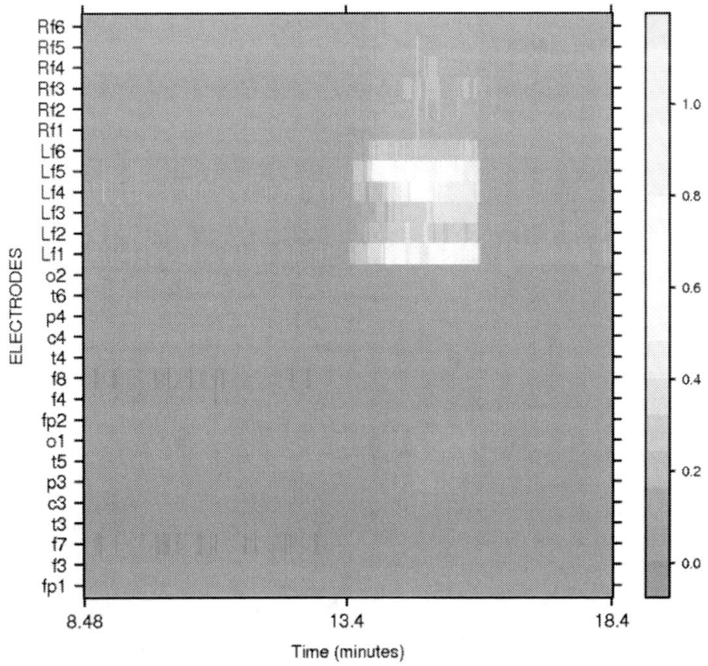

Figure 2: A typical neuro-physiological seizure of patient A. Level plot representation of the standard deviation in all electrical records, scalp and FOE. Clinical seizure begins at minute 13.4.

1.2. *Signal Analysis*

Multivariate FOE records of 12 channels (6 for each side) are divided in non-overlapping temporal windows of 512 data points (2.56 seconds). Longer values (1024 and 2048) have been used with qualitatively equivalent results. In each temporal window, cross-correlation among the 12 FOE time series was calculated, by using the Pearson correlation coefficient [18]:

$$\delta_{ij}(0) = \frac{\sum_{k=1}^{N_{win}} (x_i(k) - \overline{x_i})(x_j(k) - \overline{x_j})}{\sqrt{\sum_{k=1}^{N_{win}} (x_i(k) - \overline{x_i})^2 \sum_{k=1}^{N_{win}} (x_j(k) - \overline{x_j})^2}} \quad (1)$$

Where $\mathbf{x} = x_i(k), i = 1, N_{chan}$ *and* $k = 1, N_{win}$ is each of the 12 (N_{chan}) channels of 512 (N_{win}) data points each. We have selected this rather small number of points in each temporal window in order to obtain a suitable temporal resolution in each of the synchronization measures calculated (see below).

Correlation structure will be the base of our further calculations. No other synchronization measure will be used here. The main reason in using correlation instead of other nonlinear synchronization measures is that it performs well when it is used over neurophysiological data [10][18][19]. Another synchronization measure commonly used in recent years is phase synchronization [11][18], but as was very recently demonstrated, it must be used with extreme caution in neurophysiological electrical records [21].

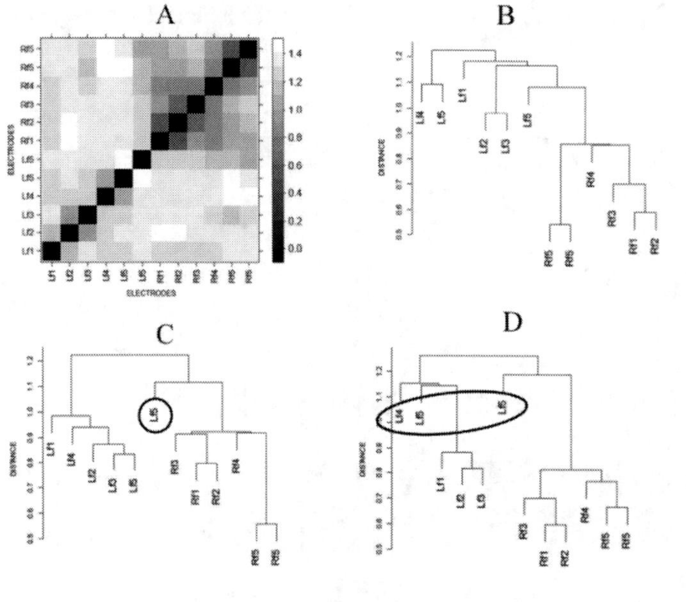

Figure 3: Synchronization and hierarchical trees. (A) Distance matrix for a particular temporal window of 512 data points in the patient A record. Only FOE electrodes are displayed. (B) Hierarchical tree (dendogram) corresponding to the distance matrix of (A). (C) and (D) two more dendogram corresponding to other temporal windows in the patient A record. Enclosed electrodes are DE (see text).

1.3. Community detection and desynchronized electrodes

In order to characterize the principal cortical interaction in the mesial area of the temporal lobe, we have implemented a recently developed algorithm aimed to detect clusters of interaction in a hierarchical tree, as it is for example the dendogram. The practical importance of hierarchical trees is that the data are not partitioned in a single step, but instead, a series of partitions are performed sequentially, accordingly with a specific rule. This is the best approach when there is no evidence in advance of how the interactions are organized. The hierarchical organization so obtained by any hierarchical clustering algorithm can be easily transformed in a community structure by simply *cutting* the hierarchical tree at a specific level.

In order to construct a distance organization from the interactions information, we shall transform correlation, as given by (1), in distances, in accordance with:

$$\rho_{ij} = |\delta_{ij}| \;\; ; \;\; d(i,j) = \sqrt{\rho_{ii} + \rho_{jj} - 2\rho_{ij}} = \sqrt{2(1-\rho_{ij})} \quad (2)$$

In this way, from the correlation matrix (1) we obtain the corresponding distance matrix. In figure 3 (A) we display a typical distance matrix for one of the temporal windows. From the distance matrix, it is now fairly easy to construct a hierarchical tree by using one of the traditional methods. We have chosen the agglomerative single-linkage [4] clustering technique. In this way, from the distance matrix, a hierarchical organization is achieved. In figure 3(B) it can be seen the dendogram corresponding to the distance matrix of figure 3(A). The correspondence between both figures is simply observed. In Figure 3(A) there exist higher correlations among the FOE of the right side than the correlations among the FOE of the left side, as displayed by the darker squares in the right side (upper-right corner), meaning shortest distances between the right electrodes. This fact is reflected in the dendogram, figure 3(B), where the right FOE are located at a deeper position in the tree that the left FOE, which are linked among them, but in an elevated position. We also display in figure 3(C) and 3(D) dendograms corresponding to other temporal windows. In these figures it is apparent that there exist equivalent correlations between the left and right electrodes, with the exception of few cases. Although both groups of electrodes are at the same level that is, left and right electrodes are grouped in individual clusters with similar height, there exists, however, some electrodes that appear outside the corresponding cluster. For instance, in figure 3(C) the left electrode Lf5 (encircled) appears desynchronized from the rest of the left electrodes because its high position in the dendogram. In the same way, left electrodes Lf4, Lf5 and Lf6 (enclosed) seems to be desynchronized from the synchronized cluster Lf1-Lf2-Lf3. These facts seem to favor the idea that there exists different cluster or community organization between both sides. Because the patient suffers from a left TLE, it seems that the areas covered by the left electrodes are less synchronized than the areas in the opposite side. In this way, we will give special attention to this kind of desynchronization.

1.4. Correlation measures

From the correlation matrix (1) we shall calculate three synchronization measures in order to compare against our results. All the calculated measures are normalized to its mean and standard deviation, for easy of comparison.

- Minimum Spanning Tree Cost (MSTC)
  From the correlation matrix it is possible to construct a tree, without loops, and with minimum total length in such a way that it represents the main interactions among the electrodes. This is called the Minimum Spanning Tree (MST). The total length of all the branches of the tree it is called "cost" of the MST. There exist a very simple way to construct the MST by using the Kruskal's algorithm. We shall use the MST cost as an

indicator of the seizure appearance because during a seizure exists an overall hyper-synchronization among different and distant areas in the brain, and in this way, the total cost of the MST must be reduced. We shall calculate the MST cost of all the FOE plus the scalp electrodes in order to consider the global synchronization produced during a clinical seizure.

- Average Synchronization (AS)
  In order to quantify the degree of synchronization among the FOE in each side, we shall simply sum the distances between pairs of FOE, for the left and for the right sides, and divide by two these quantities (due to the symmetric character of the distance matrix):

$$AS_{left} = \frac{1}{2}\sum_{i=1}^{6}\sum_{j=1}^{6} d(i,j) \quad and \quad AS_{right} = \frac{1}{2}\sum_{i=7}^{12}\sum_{j=7}^{12} d(i,j) \tag{3}$$

- Correlation eigenvalues (CE)
  Because the distribution of correlation eigenvalues (λi) is tightly linked to the synchronization characteristics of the time series, where a very unequal distribution of eigenvalues (for instance only one eigenvalue different from zero) implies a high synchronization among the time series, we shall calculate the following measure:

$$CE_{left} = \sum_{i=1}^{6} \lambda_i \log_2 \lambda_i \quad and \quad CE_{right} = \sum_{i=7}^{12} \lambda_i \log_2 \lambda_i \tag{4}$$

  When all the eigenvalues λi are equal, that is, all the time series are desynchronized we shall get a maximum of CE. On the other side, for a perfect case when exists only one eigenvalue different from zero, the case of a perfect synchronization, we shall get the minimum of CE. In this way, lower values of CE, either left or right, will mark the presence of synchronization among the involved electrodes. Note that correlation matrix must be definite positive for the proper (4) calculation. Any departure from this condition however must be due certainly produce by wrong inputs into the correlation matrix.

3. Results

We have applied the above described methods in two FOE records during typical patients' seizures. In each temporal window of 512 data points, we have calculated the correlation matrix through (1) and then transformed it into a distance matrix through (2). After that, a dendogram is constructed as explained in the Methods' section, and the identification of those electrodes that cannot be assigned to a synchronization cluster are counted. For instance, in figure 3(C) there is one electrode, namely Lf5 which seems to be desynchronized from the rest of all electrodes. Therefore, we count in this temporal window that there exists one left electrode that is not synchronized, or declusterized from the rest. In figure 3(D) instead, there exists three of these electrodes, Lf4, Lf5 and Lf6 which are declusterized because the three electrodes in the left side, Lf1, Lf2 and Lf3 seem to be tightly connected among them because its lower position in the dendogram. therefore we count 3 left DE in this temporal window. In the same way, all of the right FOE seems to be also strongly connected forming a large synchronized cluster of six electrodes.

In order to dig further into this fact we have implemented an automatic procedure to identify clusters of synchronous activity in every temporal window in records of approximately 60 minutes, yielding typically 360 temporal windows. To automatically detect synchronization clusters, we have selected a recently published algorithm [22] aimed to identify clusters in hierarchical structures as the one depicted in figure 2. Given a minimum number of electrodes the clusters must have, the algorithm automatically extract and identify all of the clusters contained in the dendogram. Those remaining electrodes not belonging to any clusters continues unassigned and are labeled as DE.

In figure 4 we show the main results in the two selected records, A and B. In each case the results are divided in both left and right counts. For instance, in figure 4(A) we show the counts of DE for the left side (upper panel) and the counts of DE for the right side (lower panel). The solid vertical line marks the beginning of the seizure as indicated by the expert physician. It is remarkable the big difference between the number of DE in the left side as compared with the right side, where there exists more DE in the left side. In this sense, this temporal lobe displays lower synchronization than the right temporal lobe because the higher number of desynchronized electrodes in this side. Also, it is interesting to note the seizure synchronization effect around minute 13.4, where the seizure begins. The left side, where it is suspected that the focus resides, seems to be "unaware" of the synchronization wave produced by the seizure, contrary to what happens in the opposite side, the right one, where the few desynchronized electrodes are still present until the moment of the seizure onset, vanish almost instantaneously. It seems that during the seizure there exist a perfect synchronization in the right side and no appreciable effect in the left one. After 2 minutes of seizure however, it seems that there exists a higher level of synchronization in the left side, looking through the less number of DE in that side, and equal level of synchronization in the right side. It must bearing in mind that this patient suffers from a left TLE, that is, the origin of its seizures is in the left temporal lobe.

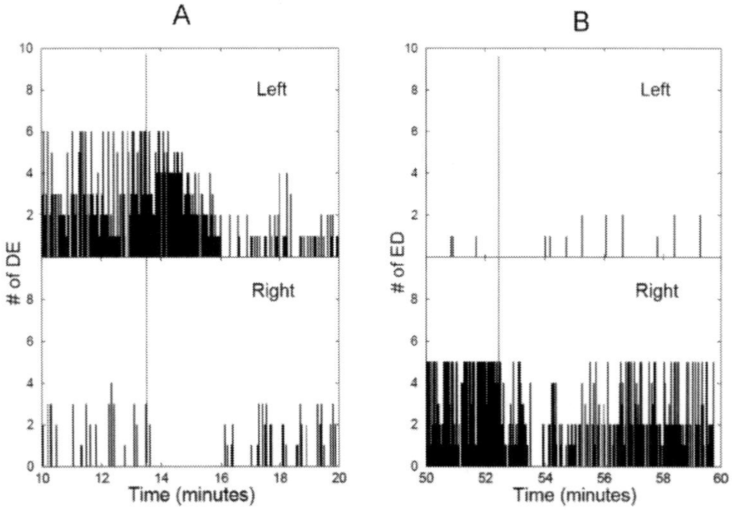

Figure 4: Number of DE as a function of time. (A) Both left and right DE counts as a function of time for patient A. (B) Both left and right DE counts as a function of time for patient B.

Something similar happens in the other patient's record, which suffers from a right TLE. The seizure onset is around minute 52.5. Also in this case, there exists a clear imbalance between right and left DE, being much more DE in the right side. After the seizure onset, it seems that there exists a lowering in the DE in the right, or epileptic, side, and none DE in the other side. But after a while after the seizure, it seems that the same levels of DE are achieved in both sides.

In order to compare our results against more classical measures, we display in figure 5 the calculations of the measures explained in the Methods section. In both cases, A and B of figure 5, we show the three measures. The seizure onset is marked by a vertical line. In the upper panel it is displayed the MST cost. In the middle panel we show the AS, for the left and right side, and in the lower panel the CE, also for the left and right sides.

The MST cost, which is a global synchronization measure, clearly shows a drop off at the beginning of the seizure, indicating that due to the higher level of correlations among all of the electrodes, the distances among them are reduced drastically.

AS are displayed in the middle panel of figure 5 (A) and (B). The solid line shows the AS_{left} and the dotted line the AS_{right}. In this case, there exists a striking difference between both sides because the left side shows a steepest drop off than the right side, at least during the seizure development, which finish when the seizure ends, approximately 2.5 minutes after the seizure onset. This effect is more intense in the patient A than in the patient B, either in intensity and in extent. The decrease in AS_{right} confirm the findings of figure 5 (A) because in the right side there exist a higher synchronization, expressed by the vanishingly number of DE as in figure 4 (A) and also by the lower values that the AS_{right} takes. Note that in the left side, there is a short period, just after the seizure onset, where the number of DE is higher than before the seizure onset (figure 4 (A)). This fact is confirmed by the short increase in the average distance in the left side in the middle panel of figure 5 (A).

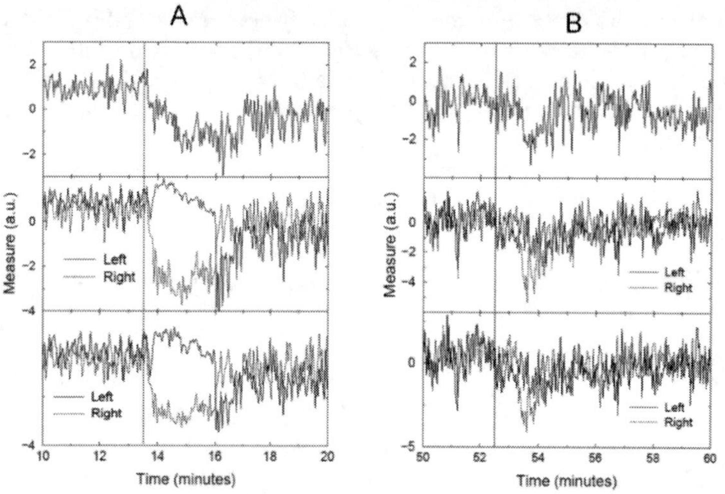

Figure 5: Correlation measures for both patients during the clinical seizure, as a function of time (10 minutes). All the measures are normalized to its mean and standard deviation. (A) Patient A, Upper panel: MST cost. Middle panel: AS for the left and right lobes. Lower panel: CE for the left and right lobes. (B). The same as in (A) but for patient (B).

Lastly, in the lower panel of figure 5 (A) and (B) we reproduce the calculations for the eigenvalues of the correlations matrix. By comparing these figures against the corresponding to the case of average synchronization, we can conclude that both measures behave in a very similar fashion.

4. Discussion

A clear imbalance between both mesial areas in TLE patients regarding its synchronization activity was shown here. Synchronization patterns have been quantified here by the property of being part of a synchronization cluster or not. Those areas belonging to any cluster have therefore its activity well synchronized with other areas within the cluster. In contrast, those regions which are not member of any cluster are, therefore, desynchronized from the rest, *i.e.* they are desclusterized regions. In both TLE patients shown here the area covered by the FOE's in the epileptic side behaves highly desclusterized as compared with the contra-lateral side, where mesial sites have a tendency to behave highly synchronized among them within one or several clusters. As we have also shown, declusterization is directly related with desynchronization in the whole side. The higher levels of declusterization displayed in the epileptic side produce lower levels of synchronization, as quantified by the synchronization measures used, AS and CE. Regarding these measures, both behaves in a similar fashion, and also are easily calculated from the information provided by the FOE records. Thus, they can be used in analyzing large amount of data.

One of the main concerns in epileptology, especially in patients reluctant to drug treatment, it is the exact location of the focus, the supposedly origin of the clinical seizures because the last alternative treatment it is the surgical removal of the brain tissue responsible of the clinical symptoms. Although there is a vast battery of diagnostic techniques, as for example Magnetic Resonance Imaging, Single Photon Emission Tomography, Positron Emission Tomography and neurophysiological EEG signal, which are aimed at localize the true position of the epileptic focus, not always they accomplish its mission. In fact, in several occasions they yield contradictory results [23], for example, locating the epileptogenic area in different hemispheres. Although v-EEG is still the most confident lateralization and localization method, it largely relies in the analysis of the ictal phase as the true indicator of the seizure origin. Because of that, patients often remain in the v-EEG room for several days in order to account for a number of seizures which allow lateralize the epileptic side in a confidently way. The methodology described here seems to give more robust results with only two or three hours of analysis, saving therefore patients's time of hospitalization.

As we have shown in this paper, tracking the number of DE in each temporal lobe seems to be a reliable technique which allows easily and quickly get insight into the synchronization phenomena in the epileptic brain during the seizure, allowing the identification of the epileptogenic side.

Acknowledgments
We would like to thanks the LAWNP 2009 organizers, Celia Anteneodo and Marcos G.E. da Luz for their very kind hospitality during the conference. This work has been funded by grants from the Fundación Mutua Madrileña (GJO, JP) and from the Plan Nacional de Investigación Científica, Desarrollo e Innovación Tecnológica (I+D+I), Instituto de Salud Carlos III, PI060349 (JP, GJO, RGS). IHP is a predoctoral investigator from the Biomedical Research Foundation, Hospital Universitario La Princesa, Universidad Autonoma de Madrid. GJO is a member of CONICET.

References
[1] Schiff SJ, So P, Chang T, Burke RE andSauer T 1996 *Phys. Rev.* E **54(6)** 6708-24.
[2] Rulkov NF, Sushchik MM, Tsimring LS and Abarbanel HDI 1995*Phys. Rev.* E **51** 980-94.
[3] Albert R and Barabási AL 2002 *Rev. Mod. Phys.* **74** 47-97.
[4] Boccaletti S, Latora V, Moreno Y, Chavez M and Hwang D 2006 *Phys. Rep.* **424(4-5)** 175-308.
[5] Wasserman S and Faust K 1994 *Social Networks Analysis* (Cambridge University Press, Cambridge).
[6] Scott J 2000 *Social Network Analysis: A Handbook 2nd ed* (Sage Publications, London).
[7] Traub RD and Wong RK 1982 *Science* **216** 4547-745.
[8] Penfield W and Jasper H 1954 *Hypersynchrony. Epilepsy and the functional anatomy of the human brain* (Boston: Little, Brown and Company) pp 193-4.
[9] Schindler K, Leung H, Elger CE and Lehnertz K. 2007 *Brain* **130** 65-77.
[10] Netoff TI and Schiff SJ 2002 *J. Neurosci.* **22** 7297-307.
[11] Mormann F, Kreuz T, Andrzejak RG, David P, Lehnertz K and Elger CE 2003 *Epilepsy Res.* **53(3)** 173-85.
[12] Gutkin BS, Lang CR, Colby CL, Chow CC and Ermentrout GB 2001 *J. Comput. Neurosci.* **11** 121-34.
[13] Dichter MA and Ayala GF 1987 *Science* **237** 157-64.
[14] Kandel ER, Schwartz JH and Jessell TM 2000 *Principles of Neural Science 4th Edition* (McGraw-Hill, New York).
[15] Li Y, Fleming IN, Colpan ME and Mogul DJ 2008 *IEEE T. Neur. Sys. Reh.* **16(1)** 62-73.
[16] Wieser HG, Elger CE and Stodieck SR 1985 *Electroencephalogr. Clin. Neurophysiol.* **61(4)** 314–22.

[17] Pastor J, Sola RG, Hernando-Requejo V, Navarrete EG and Pulido P 2008 *Epilepsia* **49(3)** 464-9.

[18] Press WH, Teukolsky SA, Vetterling WT and Flannery BP 2003 *Numerical Recipes in Fortran 77 2nd Edition: The Art of Scientific computing*. (Cambridge University Press, Cambridge).

[19] Quian Quiroga R, Kraskov A, Kreuz T and Grassberger G 2002 *Phys. Rev.* E **65** 041903.

[20] Ortega GJ, Menéndez de la Prida L, García de Sola R and Pastor J 2008 *Epilepsia* **49(2)** 269-80.

[21] Guevara R, Velazquez JL, Nenadovic V, Wennberg R , Senjanovic G and Dominguez LG 2005 *Neuroinformatics* **3** 301-14.

[22] Langfelder P, Zhang B and Horvath S 2008 *Bioinformatics* **24(5)** 719-720.

[23] Pastor J, Hernando-Requejo V, Domínguez-Gadea L, de Llano I, Meilán-Paz ML, Martínez-Chacón JL, García Sola R 2005 *Revista de Neurología* **41(12)** 709-716.

Using a one-dimensional lattice applied to the thermodynamic study of DNA.

N. F. Ribeiro, E. Drigo Filho

São Paulo State University – UNESP, Instituto de Biociências, Letras e Ciências Exatas, Rua Cristóvão Colombo, 2265, CEP 15054-000, São José do Rio Preto, SP, Brazil

E-mail: nataliafavaro@hotmail.com, elso@ibilce.unesp.br

Abstract. In this work it is analyzed a one-dimensional lattice which is composed by mass-spring systems with one additional Rosen-Morse potential on site. This kind of lattice is used to study thermodynamic properties of DNA, especially its thermal denaturation. On the context of this work, the Rosen-Morse potential simulates hydrogen bonds between double strands of the molecule. From the graphic of the average stretching of base pairs versus temperature it is possible to observe the thermal denaturation of the system. This result shows that it is possible to obtain phase transition with an asymmetric potential without an infinite barrier.

The Peyrard-Bishop (PB) model proposed in 1989 [1] has been used a lot to describe the thermal denaturation of DNA [2]. In the original model the double strand of DNA is described by two chains of particles, in which the harmonic potential simulates the stacking interactions. On the other hand, the Morse potential describes the hydrogen bonds that are one of the main interactions responsible for maintaining stable the double strand of the molecule. From this model, several dynamic and thermodynamic aspects of the DNA have been explored, for example, formation of domain wall [3, 4], energy location [5], formation and stability of breathers [6-8].

Several modifications have been proposed in the PB model in order to obtain a better description of the phase transition for a DNA molecule. One way to alter the model consists on the modification of stacking interaction. Originally, this interaction was taken as purely harmonic [1], but in more recent works [9-11], an exponential term was introduced with this harmonic term. Another way to alter the original model is to substitute the Morse potential for others potentials with different characteristics. This can be made by the addition of a new term to the Morse potential as it is done, for instance, in ref. [12]. In this reference, the additional term was interpreted as interactions of the macromolecule with the solvent.

Concerning the general behavior of non-linear lattices of the type here studied, Angelani et all [13] show that phase transitions are related to topological changes occurred in the system. One important result indicated in reference [13] is that the system does not presents phase transition when the potential on site has a symmetrical well. This happens because these potential have symmetric waves functions, obtained in the Schrödinger-type equation that emerges from the formalism, and centered in the origin. Therefore, the average value of the stretching of base pairs is always zero, what results on the impossibility of existing phase transition. This result contrasts with the one obtained for potentials with infinite barrier in one extremity of the potential well, like the Morse potential, in which the transition happens. This property is emphasized in reference [3], in which the transition is discussed both by using the dynamical method of domain wall as by using statistic mechanic. Thus, when the potential that simulates hydrogen bonds has an asymmetric well with an infinite barrier in one extremity and a finite barrier in the other one, the phase transition always occurs. This can be observed by following the evolution of one order parameter, in this case, the average stretching of base pairs, which divers for a critical temperature.

The main result here presented is that the use of Rosen-Morse potential on site in the studied lattice allows characterizing a phase transition similar to those obtained in the literature models [1-3]. As it is doing with the Morse potential [1], the use of Rosen-Morse potential permits an analytical treatment, what a priori makes easier the obtaining and interpretation of the results, besides in applications to obtain the probe design [14]. Another remarkable result is that the use of a finite and asymmetrical potential on site in a lattice used in PB model presents phase transition. The kind of potential, not yet treated in literature, can be considered as having intermediate characteristics between the symmetrical potentials [13] and the asymmetrical potential with infinite barrier in one extremity [3].

The Rosen-Morse potential is explored a lot in literature, both in the formal aspect [15-17] as in physical applications. It has been used, for example in studies of vibration on polyatomic molecules [18] and on polarons models [19]. This potential is defined as:

$$V(y_i) = B_0 \tanh(\alpha y_i) - \frac{U_0}{\cosh^2(\alpha y_i)}, \tag{1}$$

where B_0 and U_0 are parameters related, respectively, with asymmetrical and depth of the potential, and α is a parameter related with the width of the well. By adopting B_0 equals zero, the potential acquires a symmetrical shape. The variables y_i represent, in the PB model, the stretching between the i-th base pair, as shown in figure 1.

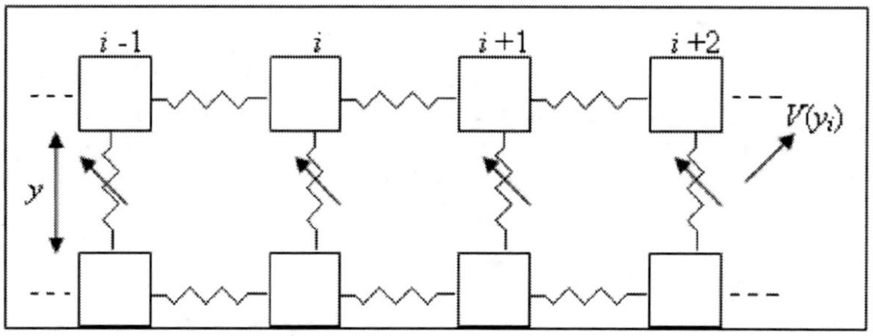

Figure 1: Representation of DNA model proposed by Peyrard and Bishop

The values of the parameters (α, B_0, U_0) here used were chosen in order to obtain a good adjustment with the Morse potential, $V_M(y) = D(1-e^{-ay})^2 - D$, with $D = 0.03eV$ and $a = 2.81\text{Å}^{-1}$, used to describe the hydrogen bonds on DNA [20]. The figure 2 shows a comparison between the originally Morse potential and the proposed Rosen-Morse one. The adjusted result, used in figure 2, points to $B_0 = -0.02eV$, $U_0 = 0.047eV$ and $\alpha = 4.3\text{Å}^{-1}$.

Following the usual PB model [1, 2] and replacing the Morse potential by the Rosen Morse potential in order to simulate the hydrogen bonds, the thermodynamic properties of the system are described by the partition function Z. The partition function of physical prominence, written in terms of the variable that represents the average stretching of hydrogen bonds, can be calculated by using the transfer integral operator method [21]. This technique relates the partition function with eigenfunctions ψ_n and energy eigenvalues ε_n obtained through a Schrödinger-type equation. Thus, the problem can be synthesized in finding the eigenfunctions and eigenvalues, expressed by the equation:

$$\left[-\frac{1}{2\beta^2 k}\frac{d^2}{dy_i^2} + V(y_i)\right]\psi_n(y_i) = \left[\varepsilon_n + \frac{1}{2\beta}\ln\left(\frac{2\pi}{\beta k}\right)\right]\psi_n(y_i), \tag{2}$$

where $\beta = \frac{1}{k_B T}$, $k_B = 8.617\times10^{-5} eV/K$ is the Boltzmann constant, T is the temperature, $k = 0.06 eV.\text{Å}^{-2}$ is the elastic constant related to the interaction alongside the chain (a purely harmonic potential is adopted to simulate the stacking interactions [2, 20]) and $V(y_i)$, in this case, is the Rosen-Morse potential which is given by the equation (1). The variables y_i represent, as usual, the stretching between the two springs [1-14], i.e., the distance between two bases that form the i-th base pair.

The partition function of the system is given by the sum $Z = \sum_n e^{-N\beta\varepsilon_n}$, where N is the particles number of the system. However, in the thermodynamically limit ($N \to \infty$) the result is dominated by the ground state and the others terms of the sum can be despised. Therefore, for ours purposes, we limit to calculate the eigenfunction and the energy eigenvalue for the ground state of equation (2).

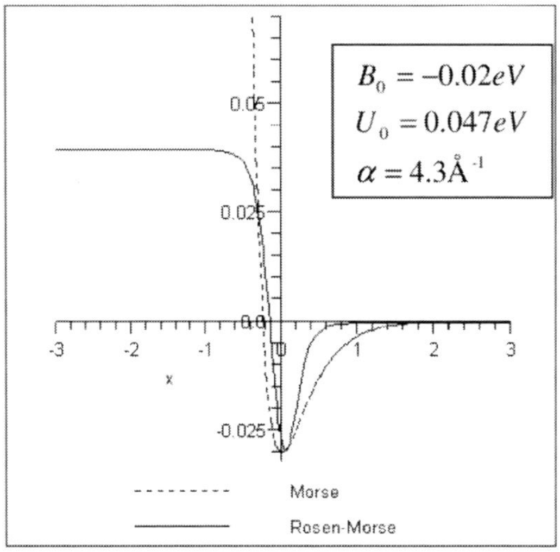

Figure 2: Graphic of the Morse potential (dotted curve) and the Rosen-Morse potential (solid curve). In the text box on the right of the graphic, there are the values obtained for the parameters of Rosen-Morse potential.

There are several methods that can be used to find the equation solutions (2), one of them is the Frobenius method [22]. Another approach is related to the factorization and supersymmetric method, as indicated in reference [23]. Anyway, for the ground state, the eigenfunction found is:

$$\psi = \frac{Ae^{az}}{\cosh^b z}, \tag{3}$$

where the new variable is $z = \alpha y$, A is the normalization constant, $a = -\frac{k\beta^2}{\alpha^2}\frac{B_0}{b}$ and $b = -\frac{1}{2} + \frac{1}{2}\left(1 + \frac{8\beta^2 k}{\alpha^2}U_0\right)^{\frac{1}{2}}$. The energy eigenvalue corresponded to the eigenfunction (3) is:

$$\varepsilon_0 = -\frac{\alpha^2}{2k\beta^2}\left(a^2 + b^2\right) - \frac{1}{2\beta}\ln\left(\frac{2\pi}{\beta k}\right). \tag{4}$$

From the eigenfunction (3) it is possible to find the average stretching between the base pairs ($< y >$) through the equation:

Journal of Physics: Conference Series **246** (2010) 012037 doi:10.1088/1742-6596/246/1/012037

$$< y >= \frac{\int_{-\infty}^{\infty} y\psi^2 dy}{\int_{-\infty}^{\infty} \psi^2 dy}. \tag{5}$$

The integrals indicated in equation (5) can be solved numerically. When the indicated parameters are used, the integration result depends only on the temperature, then, it is possible to follow the values of $< y >$ in terms of the temperature. The obtained result is showed in figure 3, where is presented a graphic of the average stretching of base pairs ($< y >$) versus temperature. For comparison effects, it is also presented in figure 3 the curve of the average stretching of base pairs versus temperature for the Morse potential on site. This curve was made by using the eigenfunction of the ground state for Morse potential, given in reference [1]. In this figure can be clearly observed that in the temperature interval between 300 and 350K there was an increase in the average stretching of base pairs. This indicates that in this temperature interval a thermal denaturation of the molecule occurred. It can be observed that in the obtained curve for Rosen-Morse potential the transition was similar to that obtained for Morse potential.

The fact that an exact/analytical solution exists for the studied system allows an alternative way to determine the temperature in which occurs the break of the double strand. This can be done through direct inspection of the wave function given by equation (3). For this function leads to on bound states, it is necessary that it be normalized in all space. By a direct inspection of the wave function (3) one observes that for ψ not divers it is necessary that a and b parameters of the wave function follow the condition $0 < \frac{a}{b} < 1$. From this condition can be established that $a < b$ to have bound estates. Thus, the critical temperature must occurs when $a = b$, what means,

$$\frac{\beta_C}{\alpha}\left(k|B_0|\right)^{\frac{1}{2}} = -\frac{1}{2} + \frac{1}{2}\left(1 + \frac{8\beta_C kU_0}{\alpha^2}\right)^{\frac{1}{2}}, \tag{6}$$

where $\beta_C = \frac{1}{k_B T_C}$, T_C in this case represents the critical temperature.

From expression (6) it is possible to obtain a critical temperature as a function of the potential parameters. The result is:

$$T_c = -\frac{k}{\alpha k_B}\frac{\left(|B_0| - 2U_0\right)}{\left(k|B_0|\right)^{\frac{1}{2}}}. \tag{7}$$

By substituting all the parameters in equation (7), the value of the critical temperature is approximately $T_C = 346K$.

This procedure also can be done to determine the critical temperature for Morse Potential. The expression for the critical temperature, was obtained in reference [2] as $T_C = \frac{2\sqrt{kD}}{ak_B}$, where a and D are the parameters of the Morse potential, k_B is the Boltzmann constant and k is the elastic constant. By substituting this parameters and constants in equation (8), the value of the critical temperature for Morse potential is approximately 350K which is close to the result obtained from (7).

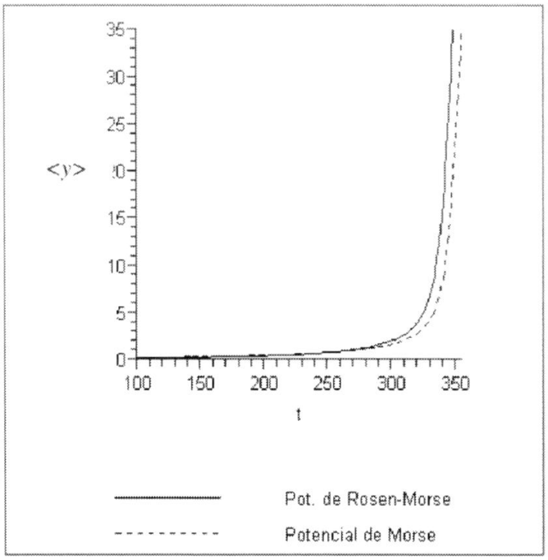

Figure 3: Graphic of the average stretching of base pairs, <y>, versus temperature, t, obtained from Morse and Rosen-Morse potential for temperature interval between 100K and 350K.

According to the reference [24] the denaturation temperature of DNA molecule varies between 318K and 372K, depending on the nucleotides involved in chain. Therefore, the obtained temperature for both potentials, Rosen-Morse and Morse, are in agreement with the values that exist in literature. In the studied scope, this result indicates that the potential used and the adopted parameters can describe both qualitative as quantitative the thermal denaturation of DNA and the similarity between the graphics of <y> for Morse and Rosen-Morse potentials indicates that it is possible to obtain a phase transition like that in ref. [1] with a asymmetrical potential that not presents an infinite barrier.

To conclude, the obtained results in this work show that is possible to characterize the thermal denaturation of DNA molecule with the use of Rosen-Morse potential to simulate the hydrogen bonds of the molecule. This result shows that it is possible to obtain a phase transition in PB-type models [1] without the potential on site having an infinite barrier. This result complements the discuss made in literature about the non-existence of phase transitions for symmetrical potential [13] and the well known fact [3] that the use of an asymmetrical potential, with wells that have an infinite barrier in one extremity and a finite barrier in another, allows the appearing of this transition. Thus, in way to be transition in one-dimensional lattices that emerges from a PB approach is necessary that the potential shows as asymmetrical well, and it is not necessary that exists an infinite barrier.

Finally, the fact of the obtained transition temperature in this system matches the values existed in literature [24] makes the adopted parameters reliable.

Acknowledgments

The authors are grateful to CAPES and FAPESP for the financial support.

References

[1] Peyrard M and Bishop A R 1989 *Phys. Rev. Lett.* **62** 2755.
[2] Peyrard M 2004 *Nonlinearity* **17** R1.
[3] Qasmi H, Barré J and Dauxois T 2004 *arXiv:cond-mat/0407662v1*.
[4] Theodorakopoulos N 2006 *Phys. D* **216** 185.
[5] Deluca J, Drigo Filho E, Ponno A and Ruggiero J R 2004 *Phys. Rev. E* **70** 026213.
[6] Cuevas J, Archilla J F R, Gaididei Yu B And Romero F R 2002 *Physica D* **163** 106.
[7] Cuevas J, Starikov E B, Archilla J F R and Hennig D 2004 *Modern Phys. Lett. B* **18** 1319.

[8] Alvarez A, Romero F R, Archilla J F R, Cuevas J and Larsen P V 2006 *European Phys. J. B* **51** 119.
[9] Dauxois T, Peyrard M and Bishop A R 1993 *Phys. Rev. E* **47** R44.
[10] Dauxois T and Peyrard M 1995 *Phys Rev. E* **51** 4027.
[11] Joyeux M and Buyukdagli S 2005 *Phys Rev. E* **72** 051902.
[12] Weber G 2006 *Europhys. Lett.* **73** 806.
[13] Angelani L, Ruocco G and Zamponi F 2005 *Phys. Rev. E* **72** 016122.
[14] Weber G, Haslam N, Essex J W and Neylon C 2009 *J. Phys.: Condens. Matter* **21** 034106.
[15] Lévai G 2008 *Phys. Lett. A* **372** 6484.
[16] Jia C S, Yi L Z and Sum Y 2008 *J. Math. Chem.* **43** 435
[17] Tezcan C and Sever R 2007 *J. Math. Chem.* **42** 387.
[18] Rosen N and Morse P M 1932 *Phys. Rev.* **42** 210.
[19] Smondyrev M A, Gerlach B and Dzero M O 2000 *Phys. Rev. B* **62** 692.
[20] Silva R A S, Drigo Filho E and Ruggiero J R 2008 *J. Biol. Phys.* **34** 511.
[21] Scalapino D J, Sears M and Ferrel R A 1972 *Phys. Rev. B* **6** 3409.
[22] Nieto M M 1978 *Phys. Rev. A* **17** 1273.
[23] Drigo Filho E 1990 *Rev. Bras. Fís.* **20** 258.
[24] Saenger W 1984 *Principles of Nucleic Acid Structure* (New York: Springer-Verlag Press).

Analysis of persistence during intracellular actin-based transport mediated by molecular motors

C. Pallavicini[1], M. A. Despósito[1,3], V. Levi[1,2,3] and L. Bruno[1,3]

[1] Departamento de Física, Facultad de Ciencias Exactas y Naturales, Universidad de Buenos Aires, 1428 Buenos Aires, Argentina.
[2] Departamento de Química Biológica, Facultad de Ciencias Exactas y Naturales, Universidad de Buenos Aires, 1428 Buenos Aires, Argentina.
[3] Consejo Nacional de Investigaciones Científicas y Técnicas, Argentina.

E-mail: lbruno@df.uba.ar

Abstract.
The displacement of particles or probes in the cell cytoplasm as a function of time is characterized by different anomalous diffusion regimes. The transport of large cargoes, such as organelles, vesicles or large proteins, involves the action of ATP-consuming molecular motors. We investigate the motion of pigment organelles driven by myosin-V motors in *Xenopus laevis* melanocytes using a high spatio-temporal resolution tracking technique. By analyzing the turning angles (ϕ) of the obtained 2D trajectories as a function of the time lag, we determine the critical time of the transition between anticorrelated and directed motion as the time when the turning angles begin to concentrate around $\phi = 0$. We relate this transition with the crossover from subdiffusive to superdiffusive behavior observed in a previous work [5]. We also assayed the properties of the trajectories in cells with inhibited myosin activity, and we can compare the results in the presence and absence of active motors.

1. Introduction

Our knowledge of intracellular transport of organelles mediated by molecular motors has markedly increased thanks to the improvement of single particle tracking (SPT) techniques. In STP experiments a probe particle or organelle is followed and tracked with millisecond temporal resolution and nanometer precision (see e.g. [1]). In this way, 2D trajectories of the motion of the particles are recovered.

Traditionally, two-dimensional trajectories are characterized by studying the statistical behavior of the mean square displacement (MSD) as a function of the time lag τ. The MSD measures the mean square distance between two points belonging to the trajectory and separated by a time lapse τ.

Although the MSD is a useful magnitude that has been used to determine the different diffusive regimes of intracellular particles motion (see e.g. [2, 3, 4, 5]), it hides relevant information about the direction or persistence of the motion. For example, in [6, 7] the cytoskeleton remodeling dynamics is studied by analyzing the motion of microbeads bound to the cytoskeleton filaments. In that system, the MSD of the microbeads undergoes a transition from subdiffusive to superdiffusive behavior with time. However, the authors were able to provide evidence of the processes that regulate the dynamics -i.e. caged dynamics at short time scales and directional memory at long time lags- only when they based their analysis on the turning angles distribution properties. Even though it is well known that molecular motors are proteins able to convert the energy form adenosine triphosphate (ATP) hydrolysis into directed motion along cytoskeleton filaments, the persistence of motor-driven cargoes motion in intracellular transport remains unclear.

Interestingly, recent theoretical works have been devoted to the origin of the competition between sub- and superdiffusive transport in different physical models [8, 9, 10]. *Xenopus laevis* melanophores are a unique system where motor-driven transport can be studied since they have pigment organelles or melanosomes containing the black pigment melanin, therefore they can be followed using bright-field transmission light microscopy [11]. Three different classes of motors are responsible for pigment organelles transport in *Xenopus laevis* melanophores cells: kinesin II and dynein move cargoes along microtubules and myosin-V carries them along actin filaments [12, 13, 14].

In a very recent paper, we study the different anomalous regimes of myosin-V driven melanosomes in cells treated with nocodazole, a drug that depolymerizes microtubules [5]. There we reported a transition from a subdiffusive to a superdiffusive regime based on the behavior of the MSD over time, computed from the 2D trajectories obtained by SPT.

In this work, we go further with the analysis of myosin-V driven melanosomes: we compute the turning angle distributions from the 2D trajectories. Using this approach, we were able to correlate the transition from sub- to superdiffusive behavior with the antipersistent-persistent change in the turning angles distributions. We compare the results in the presence and absence of active motors with the aid of parallel experiments performed in mutant cells with inhibited myosin-V activity.

2. Turning angles

We analyzed trajectories of melanosomes moving along actin filaments in *Xenopus laevis* melanophore cells obtained in single particle tracking experiments that were reported in previous works [5, 15].

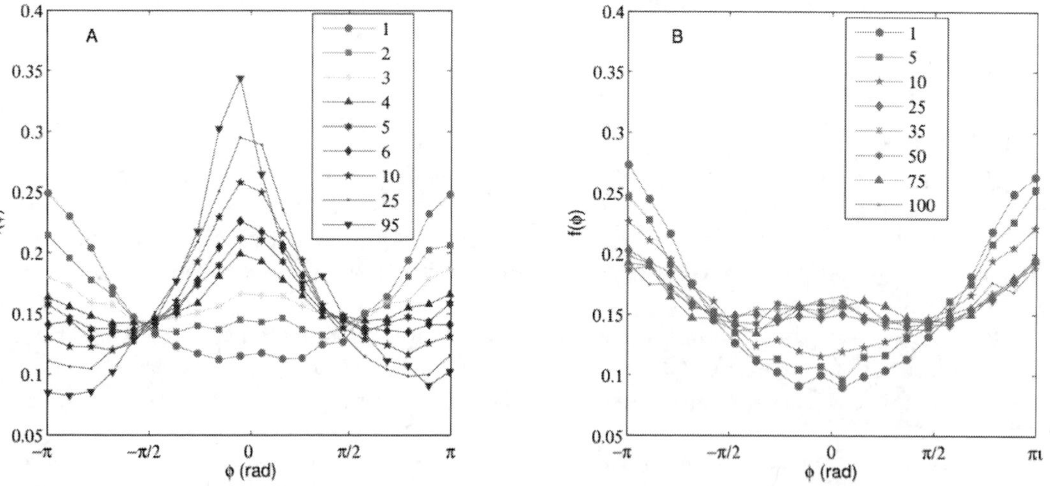

Figure 1. Turning angles at different time intervals. (A) Results obtained for WT cells during dispersion ($n = 33$). (B) Results obtained for mutated myosin-V cells ($n = 15$).

Melanosomes are black pigment granules with sizes around 500 nm. The experimental procedure followed to obtain the trajectories is described in Refs. [5, 15]. Importantly, cells were treated with nocodazole to depolymerize microtubules, so the active transport of melanosomes is only driven by the F-actin dependent motor myosin-V.

The experimental data split into two groups: one set was obtained using wild type cells (WT), while the other set was obtained using mutant cells that have a dominant-negative inhibition of myosin-V driven melanosome transport.

Given a 2D trajectory, we define the turning angle $\phi(\tau)$ as the internal angle between successive segments of duration τ. A Matlab script was used to extract the turning angle from the experimental 2D trajectories. For each τ, an angle distribution was found. We present the plots for some of these trajectories in Fig. 1.

The peaks of the turning angles distribution function $f(\phi)d\phi$ shift from $\phi = \pm\pi$ to $\phi = 0$ in WT cells (Fig. 1.A) indicating a transition from an antipersistent to a persistent motion. On the contrary, cells expressing the mutation of myosin-V do not develop this transition (Fig. 1.B).

3. Determination of the transition time lag

The transition from antipersistent to persistent behavior occurs near an essentially uniform distribution. To directly measure the transition time lag τ^* we quantify the departure of the distributions from a uniform one using:

$$\Delta_+ = \left| \int_0^{\pi/2} f(\phi)d\phi - \int_{\pi/2}^{\pi} f(\phi)d\phi \right| \tag{1}$$

$$\Delta_- = \left| \int_0^{-\pi/2} f(\phi)d\phi - \int_{-\pi/2}^{-\pi} f(\phi)d\phi \right| \tag{2}$$

We define τ^*, the time of the transition, as the time lag that minimizes this difference. A similar approach was used in [6]. This method resulted effective since we found a clear minimum for Δ_+ and Δ_- (Fig. 2).

In a previous paper [5] the transition was analyzed using a stochastic model based on the Langevin generalized equation. Within this approach, the dynamics of a melanosome of mass m, immersed in the intracellular medium and simultaneously driven by molecular motors is described as follows:

$$m\ddot{X}(t) + \int_0^t dt'\, \gamma(t-t')\, \dot{X}(t') = \xi(t) + \chi(t)\,, \tag{3}$$

where $\gamma(t)$ is the dissipative memory kernel, $\xi(t)$ is an internal noise (related to the subdiffusive passive transport) and $\chi(t)$ is an external noise related to the active transport due to the action of molecular motors. Assuming power-law correlation functions with exponents λ and α, respectively, for both types of noise, the mean square displacement (MSD) for the long time limit is given by [5]:

$$MSD(\tau) = \frac{4k_BT}{\gamma_0}\left\{\frac{1}{\Gamma(\lambda+1)}\tau^{\lambda} + \varepsilon K_{\lambda,\alpha}\tau^{2\lambda-\alpha}\right\} + 2\eta^2\,, \tag{4}$$

where

$$K_{\lambda,\alpha} = \Gamma(\alpha - 2\lambda)\,\frac{\sin(\pi(\lambda-\alpha)) - \sin(\pi\lambda)}{\pi} \tag{5}$$

is a positive constant, being $\lambda < 1$ and $1 < 2\lambda - \alpha < 2$. The last term in (4) represents the measurement errors in particle location due to SPT or biological activity [16].

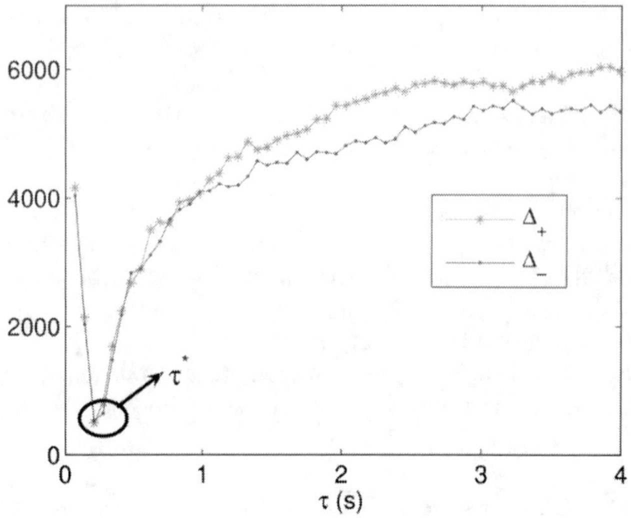

Figure 2. Δ_+ and Δ_- as a function of τ. The minimum allows to define τ^*. The transition occurs for $\tau^* = 0.21$s in the case shown in the figure.

The MSD local slope (in double log scale) is given by:

$$\beta(\tau) \;=\; \frac{\frac{\lambda}{\Gamma(\lambda+1)}\tau^{\lambda} + \varepsilon\,(2\lambda-\alpha)\,K_{\lambda,\alpha}\,\tau^{2\lambda-\alpha}}{\frac{1}{\Gamma(\lambda+1)}\tau^{\lambda} + \varepsilon\,K_{\lambda,\alpha}\,\tau^{2\lambda-\alpha} + \delta}\,, \tag{6}$$

where $\delta = \gamma_0\,\eta^2/2k_BT$.

For short time lags the dominant term is the subdiffusive one (given by λ) and for larger time lags the superdiffusive term prevails (given by $2\lambda - \alpha$). So, the transition occurs when $\beta(\tau^*) = 1$.

It can be seen in Fig. 3 that the dependence of the asymptotic slope of the MSD (given by $2\lambda - \alpha$ in the model) with the transition time obtained from the analysis of the turning angle distributions agrees with the prediction of Eq. (6) with $\beta(\tau^*) = 1$. Therefore, there is a close correlation between subdiffusion and antipersistence, and superdiffusion and persistent motion in melanosome transport driven by molecular motors.

Moreover, Fig. 3 shows that the transition occurs earlier for trajectories that have larger values of $2\lambda - \alpha$. An intuitive explanation of this is the following: assuming a given value of λ, larger $2\lambda - \alpha$ implies lower values of α. Let us recall that α is the exponent of the autocorrelation function of the force exerted by the motors and that can take values between 0 and 1 [5]. While values of $\alpha \approx 1$ are related with instantaneous pulse forces [3, 17], $\alpha \approx 0$ represents a constant force that can give rise to a persistent motion, thus the transition from sub- to superdiffusion will occur earlier. In this condition, it is harder for the thermal forces that drive the particle to randomize the direction of motion. As a consequence, the turning angles are smaller leading to a more persistent motion.

4. Summary

In this work we analyze 2D trajectories of motor-driven pigment organelles obtained by SPT in living cells. We measure the turning angle distributions and found a transition from

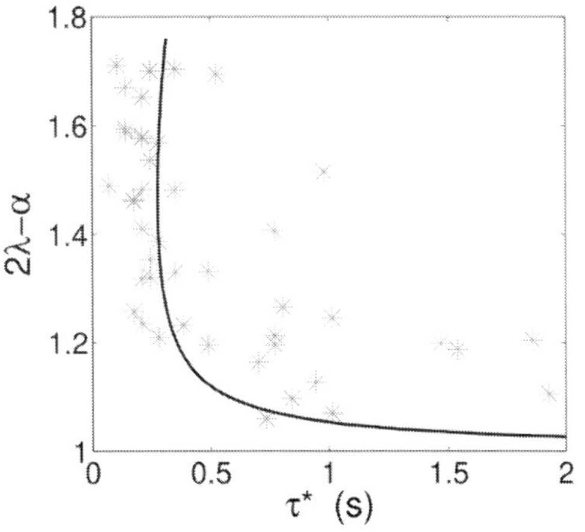

Figure 3. Asymptotic slope of log(MSD) vs. the critical time (asterisks). The solid line is the dependence of $2\lambda - \alpha$ with τ^*, given by the solution of Eq. (6) with parameter values $\lambda = 0.9$, $\epsilon = 100$ y $\delta = 3$. This parameters agree with those obtained in Ref.[5]

antipersistent to persistent motion. We were able to correlate this transition with the sub- to superdiffusive transition described in a previous work [5].

The approach followed in this paper can be extended to the analysis of other systems to extract information related with the directionality of the trajectories , which cannot be achieved by only studying the MSD.

Acknowledgments
We acknowledge support from grants PICT 928/06, PICT 31980/05 and PICT 31975/05 from Agencia Nacional de Promoción Científica y Tecnológica, Argentina.

References

[1] Levi V, Serpinskaya A S, Gratton E and Gelfand V I 2006 *Biophys. J.* **90** 318
[2] Tolic-Nørrelykke I M, Munteanu E-L, Thon G, Oddershede L and Berg-Sørensen K 2004 *Phys. Rev. Lett.* **93** 078102
[3] Wilhelm C 2008 *Phys. Rev. Lett* **101** 028101
[4] Gallet F, Arcizet D, Bohec P and Richert A 2009 *Soft Matter* **5** 2947 (2009).
[5] Bruno L, Levi V, Brunstein M and Despósito M A 2009 *Phys. Rev.* E **80** 011912
[6] Lenormand G, Chopin J, Bursac P, Fredberg J J and Butler J P 2007 *Biochem. and Biophys. Res. Comm.* **360** 797
[7] Metzner C, Raupach C, Paranhos Zitterbart D and Fabry B 2007 *Phys. Rev.* E **76** 021925
[8] Ai B, He Y 2010 *J. Chem. Phys.* **132** 094504
[9] Magdziarz M and Weron A 2007, *Phys. Rev.* E **75** 056702
[10] Dybiec B and Gudowska-Nowak E 2009 *Phys. Rev.* E **80** 061122
[11] Nascimento A A, Roland J T and Gelfand V I 2003 *Annu. Rev. Cell Dev. Biol.* **19** 469
[12] Tuma M C, Zill A, Le Bot N, Vernos I and Gelfand V I 1998 *J. Cell Biol.* **143** 1547
[13] Rogers S L , Karcher R L, Roland J T, Minin A A , Steffen W and Gelfand V I 1999 *J. Cell Sci.* **146** 1265
[14] Nilsson H and Wallin M 1997 *Cell Motil. Cytoskeleton.* **38** 397
[15] Brunstein M, Bruno L, Despósito M A and Levi V 2009 *Biophys. J.* **97** 1548
[16] Martin D S, Forstner M B and Käs J A 2002 *Biophys. J.* **83** 2109
[17] Bursac P, Lenormand G, Fabry B, Oliver M, Weitz D A, Viasnoff V , Butler J P and Fredberg J J 2005 *Nat. Mater.* **4** 557

XI Latin American Workshop on Nonlinear Phenomena IOP Publishing
Journal of Physics: Conference Series **246** (2010) 012039 doi:10.1088/1742-6596/246/1/012039

Stability of breathers in simple mechanical models for DNA

G G Slade, E Drigo Filho and J R Ruggiero

UNESP – Universidade Estadual Paulista, Instituto de Biociências, Letras e Ciências Exatas, Departamento de Física, Rua Cristovão Colombo 2265, 15054-000, São José do Rio Preto, Brazil.

Email: ggslade@hotmail.com

Abstract. This work studies through the Floquet theory the stability of breathers generated by the anti-continuous limit. We used the Peyrard-Bishop model for DNA and two kinds of non-linear potential: the Morse potential and a potential with a hump. The comparison of their stability was done in function of the coupling parameter. We also investigate the dynamic behaviour of the system in stable and unstable regions. Qualitatively, the dynamic of mobile breathers resembles DNA.

1. Introduction

The DNA molecule contains genetic information and is responsible for the transmission of hereditary information [1]. The discovery of the double helix structure [2] is considered a landmark for the development of molecular biology. However studies in the last four decades have been revealing that the DNA dynamic is complex and essential for its function. The base pairs hold the genetic information and they stay "hidden" in the helix interior. The sequences that codify the proteins must be read. This requires the display of the bases to the solution, necessitating movements of great amplitude. This implies a nonlinear dynamic.

Experiments with proton exchanges [3,4] show that the macromolecule presents fluctuations highly localized with the possibility of opening even a single base pair. They are also very dynamic because the lifetime of a base pair (the average time that the base pairs remains closed) is estimated in milliseconds. This data indicates that the DNA dynamic, in particular the study of vibrational modes that are localized in a few base pairs, is important for the comprehension of these aspects.

In nonlinear systems, breathers are solutions that are both spatially localized and time periodic [5,6]. This kind of solution is generated by the combination of two factors: the discreteness and the nonlinearity of the system [7]. The first one makes the dispersion relation (linear spectrum) limited and discrete, while the second one makes the frequency of vibration dependent on the amplitude motion. MacKay and Aubry [8] proved the existence of breathers in the one dimensional lattice when the system is composed by weakly coupled nonlinear oscillators. Their proof was based on the anti-continuous limit and showed that the solution persists in a space of periodic solutions with a fixed period and decays exponentially in space.

In this paper, we study the stability and dynamic of breathers in a simple mechanical model for DNA. The model used reduces the problem to the one-dimensional harmonic lattice with an additional non-linear potential on site, which simulates the H bond in DNA. Two kinds of non-linear potential were used: the original Morse potential and a potential with a hump. The anti-continuous limit was used to generate the breather in the lattices and the Floquet theory permits the study of its stability.

The Floquet multipliers and the energetic dynamics of the system are used to analyze the results. Tests for different frequencies of the breather and coupling constants are done in order to determine the range of stability for the breathers; these parameters are intrinsically related to the system's energy.

2. The Peyrard-Bishop (PB) model

There are several models to describe the DNA molecule [9]. Here we are interested in the localized energy structures that are provided by large amplitude movements, so we need a model whose main characteristic is to describe the base pairs separation of the nucleotides. The Peyrard-Bishop (PB) model [10] fills, in a minimalistic view, these conditions. In this model, a harmonic chain simulates each ribbon of DNA macromolecule and the interaction between them is simulated by a nonlinear potential. The three dimensional aspects of the helical structure are not considered and the motion analyzed is perpendicular to the ribbon. The positions of the nucleotides are denoted by u_j and v_j, respectively; with $j = 1, \ldots, N$. For simplicity, it is assumed that the chains are homogeneous, i.e., the masses and the elastic coupling constants are the same.

The Lagrangian of the PB model for a homogeneous chain is given by:

$$L_{\mathrm{PB}} = \sum_{j=1}^{N}\left\{\frac{m}{2}\left(\dot{u}_j^2 + \dot{v}_j^2\right) - \frac{k}{2}(u_{j+1} - u_j)^2 - \frac{k}{2}(v_{j+1} - v_j)^2 - V(u_j - v_j)\right\}. \quad (1)$$

By changing the variables to $x_j = \frac{(u_j+v_j)}{\sqrt{2}}$ and $y_j = \frac{(u_j-v_j)}{\sqrt{2}}$ it is possible to rewrite the Lagrangian as the sum of two terms, each one depending only on one coordinate, x or y. The part of the Lagrangian depending on x represents a chain of harmonic oscillators and is unimportant here. The term depending on y is written as

$$L_y = \sum_{j=1}^{N}\left\{\frac{m}{2}\dot{y}_j^2 - \frac{k}{2}(y_{j+1} - y_j)^2 - V(\sqrt{2}y_j)\right\}, \quad (2)$$

and represents a chain of harmonic oscillators with an additional potential on site.

The equations of motion obtained from Lagrangian (2) are:

$$m\ddot{y}_j + k\left(2y_j - y_{j+1} - y_{j-1}\right) + V'(\sqrt{2}y_j) = 0, \text{ with } j = 1,2,\ldots,N. \quad (3)$$

For the numerical calculation we used the periodic boundary conditions, that is, $y_0 = y_N$ and $y_{N+1} = y_1$.

3. Prescription to create the breather

The breather can be created using the principle of the anti-continuous limit suggested by MacKay and Aubry [8] and implemented by Marin and Aubry [11] and Cuevas [12]. So in the beginning we considered the uncoupled system, i.e. $k = 0$. With this limit the system is composed of isolated oscillators under the influence of the nonlinear potential on site.

Because we looking for periodic solutions, they can be written in terms of a Fourier cosine series:

$$y_j(t) = z_j^0 + 2\sum_{l=1}^{lm} z_j^l \cos(l\omega_{\mathrm{b}}t), \quad (4)$$

where ω_{b} is the breather frequency. Thus, the equations of motion can be changed through substitutions of a set of algebraic equations to obtain the $lm + 1$ coefficients of z_j^l parameters (4), these equations can be solved by the Newton method.

Once we found the breather profile for the uncoupled system, $k = 0$, we introduced the coupling into the system by small increments on k, δk. The next step is to find the numerical solution for $k=\delta k$ by using the Newton method and the original profile obtained for *k=0*. This new solution will be used like a seed for the case $k = 2\delta k$ and so on until we reach the desired value of k.

4. Breather stability

An important question concerning the breather is its stability. Using $y_j(t)$ as a periodic solution of (3) and introducing a perturbation, $\varsigma_j(t)$, it is possible to construct a perturbed solution as $\psi_j(t) =$

$y_j(t) + \varsigma_j(t)$. Considering that $\psi_j(t)$ is a solution of the system and by expanding the equations of motion (3) near the solution $y_j(t)$, we find:

$$\ddot{\varsigma}_j(t) + k\left(2\varsigma_j(t) - \varsigma_{j+1}(t) - \varsigma_{j-1}(t)\right) + V''(\sqrt{2}y_j)\varsigma_j(t) = 0, \quad (5)$$

which is a linear equation for the perturbation $\varsigma_j(t)$ (Hill's Equation).

The equations (5) can be written like a system of equations by the definition:

$$\pi_j(t) = \dot{\varsigma}_j(t) \quad (6)$$

$$\dot{\pi}_j(t) = -k\left(2\varsigma_j(t) - \varsigma_{j+1}(t) - \varsigma_{j-1}(t)\right) - V''(\sqrt{2}y_j)\varsigma_j(t), \quad (7)$$

or, in a matricial notation: $\Omega(t) = [\varsigma(t)\,\pi(t)]^T$, where $\varsigma(t) \equiv \{\varsigma_j(t)\}$ and $\pi(t) \equiv \{\pi_j(t)\}$. In the matricial form, $\Omega(t)$ is a column vector with 2N dimensions and N is the number of degrees of freedom of the system. In this notation the information contained in the equation (5) can by summarized as:

$$\dot{\Omega}(t) = A(t)\Omega(t), \quad (8)$$

where $A(t) = \begin{bmatrix} 0 & I \\ J & 0 \end{bmatrix}$. Each element of $A(t)$ is a matrix NxN, I is an identity matrix and J has the elements $J_{i,j} = -(2k + V''(\sqrt{2}y_j))\delta_{i,j} + k(\delta_{i+1,j} + \delta_{i-1,j})$ with $\delta_{i,j}$ as the delta of Kronecker. Periodic boundary conditions impose that for $i = 1$ we have $\delta_{i-1,j} = \delta_{N,j}$ and for $i = N$ we have $\delta_{i+1,j} = \delta_{1,j}$.

Since we are working with periodic equations, the system stability can be analyzed using the Floquet theory [13]. In this way, one can consider the matrix 2Nx2N $\phi(t)$, where each column represents a linearly independent solution of the system (8) corresponding to the different initial conditions. Thus, $\phi(t)$ is definite in $t = 0$ as a vector with $2N - 1$ components equal to zero, and with the component in the ith position ($i = 1,2,\dots,2N$) equal to 1. This matrix is known as the fundamental matrix and it satisfies the relation:

$$\dot{\phi}(t) = A(t)\phi(t). \quad (9)$$

Since $A(t)$ is a periodic matrix with period T, $\phi(t+T)$ must be a solution of (8) and each column of $\phi(t+T)$ can be written as a linear combination of $\phi(t)$, which means:

$$\phi(t+T) = \phi(t)M, \quad (10)$$

where M is a constant matrix. Since $\phi(0) = I$, we have that $\phi(T) = M$ and M is the monodromy matrix or Floquet operator, F.

The Floquet operator F determines the evolution of the system in a period T and it can be expressed by:

$$\Omega(T) = F\Omega(0). \quad (11)$$

For a system that obey a periodic potential, the Bloch theorem [5] guarantees eigenfunctions in the form of $\varsigma(t) = e^{i\theta\frac{t}{T}}v(t)$, where $v(t)$ is a periodic function on time, with period T. These eigenfunctions correspond to eigenvalues of F in the form of $\lambda = e^{i\theta}$, ($\theta \in \mathbb{C}$). These eigenvalues are named Floquet multipliers and θ is named as the Floquet argument. The Floquet operator is real and simpletic, that is, its physical properties are invariant for time translation. It implies that if λ is an eigenvalue, then λ^*, $\frac{1}{\lambda}$ and $\frac{1}{\lambda^*}$ are eigenvalues too.

From the Floquet theory the orbit of $y(t)$ is linearly stable if none of the Floquet multipliers have a module greater than 1. Hence a necessary condition to linear stability is that all Floquet multipliers are on the unit circle in the complex plane. In other words, θ must be real. In order to have instability, one or more pairs of eigenvalues must go out of the unit circle. If the eigenvalues go out of the unit circle in the complex plane in $\theta = 0$, we obtain an harmonic bifurcation. If they go out in $\theta = \pi$, we get a sub-harmonic bifurcation. Finally, if the eigenvalues go out of the unit circle in $\theta \neq 0, \pi$, then the bifurcation is called oscillatory.

A useful tool to study the stability is the Krein signature (κ), which is defined as the sign of the simpletic product of the real part with the imaginary part of the position-velocity vector [12]. By the

Krein criteria, for a collision between eigenvalues to generate instability, it is necessary that its Krein signature must be different.

$$\kappa(\theta) = \mathrm{sgn}([\mathrm{Re}\{\Omega(t)\}, \mathrm{Im}\{\Omega(t)\}]) = \mathrm{sgn}[\mathrm{i}\textstyle\sum_j(\varsigma_j(t)\dot{\varsigma}_j^*(t) - \varsigma_j^*(t)\dot{\varsigma}_j(t))]. \quad (12)$$

5. Results

The results reported below were obtained using chains with 21 oscillators (N = 21). The number of coefficients in the Fourier series was $lm = 17$. We have studied two different nonlinear potentials: the Morse potential and a potential with a hump. As stressed above, we have considered periodic boundary conditions. When the dynamical equations were explicitly integrated, the initial conditions were the breather profile taken with the method demonstrated in section 3, with velocities equal to zero.

The analysis were done using the Floquet multipliers and the time evolution of the energy distribution among the oscillators of the chain as a function of the coupling parameter and the frequency of the breather ω_b. These two factors determine the energy of the system calculated from the expression $E = \sum_{j=1}^N E_j$, where

$$E_j = \frac{m}{2}\dot{y}_j + \frac{k}{4}(y_{j+1} - y_j)^2 + \frac{k}{4}(y_j - y_{j-1})^2 + V(\sqrt{2}y_j) \quad (13)$$

is the energy of each oscillator and the elastic interaction energy of each spring is equally distributed between the two adjacent oscillators.

5.1 Morse potential

Originally the potential used to describe the hydrogen bonds in the PB model was the Morse potential, given by:

$$V(y) = D(\mathrm{e}^{-ay} - 1)^2, \quad (14)$$

where D is the depth of the potential well and represents the energy of dissociation, a has the dimension of inverse of length and is related to the width of the well, and y represents the displacement of the base pairs from equilibrium position ($y = 0$).

Using the Morse potential (14), the Lagrangian L_y (2) is written as:

$$L_\mathrm{y} = \textstyle\sum_{j=1}^N\left\{\frac{m}{2}\dot{y}_j^2 - \frac{k}{2}(y_{j+1} - y_j)^2 - D[\mathrm{e}^{-\sqrt{2}ay_j} - 1]^2\right\}. \quad (15)$$

It is convenient for the analytic and computational development to let the equations be depend on smallest number of parameters. Using adimensional variables $\xi_j = a\sqrt{2}y_j$ and $\tau = 2\left(\frac{Da^2}{m}\right)^{\frac{1}{2}}t$, it is possible to let the equation depend on just one parameter, C, and the Lagrangian can be expressed by

$$L = \textstyle\sum_{j=1}^N\left\{\frac{1}{2}\dot{\xi}_j^2 - \frac{1}{2}C\left(\xi_{j+1} - \xi_j\right)^2 - \frac{1}{2}\left(\mathrm{e}^{-\xi_j} - 1\right)^2\right\}, \quad (16)$$

where $\dot{\xi} = \frac{\mathrm{d}\xi}{\mathrm{d}\tau}$, $C = \frac{k}{4Da^2}$ and $L_\mathrm{y} \equiv 2DL$.

In Figure 1 we present the results for the absolute value of the Floquet multipliers as a function of the coupling parameter C for four different values of frequencies: (a) $\omega_\mathrm{b} = 0.6$, (b) $\omega_\mathrm{b} = 0.7$, (c) $\omega_\mathrm{b} = 0.8$ and (d) $\omega_\mathrm{b} = 0.9$.

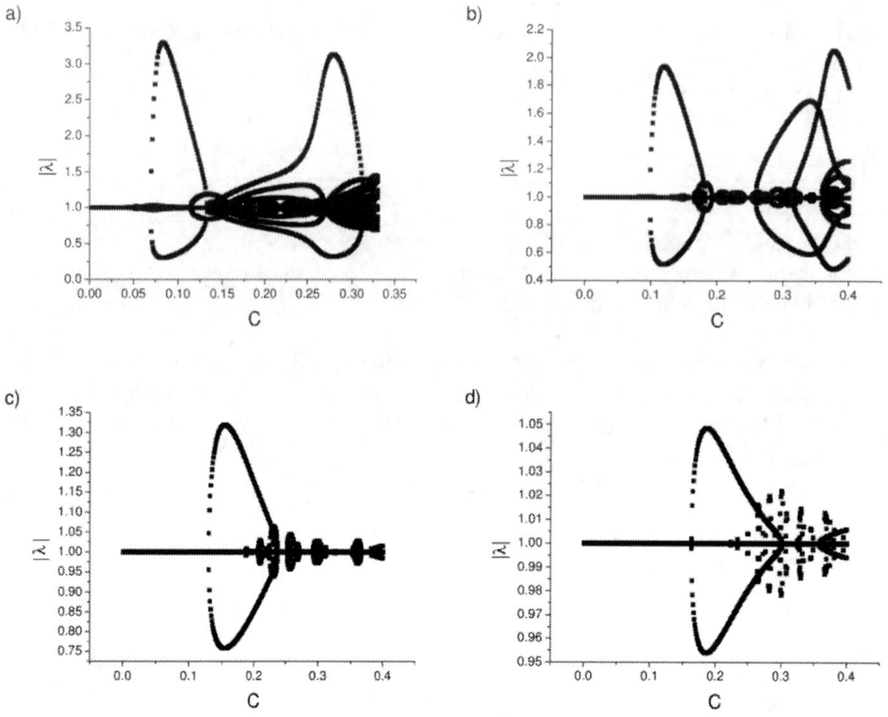

Figure 1. The absolute value of the Floquet multipliers in function of the coupling C for different frequencies of the breathers. (a) $\omega_b = 0.6$, (b) $\omega_b = 0.7$, (c) $\omega_b = 0.8$ and (d) $\omega_b = 0.9$.

A common feature in all cases is that for small values of the coupling parameter C there is not bifurcation and the stability is guaranteed. Interestingly, the range of value of C for which the system presents stability (absence of bifurcation) increases with the frequency of the breather. This behavior could be predicted since small values of C mean that the dynamics of the system are dominated by the non-linear effects. It is possible to demonstrate [12] that the adimensional energy for the system at t = 0 is given by $\sqrt{1-\omega_b^2}$. Therefore, taken $\omega_b \rightarrow 1$ the energy is small and the dynamical behavior is less sensitive to the coupling parameter C. As the values of C increase beyond a threshold depending on ω_b a series of bifurcations happen, meaning that more than one pair of eigenvalues have absolute values greater than one. For some small intervals of values of C the stability is regained, for example, figure 1b around $C \sim 0.25$. For $\omega_b = 0.8$ and $\omega_b = 0.9$ there is stability in several ranges of values of C.

It is interesting to analyze the dynamical behavior of the system, solving explicitly the equations of motion. To do this we used a Runge-Kutta fourth order method [14] to integrate the equations and the profile of the breather in the initial excitation, choosing the velocities of all oscillators equal to zero. In the sequence of figures 2-4 we present the plots in the complex plane of the Floquet multipliers with the respective Krein signature and the time evolution of the energy of each oscillator (in gray pattern) for some values of the coupling parameter C and frequencies ω_b. In figure 2 is shown a typical situation where the system is stable (C = 0.1 with $\omega_b = 0.8$). The results show that all Floquet multipliers are in the unit circle (absolute value equal to one) and the energy remains practically in the initially excited oscillator. In figure 3, on the other hand, we choose a couple of values for C and ω_b at the threshold of bifurcation (a harmonic one) shown in Figure 1c ($C = 0.13$ and $\omega_b = 0.8$). Here a pair of eigenvalues collided and left the unit circle with $\theta = 0$. The energy pattern, on the other hand, still shows the energy essentially localized in one oscillator but running on the chain. Eventually, in figure 4, we show a case where more than one pair of eigenvalues go out of the unit circle ($C = 0.23$ and $\omega_b = 0.8$). The Floquet analysis demonstrates the three kinds of bifurcation and as a result the energy spreads out the chain and the energy localization is lost.

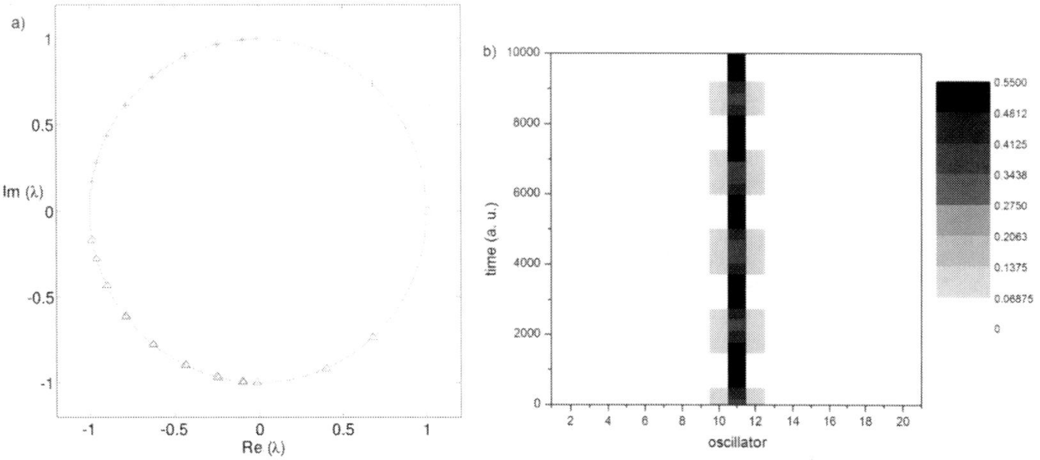

Figure 2. a) Representation of the Floquet multipliers with the respective Krein signature (+ symbol is used for $\kappa = +1$, while Δ is used for $\kappa = -1$ and O for $\kappa = 0$) and b) the energy of each oscillator in function of time. $C = 0.1$ and $\omega_b = 0.8$.

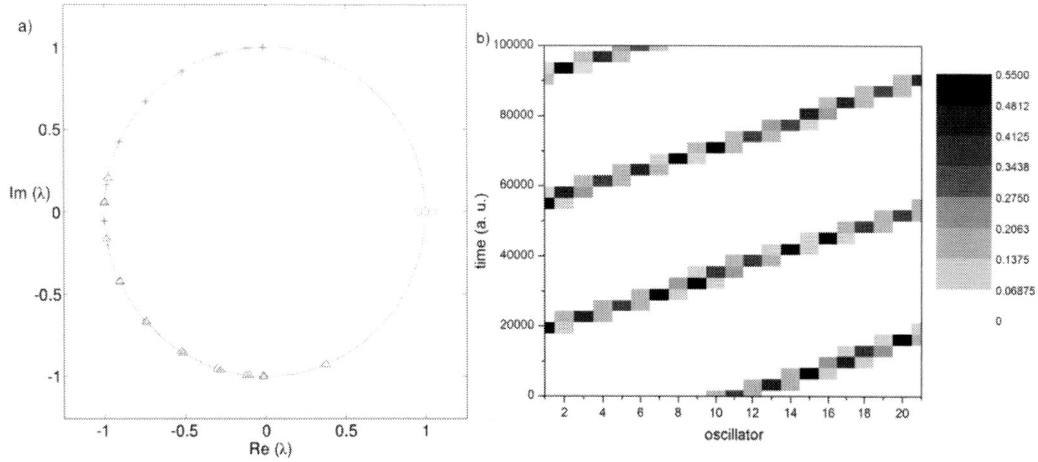

Figure 3. a) Representation of the Floquet multipliers with the respective Krein signature (same convention of figure 2) and b) the energy of each oscillator in function of time. $C = 0.13$ and $\omega_b = 0.8$.

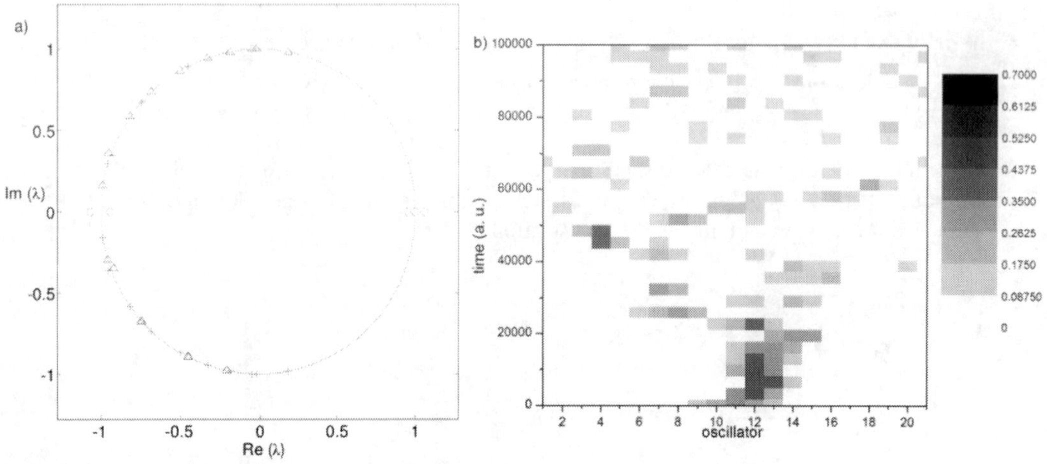

Figure 4. a) Representation of the Floquet multipliers with the respective Krein signature (same convention of figure 2) and b) the energy of each oscillator in function of time. $C = 0.23$ and $\omega_b = 0.8$.

5.2 Potential with a hump

A potential with a hump was introduced by Peyrard and co-workers [15,16], who were trying to establish conditions where the results for the opened and closed states were more compatible with the experimental observations. Essentially, they introduced a barrier of potential for y values around the distance that the hydrogen bonds can be considered broken. This resulted in an increase of time that they, once opened, stayed open. The potential was found by the intersection of the Morse potential ($y < 0$), of one quartic potential (around the origin) and one term that decays exponentially ($y > 1$).

This kind of potential is expressed by:

$$V_h(y) = \begin{cases} A[e^{-\alpha y} - 1]^2 & y < 0 \\ ay^2 + by^3 + cy^4 & 0 \le y \le 1 \\ D + Fe^{-\beta y}\left(y + \frac{1}{\beta}\right) & y > 1 \end{cases} \tag{17}$$

By replacing this potential in the lagrangian L_y and making the follow variable change, $\xi_j = \alpha\sqrt{2}y_j$ and $\tau = 2\left(\frac{A\alpha^2}{m}\right)^{\frac{1}{2}} t$, we find a similar equation to (16) expressed by:

$$L = \sum_{j=1}^{N} \left\{ \frac{1}{2}\dot{\xi}_j^2 - \frac{1}{2}C\left(\xi_{j+1} - \xi_j\right)^2 - \frac{1}{2}V_{hm}\left(\frac{\xi_j}{\alpha}\right) \right\}, \tag{18}$$

where $\dot{\xi} = \frac{d\xi}{d\tau}$, $C = \frac{k}{4A\alpha^2}$, $L_y \equiv 2AL$ and $V_{hm}\left(\frac{\xi_j}{\alpha}\right) = \begin{cases} [e^{-\xi_j} - 1]^2 & \frac{\xi_j}{\alpha} < 0 \\ \frac{a\xi_j^2}{A\alpha^2} + \frac{b\xi_j^3}{A\alpha^3} + \frac{c\xi_j^4}{A\alpha^4} & 0 \le \frac{\xi_j}{\alpha} \le 1. \\ \frac{D}{A} + \frac{F}{A}e^{-\beta\frac{\xi_j}{\alpha}}\left(\frac{\xi_j}{\alpha} + \frac{1}{\beta}\right) & \frac{\xi_j}{\alpha} > 1 \end{cases}$

As the potential function must be continuous, as soon as its first and second derivatives, the number of parameters is reduced to four: α, β, D and F. The other parameters are given by the expressions: $A = \frac{a}{\alpha^2}$, $a = 6D + \frac{1}{2}Fe^{-\beta}\left(\beta^2 + 5\beta + \frac{12}{\beta} + 12\right)$, $b = -8D - Fe^{-\beta}\left(\beta^2 + 4\beta + \frac{8}{\beta} + 8\right)$ and $c = 3D + \frac{1}{2}Fe^{-\beta}\left(\beta^2 + 3\beta + \frac{6}{\beta} + 6\right)$.

The procedure to get the results using this potential required some changes. The number of terms used in the Fourier series increased to $lm = 21$, and in order to study the energetic dynamic of the system we used the Runge-Kutta Nystron of tenth order [17]. We used the same values for the parameters as Peyrard et al [15,16], which were equivalent to: $\alpha = \beta = F = 4.0\text{Å}^{-1}$ and $D = 0.0857\text{eV}$.

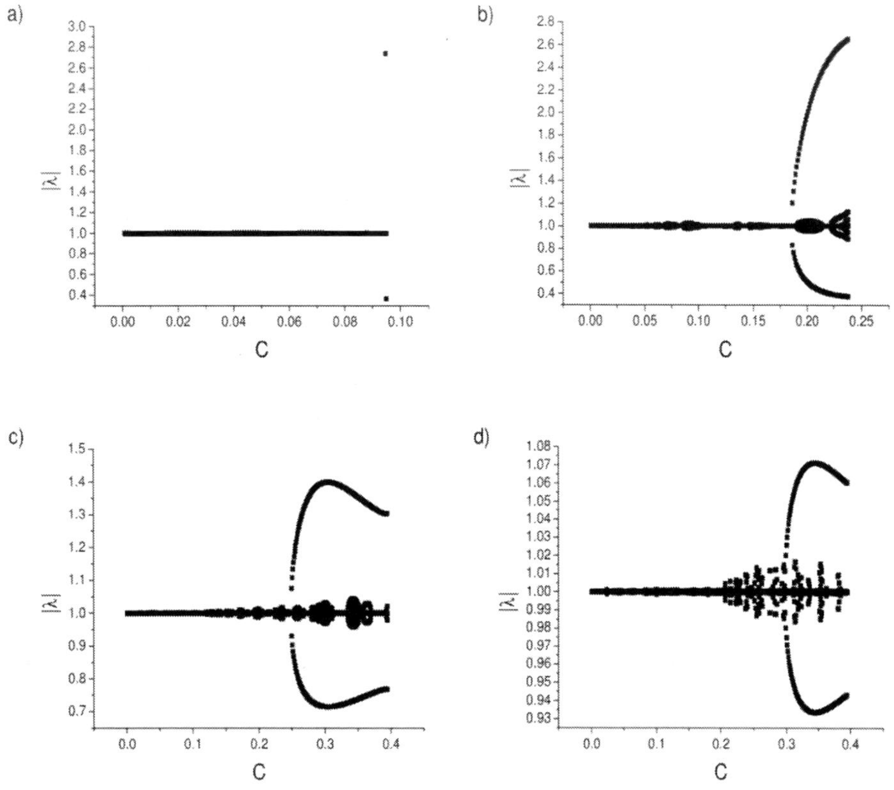

Figure 5. The absolute value of the Floquet multipliers in function of the coupling C for different frequencies of the breathers. (a) $\omega_b = 0.6$, (b) $\omega_b = 0.7$, (c) $\omega_b = 0.8$ and (d) $\omega_b = 0.9$.

In figure 5 it is shown the absolute value of the eigenvalues of the Floquet matrix as function of the coupling C, for four values of the breather frequency ω_b. Compared to the results obtained with the Morse potential the eigenvalues present greater regions of stability. However, in these regions exist some pairs of eigenvalues that have absolute values greater than 1 only by a few hundredths, and we decided to investigate their behavior. In figures 6-7 we show the plots in the complex plane of the Floquet multipliers with the respective Krein signature and the time evolution of the effective number of oscillators with significant energy n_{osc}, calculated from information entropy [18] and used as a criteria for energy localization [19,20].

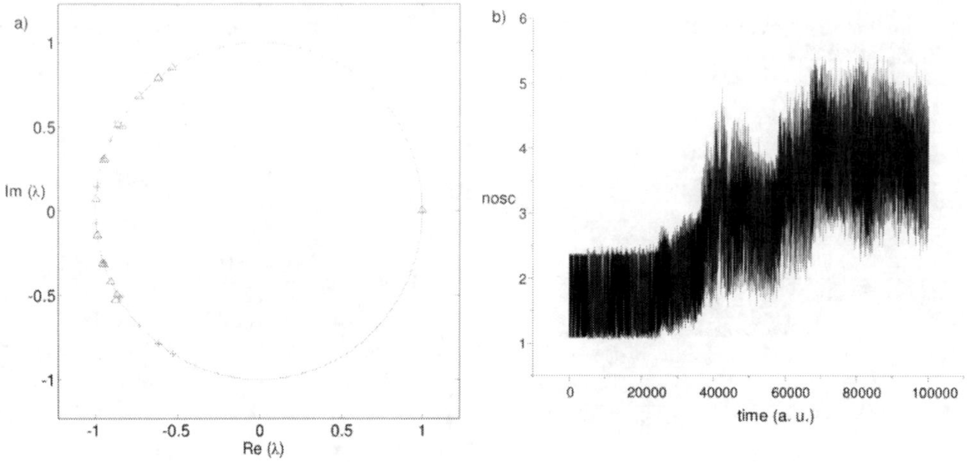

Figure 6. a) Representation of the Floquet multipliers with the respective Krein signature (same convention of figure 2) and b) the energy of each oscillator in function of time. $C = 0.09$ and $\omega_b = 0.7$.

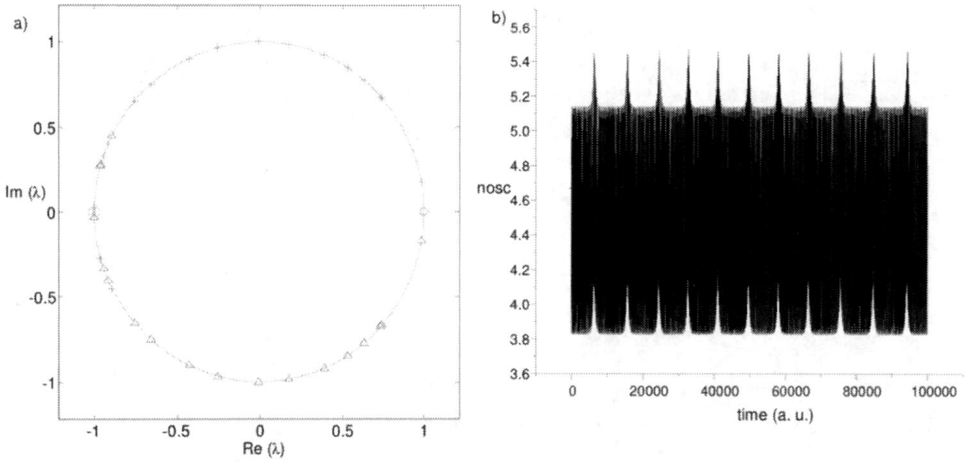

Figure 7. a) Representation of the Floquet multipliers with the respective Krein signature (same convention of figure 2) and b) the energy of each oscillator in function of time. $C = 0.255$ and $\omega_b = 0.9$.

In figure 6, the Floquet analysis shows an oscillatory bifurcation for $C = 0.09$ and $\omega_b = 0.7$. Looking at the dynamic behavior we note a gradual increase in the n_{osc} value passing from approximately 1.7 (average value) in the beginning to 3.5 by the end. In figure 7, we have a sub-harmonic bifurcation case, $C = 0.255$ and $\omega_b = 0.9$, and despite the curious periodic fluctuations in the n_{osc} values, their average value remains around 4.4. So the stability of the breather suffers the influence of these bifurcations although the energy remains localized in a small number of oscillators at least for the time interval considered.

6. Conclusion

In this paper we revisited the problem of 1-site breather formation in homogeneous nonlinear chains described by the Peyrard-Bishop model, using the numeric proceedings suggested by Marin and

Aubry. Two kinds of nonlinear potential were tested: the Morse potential and a potential with a hump. The stability analysis of these structures was done through the Floquet theory and is consistent for the Morse potential, with that found in the work of Cuevas [12].

The dynamic of stable solutions showed a localized structure where the energy is practically in one oscillator, as predicted by the theory. For instable regions, the dynamic suggests some characteristic behavior. When there is a harmonic bifurcation, the energy remains localized but it starts to walk through the chain. This case is known as a mobile breather. In the sub-harmonic bifurcation, the localized structure suffers some periodic changes that affect the effective number of oscillators with significant energy. In the oscillatory case, the energy starts to diffuse slowly through the chain, but remains localized for the time interval observed, that is relevant for the DNA dynamics. When the three bifurcations happen for one condition, their effects are joined and the energy spreads out the chain and the localized structure is completely lost.

The potential with a hump presents greater regions of stability. This fact should be related with the barrier existent in this potential, as it increases the confining region of the potential and this inhibits the loss of stability.

Qualitatively the behavior of the mobile breathers resembles that of the DNA, where a structure of localized energy walks through the chain. However, the values for the DNA amplitudes of movement and life time for open states are not equivalent to that found in our tests. So improvements in the model and the way to analyze the DNA dynamics should be made.

Acknowledgements
The authors are grateful to Capes and CNPq for financial support. The authors also are indebted to Ms. Kaylee Harrison for her assistance with the English of the manuscript.

References

[1] Alberts B, Johnson A and Lewis J 2004 *Biologia Molecular da Célula* (Porto Alegre: ARTMED)
[2] Watson G B and Crick F H 1953 *Nature* **171** 737-8
[3] Englander S W, Kallenbach N R, Heeger A J, Krumhansl J A and Litwin A 1980 *Proc. Natl. Acad. Sci.* **77** 7222-6
[4] Leroy J L, Kochoyan M, Huynh-Dinh T and Guéron M 1988 *J. Mol. Biol.* **200** 223-38
[5] Aubry S 1997 *Physica D* **103** 201-50
[6] Aubry S 2006 *Physica D* **216** 1-30
[7] Flach S and Willis C R 1998 *Physics Reports* **295** 181-264
[8] MacKay R and Aubry S 1994 *Nonlinearity* **7** 1623-43
[9] Yakushevich L 1998 *Nonlinear Physics of DNA* (Chichester: Wiley series in nonlinear science) chapter 1 pp 1-19
[10] Peyrard M, Bishop A R 1989 *Phys. Rev. Lett.* **62** 2755-8
[11] Marin J L and Aubry S 1996 *Nonlinearity* **9** 1501-28
[12] Cuevas J 2003 *Localización y Transferencia de Energia em Redes Anarmónicas No Homogéneas* Phd thesis Universidad de Sevilla
[13] Hoppensteadt F C 2000 *Analysis and Simulation of Chaotic Systems* (New York: Springer) chapter 1 pp 1-24
[14] Press W H, Teukolsky S A, Vetterling W T and Flammery B P 1992 *Numerical Recipes* (New York: Cambridge University Press)
[15] Peyrard M, Cuesta-López S and Angelov D 2009 *J. Phys.: Condensed Matter* **21** 034103
[16] Peyrard M, Cuesta-López s and James G 2008 *Nonlinearity* **21** T91-T100
[17] Tsitouras C 1999 *Celest. Mech. Dun. Astron.* **74** 223-30
[18] De Luca J, Lichtenberg A J and Liebermon 1995 *Chaos* **5** 283-97
[19] De Luca J, Drigo Filho E, Ponno A and Ruggiero J R 2004 *Phys. Rev. E* **70** 026213
[20] Silva J M, Drigo Filho E and Ruggiero J R 2009 *Eur. Phys. J. E* **29** 245-51

Journal of Physics: Conference Series **246** (2010) 012040 doi:10.1088/1742-6596/246/1/012040

Cellular automata with inertia: species competition, spatial patterns, and survival in ecotones

K. Kramer, M. Koehler, and M. G. E. da Luz*

Departamento de Física, Universidade Federal do Paraná, CP 19044, 81531-980, Curitiba-PR, Brazil

E-mail: *luz@fisica.ufpr.br

Abstract. We consider a two-dimensional CA model with three possible states for the system individual cells, 0 and $\pm$. As for the dynamical rules, only $\pm$ can exert pressure to change the cells actual states. In this way, the 0 state is neutral and in some sense competitively weaker than the other two states. We further assume an inner property, the inertia, which is an intrinsic resistance to changes in the system. We evolve an ensemble of initial configurations for the CA until reaching steady states. By calculating averages over some relevant quantities for the final stationary configurations, we discuss how certain features of the problem, namely, initial states population and degree of aggregation as well as the values of inertia, can determine the different characteristics of the spatio-temporal pattern created by the CA evolution. We finally discuss how our findings may be relevant in the understanding of structures formation due to species competition in biology, specially in the transition regions between different biomes, the so called ecotones.

1. Introduction

During the 1940's, John von Neumann, with the help of Stanislaw Ulam, introduced the idea of cellular automata (CA) [1]. The initial motivation was to mimic replication, i.e., reproduction of living organisms, by means of a simple mathematical construction: the CA. Such type of systems were then forgot until the 1970's, when John H. Conway created the famous Game of Life [2], again a CA which tries to simulate fundamental biological processes, life and death, from very straightforward rules. Around the same time, Konrad Zuse [3] probably becomes the first person to propose CA's as useful tools to model different natural phenomena. Since then, it has been an increasing interest in CA's, specially after the detailed classification of the 1D and 2D cases, made by Stephen Wolfram [4] in the 1980's.

Given their simplicity and generality (even being possible realizations for universal Turing machines [5]), nowadays CA's are considered in very broad contexts. Indeed, CA's are able to reproduce the complex behavior seen in many systems in physics, geology, social sciences, psychology, etc [6]. But an area in which CA's find most applications (perhaps, because the above mentioned historical initial developments) is still biology [7], specially when related to spatio-temporal pattern formation [8] as, e.g., in the study of spread of epidemics [9], the shape of sea shells [10], and the organization of social insects [11].

In very general terms, spatio-temporal structures often arise from driving forces acting on an extended system, formed by spatially distributed cell (or sub-system) elements. The cells evolve and interact with each other under the influence of such driving forces. Thus, the time evolution of the whole system will generate successive configurations, giving rise to the pattern, which may or may not converge to a final stationary global state. From the above description and given the way CA's are constructed (see next Section), it is not just a coincidence the great utility of CA's in generating spatio-temporal pattern, particularly in biological processes [7].

However, an aspect usually overlooked in CA models is the natural intrinsic resistance (hereafter called "inertia") which the cells or sub-systems may have to change their local states. For instance, bacterial populations can develop a certain reaction against extension of the existing colonies (growth) when harsh conditions are present [12]. It is true that some CA's with resistance have been proposed, mainly to address socio-economics problems [13]. But they have a probabilistic rather than a deterministic (mechanicist) character, with the latter being the more frequent kind of inertia in pattern formation.

Thus, in the present contribution we introduce a CA, where a term associated to inertia is explicitly considered. We moreover assume three different possible states for the CA cells. Only two of them ($+$ and $-$) are dynamically active, hence the ones playing a competitive role in the model. The third state (0) is neutral in terms of any dynamical pressure to change the other two. We study how the inertia (an inner state associated to each CA cell) can alter the emerging pattern in the system. Our analyses are based on the evolution of an ensemble of initial configurations for the CA until reaching steady states. We determine how certain relevant quantities characterizing the final structures (obtained by averaging over the evolved replicas) depend on the parameters of the initial ensemble and the inertia values. We discuss how our general findings may be of interest to the problem of maintenance and generation of biodiversity pattern [14]. Finally, we address the phenomenon of survival of less fitted species in ecotones [15], i.e., transition regions between distinct biomes. We show that the weaker species 0 can live in such transition zones (but not within the biomes), provided it has a certain small resistance to the competition pressure exerted by the stronger species $+$ and $-$.

2. The model

Next we describe our construction of a CA with inertia, as well as explain the methodology we will use to characterize the system.

2.1. CA with inertia

As previously mentioned, a CA is a system composed by K cells labeled by $k = 1, 2, \ldots, K$. To each cell k, it is ascribed a number s_k ($s = 1, 2, \ldots, S$) describing its actual local state at the discrete time instants $n = 0, 1, 2, \ldots$. There is an update rule telling how the system will evolve over time, so that $s_k(n+1)$ is function of the system configuration at the earlier time n. We will consider the 2D case and a square array of $K = \mathcal{N} \times \mathcal{N}$ elements, thus $k \equiv (i, j)$ ($i, j = 1, 2, \ldots, \mathcal{N}$). For such array, we take a wall-like (instead of the more common periodic) boundary condition, more appropriate in biological applications as in fragment pattern in ecosystems [14] and species proliferation in ecotones [16].

For the dynamics we assume the Moore neighborhood of radius 1, meaning that each cell will evolve under the influence of only its first surround neighbors, i.e., three, five and eight neighbors for, respectively, the cells at corners, edges and middle of the array. We suppose three possible states [17, 18]: $s = 0$ and $s = \pm$. The former being the neutral (passive) state, in the sense that it does not attempt to change the state of the other two. In fact, most of the CA's proposed in the literature are defined so that all states exert competitive pressure over each other, like in Ref. [19], which studies the spatio-temporal dynamics of communities. Here, as it is going to be clear in the latter applications, it is interesting to have a "weaker" player in the game, the state 0.

We add a new ingredient into the CA by considering an extra "internal state", the inertia, that quantifies a resistance to changes in the actual state of a cell. The inertia of cell k, I_k, is a number between 0 and 8, where 0 means no inertia at all and 8 is equal to the maximum possible number of first surrounding neighbors. In the extreme case of $I = 8$, the cell never changes its initial state. Intermediate I's influence the cell time evolution as described below.

Let us define

$$\mathcal{S}_k(n) = \sum_{m \,\in\, \text{neighborhood of } k} s_m(n). \tag{1}$$

Then, we define the CA rules of evolution as

- The inertia rule:
  Let $I_k(n)$ be the value of the inertia of the cell k at time n. If $I_k(n) \geq |\mathcal{S}_k(n)|$, the state s_k at $n+1$ remains the same that at n. If $I_k(n) < |\mathcal{S}_k(n)|$, we apply the dynamical rule below.
- Dynamical rule:
  If $\mathcal{S}_k(n) > 0$ ($\mathcal{S}_k(n) < 0$), the cell k state changes to $s_k(n+1) = +$ ($s_k(n+1) = -$). Otherwise it remains the same.

The above two rules are summarized by (for sign$[x]$ the signal of x)

$$s_k(n+1) = \begin{cases} \text{sign}[\mathcal{S}_k(n)] & \text{if} \quad I_k(n) < |\mathcal{S}_k(n)| \\ s_k(n) & \text{if} \quad I_k(n) \geq |\mathcal{S}_k(n)| \end{cases} \tag{2}$$

Although models where the inertia is a function of time certainly would present very rich dynamics, here we concentrate only on time independent I's.

2.2. Characterizing the system

To analyze the influence of the inertia in the formation of spatio-temporal pattern, we shall first define few quantities helping to characterize the state configurations of the CA. In fact, there are different possible sophisticated topographical analysis to identify spatial point pattern in extended systems (a nice overview is given, for instance, in Ref. [20]), some even based on fractal structures [21]. Nevertheless, here we propose a direct way to do so, which is simple to implement and capable of measuring the degree of aggregation or fragmentation found in realistic biological patterns [22].

Thus, let $p_s(n)$ be the number of cells in the state s (the population of s) at the time instant n. Of course, always $p_0 + p_+ + p_- = K = \mathcal{N}^2$. Then, we consider the "clusterization" of the full CA at n by setting the clustering functions c's, which quantify the average number of elements of a same state around each CA cell, forming small local clusters. They can be total, c, or for the state $s = 0, \pm$, c_s, given by

$$\begin{aligned} c(n) &= \frac{1}{p(n)} \sum_k [c]_k(n), \\ c_s(n) &= \frac{1}{p_s(n)} \sum_k [c_s]_k(n). \end{aligned} \tag{3}$$

Here, $[c]_k$ is the number of cells in the most populated state in the neighborhood of cell k (e.g., if around the cell k four cells have state 0, one $+$, and three $-$, then $[c]_k = 4$). If the two highest local populations have equal number of elements, then $[c]_k = 0$. $[c_s]_k = [c]_k$ if the most populated state in the neighborhood of k is s, otherwise we assume $[c_s]_k = 0$. Note then that the parameters c and c_s measure the frequency of aggregation in the CA.

Another important quantity for our CA is the convergence time τ. When an initial CA is submitted to an evolution rule, it either may converge to a certain stationary final configuration (therefore, invariant under further successive applications of the same rule) or may never entering into a steady state, changing its configuration at each time step. When it converges, we can define τ as the number of iterates necessary to reach such stationary situation.

To analyze the system dynamical evolution and the influence of varying the inertia to the formation of spatio-temporal pattern, we will discuss most of our results in terms of averages over an ensemble of initial configurations for the CA (an approach already considered in the literature but with different purposes [23]). So, we work with a large set of initial matrices, all submitted to the evolution of our CA. We will restrict our study only to those initial configurations converging to some steady state, i.e., for which τ is finite. Unless otherwise explicitly mentioned, all the calculated quantities result from proper means over the evolved ensemble.

For our simulations we set $\mathcal{N} = 22$, a good compromise of lattices already displaying rich dynamics, while keeping the computational efforts low. We also fix $p_0(0) = 161 \approx \mathcal{N}^2/3$ and $p_+ > p_-$, so that always before the evolution the majority state is $+$. In total we work with 36,146 matrices in our ensemble (randomly generated), with a fairly homogeneous distribution of parameters values in the intervals $162 \leq p_+(0) \leq 182$ and $0.12 \leq c_+(0) \leq 0.22$. To calculate the averages, we proceed in the following way. Suppose we want to determine the property γ in terms of the initial population of the state $+$, $p_+(0)$. Then, we evolve all the initial matrices of the ensemble which have different $c_+(0)$'s but a same $p_+(0)$, calculate for each of such matrices the quantity γ at the steady state, and finally obtain a simple mean. It leads to $\gamma = \gamma(p_+)$. Of course, similar procedure gives γ, say, as a function of the initial clusterization of the state $+$, $c_+(0)$.

3. Results

As already mentioned, all the initial configurations considered for our CA are such that guarantee the convergence to a certain stationary state. As we are going to see, the inertia parameter $I_k = I$ ($0 \leq I \leq 8$) – assumed to be the same for all cells k in the lattice – is a very important factor determining both the average features of the steady states reached by the CA and how long it takes for the convergence. Note that for $I = 8$, any initial configuration is already at the steady state because then the dynamical rule cannot change the system configuration (Sec. 2).

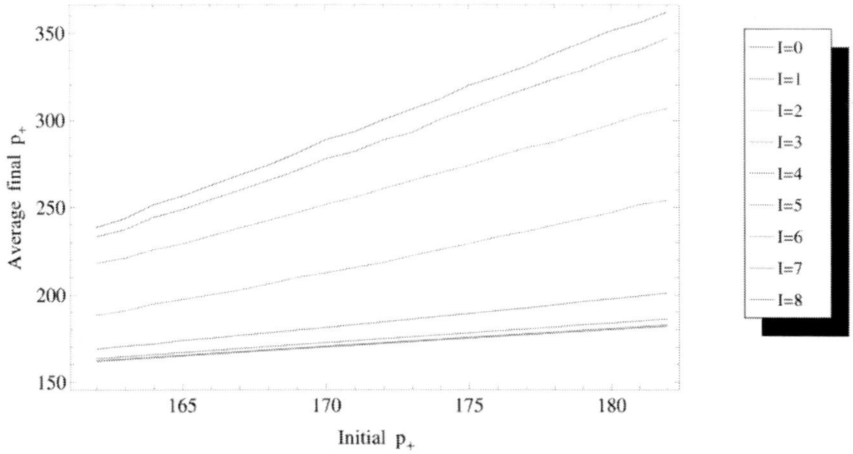

Figure 1. Average final p_+ as a function of the initial p_+ for all values of inertia.

First, as a kind of test, we consider a quantity whose dynamical evolution is intuitively easy to predict. In Fig. 1 we show the average final p_+ (i.e., already at the steady configurations) as a function of the initial p_+, assuming different values for the inertia $I_k = I$ ($I = 0, \ldots, 8$).

The observed linear increase is straightforward to understand. Indeed, in a first (mean-field) approximation, the local population of a state s is $p_s/\mathcal{N}^2$. But as the dynamical pressure to change a cell state to the value $s = \pm$ is directly proportional to the local population of s, then if the initial p_+ grows, the dynamical pressure to create new $+$'s grows in the same proportion (given that $+$ is initially the majority state). Thus, the observed results follow. Moreover, note that when I increases, we have from Fig. 1 that the difference $\Delta p_+ = p_+ - p_+(0)$ decreases. It is simply due to the stronger intrinsic resistance of each element to change its initial condition for greater I's. Actually, for $I = 6$, 7, the curves are very close to that for $I = 8$. In such cases, basically the dynamics no longer can change significantly the initial configurations.

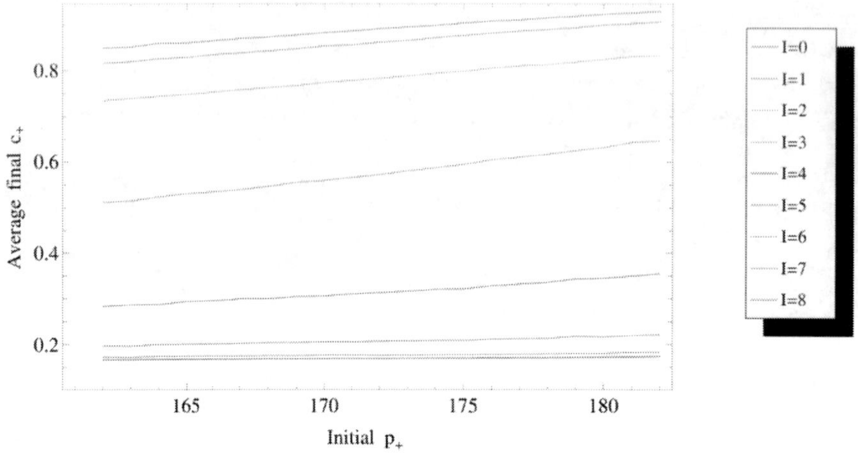

Figure 2. Average final c_+ as a function of the initial p_+ for all values of inertia.

Figure 2 shows the curves for the final average clustering of $+$ versus the initial population $+$. The behaviors are similar to those in Fig. 1. As seen, the larger the initial p_+, the higher the final p_+. But for larger final populations of the state $+$, naturally we should expect a higher aggregation of $+$, explaining the observed increase in the final c_+. Furthermore, for larger I's we also find a decrease in the final value of the average c_+, again a consequence of dynamical pressure to create the majority states $+$ to decrease with I.

The only important qualitative difference between Figures 1 and 2 – although difficult to see by direct visual inspection – is the way the lines slopes vary as a function of I. Of course they are all positive. However, in the former the slope of the curves decrease monotonically with I, whereas in the latter they increase (very slowly) for I increasing up to $I = 4$, then they start to decrease (again very slowly). To understand that, notice that if $I \neq 0$, the state of a cell k may change to a state s only if its neighborhood has at last $I + 1$ cells at state s. Hence, such neighborhood must have a relatively high clusterization to change the state of the k cell. In this way, the dynamics becomes very sensitive to the initial population and how locally agglutinated it is, leading to the observed behavior for $I \leq 4$. For $I \geq 5$, local clusterization must be really very high for any state change. Since we are not working with high initial densities for $+$, such situation is not frequently met and we start to see a very weak correlation between $p_+(0)$ and the average final c_+, until the limit of $I = 8$, for which no correlations do exist.

Figure 3 shows the average final population $+$ versus the initial clustering, $c_+(0)$. For low inertia values, $I \leq 3$, we have that the final p_+ decreases fairly linearly with the initial c_+. At first sight it may seem a contradictory result, since in principle by increasing $c_+(0)$ we could expect larger final $+$ state populations. However, an important aspect to recall is that our ensemble englobes a relatively small range for the initial $p_+(0)$: 162–182. Therefore, in this

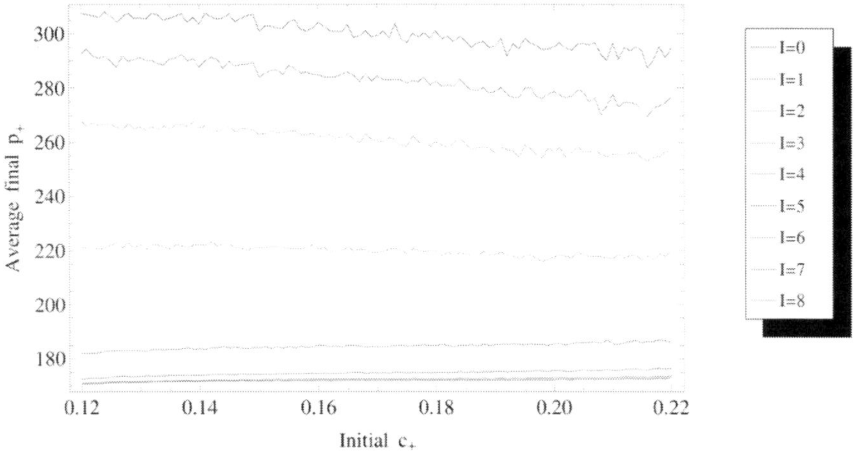

Figure 3. Average final p_+ as a function of the initial c_+ for all values of inertia.

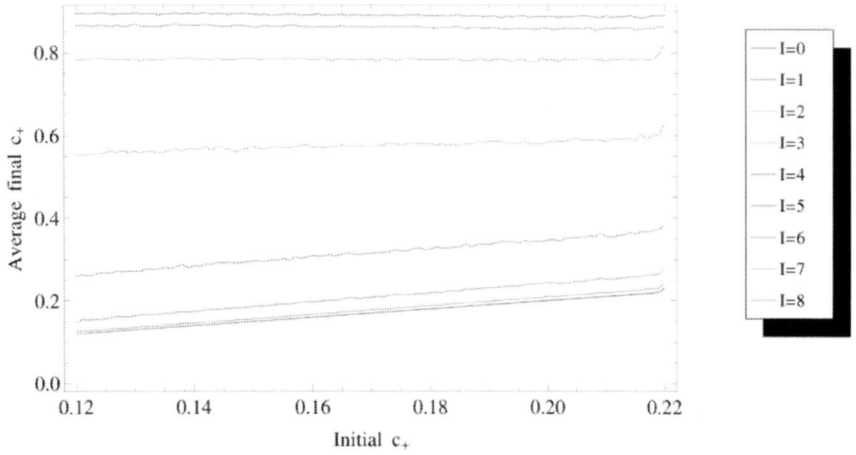

Figure 4. Average final c_+ as a function of initial c_+ for all values of inertia.

case the higher $c_+(0)$'s do not come from many small + clusters scattered off in the CA initial configurations. Rather, it is due to a small number of particular regions with relatively large populations of + (a result verified by a direct inspection of our original ensemble, but not shown here). To promote the proliferation of the + state from the CA evolution rules, it is more efficient to have many small + clusters randomly distributed (acting as nucleation centers) than to have few localized regions rich in + state. The latter can lead to a crowded region of +'s, but not allowing efficient migration to other locations of the CA lattice. It explains the behavior of the $I \leq 3$ curves in Fig. 3. On the other hand, when $4 \leq I \leq 7$ the final p_+ presents a very weak increasing with the initial c_+, in a small but positive correlation. This time, only the few regions with highly clusterized + can give rise to further creation of the + state. Thus, higher c_+'s may lead to small increments in the final p_+.

In Figure 4 we see the curves of the average final versus the initial clustering of the state +. Qualitatively these plots resemble those in Fig. 3, although here with a much weaker dependence of c_+ on $c_+(0)$. For instance, observe a very slightly negative (when $I \leq 2$) and a positive (when

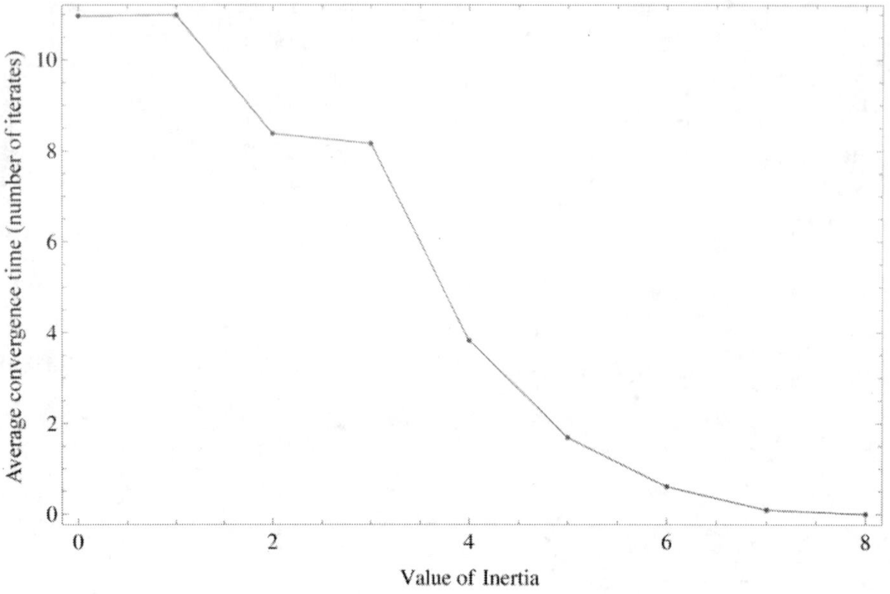

Figure 5. The mean convergence time as a function of the inertia. The same values seen for $I = 0$ and $I = 1$ is because $I = 1$ is still a too low inertia value to change such characteristic time.

$I \geq 3$) slope for the lines. Note that typically the initial clustering (recalling our procedure of averaging over the initial $p_+(0)$'s, Sec. 2.2) is not very critical to set the final clustering when the inertia is zero. As previously mentioned, $p_+(0)$ is much more important to define the final clusterization values. Indeed, for $0 \leq I \leq 2$ the average final c_+ in Fig 4 is basically equal to $c_+(p_+(0) = 172)$ in Figure 2, i.e., the value at the middle of the interval range for $p_+(0)$. Nevertheless, for intermediate values of the inertia ($3 \leq I \leq 5$), we observe a positive dependence between the initial and final clusterizations.

Finally, we present in Figure 5 the mean (over the entire evolved ensemble) convergence time as a function of I. Of course, when $I = 8$ the convergence time is zero, since for this maximum resistance any initial configuration is already a steady one. For the other values of I, the mean convergence time decreases for an increase in the inertia, a result that should be expected. Indeed, for higher I's, there are more resistance to changes, so overall fewer number of changes, resulting in a quicker relaxation to the stationary condition.

4. Discussion

As already mentioned in the Introduction, the formation and evolution of a spatio-temporal pattern is a fundamental problem in different branches of science. Although much progress in the understanding of such phenomena has been achieved [24], many important open questions remain [25]. In particular, in the context of biology, it is still not quite clear which are the main factors generating different structures in the organization and distribution of living systems [26]. For instance, competition among different species, their ability to adapt to different environmental conditions, their diffusive skills [27], among others, all play fundamental roles in the way meta-populations grow, giving rise to distinct biodiversity pattern in different ecosystems [28]. Nevertheless, the relative influence of each one of these mechanisms: competition strength; fitness; diffusiveness; etc.; are not yet properly quantified [29].

In our previous simple model of CA, we have analyzed some general factors influencing pattern formation, all them associated to realistic situations in biology [22]. In fact, as a methodology we have considered an ensemble of initial configurations and then performed averages to identify general trends, instead to focus on particular initial situations. Of course, such protocol may not be appropriated to model punctual instances, like the purple loosestrife invasion in North America [30]. Nevertheless, by properly delimiting the range of values of relevant parameters (like the ones assumed for $p_0(0)$, $p_+(0)$ and $c_+(0)$, see Sec. 2.2), a such statistical approach may be useful in pointing to typical global behavior of specific ecosystems [31]. For instance, we verified that as long as the inertia (intrinsic resistance) of the CA elements are zero, the initial population of the majority state is more important to define the final majority population than a possible spatial aggregation-clusterization of such state at $t = 0$. However, this is true provided the other active competing state (in our case, $-$) has the same degree of clusterization. We should mention that for the largest considered initial population of $+$, 182, the initial population of $-$ is 141, and thus in average $c_-(0)$ is not too radically different from $c_+(0)$. So, our present analysis would be, for example, more appropriate for homogeneous biomes.

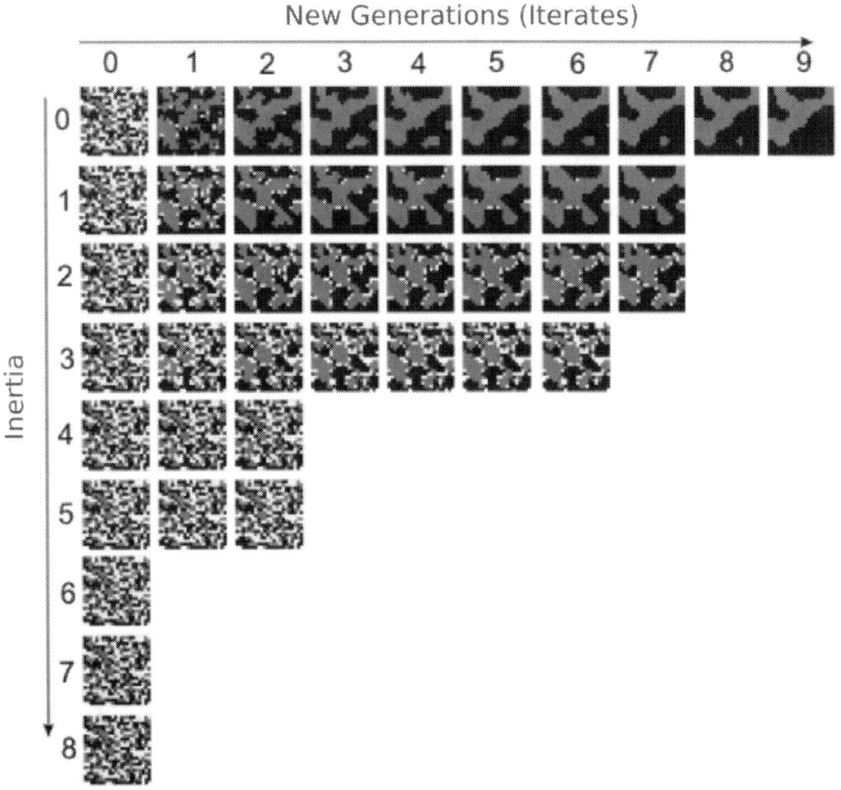

Figure 6. Typical succession of generations of a CA until reaching the steady configuration. The parameters are given in the main text and the colors black, grey, and white, represent, respectively, the states $+$, $-$ and 0.

A new factor introduced here is the inertia, quantifying the intrinsic resistance of the cell elements to change their actual states. In a biological context, such quantity could be associated to a fitness of a given species [29], allowing it to resist to other invasive species. From our results, one realizes that already very low inertia can change the final pattern of a evolutive process or even the relative influence of the other parameters for so. Indeed, depending on the I value, the relevance of $c_+(0)$ (the initial clusterizaiton) increases. This becoming clear from the curves in Figs. 3 and 4. Moreover, I can accelerate the process of reaching a final steady state for a biological pattern (see Fig. 5). Although it should be expected for larger I's, this also holds true for small and intermediate resistance values.

Actually, the above previous observations bring about another interesting point: whether a weaker less aggressive (in terms of competitiveness) species can survive in an environment with more aggressive species. For instance, models using coupled logistic maps (see, e.g., [29]) points that in fact this is possible, provided the neutral (weaker) species has minimal survival skills. However, such type of models generally does not include spatial factors, very important to determine the geographic distribution of meta-populations. With the present CA, either aspects can be considered, as exemplified in Figure 6, which displays, for different I's, a typical automaton evolving until it reaches a stationary configuration. In the example, the initial clustering and population are $c_+(0) = 0.169864$ and $p_+ = 172$, with a convergence time τ for $I = 0$ equal to 9 iterates. In Figure 7 we show the resulting number of elements in the 0 (neutral) state at the steady situation. For $I = 0$, none survives. But already for the lowest value of $I = 1$, such "weaker" species can resist to the competitive pressure to change their states (which in a biological view, could mean the death of such species and the establishment of the stronger species in that same location). Thus, the zero state can manage to survive, resisting to "attacks" in particular spatial spots of the landscape.

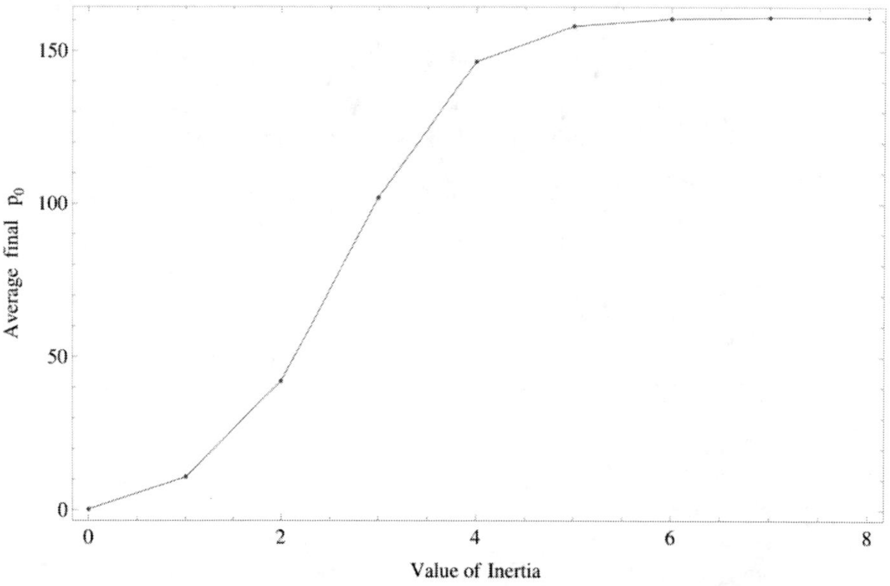

Figure 7. For the automaton of Fig.6, the final population of state 0 as a function of the inertia.

If now we come back to Fig. 6, we observe a very interesting phenomenon in biology associated to ecotones [15, 32]. Ecotones are transition zones between distinct geographic regions of different biomes. In many instances, it is not clear how groups of species coming from the different biomes can coexist in this transition region. Also how the even less fitted animals and plants (e.g., due to a less developed response to the area particular climate condition) can survive in an ecotone [33], whereas they eventually would perish in other more homogeneous environments if competing with the same stronger species [34]. In Fig. 6 we see that the neutral state, 0, with no colonization skills, will go extinct if $I = 0$. However, even for small values of I, it can survive in small "colonies", exactly at the borders of two regions dominated by the states + and −. Of course, we do not expect the present model to describe all the complexity found in such ecosystems. Nevertheless, if we interpret the states 0 and ± as meta-species, the CA with inertia may be helpful in revealing some important aspects and key mechanisms underlying ecotones biodiversity.

5. Final Remarks and Conclusion

In this contribution we have considered a CA model with three states, one being neutral (passive) in terms of dynamical pressure to change the other states. Furthermore, we have associated an inner variable to each cell, the inertia, representing an intrinsic resistance to change the cell actual state under the action of the CA dynamical rule.

By calculating averages properties of final states (after reaching the steady condition) of an ensemble of initial configurations for the CA, we have analyzed how different initial features, namely, population, aggregation of states and the inertia strength, can influence the formation and evolution of spatio-temporal pattern in the system. Moreover, based on our findings, we have presented a general discussion on how patterns of biodiversity could emerge from scenarios of competition-fitness mechanisms. We have done so assuming initial random spatial distribution for the species (the states), but also considering a certain degree of clusterization of the species involved in the process.

A particularly interesting result has been that the proposed CA naturally yields as a possible solution configurations in which the neutral (or weaker, in some evolutive sense) state can survive, provided its inertia is not null. Indeed, the 0 state can establish itself in small colonies located at the borders (transition zones or ecotones) of regions full of the stronger ± states. Such phenomenon, which does occur in real situations, is still poorly understood. So, the present can be faced as null (or even a first approximation) model to quantify such rich and complex kind of ecosystem.

Finally, this simple and exploratory implementation of our CA with inertia of course cannot explain realistic spatio-temporal pattern originated from, say, species competition. Nevertheless, it may be faced as a first step towards the incorporation of a very important feature, intrinsic resistance (or fitness), into the process. Surely, further studies are necessary for a better understanding of the problem in terms of the present model. For instance, to consider broader ranges for the initial parameters, to use clusterization measures which can distinguish between local and global state aggregations, to allow temporal as well as spatial variations for the inertia parameter, to study the influence of the CA size, etc. All such extensions would help to address important open problems in biodiversity pattern, specially in ecotones [33]. Presently, the mentioned possibilities are being investigated and hopefully will be reported in the due course.

Acknowledgments

M G E da Luz is in great debt to M C M Marques for very valuable discussions about pattern formation in biological systems, specially about ecotones. We thank CNPq and CAPES for research grants and F. Araucária, CNPq and CT-Infra (FINEP) for financial support.

References

[1] von Neumann J 1966 *Theory of Self-Reproducing Automata* (Champaign-IL: University of Illinois Press)
[2] Gardner M 1971 *Scientific American* **224** 112; Gardner M 1985 *Wheels, Life, and Other Mathematical Amusements* (New York-NY: W. H. Freeman & Company)
[3] Zuze K 1970 *Calculating Space* (Cambridge-MA: MIT University Press)
[4] Wolfram S 2002 *Cellular Automata and Complexity: Collected Papers* (Boulder-CO: Westview Press)
[5] Rendell P 2002 in *Collision-Based Computing*, Adamatzky A (ed.) (London: Springer), pp. 513-539
[6] Schiff J L 2008 *Cellular Automata: a Discrete View of the World* (New York-NY: Wiley-Interscience)

[7] Ermentrout G B and E.-Keshet L 1993 *J. Theor. Biolg.* **160** 97; Lin Y, Mynett A, and Chen Q 2009 in *Advances in Water Resources and Hydraulic Engineering - Vol. II*, Zhang C and Tang H (eds.) (Berlin: Springer) pp. 624-629
[8] Deutsch A and Dormann S 2005 *Cellular Automaton Modeling of Biological Pattern Formation* (Boston-MA: Birkhäuser)
[9] Boccara N 2004 *Modeling Complex Systems* (New York-NY: Springer-Verlag)
[10] Meinhardt H 2009 *The Algorithmic Beauty of Sea Shells - 4th Edition* (New York-NY: Springer)
[11] Miramontes O 1995 *Complexity* **1** 56; Miramontes O and de Souza O 2008 *J. Insec. Sci.* **8** 22
[12] Lin A L, Mann B A, Torres-Oviedo G, Lincoln B, Kas J, and Swinney H L 2004 *Biophys. J.* **87** 75
[13] Moldovan S and Goldenberg J 2004 *Technol. Forecast. Soc. Change* **71** 425; Zhang T, Xuan H, and Gao B 2005 in *2005 International Conference on Service System and Services Management - Proceedings of ICSSSM'05 - Vol. 2*, Chen J (ed) (Los Alamitos-CA: IEEE Computer Society); Zupan N 2007 *Technol. Forecast. Soc. Change* **75** 798
[14] Young T P 2000 *Biological Conservation* **2** 73
[15] Zang Y and Malanson G P 2006 *Geograph. Analys.* **38** 271; Favier C, Chave J, Fabing A, Schwartz D, and Dubois M A 2004 *Ecolog. Model.* **171** 85
[16] Ferreira D A C, Noguera S P, Carneiro Filho A, and Soares-Filho B 2008 *Ciência Hoje* **42**(248) 26 (in Portuguese)
[17] Platkowski T 2002 *Physica A* **311** 291
[18] Alonso-Sanz A and Martn M 2004 *Chaos, Solitons and Fractals* **21** 809
[19] Eppsteina M J, Beverb J D, and Molofskyc J 2006 *Ecolog. Model.* **197** 133
[20] Wallet F and Dussert C 1998 *Europhys. Lett.* **42** 493
[21] Lahiri T, Chakrabarti A, and Dasgupta A K 1998 *J. Struct. Biolog.* **123** 179
[22] Schneider M F 2001 *J. Appl. Ecolog.* **38** 720; Jenerette G D and Wu J G 2001 *Landscape Ecolog.* **16** 611; Kupfer J A and Runkle J R 2003 *Oikos* **101** 135; Malanson G P, Wang Q, and Kupfer J A 2007 *Ecolog. Model.* **202** 397
[23] Druzhinin O A and Mikhailov A S 1989 *Radiophys. Quant. Elect.* **32** 334; Kometer K, Zandler G, and Vogl P 1992 *Semicond. Sci. Technol.* **7** B559; Dzwinel W 2006 in *Cellular Automata: Lecture Notes in Computer Science* - **4173**, Yacoubi S E, Chopard B, and Bandini S (eds.) (Berlin: Springer-Verlag) pp. 657-666; Wissner-Gross A D 2008 *J. Cellul. Autom.* **3** 27
[24] Mocenni C, Facchini A, and Vicino A 2010 *Proc. Nat. Acad. Sci.* **107** 8097
[25] Grigoriev D and Vakulenko S 2006 *Ann. Pure Appl. Logic* **141** 412
[26] Bolliger J, Lischke H, and Green D G 2005 *Ecol. Complexity* **2** 107
[27] Santos M C, Raposo E P, Viswanathan G M, and da Luz M G E 2004 *Europhys. Lett.* **67** 734; Faustino C L, da Silva L R, da Luz M G E, Raposo E P, and Viswanathan G M 2007 *Europhys. Lett.* **77** 30002
[28] Ormerod S J and Watkinson A R 2000 *J. Appl. Ecol.* **37** 1
[29] Nowak M A 2006 *Evolutionary Dynamics* (Cambridge-MA: Harvard University Press)
[30] Welk E 2004 *Ecol. Model.* **179** 551
[31] Gower S T, Kucharik C J, and Norman J M 1999 *Remote Sens. Environ.* **70** 29
[32] Peters D P C (2002) *Ecol. Model.* **152** 5
[33] Ngai J T and Jefferies R L 2004 *J. Ecol.* **92** 1001; Chaneton E J, Mazia C N, and Kitzberger T 2010 *J. Ecol.* **98** 488
[34] Freeman D C, Wang H, Sanderson S, and McArthur E D 1999 *Evol. Ecol. Res.* **1** 487; Scarano F R 2009 *Biol. Conserv.* **142** 1201

AUTHOR INDEX

AUTHOR INDEX